AN ELEMENTARY
TREATISE ON THE CALCULUS.

AN ELEMENTARY TREATISE ON THE CALCULUS

WITH ILLUSTRATIONS FROM GEOMETRY, MECHANICS AND PHYSICS

BY

GEORGE A. GIBSON, M.A., F.R.S.E.

PROFESSOR OF MATHEMATICS IN THE
GLASGOW AND WEST OF SCOTLAND TECHNICAL COLLEGE

London
MACMILLAN AND CO., Limited
NEW YORK: THE MACMILLAN COMPANY
1901

All rights reserved

KD34760

Harvard University
Math. Dept. Library.
24 Dec. 1901

GLASGOW: PRINTED AT THE UNIVERSITY PRESS
BY ROBERT MACLEHOSE AND CO.

PREFACE.

THE rapid growth in recent years of all branches of applied science and the consequent increasing claims on the time of students have given rise in various quarters to the demand for a change in the character of mathematical text-books. To meet this demand several works have been published, addressed to particular classes of students and designed to supply them with the special kind and quantity of mathematics they are supposed to need.

With many of the arguments urged in favour of the change I am in hearty sympathy, but it is as true now as it was of old that there is no royal road to mathematics, and that no really useful knowledge can be gained except by strenuous effort.

It is sometimes alleged that a thorough knowledge of the derivatives and integrals of the simpler powers, of the exponential and the logarithmic functions, and perhaps of the sine and the cosine, is quite sufficient preparation in the Calculus for the engineer. This contention has a solid substratum of truth; but a knowledge that goes beyond the mere ability to quote results is not to be obtained by the few lessons that are too often considered sufficient to expound these elementary rules. It may be possible to state and illustrate in a few lessons a sufficient amount of the special results of the Calculus to enable a student to follow with some intelligence the more elementary treatment of mechanical and physical problems; but, though such a meagre course in the Calculus may not be without value, it is quite inadequate, both in kind and in quantity, as a preparation for the serious study of such practical subjects as Alternate Current Theory, Thermodynamics, Hydrodynamics, and the theory of Elasticity, and to a student so prepared much of the recent literature in Physics and Chemistry would be a sealed book. Besides, it should surely be the aim of every well-devised scheme of education to place the student in a position to undertake independent research in his own particular line of work, and the very complexity of the problems presented to modern science, with the vast accumulation of detail so characteristic of it, enhances in no small degree the value of a liberal training in mathematics. Subsequent specialisation makes it the more, not the less, necessary that the mathematical training in the earlier stages should be the same whether the student afterwards devotes himself to pure mathematics or to the more practical branches of science, especially as the processes of thought involved in any serious study of mechanical, physical, or chemical phenomena have much in common with those developed in the study of the Calculus.

The early text-books on the Calculus, such as Maclaurin's or Simpson's, were not written for pure mathematicians alone, but drew their illustrations largely from Natural Philosophy; the later text-books, probably in consequence of the ever-widening range of Physics, gradually dropped physical applications, and even tended to become treatises on Higher Geometry. In the present position of mathematical science, however, it is just as much out of place to make an elementary work on the Calculus a text-book of Higher Geometry as it would be to make it a text-book of Physics or of Engineering or of Chemistry. What may be reasonably required of an elementary work on the Calculus is that it should prepare the student for immediately applying its principles and processes in any department of his studies in which the Calculus is generally used. With this end in view, the subject should be illustrated from Geometry, Mechanics, and Physics while the peculiar difficulties of these branches are relegated for detailed treatment to special text-books, so that the illustrations may really serve their purpose of throwing light on general principles, and may not introduce rather than remove intellectual obscurity. As regards Chemistry, a sound knowledge of the Calculus is of special importance, since it is the properties of functions of more than one variable that are predominant in chemical investigations; the lately published book of Van Laar, *Lehrbuch der Mathematischen Chemie*, is a sign of the times that cannot be mistaken.

In this text-book an effort has been made to realise the aims just indicated. With respect to mathematical attainments, the reader is supposed to be familiar with Geometry, as represented by the parts of Euclid's Elements that are usually read, with Algebra up to the Binomial Theorem for positive integral indices, and with Plane Trigonometry as far as the Addition Theorem; but no use is made of Complex (imaginary) number, nor is a knowledge of Infinite Series presupposed. The excessive refinements of modern mathematics have been deliberately avoided, as being neither profitable nor even intelligible to the young student; constant appeal has been made to geometrical intuitions, while at the same time considerable attention has been paid to the logical development of the subject.

The early chapters may seem to contain a great deal of matter that is foreign to the book: but the theory of graphs and of units is of such importance, and is as yet so imperfectly treated in elementary teaching, that some account of it appeared to be a necessity. After considerable hesitation I have included in my plan the elements of Coordinate Geometry, so far as these were likely to be of real service in elucidating fundamental principles or important applications; but for many applications of the Calculus an extensive acquaintance with Coordinate Geometry is not necessary, and I hope that a sufficiently clear account of its principles has been given to meet the practical needs of many students. I have, however, excluded the discussion of the theory of Higher Plane Curves and of Surfaces as unsuitable for an elementary treatise.

Another innovation is the chapter on the Theory of Equations; the innovation seems to be justified, not merely as an arithmetical illustration of the Calculus, but also by the practical importance of the subject, and by the absence of elementary works that treat of transcendental equations.

The general development is that which I have followed in class-teaching for several years. The somewhat lengthy discussion of the conceptions of a rate and a limit I have found in practice to be the simplest method of enabling a student to grapple with the special difficulties of the Calculus in its applications to mechanical or physical problems; when these notions have been thoroughly grasped, subsequent progress is more certain and rapid. No rigid line is drawn between differentiation and integration, and several important results requiring integration are obtained before that branch is taken up for detailed treatment. The discussion in Chapter X. of areas and of derived and integral curves is designed, not only to furnish a fairly satisfactory basis for the geometrical definition of the definite integral, but also to illustrate a method of graphical integration that is of some importance to engineers, and that may be of some value even in purely theoretical discussions.

As in some of the more recent text-books, the discussion of Taylor's Theorem has been postponed; the Mean Value Theorem is sufficient in the earlier stages, and the somewhat abstract theorems on Convergence and Continuity of Series are most profitably treated towards the end of the course. The treatment, however, is such that teachers who prefer the usual order may at once pass from the Mean Value Theorem to Chapters XVII. and XVIII.

Functions of more than one variable are treated in less detail than functions of one variable; but I have tried to select such portions of the theory as are of most importance in physical applications. The book closes with a short chapter on Ordinary Differential Equations, designed to illustrate the types of equations most frequently met with in dynamics, physics, and mechanical and electrical engineering.

Simple exercises are attached to many of the sections; in the formal sets will be found several theorems and results for which room could not be made in the text, and which are yet of sufficient importance to be explicitly stated. I have tried to exclude all examples that have nothing but their difficulty to recommend them; and with the object of encouraging the student to put himself through the drill that is absolutely necessary for the acquisition of facility and confidence in applying the Calculus, I have freely given hints towards the solution of the more important examples.

In the preparation of the book, I have consulted many treatises, and where I am conscious of having adopted a method of exposition that is peculiar to any writer, I have been careful to make due acknowledgment. It is difficult, however, when one has been teaching a subject for years to recognise the sources of his knowledge, and it may well be that I have borrowed more largely than I am aware.

I am greatly indebted to my friends Professor Andrew Gray, F.R.S.; Mr. John S. Mackay, LL.D.; Mr. Peter Bennett; Mr. John Dougall, M.A.; and Mr. Peter Pinkerton, M.A., for help in the tedious task of the revision of proof-sheets and for useful criticism. In all matters bearing on Physics, Professor Gray's advice has been of the greatest service. To Mr. Dougall my obligations are specially great; he has taken a lively interest in the work from its inception, and has read the whole of it in manuscript, placing at my disposal, in the most generous way, his great knowledge of the subject and the fruits of his experience as a teacher; to him, too, I owe the verification of the examples.

I desire to thank Professor R. A. Gregory for his constant and kindly advice on matters relating to the passage of the book through the press. I am also grateful to the printers for the excellence of their share of the work.

GEORGE A. GIBSON.

GLASGOW, *September*, 1901.

CONTENTS.

CHAPTER I.
COORDINATES. FUNCTIONS.

ART.		PAGE
1.	Directed Segments or Steps,	1
2.	Addition of Steps,	2
3.	Symmetric Steps and Subtraction of Steps,	3
4.	Abscissa of a Point. Fundamental Axiom,	4
5.	Measure of a Step,	5
6.	Axes of Coordinates. Squared Paper,	6
7.	Distance between two Points,	9
8.	Polar Coordinates,	10
9.	Variable. Continuity,	12
10.	Geometrical Representation of Magnitudes,	13
11.	Function. Dependent and Independent Variables,	13
12.	Notation for Functions,	16
13.	Explicit and Implicit Functions,	16
14.	Multiple-valued and Inverse Functions,	17
	Exercises I.,	19

CHAPTER II.
GRAPHS. RATIONAL FUNCTIONS.

15.	Object of the Calculus. Graphs,	20
16.	Graph of x^2,	20
17.	Equation of a Curve. Symmetry. Turning Values,	23
18.	Graph of cx^2,	24
19.	Scale Units,	26
20.	Coordinate Geometry,	27
	Exercises II.,	29

ART.		PAGE
21.	The Linear Function. Intercepts,	30
22.	Gradient,	32
	Exercises III.,	32
23.	Rational Functions. Point of Inflexion,	34
24.	Asymptotes,	37
	Exercises IV.,	41

CHAPTER III.

GRAPHS. ALGEBRAIC AND TRANSCENDENTAL FUNCTIONS. CONIC SECTIONS.

25.	Algebraic Functions. Graph of Inverse Functions. Cusps,	43
26.	Conic Sections,	47
27.	Change of Origin and of Axes,	52
	Exercises V.,	54
28.	Transcendental Functions. Trigonometric Functions,	56
29.	The Exponential Function and the Logarithmic Function,	57
30.	General Observations on Graphs,	59
	Exercises VI.,	60

CHAPTER IV.

RATES. LIMITS.

31.	Rates,	65
32.	Increments,	65
33.	Uniform Variation. Measure of a Uniform Rate,	67
34.	Dimensions of Magnitudes,	68
35.	Variable Rates,	71
36.	Average Rate,	71
37.	Measure of a Variable Rate,	73
38.	Limits,	74
39.	Examples of Limits. Definition of Tangent,	74
40.	General Explanation of a Limit,	79
41.	Definition of a Limit. Notation. Distinction between Limit and Value,	79
42.	Theorems on Limits,	81
43.	Examples. Mensuration of Cylinder and Cone,	83

CHAPTER V.

CONTINUITY OF FUNCTIONS. SPECIAL LIMITS.

ART.		PAGE
44.	Continuity of a Function,	87
45.	Theorems on Continuous Functions,	89
46.	Continuity of the Elementary Functions,	90
47.	$\underset{x=a}{L} (x^n - a^n)/(x-a)$,	91
48.	$\underset{m=\infty}{L} \left(1 + \frac{1}{m}\right)^m$. The number e,	92
49.	The Function e^x,	96
50.	Compound Interest Law,	97
	Exercises VII.,	98

CHAPTER VI.

DIFFERENTIATION. ALGEBRAIC FUNCTIONS.

51.	Derivatives. Differentiation,	101
52.	Increasing and Decreasing Functions. Stationary Values,	103
53.	Geometrical Interpretation of a Derivative,	105
54.	Derivative as an Aid in Graphing a Function,	106
55.	Derivative not Definite,	107
56.	Fluxions. Velocity,	109
57.	Derivative of a Power,	111
58.	General Theorems,	111
	Exercises VIII.,	115
59.	Derivative of a Function of a Function and of Inverse Functions,	116
	Exercises IX.,	119
60.	Differentials,	120
61.	Geometrical Applications. Tangent. Subtangent, etc.,	122
62.	Derivative of Arc,	124
	Exercises X.,	125

CHAPTER VII.

DIFFERENTIATION (*continued*). TRANSCENDENTAL FUNCTIONS. HIGHER DERIVATIVES.

63.	Derivatives of the Trigonometric Functions,	129
	Exercises XI.,	131
64.	Inverse Trigonometric Functions,	133
	Exercises XII.,	134

ART.		PAGE
65.	Exponential and Logarithmic Functions,	135
	Exercises XIII.,	139
66.	Hyperbolic Functions,	139
67.	Higher Derivatives,	142
68.	Leibniz's Theorem. Examples,	144
	Exercises XIV.,	146

CHAPTER VIII.

PHYSICAL APPLICATIONS.

69.	Applications of Derivatives in Dynamics. Simple Harmonic Motion. Potential,	149
70.	Coefficients of Elasticity and Expansion,	156
71.	Conduction of Heat,	157
	Exercises XV.,	159

CHAPTER IX.

MEAN VALUE THEOREMS. MAXIMA AND MINIMA. POINTS OF INFLEXION.

72.	Rolle's Theorem and the Theorems of Mean Value,	161
73.	Other Forms of the Theorems of Mean Value,	164
74.	Maxima and Minima,	166
75.	Examples. Graph of $e^{-ax}\sin(bx+c)$,	168
76.	Elementary Methods,	171
77.	Variation near a Turning Value,	174
	Exercises XVI.a, XVI.b, XVI.c,	176-180
78.	Concavity and Convexity. Points of Inflexion,	180
	Exercises XVII.,	182

CHAPTER X.

DERIVED AND INTEGRAL CURVES. INTEGRAL FUNCTION. DERIVATIVES OF AREA AND VOLUME OF A SURFACE OF REVOLUTION. POLAR FORMULAE. INFINITESIMALS.

79.	Derived Curves,	183
80.	Derivative of an Area,	185
81.	Interpretation of Area,	187

CONTENTS.

ART.		PAGE
82.	Integral Function,	188
83.	Integral Curve,	190
84.	Graphical Integration,	192
85.	Surfaces of Revolution,	193
86.	Infinitesimals,	195
87.	Fundamental Theorems,	197
88.	Polar Formulae,	200
	Exercises XVIII.,	201

CHAPTER XI.

PARTIAL DIFFERENTIATION.

89.	Partial Differentiation. Continuity of a Function of two or more Independent Variables,	204
89a.	Coordinate Geometry of Three Dimensions. Direction Cosines. Equations of Line and Plane. Equation of Surface,	205
90.	Total Derivatives. Complete Differentials,	211
91.	Geometrical Illustrations. Tangent Plane. Normal,	214
92.	Rate of Variation in a given Direction. Note on Angles,	218
93.	Derivatives of Higher Orders. Commutative Property. Laplace's Equation,	220
94.	Complete Differentials,	224
95.	Application to Mechanics. Potential,	225
96.	Application to Thermodynamics,	228
97.	Four Thermodynamic Relations,	231
98.	Change of Variable. Differentials of Higher Orders,	233
99.	Transformation of $\nabla^2 u$,	235
	Exercises XIX.,	238

CHAPTER XII.

APPLICATIONS TO THE THEORY OF EQUATIONS.

100.	Rational Integral Functions. Zeroes,	242
101.	Any Continuous Function,	243
102.	Newton's Method of approximating to the Roots of an Equation,	244
103.	Tests for Degree of Approximation,	245
104.	Examples,	246
105.	Successive Approximations,	247

ART.		PAGE
106.	Expansion of a Root in a Series. Reversion of Series,	249
107.	The Equation $x = \tan x$,	251
	Exercises XX.,	253
108.	Proportional Parts,	255
109.	Small Corrections,	258
	Exercises XXI.,	260

CHAPTER XIII.

INTEGRATION.

110.	Integration. Indefinite and Definite Integral. Constant of Integration,	262
111.	Standard Forms,	264
112.	Algebraic and Trigonometric Transformations,	267
	Exercises XXII.,	269
113.	Change of Variable,	271
114.	Examples of Change of Variable,	272
115.	Quadratic Functions,	274
116.	Trigonometric and Hyperbolic Substitutions,	277
117.	Some Trigonometric Integrands,	278
	Exercises XXIII.,	280
118.	Integration by Parts,	281
119.	Successive Reduction. The Integral $\int_0^{\frac{\pi}{2}} \sin^m x \cos^n x \, dx$,	284
	Exercises XXIV.,	288
120.	Partial Fractions,	290
121.	Integration of Rational Functions,	292
122.	Irrational Functions,	294
123.	General Remarks,	295
	Exercises XXV.,	296

CHAPTER XIV.

DEFINITE INTEGRALS. GEOMETRICAL APPLICATIONS.

124.	Definite Integrals. Theorems,	298
125.	Related Integrals,	301
126.	Infinite Limits. Infinite Integrand,	304
	Exercises XXVI.,	306

CONTENTS.   xvii

ART.		PAGE
127.	Some Standard Areas and Volumes. Curve Tracing,	309
	Exercises XXVII.,	312
128.	Area of Closed Curves,	316
129.	Area Swept out by a Moving Line,	319
130.	Planimeters,	321
	Exercises XXVIII.,	322

CHAPTER XV.

INTEGRAL AS LIMIT OF A SUM. DOUBLE INTEGRALS.

131.	Integral as the Limit of a Sum,	324
132.	Examples,	327
133.	Approximations. Simpson's Rule,	328
	Exercises XXIX.,	331
134.	Mean Values,	333
135.	Double integrals,	334
136.	Notations for Double Integrals. Polar Elements. Triple Integrals,	337
137.	Centres of Inertia,	341
138.	Moments of Inertia,	343
139.	Polar Element of Volume. Definition of Line Integral and of Surface Integral,	346
	Exercises XXX.,	347
	Gamma and Beta Functions,	349

CHAPTER XVI.

CURVATURE. ENVELOPES.

140.	Curvature. Fundamental Formula,	352
141.	Circle, Radius, and Centre of Curvature,	354
142.	Various Formulae for the Curvature. Intrinsic Equation of a Curve,	355
	Exercises XXXI.,	359
143.	Evolute, Involute, Parallel Curves,	361
144.	Envelopes,	364
145.	Equation of Envelope. Contact Theorem,	365
146.	Cycloids. Epicycloids. Hypocycloids,	368
	Exercises XXXII.,	371

CHAPTER XVII.

INFINITE SERIES.

ART.		PAGE
147.	Infinite Series—Convergent, Divergent, Oscillating,	375
148.	Existence of a Limit. Theorems,	377
149.	Tests of Convergence. Fundamental Test; Comparison Test; Test Ratio. Remainder,	379
150.	Absolute Convergence. Power Series,	382
151.	Uniform Convergence. Continuity of Series,	385
	Exercises XXXIII.,	387

CHAPTER XVIII.

TAYLOR'S THEOREM.

152.	Taylor's Theorem. Maclaurin's Theorem. Remainder,	390
153.	Examples of Expansions: $\sin x$, $\cos x$, e^x, $(1+x)^m$, $\log(1+x)$,	393
154.	Calculation of the nth Derivative. Examples,	397
155.	Differentiation and Integration of Series,	399
156.	Expansions. Approximations. Examples of Integration of Series,	401
	Exercises XXXIV.,	403

CHAPTER XIX.

TAYLOR'S THEOREM FOR FUNCTIONS OF TWO OR MORE VARIABLES. APPLICATIONS.

157.	Taylor's Theorem for Functions of two or more Variables,	408
158.	Examples:—Tangent Plane, Euler's Theorems of Homogeneous Functions,	411
159.	Maxima and Minima of a Function of two or more Variables,	412
160.	Examples. Undetermined Multipliers,	414
	Exercises XXXV.,	416
161.	Indeterminate Forms. Elementary Methods,	418
162.	Method of the Calculus,	419
	Exercises XXXVI.,	422

CHAPTER XX.

DIFFERENTIAL EQUATIONS.

ART.		PAGE
163.	Differential Equations. Definitions. Examples,	424
164.	Complete Integral,	426
	Exercises XXXVII.,	427
165.	Equations of the First Order and of the First Degree. Variables Separable. Homogeneous Equations. Linear Equations. Exact Equations,	428
166.	Equations of First Order, but not of First Degree. Clairaut's Equation. Singular Solutions,	431
167.	Equations of the Second Order. Simple Pendulum,	432
168.	Linear Equations. General Property,	433
169.	The Complementary Function,	434
170.	The Particular Integral,	436
171.	Simultaneous Equations. Example from Electric Circuits,	437
	Exercises XXXVIII.,	439
	ANSWERS,	442
	INDEX,	454

AN ELEMENTARY TREATISE ON THE CALCULUS.

CHAPTER I.

COORDINATES. FUNCTIONS.

§ 1. Directed Segments or Steps. Let A, B (Fig. 1) be any two points on a straight line. In Elementary Geometry it is customary to denote the segment of the line between A and B by AB or by BA indifferently, the order of the letters being of no consequence. It is useful, however, for many purposes to distinguish the segment traced out by a point which moves along the line from A to B from that traced out by a point which moves from B to A. When this distinction is made, the segment is called a *directed segment* or *vector* or *step*, and the distinction is represented in the symbol for the segment by the order of the letters; thus, AB denotes the segment traced out by a point which moves from A to B, while BA denotes the segment traced out by a point which moves from B to A. The length of the step AB is the same as that of the step BA, but the steps have opposite directions.

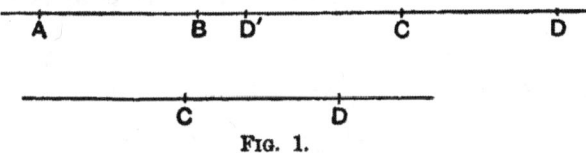

Fig. 1.

Two steps AB, CD are defined to be equal if (1) they are on the same straight line or on parallel straight lines, (2) the lengths of AB and CD are equal, and (3) D is on the same side of C as B is of A. Thus, if D' be at the same distance from C as D is, but on the opposite side, AB is not equal to CD' but to $D'C$. The step AB has the same length and the same direction as CD or $D'C$, but though it has the same length as CD', it has not the same direction and is therefore not equal to CD' in the sense in which "equal" has been defined for steps.

§ 2. Addition of Steps. Let A, B, C be any three points on a straight line. Whatever be the relative position of the points A, B, C, a point which moves along the line from A to B, and then from B to C, will be at the same distance from A and on the same side of A as if it had moved directly from A to C. AC is therefore taken as the *sum* of the steps AB and BC, and the operation of addition of steps is defined by the equation

$$AB + BC = AC.$$

When B lies between A and C, the sum of the lengths of the steps AB and BC is equal to the length of the step AC, and therefore in this case addition of steps agrees with the usual geometrical meaning of addition of segments in which length alone is considered. But when B does not lie between A and C, the sum of the lengths of the steps AB and BC is not equal to the length of the step AC. It will be seen immediately that steps can be represented as positive or negative, and that addition of steps corresponds to algebraical addition.

If D be any fourth point on the line

$$AB + BC + CD = AC + CD = AD,$$

and in the same way the sum of any number of steps may be defined.

To find the sum of AB and CD when B and C are not coincident, take the step BE equal to the step CD; then

$$AB + CD = AB + BE = AE.$$

If x be any positive number, xAB is a step in the same direction as the step AB, and of a length which is to the length of AB in the ratio of x to 1; thus, $3AB$ is a step thrice as long, and in the same direction as the step AB;

$\frac{5}{3}AB$ is a step five-thirds of the length of AB and in the same direction.

The student will have no difficulty in showing that the commutative and associative laws for the addition of numbers hold for the addition of steps.

§ 3. Symmetric Steps and Subtraction of Steps. If in the first case of the preceding Article the point C be supposed to coincide with A, the step AC becomes the *zero-step* AA, which is denoted by 0. Hence, in symbols,

$$AB + BA = AA = 0.$$

Similarly, $\quad AB + BC + CA = AC + CA = 0.$

In Algebra the negative number $-a$ is defined by the equation

$$a + (-a) = 0.$$

In the same way the negative step $-AB$ may be defined by the equation

$$AB + BA = 0$$

as being the step BA; that is, the step $-AB$ is the step BA of the same length in the opposite direction. The symbol $+$ may now be attached to a step AB, and $+AB$ may be called a positive step. The two steps $+AB$ and $-AB$ (or BA) are called *symmetric* steps. Obviously, if two steps are equal, so also are their symmetric steps.

The operation of subtraction of a step is defined as the addition of the symmetric step; in symbols,

$$AC - BC = AC + CB = AB;$$

or, $\quad AB - CD = AB + DC = AC'$, if $BC' = DC.$

Precisely as in Algebra, the commutative and associative laws may be shown to hold for subtraction of steps, and there will be no confusion caused by the use of the symbols $+$ and $-$ to indicate symmetric steps as well as the operations of addition and subtraction.

By the definition of subtraction, if A, B be any two points on a line and O any third point,

$$AB = AO + OB = OB + AO = OB - OA.$$

§ 4. Abscissa of a Point. Let O be a fixed point on a line $X'OX$ and P, P' two points on opposite sides of O but at the same distance from it (Fig. 2); let U be another point on the line on the same side of O as P is, say to the right of O.

The steps OU, OP have the same sign; the steps OU, OP' have opposite signs.

Let OU be taken as a standard of length, say 1 inch, and as a standard of direction; it may therefore be called the unit step. Steps measured like OU to the right will be called positive steps, while those measured to the left will be called negative. Thus OP, $P'P$ are positive, OP', PP' negative steps.

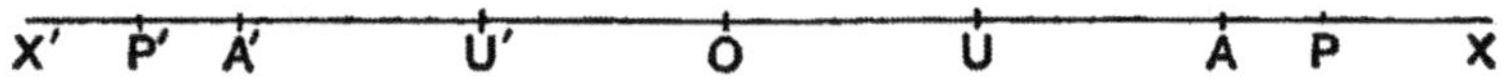

```
X'   P'   A'        U'        O        U        A    P    X
```
FIG. 2.

If OP is equal to xOU, then

$$OP' = -P'O = -OP = -xOU.$$

The positive number x is called the *abscissa* of P with respect to the *origin* O; the negative number $-x$ is called the abscissa of P' with respect to the same origin, and the line $X'OX$ is called the *axis of abscissae*. Every point of the line to the right of O will have a positive number for abscissa, and every point to the left of O a negative number; the abscissa of O itself is zero. Thus if $OA = 2OU$, the abscissa of A is 2; the abscissa of U is 1; the abscissae of U' and A', the points symmetric to U and A, are -1 and -2 respectively.

As thus defined, the abscissa of a point P is the ratio of OP to the unit step OU, taken with the positive or negative sign according as P is to the right or to the left of O, U being supposed to be to the right of O. When a point P has the abscissa x, it is convenient to say that the point P and the number x *correspond* to each other. Thus the point A and the number 2, the point U' and the number -1, the point O and the number 0 correspond to each other.

AXIOM.—**The fundamental axiom** on which the application of Algebra to Geometry rests is that, when the origin O and the unit step OU have been fixed, there is a one-to-one correspondence between the points of the axis and the system of real numbers; that is, to every point on the axis corresponds a definite number, namely the abscissa of the point, and to every number corresponds a point on the axis, namely the point which has the number for abscissa.

When the ratio of OP to OU is a *rational* number, that is, a positive or negative integer or fraction, P is determined by laying off OU, or a submultiple of OU, a certain number of times along the axis, to the right or to the left, according as the number is positive or negative. Thus if the number be $-\frac{7}{3}$, we lay off to the left a line equal to 7 times the third part of OU. When, however, the ratio of OP to OU is an *irrational* number, such as $\sqrt{2}$ or π, the position of P may be determined in practice by taking a rational approximation to the irrational number. Thus for π we may take 3·1 or 3·14 or 3·142, etc., according to the size of the unit line. Of course, whatever size the unit line may be, a stage is soon reached when the closer approximations become indistinguishable in the diagram; if the unit be 1 inch it would be difficult to distinguish the points whose abscissae are 3·14 and 3·142 from each other. Irrational numbers are, however, subject to the same laws of operation as rational numbers, and though in a diagram it may be impossible to distinguish the points corresponding say to π and 3·142 from each other, yet in our reasoning they are to be considered distinct, just as in reasoning about a straight line we consider it to have no breadth, although we cannot represent such a line in a diagram.

Ex. 1. Mark the points whose abscissae are:

$$2\tfrac{1}{2}\,;\ -\tfrac{4}{3}\,;\ \sqrt{2}\,;\ -\sqrt{3}\,;\ -\tfrac{1}{\sqrt{2}}\,;\ \pi\,;\ -\tfrac{\pi}{2}\,;\ -\sqrt{\pi}.$$

Ex. 2. If x be the abscissa of a point, mark the points which are determined by the equations:

$$2x-3=0\,;\ 3x+5=0\,;\ x^2-4=0\,;\ 3x^2-4x-1=0.$$

§ 5. Measure of a Step. If the abscissae of A, B are a, b respectively, then

$$AB = OB - OA = bOU - aOU = (b-a)OU.$$

The number $b-a$ may be taken as the measure of AB; the *numerical* value of $b-a$ gives the ratio of the length of AB to the length of the unit step OU, and the sign of $b-a$ gives the *direction* of AB. Thus if OU be 1 inch, $b=5, a=2$, AB will be 3 inches and B will be to the right of A; if $b=-5$, $a=-2$, AB will be 3 inches long, and since $-5+2$ is negative B will be to the left of A. The unit step OU is generally omitted, and AB is said to be equal to $b-a$.

By the definition of the expression "algebraically greater," b is algebraically greater than a when $b-a$ is positive; therefore when b is algebraically greater than a, B lies to the right of A. Similarly when b is algebraically less than a, B lies to the left of A. We have, therefore, the convenient relation that the number b is algebraically greater or less than the number a, according as the point whose abscissa is b lies to the right or to the left of the point whose abscissa is a. Instead of the expression "the point whose abscissa is a," it will be more compact and equally clear to use the phrase "the point a."

Ex. 1. Determine in sign and magnitude the step AB for the cases:

$a=\frac{1}{2}, b=4$; $a=-1, b=1$; $a=-2, b=-5$; $a=-\sqrt{2}, b=\pi$.

Ex. 2. Show that the abscissa of the middle point of AB is $\frac{1}{2}(a+b)$.

Ex. 3. If $AP:PB=k:1$ show that the abscissa of P is $(a+kb)/(k+1)$.

For if x is the abscissa of P

$$AP=x-a, \quad PB=b-x \text{ and } x-a=k(b-x).$$

What is the sign of k (i) when P lies between A and B, (ii) when P does not lie between A and B?

§ 6. **Axes of Coordinates.** Let $X'OX$, $Y'OY$ (Fig. 3) be two unlimited straight lines at right angles to each other, and P any point in the plane of the diagram; draw PM, PN perpendicular to $X'X$, $Y'Y$ respectively.

When P is given, the steps OM, ON are definitely determined; and conversely when the steps OM, ON are given, P is definitely determined as the point of intersection of the perpendiculars MP, NP.

Let OU be the unit step for the direction $X'X$, OV the unit step for the direction $Y'Y$, and for the present suppose

RECTANGULAR COORDINATES.

these two steps to be of the same length, say an inch. The step OM or its equal, the step NP, will be considered positive when P lies to the right of $Y'Y$, but negative when P lies to the left of $Y'Y$; the step ON or its equal, the step MP, will be considered positive when P lies above $X'X$, but negative when P lies below $X'X$.

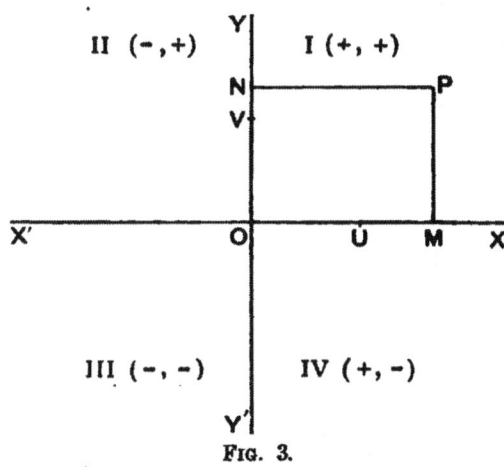

Fig. 3.

Of course, the direction which is to be considered positive may be chosen at pleasure, but unless the contrary is stated, the positive directions will be assumed to be from left to right and from below upwards respectively. Again OM and MP will only be compared as to their *lengths*; we only compare steps with each other when they are on the same straight line or on parallel straight lines. Obviously the theorems that hold for the comparison of steps with each other are true, whatever be the particular line on which the steps are taken, but we have given no definition of equality or of sum or of difference, except when the steps compared are on the same straight line or on parallel straight lines.

Suppose now that

$$OM = NP = aOU; \quad ON = MP = bOV;$$

the numbers a, b are called the *coordinates* of P with respect to the *axes* $X'X, Y'Y$; a is the *abscissa*, b the *ordinate*, and P is described shortly as "the point (a, b)." In thus describing the point the first coordinate is understood to be the abscissa, the second the ordinate. The axes are at right angles to each other, and it will be assumed,

8   AN ELEMENTARY TREATISE ON THE CALCULUS.

unless the contrary is stated, that the axes are always rectangular. O is called the *origin* of coordinates, and its coordinates are 0, 0.

The axes divide the plane into four quadrants; the first quadrant is that bounded by OX, OY, the second by OY, OX', the third by OX', OY', and the fourth by OY', OX.

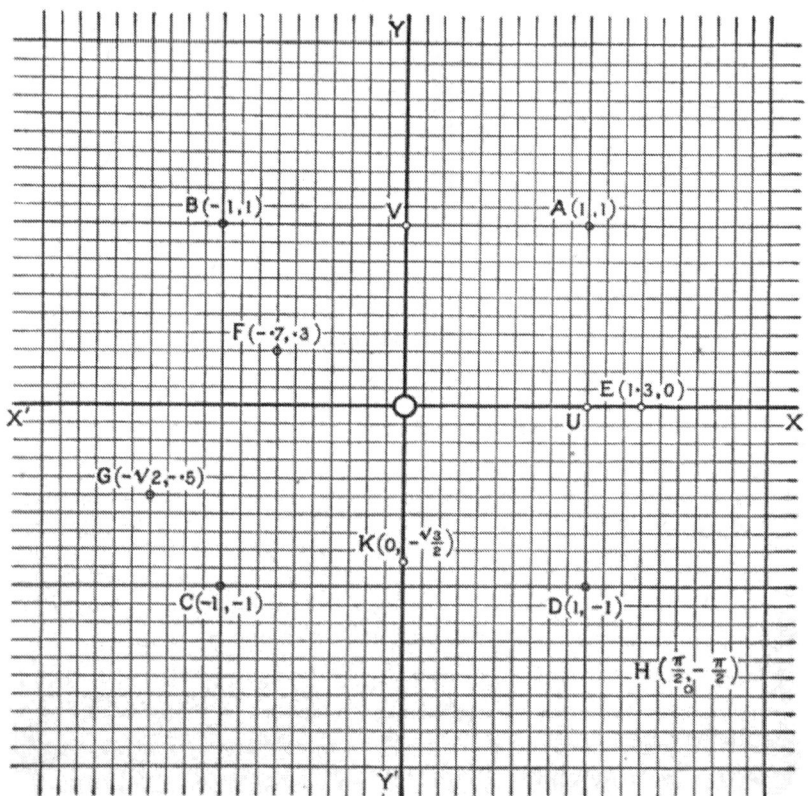

Fig. 4.

The *signs* of the coordinates show at once the *quadrant* in which a point lies: in the first quadrant XOY the signs (the first being that of the abscissa) are $+$, $+$; in the second, YOX', $-$, $+$; in the third, $X'OY'$, $-$, $-$; in the fourth, $Y'OX$, $+$, $-$.

Paper called "squared paper," ruled twice over with two sets of equidistant parallel lines, can be readily purchased, and its use greatly facilitates the plotting of points.

Fig. 4 shows several points referred to the axes $X'X$, $Y'Y$. The four points A, B, C, D are each at unit distance from both axes, but no two of them are in the same quadrant, since no two pairs of coordinates agree both in sign and in magnitude.

E lies on $X'X$, and its ordinate is therefore zero; the abscissa of K is zero, since K lies on $Y'Y$.

Since OU is divided by the faintly ruled lines into 10 equal parts, each of these parts will represent ·1; it is easy, therefore, to mark off a length such as 1·3 or −·7. In the same way $-\sqrt{2}$, $-\tfrac{1}{2}\sqrt{3}$ are represented by -1·41, $-$·87, though the second decimal can only be roughly indicated.

Ex. 1. Plot the points $(1, -2)$; $(-\tfrac{2}{3}, 0)$; $(-3, -2)$; $(0, \tfrac{2}{3})$; $(1, 0)$; $(-1, 0)$; $(0, 1)$; $(0, -1)$; $(\pi, \tfrac{1}{2}\pi)$; $(\sqrt{2}, \sqrt{3})$; $(-\sqrt{2}, -\sqrt{3})$.

Ex. 2. What is the locus of a point whose abscissa is (i) 2, (ii) -2, (iii) 0, (iv) a? What is the locus of a point whose ordinate has these values?

Ex. 3. Two points P, Q are said to be *symmetric with respect to a line* when the line bisects PQ, and is perpendicular to PQ; two points P, Q are said to be *symmetric with respect to a point O*, when O is the middle point of the line PQ. If P is the point (a, b) show

(i) that the point $(a, -b)$ is symmetric to P with respect to $X'X$.

(ii) that the point $(-a, b)$ is symmetric to P with respect to $Y'Y$.

(iii) that the point $(-a, -b)$ is symmetric to P with respect to O.

For simplicity take first the case $a=1$, $b=2$.

Ex. 4. If A is the point (x_1, y_1), B the point (x_2, y_2), and P the point dividing AB in the ratio of k to 1, show, as in § 5, Ex. 3, that the coordinates of P are

$$\frac{x_1+kx_2}{1+k}, \quad \frac{y_1+ky_2}{1+k}.$$

What is the sign of k (i) when P lies between A and B, (ii) when P does not lie between A and B?

§ 7. Distance between two points. Let P (Fig. 5) be the point (x_1, y_1), Q the point (x_2, y_2); draw PM, QN perpendicular to $X'X$, and let PR be drawn parallel to $X'X$ to meet NQ (or NQ produced) at R.

Whatever be the relative position of P and Q, we have for the measures of PR, RQ
$$PR = MN = x_2 - x_1; \quad RQ = y_2 - y_1.$$
As regards magnitude we have, by Euclid I. 47,
$$PQ^2 = PR^2 + RQ^2,$$
and whether the signs of $x_2 - x_1$ and $y_2 - y_1$ be positive or negative the squares of these numbers will give the number of square units in the squares described on PR and RQ. Hence

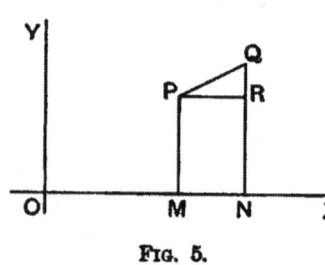

Fig. 5.

$$PQ^2 = (x_2 - x_1)^2 + (y_2 - y_1)^2,$$
and therefore the length of PQ is
$$\sqrt{\{(x_2 - x_1)^2 + (y_2 - y_1)^2\}}$$
where the positive sign must be given to the root.

If Q coincide with O, x_2 and y_2 are both zero, and the length of OP is $\sqrt{(x_1^2 + y_1^2)}$.

The student should verify the result for different positions of P and Q.

Ex. 1. Find the distance between the points (3, 7), (9, 6), the length of the unit being 1 inch.

Let the distance be r inches; then
$$r^2 = (3 - 9)^2 + (7 - 6)^2 = 37; \quad r = \sqrt{37} = 6\cdot083,$$
so that the distance is 6·083 inches.

Ex. 2. Find the distances between the following pairs of points: plot the points in each case.

i. (1, 1), (3, 2). ii. (−1, 1), (3, 2). iii. (−1, 0), (0, 2).

iv. (−2, −3), (2, 3). v. (π, −π), $\left(-\dfrac{\pi}{2}, \dfrac{\pi}{2}\right)$.

Ex. 3. Show that if the point (x, y) be any point on the circle whose radius is 3, and whose centre is the point (2, 1),
$$x^2 + y^2 - 4x - 2y - 4 = 0.$$

§ 8. Polar Coordinates. The position of the point P (Fig. 6) would clearly be determined by the angle which OP makes with the fixed line OX, and by the length of the radius OP. We must be clear, however, as to the meaning of the word "angle." Following the usual convention in Trigonometry, we consider the radius OP to be always positive, and define the angle that OP makes with the *positive* direction of OX as the angle through which a line coinciding with OX (not with OX') has to be turned till it passes through the point P. The angle will be considered positive when the rotation is counter-clockwise.

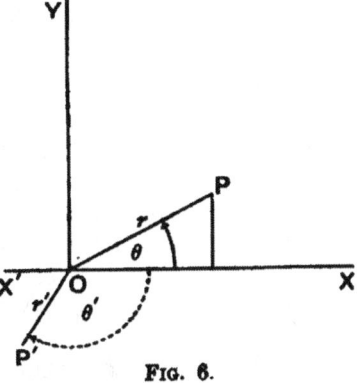

Fig. 6.

If OP be r units of length and the angle XOP θ degrees or radians according to the unit of angle adopted, the two numbers r, θ are called the polar coordinates of P, and P is described as the point (r, θ). Similarly, P' is the point (r', θ'); θ' is negative.

With the usual system of rectangular axes in which OX has to be rotated counter-clockwise through $90°$ till it coincides with OY, the positive direction of the axis $Y'Y$, we see that the polar coordinates (r, θ) of P are connected with the rectangular coordinates (x, y) by the equations

$$x = r \cos \theta, \quad y = r \sin \theta.$$

These equations, when solved for r and θ in terms of x and y, give

$$r = +\sqrt{(x^2 + y^2)}, \quad \tan \theta = \frac{y}{x}.$$

It must be noted, however, that $\tan \theta$ does not definitely determine the angle θ. For if $\tan \theta$ be positive we can only infer that P lies in the *first* or *third* quadrant, while if $\tan \theta$ be negative that P lies in the *second* or *fourth* quadrant. We must consider also the signs of x and y or of $\cos \theta$ and $\sin \theta$.

It is usually most convenient to suppose θ to vary from $-180°$ to $+180°$ so that a point above the axis $X'X$ has a positive angle, and a point below that axis a negative angle.

Ex. 1. If P is the point $(-3, 4)$, find its polar coordinates (r, θ).

$$r = \sqrt{(9+16)} = 5\,;\ \tan\theta = \frac{4}{-3} = -1{\cdot}3333\,;\ \theta = 126°\ 52'.$$

Since $\tan\theta$ is negative, θ is in the second or fourth quadrant; but x or $\cos\theta$ is negative, and therefore θ is in the second.

Ex. 2. If P is the point $(3, -4)$, show that its polar coordinates are $(5, -53°\ 8')$.

§ 9. Variable. Continuity. Let A be a fixed point on a line, say, on the x-axis $X'X$, and let a point P start from the position A and move steadily along the axis, say to the right, till it reaches another position B. The segment AB described by the point P is the most perfect type of a *continuous* magnitude; there is no gap or break in it. As P moves from A to B, the step AP steadily increases; AP is a *continuously varying* magnitude during the motion of P.

If a, b are the abscissae of A, B, and x the abscissa of P at any stage of the motion, then, as P moves from A to B, since $AP = x - a$,* x steadily increases (algebraically) from a to b; x is a *continuously varying number* or, more briefly stated, x is a *continuous variable*.

Again, since P coincides in succession with every point lying between A and B, so x assumes in succession every value lying between a and b. If a be negative and b positive, A will be to the left and B to the right of the origin O, and when P passes through O, x will be zero so that as x passes from negative to positive values it passes through the value zero. Had P instead of moving always to the right moved sometimes forward, sometimes backward, then every time it passed through O the value of x would have been zero, so that x would only change from negative to positive or from positive to negative by passing through zero.

We will assume then, as characteristic of a continuous variable, that as it varies continuously from a value a to a value b it assumes once at least every value inter-

---

*Here, and in similar cases, it is *the measure* of the step AP that is of importance; it will cause no confusion to let AP stand for the step, and also for the measure of the step as is usually done in all applications of geometrical theorems.

mediate to a and b; if one of these values, say a, be negative and the other positive, one of the values the variable takes will be zero.

§ 10. **Geometrical Representation of Magnitudes.** The measure x of any magnitude A is the ratio of A to another magnitude U of the same kind that is chosen as the unit. If then on any axis a unit step OU is taken as representing the unit magnitude U, the step OM where OM is equal to xOU will represent the magnitude A. There is thus established a correspondence between the magnitudes of the particular kind considered and the points of the axis; the point 1 corresponds to the unit magnitude U, the point 2 to the magnitude $2U$, and so on.

Many of the magnitudes considered in Geometry and Physics, for example, lines, angles, velocities, forces, are often treated as *directed* magnitudes, and their measures may then be either positive or negative; when the measures are negative, the points that correspond to the magnitudes will lie on the opposite side of O from that on which U lies.

A variable magnitude P will be represented by a variable segment OP, and when the magnitude varies continuously the point P will trace out a continuous segment of the axis.

For purposes of calculation it is the measure of the magnitude that is of importance, and, to avoid a tedious prolixity of statement, such an expression as "a velocity v" will often be used in the sense "a velocity whose measure is v units of velocity." Of course in all cases care should be taken to prevent ambiguity as to the units employed.

§ 11. **Function. Dependent and Independent Variables.** In any problem the magnitudes dealt with will usually be of two classes, namely, those that retain the same value all through the investigation and those that are supposed to take different values: the former are called *constants*, the latter *variables*. It has become customary to denote constants by the earlier letters of the alphabet, $a, b, c, \ldots$, and variables by the later letters, $z, y, x, \ldots$. Of course when there is any advantage in denoting a variable by a or a constant by z there is no reason against doing so.

Again, taking first the case of only two variables, it will usually happen that when one of the variables is given a series of values the other variable will take a series of definite values, one for each that the first is supposed to have been given. The second variable is then said to be a *function* of the first, or to be a variable *dependent* on the first, which is distinguished as the *independent* variable. Instead of the phrase "independent variable," the word *argument* is often used, and the dependent variable is then called a function of its argument.

Thus, if we consider a series of triangles, all of the same altitude, the area of any triangle is a function of its base. The distance travelled by a train which moves at a constant speed is a function of the time during which it has moved at that speed. The pressure of a given quantity of gas which is maintained at a constant temperature is a function of its volume. In these examples the independent variable or argument is the base, the time, the volume; and the dependent variable or function is the area, the distance, the pressure respectively.

It is usually a mere matter of convenience which of the two variables is considered as independent. Thus if the time at which the train passed certain stations on the railroad were the subject of inquiry, the distance would be taken as the independent variable and the time as the dependent.

When there are more than two variables it may happen that when definite values are assigned to all but one of them the value of that one becomes determinate; this one variable is then said to be a function of or to be dependent on the other variables which are called the independent variables of the problem.

Thus the area of a triangle is a function of the base and of the altitude when both base and altitude vary. The pressure of a given quantity of gas is a function of the volume and of the temperature when both volume and temperature vary.

Generally, a variable y is said to be a function of another variable x when to every value of x there corresponds a definite value of y; a variable y is said to be a function of two or more

variables, x, u, ..., **when to each set of values of the variables x, u, ... there corresponds a definite value of y.**

While it is important to keep this general notion of functional dependence in mind, it will, however, be usually assumed that a function is defined by an equation (see §§ 13, 26, 27, 28), and that it can be represented by a graph (§ 16). This assumption implies (i) that as the argument varies continuously, in the sense explained in § 9, from a value a to a value b, the function also varies continuously from a value, A say, to a value B; (ii) that to a small change in the argument corresponds also a small change in the function. The assumption implies a good deal more than what is here stated, but at this stage the student is earnestly urged to pass lightly over the purely theoretical difficulties and to try to get a thorough grasp of the fundamental conceptions of variation and functional dependence by working out for himself the graphical exercises in the next chapter. He will find by trial that, except for special values of the argument, the property (ii) is actually found in all the ordinary functions; the property (i), though apparently simpler, is really much harder to demonstrate mathematically. A mathematical definition of the continuity of a dependent variable will be given in Chapter V., § 44.

The student should notice the phrase "definite value" or "determinate value." It may happen that the analytical expression for a function ceases to have meaning for certain values of the argument; for these values, therefore, the function is *not defined*. Thus the function $(x^2-1)/(x-1)$ is defined for all values of x, *except the value* 1; because when $x=1$ the expression takes the form $0/0$, which is absolutely meaningless. We should not get out of the difficulty by first dividing numerator and denominator by $x-1$ and then putting 1 for x; because in dividing by $x-1$ we assume that $x-1$ is not zero, division by zero being excluded by the fundamental laws of algebra.

Again, such a function as $\sqrt{(1-x^2)}$ is only defined for values of x that are numerically less than or equal to 1; in this case we may say that the function is defined for values of the argument in the range from -1 to $+1$ inclusive.

It has always to be understood in reasoning about a function that only those values of the argument are to be considered for which the function has a definite value, or, in other words, for which the function is *well-defined*.

§ 12. Notation for Functions. A function of a variable is often denoted by enclosing the variable in a bracket and prefixing a letter; thus, $f(x)$, $F(x)$, $\phi(x)$ denote functions of x. The letters f, F, ϕ are functional symbols, not multipliers; the symbol $f(x)$ must be taken as a whole, and means simply "some function of x," the context or some explicit statement determining which particular function is meant. For different functions occurring in the same investigation different functional symbols must of course be used.

$f(a)$ means "the value of the function $f(x)$ when x has the value a," or "the value of the function $f(x)$ when x is replaced by a." Thus, if $f(x)$ denote the function
$$x^2 - 3x - 1,$$
then $f(0) = -1$; $f(1) = -3$; $f(a+b) = (a+b)^2 - 3(a+b) - 1$;
$$f(x^2) = (x^2)^2 - 3x^2 - 1 = x^4 - 3x^2 - 1.$$

A similar notation is used for functions of two or more variables; thus, $f(x, y)$, $F(p, v)$, $\phi(x, y, z)$ denote functions of x and y, of p and v, of x, y, and z respectively.

If $\qquad f(x, y) = 3x^2 - 2xy - y^2 + 4,$
then $\qquad f(1, -1) = 3 + 2 - 1 + 4 = 8;$
$\qquad\qquad f(a, b) = 3a^2 - 2ab - b^2 + 4.$

The letters should be separated by a comma to indicate that there are two or more variables, and thus distinguish the function from one in which the argument is the product of two or more variables. Thus, $f(xy)$ is a function whose argument is the product xy, and if $f(x)$ be $ax + b$, then $f(xy)$ is $axy + b$.

§ 13. Explicit and Implicit Functions. One variable is usually defined as a function of another by an equation. The dependent variable is called an *explicit* function of its argument, or is said to be given *explicitly* when the equation is solved for the dependent variable in terms of the argument. Thus

$$y = x^3 - 2x + 3; \quad s = \cos(nt + e); \quad p = \frac{c}{v};$$

$$y = f(x); \quad s = \phi(t); \quad p = F(v)$$

are equations which give y, s, p explicitly as functions of x, t, v respectively.

When the equation is not solved, the dependent variable is called an *implicit* function of its argument, or is said to be given *implicitly*. Thus y is given as an implicit function of x by the equation

$$axy + bx + cy + d = 0.$$

This equation when solved for y in terms of x gives

$$y = -\frac{bx + d}{ax + c},$$

and y is now an explicit function of x.

§ 14. **Multiple-valued and Inverse Functions.** When a function is given implicitly by an equation, it may happen that to one value of the one variable there correspond two or more values of the other. The definition of a function given in § 11 assumes that to each value of the argument there corresponds but one value of the function, and in reasoning about a function we must always suppose that it has but one value for each value of its argument; in other words, that the function is *single-valued*. When the defining equation gives more than one value of the one variable for one value of the other, we can usually consider the equation as defining a function that is made up of two or more functions each of which is single-valued; such a function is called a *multiple-valued function*.

Thus, if y is given as a function of x by the equation

$$x^2 + 2xy - y^2 - 1 = 0,$$

then
$$y = x \pm \sqrt{(2x^2 - 1)},$$

and to each value of x there correspond *two* values of y; y is a two-valued function of x. The equation really gives two functions of x, namely,

$$y = x + \sqrt{(2x^2 - 1)}, \quad y = x - \sqrt{(2x^2 - 1)},$$

each of which is single-valued, and defined for those values of x for which $2x^2$ is greater than or equal to 1.

Again, the equation

$$x^2 - y + 1 = 0$$

defines y as a single-valued function of x, but x as a two-valued function of y, namely x is either $\sqrt{(y-1)}$ or $-\sqrt{(y-1)}$.

When the graphical representation of functions is considered, it will be seen that the separate functions represent different parts of the one curve (*e.g.* § 20).

The equation $x^2 - y + 1 = 0$, as we have just seen, not only defines y as a function of x but also defines x as a function of y. More generally, the equation $y = f(x)$, which defines y explicitly as a function of x, also defines x implicitly as a function of y; the two functions thus defined by the one equation are said to be *inverse* to each other.

For example the equation $y = x^3$ when solved for x gives $x = \sqrt[3]{y}$ and thus defines two functions which are inverse to each other, namely *the cube* and *the cube root*.

It is usual in English books to employ f^{-1} as the symbol of the function inverse to that denoted by the symbol f so that

$$x = f^{-1}(y) \text{ when } y = f(x).$$

The student will be already familiar with this notation in the case of angles. Thus $\sin^{-1} y$ means, not $1/\sin y$ but, the angle (within a certain range) whose sine is y; and just as we have the identity, $\sin(\sin^{-1} y) = y$, so we have

$$f\{f^{-1}(y)\} = y$$

or, as the identity is usually written,

$$ff^{-1}(y) = y.$$

Again it may well happen that the inverse function is not single-valued. Thus, $\sin^{-1} x$ may, unless some restriction be imposed, be any one of an infinite number of angles. To secure definiteness some restriction has in such cases to be placed on the range of the variable; for example, $\sin^{-1} x$ may be restricted to angles lying between $-\pi/2$ and $+\pi/2$ (inclusive of $-\pi/2$ and $+\pi/2$), and then $\sin^{-1} x$ is single-valued. For further information, see §§ 25, 27, 28.

EXERCISES I.

1. If $f(x)=x^3-x+1$, find $f(0)$, $f(1)$, $f(-1)$, and show that
$$f(x+h)=f(x)+(3x^2-1)h+3xh^2+h^3.$$

2. If $f(x)=x^2-x-2$, write down $f(ax+b)$.

3. If $f(x)=x^2-5x+1$, write down $f(x^2)$, $f(x^3)$, $f(\sin x)$.
What is the value of $f\left(\sin\dfrac{\pi}{4}\right)$?

4. If $f(x)=\log x$, show that
$$f(xy)=f(x)+f(y)\,;\quad f\left(\frac{x}{y}\right)=f(x)-f(y).$$

5. If $f(x)=ax^6+bx^4+cx^2+d$, show that $f(-x)$ is equal to $f(x)$.
When $f(-x)=f(x)$, the function $f(x)$ is called an *even* function of its argument.

6. If $f(x)=ax^7+bx^5+cx^3+dx$, show that $f(-x)$ is equal to $-f(x)$.
When $f(-x)=-f(x)$, the function $f(x)$ is called an *odd* function of its argument.

7. Show that $\sin x$, $\operatorname{cosec} x$, $\tan x$, $\cot x$ are odd functions of x, and that $\cos x$, $\sec x$ are even functions of x.

8. Show that $(e^x-e^{-x})/x$ is an even function of x.

9. If $f(x, y)=ax^2+bxy+c$, write down $f(y, x)$, $f(x, x)$, and $f(y, y)$.

10. If $y=f(x)=\dfrac{x+2}{3x+4}$, show that $f(y)=\dfrac{7x+10}{15x+22}$.

11. If $y=f(x)=\dfrac{ax+b}{cx-a}$, show that $x=f(y)$.

12. If $f(x, y)=x^2-y^2$, show that $f(\cos\theta,\sin\theta)=\cos 2\theta$, and that $f(\sec\theta,\tan\theta)=1$.

CHAPTER II.

GRAPHS. RATIONAL FUNCTIONS.

§ 15. Object of the Calculus. Graphs. Stated in the most general terms the object of the Calculus may be said to be the study of the changes of a continuously varying function. The investigation of the *rate* at which a given function is changing for any specified value of its argument belongs to the Differential Calculus; the converse problem of determining the *amount* by which a function changes for a specified change in its argument, when the rate of change of the function is known, belongs to the Integral Calculus.

An almost indispensable aid to this study is furnished by the graphical representation of a function, and for the sake of those students who may have had little or no experience in graphical work a few hints will now be given that may be of service to them. At times the tracing of a graph involves a good deal of tedious calculation, but the student will be well repaid for his labour by the insight he will obtain into the fundamental conceptions of *variation* and *continuity* of a function. When he has made but a little progress in the differential calculus he will find several methods of reducing the necessary calculations. An extremely good discussion of graphs from an elementary standpoint will be found in Professor Chrystal's *Introduction to Algebra*. (London: A. & C. Black.)

§ 16. Graph of x^2. In geometry and physics we frequently find a function defined by an equation of the form $y = cx^2$ where c is a constant. Thus the area of a circle varies as the square on the radius; the distance that a body falls from rest, the resistance of the air being neglected, varies as the square of the time of fall; the heat generated by an electric current in a given time varies as the square of the current in the circuit and so on. These statements when expressed in the usual algebraical way all lead to an equation of the above form; x denotes the number of units of the one kind of quantity, for example the number of feet, or the number of seconds, or the number of amperes; y denotes the number of units of the second kind, for example the number of square feet, or the number of linear feet, or the number of ergs (or other heat units). The number c is a constant, that is, does not change when x changes; it is not, however, the same constant in the different problems; thus for the area of the circle $c = \pi$, for the falling body $c = \frac{1}{2}g$, for the electric circuit c depends on the resistance and on the heat unit.

Suppose for simplicity that $c = 1$; the more general case can be deduced from this one. Let $X'X$, $Y'Y$ be two rectangular axes (Fig. 7), OU, OV unit segments on these axes. Give to x a series of values, and from the equation $y = x^2$ deduce the corresponding values of y. Associating each value of x with the corresponding value of y, we obtain a series of pairs of numbers, and each pair may be taken as the coordinates of a point in the plane of the diagram, the value of x being the abscissa and the corresponding value of y the ordinate of the point. If the values given to x form an increasing or a decreasing series of numbers, and if the difference between any two consecutive values be small it will be found that the consecutive points determined on the diagram lie pretty close to each other; the curve drawn through these points with a free hand is called the *graph* of the function x^2.

Tabulating values, we have

x	0, ·1, ·2, ·3, ... 1, 1·1 ...
y	0, ·01, ·04, ·09, ... 1, 1·21 ...
x	−·1, −·2, −·3, ... −1, −1·1 ...
y	·01, ·04, ·09, ... 1, 1·21 ...

Take OU, OV each, say, 1 inch and plot the points

22   AN ELEMENTARY TREATISE ON THE CALCULUS.

(0, 0), (·1, ·01) ... (−·1, ·01), (−·2, ·04) ... ; by drawing a curve through the points we get the graph of x^2 (Fig. 7).

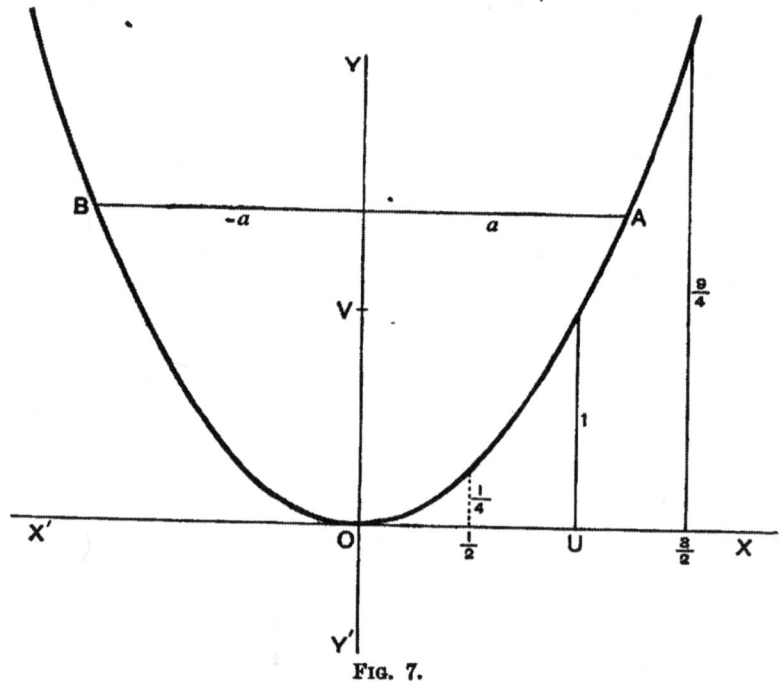

Fig. 7.

Of course only a comparatively small number of points can be plotted, but by actual calculation we find that a small change in x produces but a small change in y; we are therefore warranted in concluding that an ordinate corresponding to a value of x that has not been used in plotting the points but that lies between two values that have been used can differ but little from the ordinate of the graph corresponding to that value of x. When there is any room for doubt, a few more values of y at closer intervals may be calculated.

When x is at all large, y will be much larger and it becomes impossible to plot the points in the diagram; we must then try to follow in imagination the course of the graph or if it be of importance to know the form of the graph for such values we may take the unit lines OU, OV smaller. See further § 19.

§ 17. **Equation of a Curve. Symmetry. Turning Values.**
Let us now consider the graph of x^2 from the purely geometrical point of view.

(i) *Equation of the Curve.* A point in the plane will or will not be on the graph of x^2 according as the ordinate of the point is or is not equal to the square of its abscissa; in other words, *the condition that a point should lie on the graph* is that the coordinates of the point should satisfy the equation $y = x^2$ which states the law according to which the curve was constructed. This equation is generally called *the equation of the curve*, and the curve is said to be represented by the equation; the two expressions "the graph of the function x^2" and "the curve whose equation is $y = x^2$" (or "the curve represented by the equation $y = x^2$") mean the same thing.

More generally, "the graph of the function $f(x)$" and "the curve whose equation is $y = f(x)$" mean the same thing, and the condition that a point should lie on the curve or graph is that its coordinates should satisfy the equation $y = f(x)$. Thus the point $(-\frac{1}{2}, \frac{1}{4})$ does, and the point $(-\frac{1}{3}, \frac{1}{4})$ does not, lie on the graph of x^2; the origin lies on the graph of x^2 but not on that of $x^2 + 1$.

(ii) *Symmetry.* The ordinate of the point on the graph of x^2 which has $-a$ for its abscissa is equal to the ordinate of the point which has a for its abscissa, since each ordinate is a^2. If A is the point (a, a^2) and B the point $(-a, a^2)$ AB will be perpendicular to OY and will be bisected by OY; that is, since a may be any number whatever, the graph is symmetrical about OY, or OY is an axis of symmetry (cp. § 6, ex. 3). In plotting the graph by points therefore, it would be sufficient to calculate y from positive values of x alone; the part of the curve to the left of OY is simply the reflection in OY of the part to the right. We might imagine the plane of the diagram turned through two right angles about OY and the part of the curve originally to the right of OY would after rotation form the part to the left of OY.

The graph of a function $f(x)$ is not, as a rule, symmetrical about the y-axis or about any other line; but the function should always be examined for symmetry since the presence of symmetry saves labour. The graph of $f(x)$ will be symmetrical about OY if $f(x)$ is an *even* function (Exer. I., ex. 5), for in that case the ordinate $f(-a)$ of the point whose abscissa is $-a$ is equal in sign and in magnitude to the ordinate $f(a)$ of the point whose abscissa is a.

(iii) *Variation of the Function.* Suppose a point to start from O and move along the graph. At first the ordinate of the point increases very slowly; as the point gets nearer to the point (1, 1) its ordinate grows more rapidly; when it has passed (1, 1) its ordinate grows still more rapidly. As x increases from 0 to $\frac{1}{2}$ the ordinate increases from 0 to $\frac{1}{4}$; as x increases from $\frac{1}{2}$ to 1 the ordinate increases from $\frac{1}{4}$ to 1; as x increases from 1 to $\frac{3}{2}$ the ordinate increases from 1 to $\frac{9}{4}$. Thus for the same increase of $\frac{1}{2}$ in x the ordinate increases by the amounts $\frac{1}{4}, \frac{3}{4}, \frac{5}{4}$ respectively. The course of the graph shows very clearly that after a certain point has been reached the ordinate grows more rapidly than the abscissa while near the origin it grows less rapidly; the graph thus gives a vivid picture of the variation of the function x^2 represented by the ordinate.

(iv) *Turning Values.* If a point move along the graph from any position on the left of OY to any position on the right the ordinate of the point decreases till the point reaches O and then increases. The point O where the ordinate ceases to decrease and begins to increase is called a *turning point* of the graph, and by analogy the value of the function x^2 at O, namely zero, is called a *turning value* of the function. The turning value is in this case a *minimum* value of the function or ordinate.

In general those points on a graph at which the ordinate ceases to decrease and begins to increase, or else ceases to increase and begins to decrease are called *turning points* of the graph, and the corresponding values of the function *turning values*; the turning values are respectively *minima* and *maxima* values of the function, that is values respectively less and greater than any other values of the function in their neighbourhood.

§ 18. Graph of cx^2. We might by assigning values to x, and calculating the corresponding values of y from the

SYMMETRY. TURNING VALUES. 25

equation $y = cx^2$ construct the graph of cx^2; it will be instructive to consider another method of deriving the graph.

First let c be positive. Let any ordinate of the graph of x^2 be denoted by y_1 and the ordinate of the graph of cx^2 *for the same value of x* by y_2; then $y_2 = cy_1$, because $y_1 = x^2$, $y_2 = cx^2$ and x is the same number in both equations. The two ordinates may be called "corresponding ordinates."

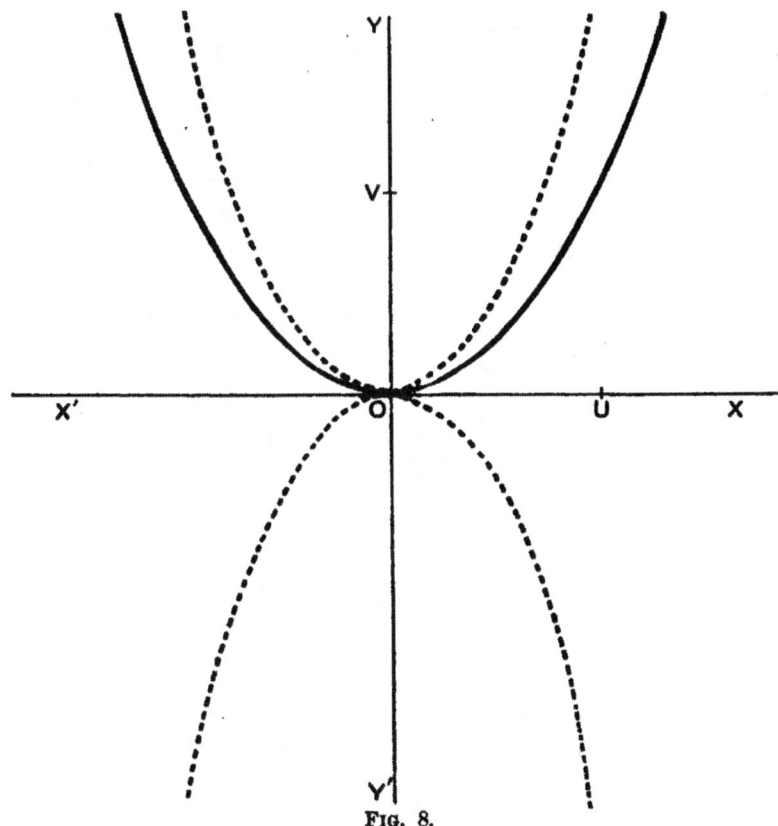

Fig. 8.

Hence to obtain any ordinate of the graph of cx^2 we have only to multiply the corresponding ordinate of that of x^2 by c; in other words, if MP is any ordinate of the graph of x^2 divide MP or MP produced at P' so that MP'

is to MP as c to 1 and P' will be a point on the graph of cx^2.

The upper dotted curve (Fig. 8) is the graph of $2x^2$, and is obtained by doubling each ordinate of the graph of x^2 (full curve). It will be noticed that the general character of the two graphs is the same; the graph of $2x^2$ however recedes more rapidly from $X'X$ than does that of x^2 and is steeper. In general, the graph of cx^2 lies above or below that of x^2 according as c is greater or less than 1.

Next let c be negative, say -2. The graph of $-2x^2$ may be got from that of $2x^2$ by reflection in $X'X$, or by rotating the graph of $2x^2$ through two right angles about $X'X$; for the ordinates of the graph of $-2x^2$ are simply those of the graph of $2x^2$ with signs changed. The lower dotted curve is the graph of $-2x^2$; O is a turning point of the graph and zero a *maximum* value of the function $-2x^2$, the value being taken *algebraically*.

§ 19. Scale Units. Let us now consider the graph of cx^2 as the geometrical representation of the law of falling bodies; c may be taken as 16 when the foot and the second are the units of space and time. The graph shows clearly how rapidly the distance fallen increases with the time, for the curve moves rapidly away from the axis OX; in this case the part of the curve to the left of OY does not belong to the representation since negative values of x are not considered.

But if OU and OV are, as has been supposed, of the same length it will be impossible to represent the connection between the distance fallen and the time of fall, even for values of x up to 1, within the limits of an ordinary sheet unless OU and OV are both very small. The remedy is to choose these segments of different lengths. The foot and the second are magnitudes of different kinds and there is no necessity therefore that the segment which represents 1 second should be of the same length as that which represents 1 foot, nor is it implied in the definition of the coordinates of a point that OU and OV should be of the same length. M being the foot of the perpendicular from P on $X'X$, the coordinates of P are x, y if $OM = xOU$, $MP = yOV$

and P is definitely determined whether OU, OV are of the same length or not.

In the case of $y = 16x^2$ we might therefore take OU equal to 1 inch and OV equal, say to $\frac{1}{16}$th of an inch; an abscissa 1 inch long would therefore represent 1 second while an ordinate 1 inch long would represent 16 feet; an abscissa 2 inches long would represent 2 seconds, an ordinate 2 inches long would represent 32 feet and so on. A similar choice would in other cases bring the graph within manageable size.

But even when the two magnitudes whose connection is represented by a graph are of the same kind it is often advisable to have units of different lengths. The value of the graph will not be thereby impaired; the purpose of the graph is to show to the eye how one magnitude changes as another with which it is connected changes, and the ratio of the two lines, say MP and NQ, which represent any two values of the first magnitude is independent of the size of the line which represents the unit magnitude. For

$$MP : NQ = y_1 OV : y_2 OV = y_1 : y_2$$

where OV represents the unit magnitude and $y_1 OV$, $y_2 OV$ the two values considered.

Thus, in a contour road map, if the heights were represented on the same scale as the horizontal distances, it would be difficult to trace the character of the road; hence the heights are exaggerated by using a much larger unit for the vertical than for the horizontal distances. If the graph is to be used to determine actual heights, the scale of the drawing must of course be given.

§ 20. **Coordinate Geometry.** Many of the properties of a curve can be most simply investigated by using the equation of the curve; the study of curves from this point of view is the subject of coordinate geometry.

On the one hand the curve may be defined by some geometrical property; the law of the curve is then expressed in the equation of the curve. Thus the law of the circle is that every point on it is at the same distance from the centre. Now, taking rectangular axes, let O be the centre of the circle, c its radius and $P(x, y)$ any point

on it. Then (§ 7) OP^2 is equal to x^2+y^2; also OP is equal to c. Hence
$$x^2+y^2=c^2 \quad \ldots\ldots\ldots\ldots\ldots\ldots\ldots\ldots\ldots(1)$$
and this equation is true for the abscissa and the ordinate of every point on the circle but of no other point. As P moves round the circle, x and y change in value, but *always* the sum of their squares is equal to c^2. Equation (1) is therefore called the equation of the circle with radius c.

On the other hand an equation between x and y defines y as a function of x, and the graph of this function may be plotted point by point; numerous examples will be found in later articles. As a simple case we might consider the equation $y=x^2$ which gives the graph of § 16; or we might take equation (1). In that case y is defined as a two-valued function of x, $y=\pm\sqrt{(c^2-x^2)}$, for values of x from $x=-c$ to $x=+c$; clearly if x is numerically greater than c y is imaginary. The graph will be symmetric about the axis $X'X$, and by considering the inverse function $x=\pm\sqrt{(c^2-y^2)}$, we see that the graph is also symmetric about $Y'Y$. We might then plot points for which x and y are both positive and thus arrive at the form of the graph. The two functions $+\sqrt{(c^2-x^2)}$ and $-\sqrt{(c^2-x^2)}$ are represented respectively by the semicircles above and below the x-axis.

In later sections it will be seen how the geometrical properties of the graphs of the simpler functions can be deduced from the equations (see § 26).

If in plotting the graph of the function defined by equation (1) the units OU, OV are of different lengths the graph will *seem* to be not a circle, but an ellipse (Exer. V. 4); if OV be, say, half of OU, each ordinate will be only half the actual length of the ordinate of the circle. So long as OU, OV are of the same length the *shape* will not be altered; a change in the size of the units, so long as the units remain of equal length, only enlarges or reduces the figure since all lines are altered in the same proportion.

Even in studying the geometrical properties of curves, however, it is often necessary to choose units of different lengths in order to get the curve represented on a sheet of reasonable size; it must then be borne in mind that the graph will only show the ratios and not the actual lengths of the lines whose measures are the numbers taken as the ordinates.

In all cases the units should be chosen so as to make the graph as large as possible; a diminutive graph usually defeats the end of its existence.

EXERCISES II.

1. Are the points $A(\frac{1}{2}, 1)$, $B(\frac{1}{3}, \frac{4}{9})$, $C(-\frac{1}{4}, \frac{1}{4})$, $D(5, 100)$, $E(3, 40)$ on the curve whose equation is $y = 4x^2$?

2. Is the y-axis $Y'OY$ an axis of symmetry for the graph of any of the functions—

   (i) $2x^2 - 3x^4$;   (ii) $2x^3 - 3x^5$;   (iii) x^{2n};   (iv) x^{2n+1} (n integral);

   (v) $(x+1)/(x^2+1)$;   (vi) $1/(x^2+1)$;   (vii) $a + bx^2 + cx^4 + dx^6$?

   Does the point $(1, -1)$ lie on any of the graphs? What must be the value of a if the origin lies on the graph of (vii)?

3. Trace the graphs of the following functions for values of x between -2 and $+2$, and find the turning points of the graphs and the abscissae of the points where the graphs cross the axis of abscissae—

   (i) $x^2 - 1$;   (ii) $2x^2 - 1$;   (iii) $-2x^2 + 1$;

   (iv) $3x - 2x^2$;   (v) $1 - x - x^2$;   (vi) $-1 + 3x - 2x^2$.

   How may the graphs (i), (ii), (iii) be derived without calculation from the graphs of x^2, $2x^2$, $-2x^2$ respectively? How may the graph of (iii) be derived from that of (ii), and the graph of (vi) from that of (iv)?

4. Having given the graph of *the function* $f(x)$, show how to obtain the roots of *the equation* $f(x) = 0$. Illustrate from the graphs of ex. 3.

   [Let a be the abscissa of any point A on the graph; by the nature of a graph the ordinate of A is $f(a)$. Hence, if $f(a) = 0$, A must be on the axis of abscissae; but if $f(a) = 0$, then a is a root of *the equation* $f(x) = 0$. Therefore the roots of the equation $f(x) = 0$ are the abscissae of the points where the graph of $f(x)$ crosses the axis of abscissae.]

5. Trace the curve whose equation is $y = x^3$.

   To every point P on the curve there corresponds another point P' on the curve which is symmetric to P with respect to the origin (§ 6, ex. 3.); for if P is the point (a, b), P' is the point $(-a, -b)$, and when $b = a^3$ then also $-b = (-a)^3$. When, as in this case, the equation is not altered by replacing x and y by $-x$ and $-y$ respectively, the origin is called *a centre of symmetry of the curve*.

6. On which of the curves given by the following equations is the origin a centre of symmetry—

   (i) $y = ax^3 + bx^5$;   (ii) $y = x^{\frac{5}{3}}$;   (iii) $y = x^{\frac{3}{2}}$;   (iv) $ax^2 + by^2 = c$?

§ 21. The Linear Function.

If any point be taken on the bisector of the angle XOY, the ordinate of the point will be equal both numerically and in sign to the abscissa of the point; but if any point not on that bisector be taken its ordinate will not be equal *both numerically and in sign* to its abscissa. Hence the bisector has for equation $y = x$; the bisector is the graph of the function x.

Similarly $y = -x$ is the equation of the bisector of the angle YOX'.

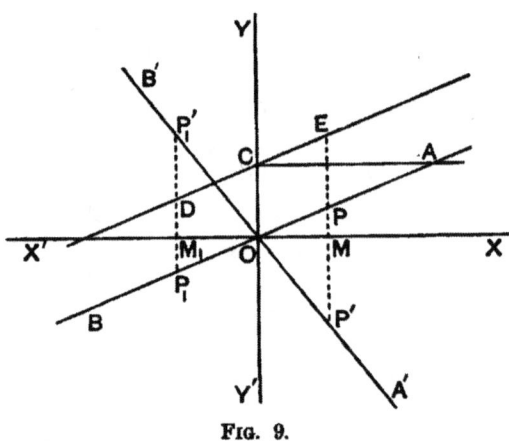

Fig. 9.

If P is any point on the straight line BOA (Fig. 9) and if x, y are the coordinates of P, then $y = x \tan XOA$; this equation is true whether the coordinates of P are both positive or both negative as when P has the position P_1. Conversely, if the point is not on BOA the equation $y = x \tan XOA$ will not be true for the coordinates of the point. Hence the straight line BOA has for its equation $y = x \tan XOA$; BOA is the graph of the function $x \tan XOA$.

Similarly $y = x \tan XOA'$ is the equation of the straight line $B'OA'$; the angle XOA' and $\tan XOA'$ are both negative.

Hence the equation $y = ax$ always represents a straight line through O, the origin of coordinates, and a is the tangent of the angle which the line makes with OX.

THE LINEAR FUNCTIONS. 31

If through C a line DCE be drawn parallel to BOA then
$$ME = MP + PE = OM \tan XOA + OC$$
$$M_1D = M_1P_1 + P_1D = OM_1 \tan XOA + OC$$
by the rule for addition of steps (§ 2).

Hence if x, y are the coordinates of E and OC is equal to b
$$y = x \tan XOA + b$$
and the same equation holds if x, y instead of being the coordinates of E are the coordinates of D or of any other point on DE.

If C were taken on OY', the only difference would be that its measure b would be a negative number.

The graph of any function of the form $ax + b$ is therefore a straight line; a is the tangent of the angle which the line makes with OX and b is the distance from O of the point where the line crosses the axis OY, or as it is usually called the *intercept* on OY. (See also Exer. III., ex. 2.)

If $a = 0$, the line is parallel to the axis OX if b is not also zero; if both a and b are zero the line is the axis itself.

The equation $x = c$ represents a line parallel to the axis OY if c is not zero; if $c = 0$, the equation represents the axis $Y'Y$. In this case, the line is perpendicular to OX and the tangent of the angle it makes with OX is infinite.

Since the graph of $ax + b$ is a straight line, $ax + b$ is often called a *linear* function of its argument x.

It is important that the student should attach a definite meaning to the phrase "the angle that a straight line makes with the axis of abscissae." We make the following convention which will save constant repetitions; the line is understood not to be perpendicular to OX. Through O draw a parallel to the given line; by *the angle which the given line makes with OX* is meant the acute angle (positive or negative) through which a line coinciding with OX (not OX') must be turned till it coincides with the parallel through O: or, what amounts to the same thing, it is the acute angle (positive or negative) through which a line drawn from any point on the given line parallel to OX (not OX') must be turned till it coincides with the given line.

Thus the angle which DE makes with OX is XOA or ACE, if CA be parallel to OX; this angle is positive. The angle which $B'A'$ makes with OX is XOA' and is negative.

§ 22. Gradient. The *gradient* of a line is the tangent of the angle the line makes with the axis of abscissae OX; the gradient is therefore positive or negative according as the angle is positive or negative. Instead of "gradient" the word "slope" is used by some writers; but the term "gradient" is already well established in this meaning.

If we suppose the axis of abscissae OX to be horizontal and the axis of ordinates OY vertical, the positive directions being to the right and upwards respectively, we can describe the motion of a point which moves along the line briefly thus: as the projection of the point on $X'X$ moves to the right or to the left the point itself moves upwards or downwards; or, if the coordinates of the point be (x, y), we may say, as the *point* x moves to the right or left the *point* (x, y) moves upwards or downwards.

When, as on the straight line DE, the gradient is *positive* we see that as the point x moves to the right the point (x, y) on the line moves *upwards*; but when, as on the straight line $B'A'$, the gradient is *negative*, as the point x moves to the right the point (x, y) on the line moves *downwards*. Of course if the direction of motion of the point x be reversed so is that of the point (x, y). Instead of "the point (x, y) on a line or curve" we shall sometimes say simply "the graphic point" meaning the point supposed to be describing the graph.

EXERCISES III.

1. Find the gradients of and the intercepts on the axis of y made by the lines whose equations are

(i) $y = -x + 2$; (ii) $y = \frac{3}{2}x - 1$; (iii) $y = -\frac{3}{2}x - 1$.

Trace the lines on a diagram.

2. Show that the equation
$$2y + 3x - 1 = 0$$
represents a straight line, and find its gradient.

The equation may be written $y = -\frac{3}{2}x + \frac{1}{2}$; it therefore represents a straight line with the gradient $-\frac{3}{2}$.

EXERCISES III.

In the same way it may be seen that the equation

$$ax + by + c = 0 \quad \text{................................} \quad (i)$$

represents a straight line. If b is not zero, the gradient is $-a/b$. If b is zero, the equation becomes $x = -c/a$ and represents a straight line perpendicular to the x-axis; in this case the gradient is infinite.

If a, b are both different from zero, and if the line cut the x-axis at A and the y-axis at B, then $OA = -c/a$, $OB = -c/b$. For the coordinates of A are $(OA, 0)$, and since these satisfy (i), we must have

$$aOA + c = 0 \quad \text{or} \quad OA = -c/a.$$

Similarly the coordinates of B are $(0, OB)$, and therefore $bOB + c = 0$.

OA, OB are called *the intercepts* made by the line on the coordinate axes; of course, the simplest method of graphing the straight line is to find the intercepts OA, OB, and to join AB.

3. Determine whether any or all of the points $A(1, 1)$, $B(2, -1)$, $C(9, -4)$ lie on the straight line given by the equation

$$2x + 3y = 6.$$

4. Show that whatever constant value a may have the point (x_1, y_1) will lie on the line given by the equation

$$y - y_1 = a(x - x_1).$$

The equation is true when for x we put x_1 and for y we put y_1, and this is the only condition required.

5. Determine the constant a in Ex. 4 so that the point (x_2, y_2) may lie on the line.

Since the coordinates (x_2, y_2) must satisfy the equation, we find

$$y_2 - y_1 = a(x_2 - x_1) \quad \text{or} \quad a = (y_2 - y_1)/(x_2 - x_1),$$

and therefore the equation of the line through the points (x_1, y_1), (x_2, y_2) is

$$y - y_1 = \frac{y_2 - y_1}{x_2 - x_1}(x - x_1).$$

6. Find the equations of the lines through the following pairs of points—

    (i) $(1, 2), (2, 1)$;   (ii) $(-1, 2), (2, -1)$;
    (iii) $(0, 0), (1, -1)$;   (iv) $(0, 3), (-2, 0)$.

7. Find the equation of the line with the gradient 2 passing through the point $(3, 1)$.

8. Find the equation of the line with the gradient c passing through the point (a, b).

9. Find the coordinates of the point of intersection of the two lines given by the equations

    (i) $x + 2y = 3$;   (ii) $3x + y = 4$.

Since the point of intersection lies on *both* lines, its coordinates must satisfy *both* equations (i) and (ii). Solving these as simultaneous equations, we get for the required coordinates $x=1$, $y=1$. Verify the result by means of a diagram.

10. Draw on one diagram the curves whose equations are
$$2x+y-3=0, \quad y=x^2,$$
and find by measurement the coordinates of the points of intersection. Verify by solving the equations as simultaneous equations.

11. Show that the roots of the equation $2x+x^2-3=0$ are the abscissae of the points of intersection of the curves of ex. 10.

12. Show that the roots of the equation $f(x)=c$ are the abscissae of the points of intersection of the curves given by
$$y=c, \quad y=f(x).$$
Compare Exer. II. ex. 4.

§ 23. Rational Functions.
An expression of the form
$$a+bx+cx^2+\ldots+kx^{n-1}+lx^n\ldots\ldots\ldots\ldots(1)$$
where the coefficients a, b, c,... are constants and the indices of the powers of x are all positive integers of which n is the greatest is called a *Rational Integral Function* of x of degree n.

The quotient of two rational integral functions of x is called a *Rational Fractional Function* of x.

It is known from the theory of equations that an expression of the form (1) will in general vanish for n values of x; hence the graph of the function (1) will in general cross the x-axis n times. (See Exer. II. ex. 4). Some of the values of x for which (1) vanishes may however be imaginary and for such values of the abscissa there are no real points on the axis so that the graph may not have as many as n crossings. When two of the values of x for which (1) vanishes are equal, the student will find that the graph *touches* the x-axis at the corresponding point.

Graphs of the even powers. The graphs of the *even* powers of x, x^2, x^4 ... are all of the same general character; they touch the x-axis at O and have the y-axis as an axis of symmetry. The greater the index however, the slower does the graph recede from the x-axis near the origin; on the other hand, the greater the index the more rapidly does the graphic point move upwards when x is greater than 1. The general shape of the graphs of ax^2, ax^4, ...

GRAPHS OF POWERS. INFLEXION.

can be seen by dividing the corresponding ordinates of the graphs of x^2, x^4 ... in the ratio of a to 1, as in § 18.

Graphs of the Odd Powers. The graphs of the *odd* powers higher than the first, x^3, x^5, ... touch the x-axis at the origin but they do not have the y-axis as an axis of symmetry. For these the origin is a *centre of symmetry*. (Exer. II. 5). For positive values of x the graphs resemble those of the even powers; near the origin the graph of x^3 is flatter than that of x^2, not so flat as that of x^4, while for values of x greater than 1 the graph of x^3 lies above that of x^2, below that of x^4.

To construct the graph of x^3 for negative values of x, take a point P on the graph of the positive values of x, produce PO backwards its own length to P', and P' will be the point on the graph symmetric to P (Fig. 10).

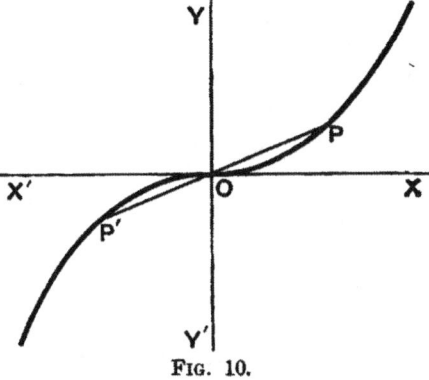

FIG. 10.

The same construction holds for any curve that has the origin for a centre of symmetry.

The graphs of the odd powers thus *both touch and cross* the x-axis at O, bending away from the axis in opposite directions on opposite sides of O (Fig. 10).

DEFINITION. A point such as O where the curve crosses its tangent and bends away from it in opposite directions on opposite sides is called a *Point of Inflexion*, and the tangent at the point is called an *Inflexional Tangent*.

The student should plot on the same diagram for values of x between -1 and $+1$, using a pretty large unit, the graphs of x^2, x^3, x^4, x^5. He will gain useful ideas of the relative magnitude of the powers of x when x is a proper fraction. He will also be able to deduce the general course of the graph of such a function as $x^{\frac{3}{2}}$ for values of x between 0 and 1; the graph will lie below that of x^2, but above that of x^3. If x be negative $x^{\frac{3}{2}}$ is imaginary, and there is no part of the graph to the left of the y-axis.

In the same way by plotting the graphs of the same functions for values of x between 1 and 3, using a small unit, he will see how rapidly the higher powers of x increase when x is greater than 1. He can readily verify the important principle that the term of highest degree in a rational integral function will for sufficiently large values of x be numerically greater than the sum of all the other terms, and will therefore determine the *sign* of the function for large values of x.

The construction of the graph of the general rational integral function is usually laborious; when the student is able to differentiate a function he will find that the labour may be considerably reduced.

As an example take the function $f(x)$, where

$$f(x) = x^3 - 3x + 1.$$

Write
$$f(x) = x^3\left(1 - \frac{3}{x^2} + \frac{1}{x^3}\right).$$

Now, if x is *numerically* equal to or greater than 2 the expression within the bracket will be positive, as a little consideration shows. Hence if x is positive and equal to or greater than 2, $f(x)$ will be positive; if x is negative and *numerically* equal to or greater than 2, $f(x)$ will be negative, since x^3 will be negative and the expression within the bracket positive. The graph must therefore cross the x-axis once at least between the points on that axis at which x is -2 and 2 respectively.

Examining further, we find

$$f(-2) = -1;\ f(-1) = +3;\ f(1) = -1;\ f(2) = +3,$$

and therefore the graph must cross *thrice*, namely, between the points -2 and -1, -1 and 1, 1 and 2; since the equation is of the third degree, the graph cannot cross more than thrice. There will thus be two turning points.

Again,
$$f(-1\cdot9) = -\cdot159,$$
$$f(-1\cdot8) = +\cdot568,$$

so that the graph crosses between $-1\cdot9$ and $-1\cdot8$.

GRAPHS OF RATIONAL FUNCTIONS. 37

When $x = -1\cdot 88$, $f(x) = -\cdot 005$,

so that the graph crosses very nearly where $x = -1\cdot 88$, and this value is an approximate root of the equation

$$x^3 - 3x + 1 = 0.$$

In the same way it may be found that the other two roots are approximately $\cdot 35$ and $1\cdot 53$.

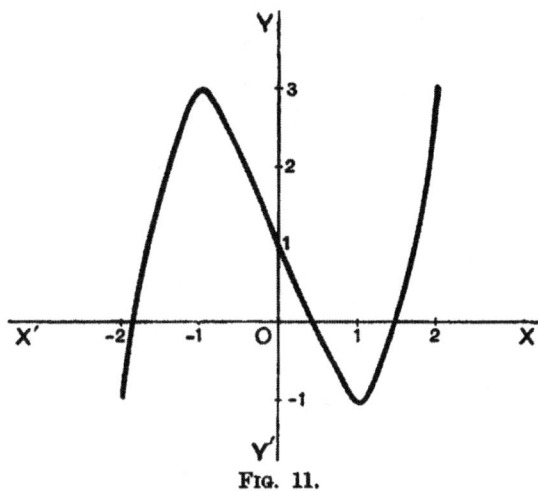

Fig. 11.

The turning points occur where $x = -1$ and $x = 1$, and the calculation of a few values of $f(x)$ shows that the graph is of the form shown in Fig. 11.

§ 24. Asymptotes.

The simplest example of a rational fractional function is $1/x$.

When x is small and positive, $1/x$ is large and positive, and as x tends towards zero $1/x$ becomes extremely large or, in the usual language, $1/x$ tends toward infinity; thus when x takes the values $\cdot 1$, $\cdot 01$, $\cdot 001$, ... $1/x$ takes the values 10, 100, 1000, ... respectively. Hence as the point x moves from the right toward O till it all but coincides with O the graphic point moves upward and recedes to a very great distance from the x-axis while approaching very close to the y-axis; when x is zero, that is when the point x coincides with O, the graphic point may be said to be at infinity. In this case the graph is said to approach the positive end of the *y*-axis *asymptotically*, or to have the *y*-axis as an *asymptote*.

In the same way it may be seen that when x is very large and positive $1/x$ is very small and positive; the graph approaches the positive end of the *x*-axis asymptotically.

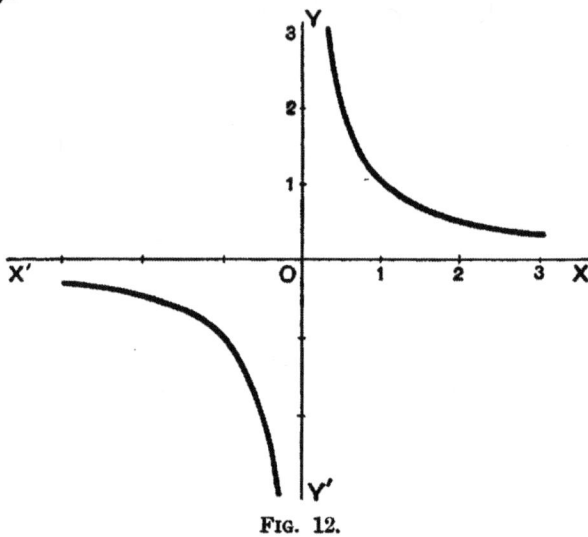

Fig. 12.

The graph is obviously symmetrical with respect to the origin, and approaches both ends of both coordinate axes asymptotically (Fig. 12).

DEFINITION. In general, when a curve has a branch extending to infinity, the branch is said to approach a straight line *asymptotically*, or to have the straight line for an *asymptote*, if as a point moves off to infinity along the branch the distance from the point to the straight line tends towards zero as a limit, that is, if as the point moves off to infinity the distance becomes and remains less than any given length.

If $x-a$ be a factor of the denominator of a rational fractional function of x in its lowest terms, the function will tend towards infinity as x tends towards a and the line whose equation is $x=a$ will be an asymptote. If as x tends towards infinity the function y tends to a finite value

ASYMPTOTES.

β then $y=\beta$ will be the equation of an asymptote. These asymptotes are parallel to or coincident with the coordinate axes, as in the example just considered; but there may be asymptotes that are not parallel to either axis, as in the following example:
$$y = \frac{x^3+x^2+1}{x^2}.$$

Here we may write
$$y = x+1+\frac{1}{x^2}.$$

If we denote by y_1 the ordinate of the graph and by y_2 the corresponding ordinate of the straight line whose equation is $y=x+1$, we see that
$$y_1 = y_2 + \frac{1}{x^2}.$$

Hence whether x be positive or negative y_1 is greater than y_2 and therefore the graph of the function is always above the straight line.

Again when x is numerically very large $1/x^2$ is very small, and the difference between y_1 and y_2 will as the point x moves either to the extreme right or to the extreme left of the x-axis become less than any given fraction; hence the graph approaches both ends of the line whose equation is $y=x+1$ asymptotically.

The y-axis is also an asymptote; y is *positive* when x is either a small positive or a small negative number and therefore the graph does not approach the negative end of the y-axis but it approaches the positive end both from the right and from the left.

The graph will cross the x-axis for those values of x which make the numerator x^3+x^2+1 zero; a few trials will show that the numerator vanishes only once, namely when $x=-1\cdot 47$ approximately. When x is algebraically less than $-1\cdot 47$, y is negative; for all other values of x the ordinate is positive.

When $\quad\quad\quad\quad x=1,\ y=3$;
when $\quad\quad\quad\quad x=2,\ y=3\frac{1}{4}$,
and there is a turning point when $x=1\cdot 3$ approximately.

40   AN ELEMENTARY TREATISE ON THE CALCULUS.

The graph is shown in Fig. 13. The unit for the abscissae is double that for the ordinates; if the units were equal the portion ABC would be at a considerable distance above $X'X$ and the diagram would have to be very large

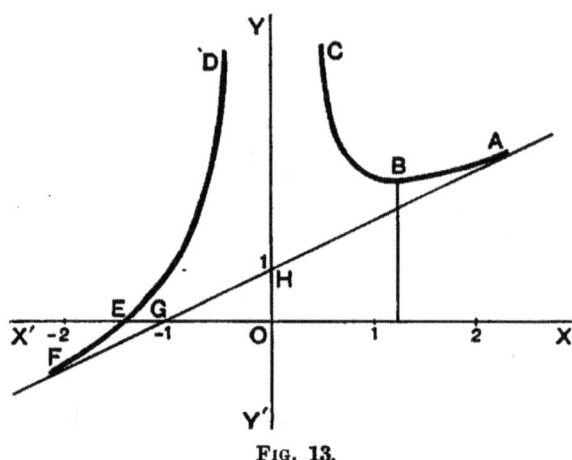

FIG. 13.

to show that part clearly. The curve approaches the asymptote GH very rapidly but the asymptote OY more slowly.

In plotting the graph of a fractional function it will be frequently found convenient to split the function up into partial fractions as has been done above. Thus, if

$$y = \frac{x^2+1}{(x-1)(x-2)},$$

we can write

$$y = 1 - \frac{2}{x-1} + \frac{5}{x-2},$$

and we see that there are three asymptotes whose equations are

$$y = 1, \quad x = 1, \quad x = 2.$$

In this case the graph crosses the horizontal asymptote at the point whose abscissa is $\frac{1}{3}$, because when $y = 1$ we have

$$1 = \frac{x^2+1}{(x-1)(x-2)}, \quad \text{or} \quad x = \frac{1}{3}.$$

For the equation $\quad y = \dfrac{x^3+1}{(x-1)(x-2)},$

we should have $\quad y = x + 3 - \dfrac{2}{x-1} + \dfrac{9}{x-2},$

and there would again be three asymptotes, two of which are parallel to the y-axis while the third has for equation
$$y = x+3,$$
and this third asymptote cuts the graph again at the point whose abscissa is $\tfrac{4}{3}$.

EXERCISES IV.

Graph the functions 1—6 :

1. $3x^2 - 5x - 1$; 2. $x^2 + 2x + 2$; 3. $x^3 - x$;

4. $x^3 - x + 2$; 5. $\dfrac{2x-3}{x-1}$; 6. $\dfrac{2x^2 + x - 1}{x-1}$.

7. Show that the roots of the equation $x^3 - ax - b = 0$ are the abscissae of the points of intersection of the graphs of x^3 and of $ax + b$.

8. Find to two decimals the roots of the equations

   (i) $x^3 - 7x + 3 = 0$;   (ii) $x^3 - 7x + 9 = 0$.

Graph the functions.

9. If $f(x) = x^4 - 4x^3 - 4x^2 + 16x + 1$, show that the equation $f(x) = 0$ has four real roots, and find these to two decimals.

[Find the values of $f(x)$ for x equal to $-2, -1, 0, 3, 4$ respectively. The ordinate $f(-2)$ is positive and the ordinate $f(-1)$ negative, so that the graph crosses the axis of abscissae between the point -2 and the point -1. Proceed in the same way with the other numbers.]

10. A point is moving in a plane and at time t seconds reckoned from a fixed instant, its coordinates with respect to two rectangular axes in the plane are x and y feet. Construct the path of the point in the following cases :

   (i) $x = t + 1, y = 2t$;   (ii) $x = a + bt, y = c + dt$;

   (iii) $x = 2t, y = 8t^2$;   (iv) $x = t, y = t^3$.

[The position of the point at any instant may be found by calculating the values of x and y for the value of t at that instant; having found the position of the point for a number of values of t, the graph can be drawn in the usual way. Or, the equation of the path may be found by eliminating t. Thus in (i) t may be considered a function of x, namely $t = x - 1$; but y is always $2t$, and therefore y and x are always connected by the equation $y = 2(x - 1)$. In this case therefore the path is a straight line. In (ii) the path is also a straight line. The equations of the paths in (iii), (iv) are $y = 2x^2$, $y = x^3$. This method of representing the path of a point by means of two equations is of frequent occurrence both in Geometry and in Mechanics.]

11. The angle θ between the two straight lines whose equations are

(i) $y = mx + c$, (ii) $y = m'x + c'$

may be found from the equation

$$\tan \theta = \frac{m - m'}{1 + mm'}.$$

[Let (i) make the angle α, (ii) the angle β with $X'OX$; suppose $\alpha > \beta$, then $\theta = \alpha - \beta$ and

$$\tan \theta = \frac{\tan \alpha - \tan \beta}{1 + \tan \alpha \tan \beta} = \frac{m - m'}{1 + mm'}.$$

If the *numerical* value of $(m - m')/(1 + mm')$ be taken, the acute angle between the lines will be obtained whether $\alpha > \beta$ or $\alpha < \beta$.]

12. The angle between the lines given by

$$ax + by + c = 0 \quad \text{and} \quad a'x + b'y + c' = 0$$

is given by $\quad \tan \theta = (ab' - a'b)/(aa' + bb')$.

13. Show that the lines of ex. 12 are

(i) parallel if $a/b = a'/b'$,

(ii) perpendicular if $aa' + bb' = 0$.

CHAPTER III.

GRAPHS. ALGEBRAIC AND TRANSCENDENTAL FUNCTIONS. CONIC SECTIONS.

§ 25. Algebraic Functions. y is called an *Algebraic Function* of x when it is determined by an equation of the form

$$Ay^n + By^{n-1} + \ldots + Ky + L = 0,$$

in which the indices of the powers of y are positive integers and the coefficients $A, B, \ldots K, L$, are rational integral functions of x. Manifestly, rational functions are special cases of algebraic functions.

y will usually be multiple-valued and its graphical representation is much more difficult than that of the rational function except in particular cases of which the following are of special importance[1]:

Type I. $\qquad y^n - x = 0 \ \text{ or } \ y = x^{\frac{1}{n}}.$

When n is an even integer, x must be positive and y will be two-valued; when n is an odd integer, x may have any value and y will be single-valued. The graph of $x^{\frac{1}{n}}$ is readily found from that of x^n.

Let QOP (Figs. 14, 15) be the graph of x^n, and let PN be perpendicular to $Y'Y$; then $ON = NP^n$, or $NP = ON^{\frac{1}{n}}$.

Hence if OY be taken as the axis of abscissae, that is, as the axis of the argument, and OX as the axis of ordi-

---

[1] The beginner may find this article somewhat difficult; he should work out the simple examples of the various cases that are set down at the end of the article and the discussion will become more definite. He need not however spend much time on this article at a first reading of the subject.

nates, that is, as the axis of the function, the curve QOP will be the graph of the function $ON^{\frac{1}{n}}$. It is desirable however to have OX as the axis of abscissae and OY as the axis of ordinates, that is, the figure has to be turned so that OY becomes horizontal and coincides with the present position of OX, while OX becomes vertical and coincides with the present position of OY. The simplest

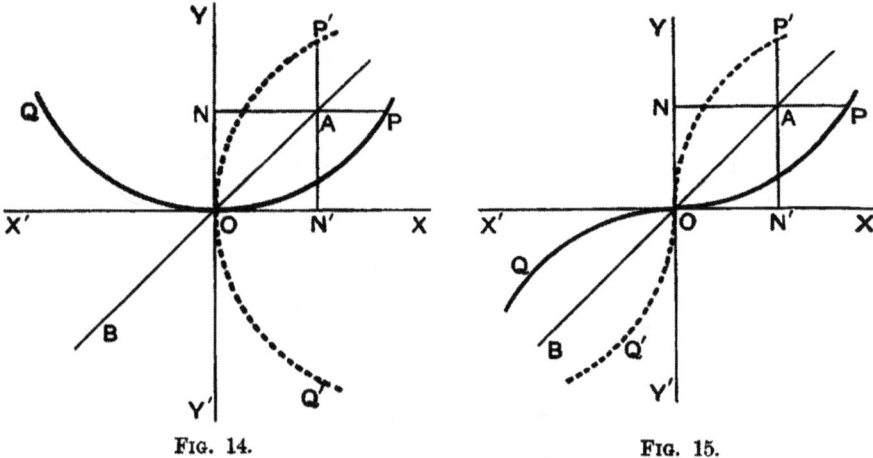

FIG. 14.   FIG. 15.

way of securing this is to suppose the whole figure rotated through two right angles about the bisector BOA of the angle XOY as axis; NAP will thus come into the position $N'AP'$, and QOP will come into the position $Q'OP'$. $Q'OP'$ will be the graph of $x^{\frac{1}{n}}$, because $N'P' = ON'^{\frac{1}{n}}$, since $N'P' = NP$ and $ON' = ON$.

Fig. 14 is the graph when n is even and when, therefore, for one value of x there are two values of y; on the other hand, when n is odd, as shown in Fig. 15, to one value of x there is but one value of y.

Construction of Graph of an Inverse Function. The same transformation gives the graph of the function inverse to a given function. If $y = x^n$ and if x be taken as the argument, QOP is the graph of x^n; the function inverse to y is x where $x = y^{\frac{1}{n}}$, and, when y is taken as the argument, QOP

is the graph of $y^{\frac{1}{n}}$, that is, of the function inverse to y or x^n. It is convenient however to represent the argument in all cases by lines measured along $X'OX$ and to denote the argument of the inverse function by the *same* letter as is used for the argument of the original function; that is when the inverse function has been formed we then replace y by x and x by y, and the graph of the inverse function, when this replacement has been made, will be the original graph rotated through two right angles about the bisector of the angle XOY.

In this notation the graph of x^n is QOP; the inverse function, which as first stated is $y^{\frac{1}{n}}$, is now $x^{\frac{1}{n}}$, and its graph is $Q'OP'$.

Again, when the graph of a function has been constructed, we see how to choose the range of the variables so that the inverse function may be single-valued. When n is even OP' is the graph of $+x^{\frac{1}{n}}$ and OQ' that of $-x^{\frac{1}{n}}$; that is, OP' and OQ' are the two branches of the two-valued function inverse to x^n when n is even.

Type II. $y^n - x^m = 0$ **where m, n are unequal and not both even.**

If m, n were both even the equation would be equivalent to the two equations $y^{\frac{n}{2}} - x^{\frac{m}{2}} = 0$, $y^{\frac{n}{2}} + x^{\frac{m}{2}} = 0$, and there would therefore be two graphs, each of which would come under one of the following groups.

The student should notice the remark in § 23 about the graph of such a function as $x^{\frac{5}{2}}$; it will be found useful in the discussion of the groups contained in the general equation.

(A) $m > n$; $y = x^{\frac{m}{n}}$ where $\dfrac{m}{n}$ is an improper fraction.

(A_1) m, n **both odd.** The graph is of the form QOP (Fig. 15); O is a point of inflexion and $X'OX$ a tangent at O.

(A_2) m **even,** n **odd.** The graph is of the form QOP (Fig. 14); OY is an axis of symmetry and $X'OX$ a tangent at O.

46   AN ELEMENTARY TREATISE ON THE CALCULUS.

(A_3) *m* **odd,** *n* **even.** *y* is imaginary when *x* is negative; OX is an axis of symmetry, and both branches touch OX (and each other) at O. (Fig. 16.)

DEFINITION. A point on a curve such as O, at which two branches OA, OB have the same tangent, but *beyond which they do not pass*, is called a *Cusp*. It must be observed that neither branch passes beyond O; a point moving from A along the curve to O reverses its direction in order to proceed along the other branch OB.

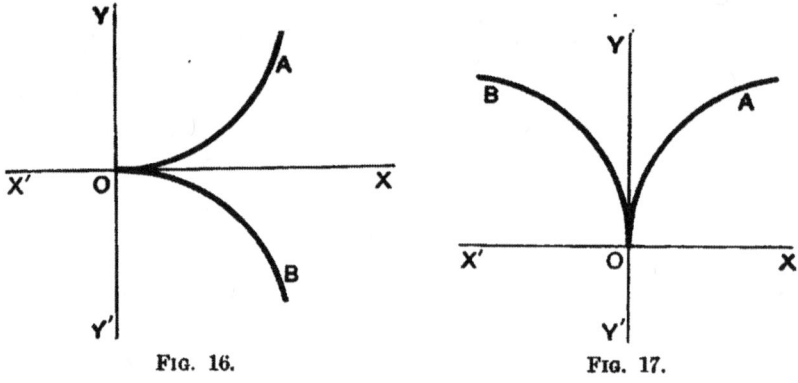

FIG. 16.   FIG. 17.

(B) $m < n$; $y = x^{\frac{m}{n}}$ where $\frac{m}{n}$ **is a proper fraction.**

(B_1) *m, n* **both odd.** The graph is of the form $Q'OP'$ (Fig. 15); O is a point of inflexion and $Y'OY$ a tangent at O.

(B_2) *m* **odd,** *n* **even.** *y* is imaginary when *x* is negative; OX is an axis of symmetry and $Y'OY$ a tangent at O. The graph is of the form $Q'OP'$. (Fig. 14.)

(B_3) *m* **even,** *n* **odd.** OY is an axis of symmetry and is a tangent at O; O is a cusp. (Fig. 17.)

Thus if $m=2$, $n=5$, since $\frac{2}{5}$ lies between $\frac{2}{4}$ or $\frac{1}{2}$ and $\frac{2}{6}$ or $\frac{1}{3}$, the graph of $x^{\frac{2}{5}}$ will, when x is positive, lie between those of $x^{\frac{1}{2}}$ and $x^{\frac{1}{3}}$, each of which has the form OP' (Fig. 15). The branch OB is present because OY is an axis of symmetry.

The student will have no difficulty in deducing the graphs when the equation is $y^n + x^m = 0$; they are deduced from those of $y^n - x^m = 0$ by rotation about one of the coordinate axes. Thus the graphs corresponding to (A_1) and (B_1) are obtained by rotation about $X'X$. More generally, the graphs of $y^n - ax^m = 0$ can be deduced by dividing the ordinates of $y^n - x^m = 0$ in the ratio of $a^{\frac{1}{n}}$ to 1.

Ex. 1. Draw the graphs of the following cases of Type I.:
    (i) $y^3 = x$;   (ii) $y^3 = x$;   (iii) $y^2 = -x$;   (iv) $y^3 = -x$.

Ex. 2. Draw the graphs of the following cases of Type II. (A):
    (i) $y^3 = x^5$;   (ii) $y^3 = x^4$;   (iii) $y^2 = x^3$;
    (iv) $y^3 = -x^5$;   (v) $y^3 = -x^4$;   (vi) $y^2 = -x^3$.

Ex. 3. Draw the graphs of the following cases of Type II. (B):
    (i) $y^5 = x^3$;   (ii) $y^3 = x^2$;   (iii) $y^4 = x^3$;
    (iv) $y^5 = -x^3$;   (v) $y^3 = -x^2$;   (vi) $y^4 = -x^3$.

Ex. 4. Draw the graphs of
    (i) $y^2 = 9x^3$;   (ii) $y^2 = -9x^3$;   (iii) $y^3 = 27x^2$.

Ex. 5. Graph the functions
    (i) $\dfrac{1}{x^{\frac{3}{2}}}$;   (ii) $\dfrac{1}{x^{\frac{2}{3}}}$;   (iii) $\dfrac{1}{x^{\frac{5}{3}}}$.

§ 26. Conic Sections. For the sake of readers unfamiliar with the conic sections we give in this article the equations of the conic sections and define the most frequently occurring technical terms connected with them.

DEFINITION.—A conic section is the locus of a point which moves in a plane so that its distance from a fixed point is in a constant ratio to its distance from a fixed straight line.

The fixed point is called *the focus*, the constant ratio *the eccentricity*, and the fixed line *the directrix*.

Let S (Fig. 18) be the focus, KN the directrix and SK perpendicular to KN.

Let e be the eccentricity and on KS take A so that $AS = eKA$; then A is a point on the conic.

As axes of coordinates take KAS and the perpendicular through A to KAS. Let P be any point (x, y) on the conic and draw PM perpendicular to KS; then $x = AM$, $y = MP$.

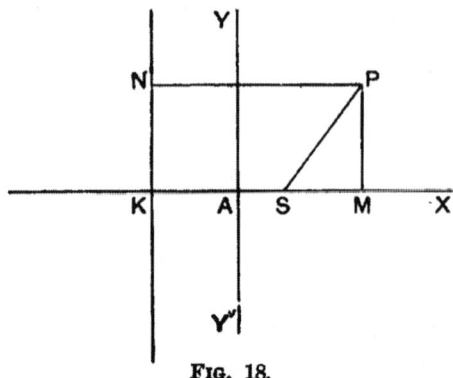

Fig. 18.

Let $KA = p$; then
$$AS = ep.$$
Now
$$SM = AM - AS = x - ep;$$
$$NP = KM = p + x.$$
But $SP = eNP$ by the definition of the conic; hence
$$SP^2 = e^2 NP^2$$
or $SM^2 + MP^2 = e^2 NP^2$,

so that, inserting the values of SM, MP, NP, we get
$$(x - ep)^2 + y^2 = e^2(x + p)^2,$$
or after reduction
$$(1 - e^2)x^2 - 2e(1 + e)px + y^2 = 0 \dots\dots\dots\dots(1).$$

Every point whose coordinates satisfy equation (1) will be a point on the conic section; for different values of the constants e, p there will be different conics. Evidently AS is an axis of symmetry.

If AK were taken as the positive direction of the axis of abscissae, then in equation (1) we should have $+ 2e(1+e)px$, for the change in the direction of the axis is equivalent to writing $-x$ in place of x.

Special Forms of the Conic Section.—I. If $e = 1$, the conic is called a **parabola**. In this case equation (1) reduces to
$$y^2 = 4px \dots\dots\dots\dots\dots\dots\dots\dots\dots(\text{P})$$

A is called *the vertex*, AX *the axis* of the parabola.

When $e = 1$, $AS = p$ and if SL is the ordinate at S equation (P) shows that $SL = 2p = KS$. SL is called the *semi-latus-rectum* of the parabola; in every conic section the double ordinate through the focus is called the *latus rectum*. Sometimes $4p$ is called *the parameter* of the parabola.

It is easily seen that the curve is of the form of Fig. 19, extending to infinity towards the right. The graph of x^2 is a parabola with its axis *vertical* (see § 16); its latus rectum is 1, its focus the point $(0, \frac{1}{4})$ and its directrix is the line through $(0, -\frac{1}{4})$ parallel to the axis of abscissae.

II. If e is less than unity the conic is called **an ellipse**. In this case equation (1) takes the form

$$x^2 - \frac{2ep}{1-e}x + \frac{y^2}{1-e^2} = 0;$$

or, putting a for $ep/(1-e)$ and b^2 for $a^2(1-e^2)$,

$$\frac{x^2}{a^2} - \frac{2x}{a} + \frac{y^2}{b^2} = 0. \dots\dots\dots\dots\dots\dots\text{(E)}$$

III. If e is greater than unity, the conic is called a **hyperbola**. In this case, if $a = ep/(e-1)$ and $b^2 = a^2(e^2-1)$, equation (1) becomes

$$\frac{x^2}{a^2} + \frac{2x}{a} - \frac{y^2}{b^2} = 0. \dots\dots\dots\dots\dots\dots\text{(H)}$$

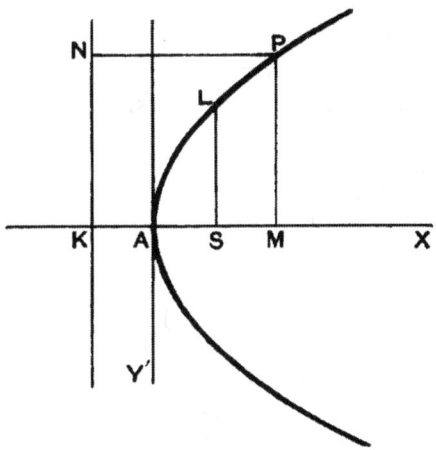

Fig. 19.

A more convenient form for the equations of the ellipse and the hyperbola is got as follows:

In (E) let $y = 0$; then $x = 0$ or $2a$. The ellipse therefore cuts the x-axis at two points, namely at A where $x = 0$, and at another point, A' say, to the right of A where $x = AA' = 2a$. AA' is called the *major axis* and A, A' the *vertices* of the ellipse.

Similarly from (H) it will be found that the hyperbola cuts the x-axis at A and at another point, A' say, to the *left* of A where AA' is equal in length to $2a$. AA' is called the *transverse axis* and A, A' the *vertices* of the hyperbola.

To find the shape of the ellipse take the origin of coordinates at C, the middle point of AA' (Fig. 20).

Attending to the sign of the segments we have in all cases
$$AM = AC + CM.$$
Let $CM = x'$; $AM = x$; then since $AC = a$
$$x = a + x'.$$
Replacing x in (E) by $a + x'$ and reducing we get
$$\frac{x'^2}{a^2} + \frac{y^2}{b^2} = 1. \ldots\ldots\ldots\ldots\ldots\ldots\ldots\ldots(\text{E}')$$
In exactly the same way we find, in place of (H),
$$\frac{x'^2}{a^2} - \frac{y^2}{b^2} = 1. \ldots\ldots\ldots\ldots\ldots\ldots\ldots\ldots(\text{H}')$$

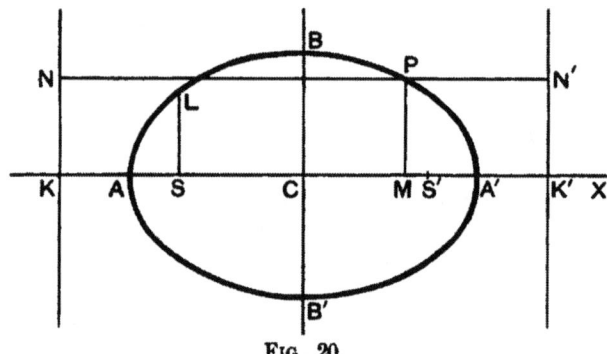

Fig. 20.

If we remember that the abscissae are now measured from C and not from A we may drop the accent; the equations are then
$$\frac{x^2}{a^2} + \frac{y^2}{b^2} = 1 \quad \text{and} \quad \frac{x^2}{a^2} - \frac{y^2}{b^2} = 1, \ldots\ldots\ldots\ldots(\text{c})$$
and these may be considered the standard forms.

From these equations we see that both curves are symmetrical about both axes. The origin C is a centre of symmetry; C is called the *centre* of the conics, and the ellipse and the hyperbola are called *central conics*. The parabola has no centre.

The axis of ordinates meets the ellipse (Fig. 20) at two

points B, B'; BB' is called the *minor axis*. From the equation
$$x^2/a^2 + y^2/b^2 = 1$$
it is easy to see that x is never numerically greater than a nor y greater than b. The ellipse is therefore a closed curve.

The circle is the particular case of the ellipse in which $b = a$ and $e = 0$.

The axis of ordinates does not meet the hyperbola because when $x = 0$, $y^2 = -b^2$ and therefore y is imaginary. It will be seen further that y is imaginary if x is numerically less than a, so that no part of the hyperbola lies between the lines through A, A' perpendicular to AA'.

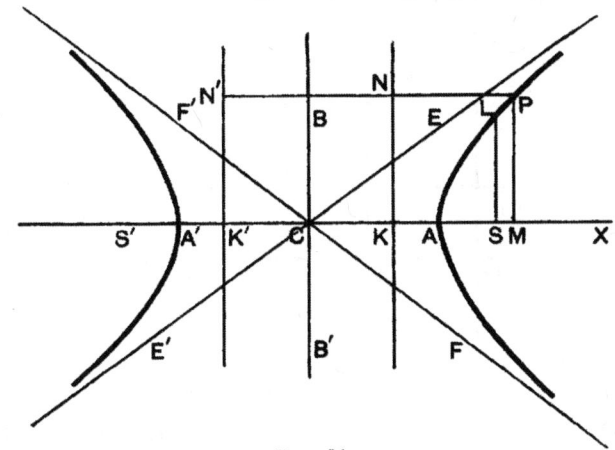

Fig. 21.

The curve consists of two branches extending to infinity to the right of A and to the left of A' respectively. It will be a good exercise for the student to prove that the lines $E'E$, $F'F$ whose equations are
$$y = bx/a, \; y = -bx/a,$$
are asymptotes (Fig. 21).

If $b = a$ the hyperbola is said to be *equilateral*; since the asymptotes are in that case at right angles the hyperbola is also said to be *rectangular*.

From the symmetry of the central conics about the axis of ordinates through C it may be inferred that they have a second focus S' and a second directrix $K'N'$ symmetrical to S and KN with respect to C; the curves might be constructed from S' and $K'N'$ in the same way as from S and KN, the eccentricity being the same.

Some useful properties of the Conic Sections will be found in *Exercises V., VI.*

§ 27. Change of Origin and of Axes. The device of changing the origin of coordinates is often useful in simplifying the equation of a curve and thus making the construction of the curve more simple.

I. *New Axes parallel to Old Axes.* In Fig. 22 let B be the new origin, and let X'_1BX_1, Y'_1BY_1 be parallel to $X'OX$, $Y'OY$ respectively.

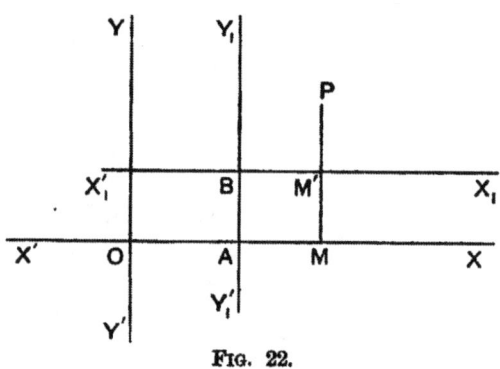

Fig. 22.

Let (a, b), (x, y) be the coordinates of B and of any other point P with respect to the old axes $X'OX$, $Y'OY$; and let (x', y') be the coordinates of P with respect to the new axes X'_1BX_1, Y'_1BY_1. Then

$OA = a, AB = b$; $OM = x, MP = y$; $BM' = x'$; $M'P = y'$;

and $\qquad OM = OA + AM = OA + BM'$;

$\qquad MP = MM' + M'P = AB + M'P$;

and therefore $\qquad x = a + x'; \quad y = b + y'$ ....(1)

Conversely $\qquad x' = x - a; \quad y' = y - b.$ ....(1')

When x and y have been replaced by $a + x'$ and $b + y'$ the accents may be dropped, it being remembered that the origin is then B, so that x will mean not OM but BM', and y not MP but $M'P$.

CHANGE OF AXES.

EXAMPLE. The equation $y^2 - 4x - 2y - 1 = 0$ may be written
$$(y-1)^2 = 4(x+\tfrac{1}{2}).$$
Put $x + \tfrac{1}{2} = x'$, that is, $x = -\tfrac{1}{2} + x'$ and $y - 1 = y'$, that is, $y = 1 + y'$, which means transferring the origin to the point $(-\tfrac{1}{2}, 1)$, and the equation becomes
$$y'^2 = 4x'.$$
This equation, and therefore also the given one, represents a parabola with its vertex at the new origin and with the new axis of abscissae as its axis. The latus rectum is 4; the focus is the point (1, 0) with respect to the new axes, and therefore the point $(\tfrac{1}{2}, 1)$ with respect to the old because the coordinates of any point with respect to the old axes are equal to those with respect to the new increased by the coordinates of the new origin.

II. *The origin not changed, but the New Axes obtained by turning the Old Axes through a positive or negative angle θ.* In Fig. 22a let P be the point (x, y) when referred to the old axes $X'X$, $Y'Y$, and the point (x', y') when referred to the new axes X'_1X_1, Y'_1Y_1, so that

$x = OM$, $y = MP$;
$x' = OM'$, $y' = M'P$;
$\angle XOX_1 = \theta = \angle YOY_1$.

By elementary trigonometry,

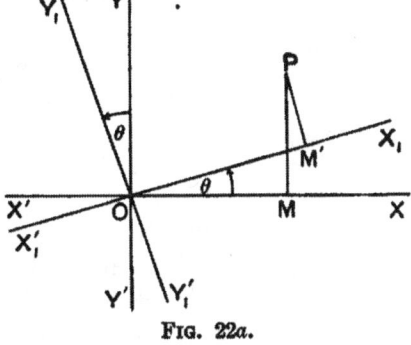

FIG. 22a.

$OM = OM' \cos\theta - M'P \sin\theta$; $MP = OM' \sin\theta + M'P \cos\theta$;
that is,

$x = x' \cos\theta - y' \sin\theta$; $\qquad y = x' \sin\theta + y' \cos\theta$. ...... (2)

Conversely, solving for x' and y' in terms of x and y,

$x' = x \cos\theta + y \sin\theta$; $\qquad y' = -x \sin\theta + y \cos\theta$ ..... (2')

It may be possible to choose θ, so that the new equation is simpler than the old or even is an equation of which the graph is known.

EXAMPLE. By turning the axes through 45° the equation $xy = c^2$ becomes
$$\frac{x'-y'}{\sqrt{2}} \cdot \frac{x'+y'}{\sqrt{2}} = c^2 \quad \text{or} \quad x'^2 - y'^2 = 2c^2,$$
since, by (2), $\quad x = \frac{1}{\sqrt{2}}x' - \frac{1}{\sqrt{2}}y', \quad y = \frac{1}{\sqrt{2}}x' + \frac{1}{\sqrt{2}}y'.$

The new form shows that the curve is a rectangular hyperbola; half the transverse axis, denoted in § 26 by a, is $\sqrt{2c}$. Hence the graph of c^2/x is a rectangular hyperbola referred to its asymptotes as coordinate axes.

III. *The origin changed to (a, b) and the axes turned through an angle θ.* Combining cases I., II. we get the more general transformation

$$x = a + x' \cos \theta - y' \sin \theta;$$
$$y = b + x' \sin \theta + y' \cos \theta \quad \ldots\ldots\ldots\ldots\ldots (3)$$
$$x' = (x - a) \cos \theta + (y - b) \sin \theta;$$
$$y' = -(x - a) \sin \theta + (y - b) \cos \theta \ldots\ldots (3')$$

EXERCISES V.

Unless otherwise stated the equations of the conic sections in this set of Exercises are supposed to be in the standard forms (P), (C) of § 26.

1. In the central conics prove $\quad CS = eCA, \quad CA = eCK$.

For the ellipse, $\quad\quad AS : AK = e = A'S : A'K$,

and therefore $\quad e = A'S - AS : A'K - AK = S'S : A'A = CS : CA$,
$$e = A'S + AS : A'K + AK = A'A : K'K = CA : CK.$$

For the hyperbola, $\quad A'S - AS : A'K - AK = CA : CK$,
$$A'S + AS : A'K + AK = CS : CA.$$

2. In Fig. 20, S is the point $(-ea, 0)$, S' the point $(ea, 0)$.

In Fig. 21, S is the point $(ea, 0)$, S' the point $(-ea, 0)$.

In Fig. 19, S is the point $(p, 0)$.

3. Show that the latus rectum (or *parameter*) of a central conic is $2b^2/a$.

4. On AA' (Fig. 20) as diameter a circle is described; if MP is produced to meet the circle at Q, show that

$$MP : MQ = b : a = \text{constant}.$$

For $\quad MQ^2 = CA^2 - CM^2 = a^2 - x^2; \quad MP^2 = \dfrac{b^2}{a^2}(a^2 - x^2)$.

The circle is called the *Auxiliary Circle* of the ellipse. The theorem shows that if the ordinate MQ of a circle to any diameter is divided internally at P, so that $MP : MQ = $ constant, the locus of P is an ellipse whose major axis is the diameter of the circle.

The student may prove that if P is in MQ produced, the locus is an ellipse whose *minor* axis is the diameter of the circle.

5. Show that the point $(a\cos\theta, b\sin\theta)$ lies on the ellipse, whatever be the value of θ.

For the equation of the ellipse is satisfied by $x = a\cos\theta, y = b\sin\theta$.

As θ varies from $0°$ to $360°$ the point travels round the ellipse. In the notation of ex. 4, if P is the point $(a\cos\theta, b\sin\theta)$ θ is the angle $A'CQ$ and is called the *Eccentric Angle* of P.

6. Show that the point $(pt^2, 2pt)$, p being a constant, lies on a parabola whatever be the value of t.

7. In Fig. 20, if $CM = x$, prove that $SP = a + ex$, $S'P = a - ex$, $SP + S'P = 2a$.

For
$$SP = eNP = eKC + eCM = a + ex,$$
$$S'P = e \cdot PN' = e \cdot CK' - eCM = a - ex.$$

$SP, S'P$ are called the focal distances of P, and therefore in the ellipse the *sum* of the focal distances is constant, the constant being the major axis.

8. In Fig. 21, if $CM = x$, prove $SP = ex - a, S'P = ex + a, S'P - SP = 2a$. Hence the *difference* of the focal distances of a point on a hyperbola is constant.

9. In the parabola (Fig. 19) prove
$$SP = KA + AM = AS + AM = p + x,$$

x being the abscissa of P.

10. On any of the conics (Figs. 19, 20, 21) a point Q is taken and the chord PQ (produced if necessary) meets the directrix KN at Z. Prove that SZ bisects the exterior angle PSQ, except when P and Q are on different branches of the hyperbola when SZ bisects the interior angle.

Draw QR perpendicular to KN; then
$$SP : PN = e = SQ : QR,$$
therefore
$$SP : SQ = PN : QR = PZ : QZ,$$
and the theorem follows by Euc. vi. 3 or A.

11. Trace the conics given by the equations,

(i) $x^2 + 4y^2 = 4$; (ii) $2x^2 - 3y^2 = 6$,

and find the eccentricity of each.

In (i) $a^2 = 4, b^2 = 1$, and $b^2 = a^2(1 - e^2)$, so that $e^2 = (a^2 - b^2)/a^2$ etc.

12. Show by transferring the origin to $(0, -b)$ that the equation of the ellipse when B' is the origin and $B'B$ the axis of ordinates is
$$x^2/a^2 + y^2/b^2 - 2y/b = 0.$$

If B is the origin and $B'B$ the axis of ordinates the equation is
$$x^2/a^2 + y^2/b^2 + 2y/b = 0.$$

13. Show by finding the values of A and B in terms of p, a, b that when A is positive the equation
$$y^2 = Ax + Bx^2$$
represents (i) a parabola if $B=0$; (ii) an ellipse if B is negative, the ellipse becoming a circle if $B=-1$; (iii) a hyperbola if B is positive.

Show that when B is negative and numerically greater than 1 the major axis of the ellipse lies along the axis of ordinates. Show that all the results hold also when A is negative. (See note on sign of term in x in equation (1) § 26.)

14. Graph the ellipses given by

(i) $y^2 = 4x - \tfrac{1}{2}x^2$; (ii) $y^2 = 4x - 2x^2$,

and find their eccentricity.

15. Show that the equation
$$4y^2 - x^2 + 8y + 6x - 5 = 0$$
represents two straight lines through the point $(3, -1)$.

§ 28. Transcendental Functions. All functions that are not algebraic are classed as *Transcendental functions.*

The elementary transcendental functions are (i) the Trigonometric Functions, Direct and Inverse, (ii) the Exponential Function and its Inverse the Logarithmic Function.

Graphs of the direct trigonometric functions, $\sin x$, $\cos x$, $\tan x$, $\csc x$, $\sec x$, $\cot x$ will be found in most textbooks of Trigonometry. The characteristic property of these functions is that they are *periodic*; that is, if $f(x)$ denote any one of these functions and if n be any positive or negative integer
$$f(x + 2n\pi) = f(x).$$

In other words the function is not altered if its argument be increased or diminished by any multiple of 2π. This number 2π is called *the period* of the function. The tangent and cotangent have also the shorter period π.

The graphs of the Inverse Functions can be constructed as explained in § 25 by rotation about the bisector of the angle XOY. To make the inverse functions single-valued we shall always suppose the angle denoted by $\sin^{-1}x$, $\csc^{-1}x$, $\tan^{-1}x$, $\cot^{-1}x$ to lie between $-\tfrac{\pi}{2}$ and $\tfrac{\pi}{2}$ and the angle denoted by $\cos^{-1}x$, $\sec^{-1}x$ to lie between 0 and π.

TRIGONOMETRIC FUNCTIONS.

Thus

$$\sin^{-1}\left(\frac{1}{2}\right) = \frac{\pi}{6} = \cos^{-1}\left(\frac{\sqrt{3}}{2}\right); \quad \sin^{-1}\left(-\frac{1}{2}\right) = -\frac{\pi}{6},$$

$$\cos^{-1}\left(-\frac{\sqrt{3}}{2}\right) = \frac{5\pi}{6}; \quad \sin^{-1}x + \cos^{-1}x = \frac{\pi}{2};$$

$$\tan^{-1}x + \cot^{-1}x = \frac{\pi}{2} \text{ if } x \text{ be positive,}$$

$$= -\frac{\pi}{2} \text{ if } x \text{ be negative.}$$

Ex. 1. Plot to the same axes the graphs of $\sin x$, $2\sin x$, $3\sin x$, $\frac{1}{2}\sin x$, $\frac{1}{3}\sin x$ between -2π and 2π.

Ex. 2. Plot to the same axes the graphs of $\sin \frac{1}{2}x$, $\sin x$, $\sin 2x$ between -2π and 2π.

Ex. 3. Plot the graph of $\sin \frac{1}{2}x + \sin x + \sin 2x$, making use of the graphs of ex. 2. $\left(-\frac{\pi}{2} \leqq x \leqq \frac{\pi}{2}\right).$

Ex. 4. From the graph of $\sin x$ deduce without calculation the graph of $\sin(x+a)$ where a is any positive or negative number. Deduce the graph of $\cos x$.

[Shift the origin to the point (a, o).]

Ex. 5. Plot the graph of $\sin x + \cos x$.

$$\left[\sin x + \cos x = \sqrt{2}\sin\left(x+\frac{\pi}{4}\right), \text{ etc.}\right]$$

Ex. 6. With the notation of ex. 10, Exer. IV. construct the path of the point when

(i) $x = 2t$, $y = 3\sin 4t$;   (ii) $x = 2t$, $y = 3\tan^{-1}t$;

(iii) $x = a\cos nt$, $y = b\sin nt$.

§ 29. The Exponential Function and the Logarithmic Function. The power a^x is called an *Exponential Function* of x; here the base a is any *positive constant*, and the index or exponent x is the argument of the function.

a^x is always positive. If x be a positive fraction $\frac{m}{n}$ (m, n integers), a^x means the (positive) n^{th} root of a^m; if x be negative, say $-\frac{m}{n}$ (m, n positive integers), a^x means the reciprocal of the (positive) n^{th} root of a^m; if x is zero, a^x is 1. If x be an irrational number we may for the present suppose it to be replaced by a rational approximation.

(i) $a > 1$. As x increases from $-N$ to $+N$, where N is a large positive number, a^x will increase from a very small positive number a^{-N} through 1, the value of a^x when $x=0$, to a very large positive number a^{+N}.

(ii) $a = 1$. In this case a^x is always 1.

(iii) $a < 1$, say $a = 1/b$ where b is greater than 1. As x increases from $-N$ to $+N$, a^x will decrease from a large positive number b^{+N} to a very small positive number b^{-N}.

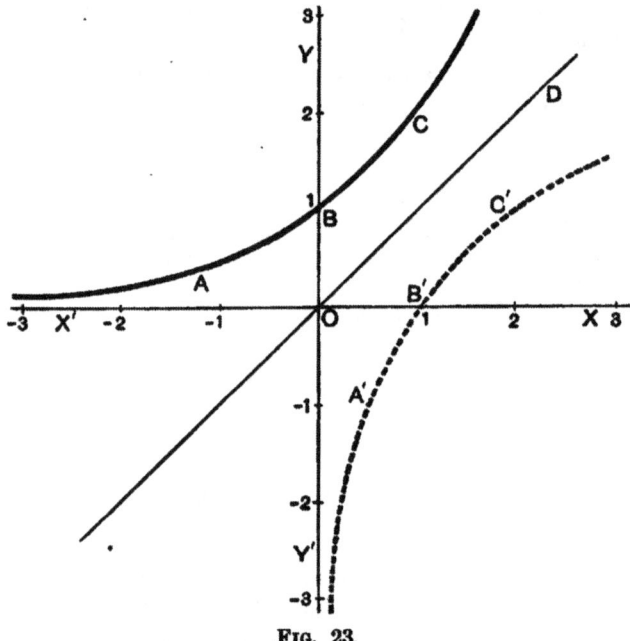

FIG. 23.

ABC in Fig. 23 shows the graph of a^x when $a=2$. The graph approaches the negative end of the x-axis asymptotically.

If a is greater than 1, $1/a$ is less than 1, and since $(1/a)^x = a^{-x}$ it is evident that the graph of $(1/a)^x$ can be found from that of a^x by rotating the latter about the axis $Y'Y$. Hence when a is less than 1, the graph of a^x will approach the *positive* end of the x-axis asymptotically.

EXPONENTIAL AND LOGARITHMIC FUNCTIONS. 59

By the definition of a logarithm, $x = \log_a y$ if $y = a^x$. Hence the logarithm x is the function inverse to the exponential function a^x. By the method given in § 25 for finding the graph of an inverse function, we get the graph of $\log_a x$ by rotating the graph of a^x about OD the bisector of the angle XOY. The curve $A'B'C'$ in Fig. 23 is the graph of $\log_2 x$.

The most convenient base for the exponential function is an irrational number, usually denoted by e and called Napier's base; approximately $e = 2\cdot71828$. Logarithms to the base e will throughout the book be denoted by the symbol "log" (without suffix), unless the contrary is expressly stated; they can be converted into logarithms to the base 10 by the ordinary rule.

$$\log_{10} x = \log_e x \times \log_{10} e = \log_e x \div \log_e 10,$$
and $\quad \log_{10} e = \cdot 434\ 294 \quad \log_e 10 = 2\cdot 302\ 585.$

The exponential function will be considered more fully when the number e is defined (§ 48).

§ 30. General Observations on Graphs. The graphs that have been discussed up to this point have been those of functions defined by equations of the kind that occur in elementary algebra and trigonometry, and it has been assumed that the functions are *Continuous*. It is only on this assumption that we are justified in joining the points whose coordinates satisfy an equation and concluding that the coordinates of the points which lie on the short lines or arcs that we draw will actually satisfy the equation. In other words we assume that when the argument x changes by a small amount the function y will also change only by a small amount. The only exception we have found has been in those cases in which as x tended towards a special finite value y tended to a very large value (numerically). See § 24. Thus if $y = 1/x$, as x changes say from 1/1000 to 1/1001, y changes from 1000 to 1001, that is, an extremely small change in x produces a change of 1 unit in y; as x gets nearer still to 0 a change of the same amount as before would produce a still larger change in y. Hence as x approaches 0, y or $1/x$ ceases to be continuous, or, as it is usually expressed, becomes discontinuous.

Since division by zero is expressly excluded in stating the rules of division in algebra, the symbol $\frac{1}{0}$ is really meaningless; but since it is possible by taking x nearer and nearer to 0 to make $\frac{1}{x}$ greater than any given finite number, it is usual *to define* $\frac{1}{0}$ as "infinite," or "an infinite number."

Hence a function becomes discontinuous for those values of its argument that make it, in the above sense, infinite.

The question of continuity will be taken up in Chap. V.

When, as frequently occurs in practical work, the relation between a function and its argument is determined by measurements, it is only possible to calculate a comparatively small number of corresponding values of function and argument. In such a case it would be possible to find a great variety of curves which would be continuous in the mathematical sense and would pass through all the points that are plotted. In practice the points are not joined by straight lines; but the simplest curve on or near which the points seem to lie is usually taken as the graph of the function. The broken line or curve which would be obtained by joining the plotted points by straight lines would have this disadvantage, that its *curvature* would not be continuous; in the language of the Calculus, the *derivative* of the function as represented by the graph would change abruptly, as a rule, for the values of the function actually calculated.

Of course considerable care must be taken in selecting the curve and no inference should be drawn, as a rule, from the form of the graph outside the range of the argument for which the values of the function have been calculated. Examples of such graphs will be found in most text books of mechanics, physics or chemistry.

EXERCISES VI.

1. From the graph of $f(x)$ derive that of $f(kx)$, k being a constant.

Denote the graph of $f(x)$ by G_1 and that of $f(kx)$ by G_2. When $x = a$, the ordinate of G_2 is $f(ka)$; but $f(ka)$ is the value of $f(x)$ when $x = ka$. Hence the ordinate of G_2 when $x = a$ is equal to that of G_1

GENERAL OBSERVATIONS. EXERCISES VI.

when $x = ka$. Since a may be any value whatever of the abscissa, G_2 may be derived from G_1 without further calculation of ordinates; we may say that every line parallel to the x-axis is contracted in the ratio k to 1, while every line parallel to the y-axis is unaltered.

2. Apply the principle of ex. 1 to construct (i) the graph of $\sin \tfrac{1}{2}x$ and of $\sin 2x$ from that of $\sin x$; (ii) the graph of 2^{kx} from that of 2^x; (iii) the graph of 2^{-kx} from that of 2^{-x}.

3. From the graph of $f(x)$ derive that of $cf(kx)$, c and k being constants.

Deduce the graph of the ellipse $x^2/a^2 + y^2/b^2 = 1$

(i) from that of the circle $x^2 + y^2 = a^2$;

(ii) from that of the circle $x^2 + y^2 = 1$.

4. A point moves in a plane and at time t its coordinates are

$$x = Vt \cos a, \quad y = Vt \sin a - \tfrac{1}{2}gt^2;$$

show that the path of the point is a parabola with its axis vertical downward, that its vertex is the point $(V^2 \sin a \cos a/g,\ V^2 \sin^2 a/2g)$, and that its latus rectum is equal to $2V^2 \cos^2 a/g$. (Compare Exer. IV., 10.)

Eliminate t, then the equation between x and y may be written

$$\left(x - \frac{V^2 \sin a \cos a}{g}\right)^2 = -\frac{2V^2 \cos^2 a}{g}\left(y - \frac{V^2 \sin^2 a}{2g}\right).$$

5. Show that the equation of the directrix of the parabola in ex. 4 is

$$y = V^2/2g$$

6. If the coordinates of a point are given by

$$x = a + bt, \quad y = A + Bt + Ct^2$$

where t is variable and $a, b, \ldots C$ constants, show that the locus of the point is in general a parabola whose vertex is the point

$$\left(a - \frac{bB}{2C},\ A - \frac{B^2}{4C}\right),$$

and whose latus rectum is equal to b^2/C.

7. Apply the transformations of § 27, (2) to the equation

$$Ax^2 + 2Bxy + Cy^2 = D, \quad \ldots\ldots\ldots\ldots\ldots\ldots\ldots\ldots\text{(i)}$$

and show that the new equation will be

$$Lx'^2 + 2Mx'y' + Ny'^2 = D, \quad \ldots\ldots\ldots\ldots\ldots\ldots\text{(ii)}$$

where
$L = A \cos^2\theta + 2B \sin\theta \cos\theta + C \sin^2\theta,$

$M = (C - A)\sin\theta \cos\theta + B(\cos^2\theta - \sin^2\theta),$

$N = A \sin^2\theta - 2B \sin\theta \cos\theta + C \cos^2\theta.$

8. Show that equation (i) of ex. 7 represents, in general, a central conic.

For equation (ii) will become $Lx'^2 + Ny'^2 = D$, which is of the form (c), § 26, if $M=0$. It is always possible to choose θ so that M shall vanish, because

$$(C-A)\sin\theta\cos\theta + B(\cos^2\theta - \sin^2\theta) = 0, \text{ if } \tan 2\theta = \frac{2B}{A-C},$$

and whatever be the values of A, B, C, an angle can always be found to satisfy this equation. The values of $\cos\theta$, $\sin\theta$ found from this equation have to be inserted in the values of L and N.

9. Show by turning the coordinate axes through 45° that the equation
$$13x^2 - 10xy + 13y^2 = 72$$
represents an ellipse whose axes are 6 and 4. Sketch the curve.

10. The coordinates of a point are given by
$$x = a\cos\frac{2\pi t}{T}, \quad y = b\cos 2\pi\left(\frac{t}{T} + a\right),$$
where t is a variable, say the time. Show that the point describes the ellipse given by the equation
$$\frac{x^2}{a^2} - 2\frac{xy}{ab}\cos 2\pi a + \frac{y^2}{b^2} = \sin^2 2\pi a.$$

11. The coordinates of a point are given by
$$x = a\cos(2\pi t/T), \quad y = b\cos(4\pi t/T);$$
show that the point describes the parabola given by
$$\frac{y}{b} + 1 = \frac{2x^2}{a^2}.$$

12. Find the coordinates of the centre and the lengths of the axes of the central conics given by the equations

(i) $4x^2 + 9y^2 - 24x + 72y + 144 = 0$;

(ii) $3x^2 - 4y^2 + 66x + 40y + 251 = 0$.

Equation (i) may be written
$$\frac{(x-3)^2}{9} + \frac{(y+4)^2}{4} = 1.$$

13. Show by turning the axes through 45° that the equation
$$x^3 + y^3 = 3\sqrt{2}kxy \quad\ldots\ldots\ldots\ldots\ldots\ldots\ldots\ldots\ldots\ldots\ldots\text{(i)}$$
becomes, the accents being dropped,
$$(x+k)y^2 = (k - \tfrac{1}{3}x)x^2 \quad\ldots\ldots\ldots\ldots\ldots\ldots\ldots\ldots\ldots\text{(ii)}$$

From the form (ii) show that there is an asymptote perpendicular to the new axis of abscissae. Show further that the new x is not greater than $3k$ nor less (algebraically) than $-k$, it being assumed that k is positive.

14. In Fig. 20, taking S as origin, SK as initial line, $SP=r$, $\angle KSP=\theta$, show that for this system the polar equation of an ellipse is
$$r=l/(1+e\cos\theta),$$
where $l=eKS=SL=$ semi-latus rectum.

By the definition of a conic $SP=eNP$; hence, since $NP=KS+SM$,
$$r=eKS+eSM=l+er\cos(\pi-\theta)=l-er\cos\theta,$$
and therefore $r(1+e\cos\theta)=l$.

The equation is the same if S is origin, $S'K'$ initial line, and $\angle K'SP=\theta$.

15. Show that the polar equation of a hyperbola is
$$r=l/(1+e\cos\theta).$$

16. Show that the polar equation of the parabola (Fig. 19) is
$$r=2p/(1+\cos\theta),$$
where $2p=SL$ and $\angle KSP=\theta$, the origin being S and the initial line SK. If $\angle XSP=\theta$, we shall have $1-\cos\theta$ instead of $1+\cos\theta$.

17. Show that the length of the perpendicular from the point (x_1, y_1) to the line $y-x\tan\theta=0$ is
$$(y_1-x_1\tan\theta)/\sqrt{(1+\tan^2\theta)}.$$

In Fig 22a, § 27, let P be the point (x_1, y_1); then $M'P$ is the required perpendicular, since $X_1'OX_1$ is the line $y-x\tan\theta=0$. But by (2'), § 27,
$$M'P=y'=y_1\cos\theta-x_1\sin\theta=\cos\theta(y_1-x_1\tan\theta)$$
$$=(y_1-x_1\tan\theta)/\sqrt{(1+\tan^2\theta)}.$$

By putting $-a/b$ for $\tan\theta$, we see that the perpendicular on the line whose equation is $ax+by=0$ is
$$(ax_1+by_1)/\sqrt{(a^2+b^2)}.$$

Hence to find **the length of the perpendicular** from the point (x_1, y_1) on the line whose *equation* is $ax+by=0$, substitute x_1, y_1 for x, y in the *expression* $ax+by$ and divide by the square root of the sum of the squares of the coefficients of x and y.

18. By the method of ex. 17 show that **the length of the perpendicular** from the point (x_1, y_1) to the line whose equation is
$$ax+by+c=0$$
is $\qquad (ax_1+by_1+c)/\sqrt{(a^2+b^2)}.$

The *sign* of the expression for the perpendicular will be positive or negative according as $ax_1 + by_1 + c$ is positive or negative if the root have always the positive sign; the numerical value however always gives the length.

19. Find the length of the perpendicular in the cases:

  (i) point (2, 1);   line, $3x - 4y + 5 = 0$.
  (ii) point (2, -1);   line, $12x - 13y - 10 = 0$.

20. Find the length of the perpendicular on the line given by

$$\frac{x}{a}\cos\theta + \frac{y}{b}\sin\theta = 1,$$

from the points   (i) (0, 0);   (ii) (ea, 0);   (iii) ($-ea$, 0).

If $e^2 = (a^2 - b^2)/a^2$ show that the product of the perpendiculars from the points (ii) and (iii) is equal to b^2.

21. Show that the straight line in ex. 20 meets the ellipse given by

$$x^2/a^2 + y^2/b^2 = 1$$

at only *one* point, namely the point ($a\cos\theta$, $b\sin\theta$). (Compare Exercises III. 9, 10.)

The line is therefore a tangent to the ellipse; the three perpendiculars are those from the centre of the ellipse and the two foci. (See Exercises X. 9.)

22. If MP (Fig. 21) is produced to meet the hyperbola again at P' and the lines CE, CF, at Q, Q' show that

$$Q'P \cdot PQ = MQ^2 - MP^2 = b^2 = Q'P' \cdot P'Q.$$

From these equations prove that CE and CF are asymptotes; also that PQ and $Q'P'$ are equal.

CHAPTER IV.

RATES. LIMITS.

§ 31. Rates. The fundamental problem of the Differential Calculus may be considered as the investigation of the *rate* at which a function changes with respect to its argument.

The element of *time* does not necessarily enter into the conception of a rate. Whatever be the nature of the magnitudes under consideration a change in the one which is taken as the independent variable or argument will usually produce a change in that which is taken as the dependent variable or function, and by comparing the change in the function with the change in the argument we can determine the rate at which the function changes with respect to its argument. Many problems in pure and applied mathematics depend on such a comparison, so that their solution reduces to the determination of a rate; for example, the problem of drawing a tangent to a curve is equivalent to that of determining the rate at which the ordinate varies with respect to the abscissa.

§ 32. Increments. When a variable x changes from a value x_1 to a value x_2 the difference $x_2 - x_1$ (not $x_1 - x_2$) is called the *increment* that x has taken, and is often denoted by the symbol δx_1 or Δx_1, read "delta x_1"; δ, Δ are the Greek forms of the small d and capital d, the initial letter of the word "difference." The symbol δx_1 must be taken as a whole; δ by itself in this use of the letter is meaningless.

If $x_2 > x_1$ the increment is positive, so that x has increased algebraically; if $x_2 < x_1$ the increment is negative, so that x has decreased algebraically. In both cases the one word "increment" is used, so that a negative increment is an algebraic decrease.

Since $x_2 - x_1 = \delta x_1$ we have $x_2 = x_1 + \delta x_1$, so that if x change from the value x_1 to another value and if the increment that x takes is δx_1 that other value is $x_1 + \delta x_1$; the student must accustom himself to this method of denoting the value *to* which x changes, for although $x_1 + \delta x_1$ seems more cumbrous than x_2 its form is more suggestive and is really simpler in many investigations.

Let y be a function of x, say $5x - 3$, and let x_1, y_1 be corresponding values of x and y. When x changes from x_1 to $x_1 + \delta x_1$ let y take the increment δy_1, so that the value of y corresponding to $x_1 + \delta x_1$ is $y_1 + \delta y_1$; then

$$y_1 = 5x_1 - 3; \quad y_1 + \delta y_1 = 5(x_1 + \delta x_1) - 3,$$

and therefore $\delta y_1 = 5\delta x_1$.

If $y = 3x^2 + 7x - 2$, we find, using the same notation,

$$y_1 = 3x_1^2 + 7x_1 - 2; \quad y_1 + \delta y_1 = 3(x_1 + \delta x_1)^2 + 7(x_1 + \delta x_1) - 2,$$

and therefore, by subtracting the left side of the first equation from the left side of the second, and the right side of the first from the right side of the second,

$$\delta y_1 = 6x_1 \delta x_1 + 3(\delta x_1)^2 + 7\delta x_1$$
$$= (6x_1 + 7)\delta x_1 + 3(\delta x_1)^2.$$

In general, if $y = f(x)$, we have

$$\delta y_1 = f(x_1 + \delta x_1) - f(x_1) = \delta f(x_1).$$

The same notation is used whatever letters denote the variables, so that if $s = \phi(t)$,

$$\delta s_1 = \phi(t_1 + \delta t_1) - \phi(t_1) = \delta \phi(t_1),$$

and so on.

As this process of finding increments is of constant occurrence the student should make himself quite familiar with it. The following examples should be worked through.

Ex. 1. If $y = \dfrac{1}{x^2}$, show that $\delta y_1 = -\dfrac{2x_1 \delta x_1 + (\delta x_1)^2}{(x_1 + \delta x_1)^2 x_1^2}$.

Ex. 2. If $f(x) = x^3 - 1$, show that
$$\delta f(x_1) = 3x_1^2 \delta x_1 + 3x_1(\delta x_1)^2 + (\delta x_1)^3.$$

INCREMENTS. UNIFORM RATE. 67

Ex. 3. If $y = \log x$, show that
$$\delta y_1 = \delta \log x_1 = \log \frac{x_1 + \delta x_1}{x_1} = \log\left(1 + \frac{\delta x_1}{x_1}\right).$$

Ex. 4. If $y = x^3$, $\delta y_1 = (x_1 + \delta x_1)^3 - x_1^3$, calculate δy_1 and $\delta y_1/\delta x_1$ when
$$x = 10, \quad \delta x_1 = 1, \cdot 5, \cdot 1, \cdot 01, \cdot 001.$$

Ex. 5. If $y = \sin x$, show that
$$\delta y_1 = \delta \sin x_1 = \sin(x_1 + \delta x_1) - \sin x_1.$$

From the Tables calculate δy_1 and $\dfrac{\delta y_1}{\delta x_1}$ for the following values of x_1 and δx_1, the numbers denoting the value of the angles in degrees:

(i) $x_1 = 30$, $\delta x_1 = 1, \cdot 5, \cdot 2, \cdot 1$;

(ii) $x_1 = 60$, $\delta x_1 = 1, \cdot 5, \cdot 2, \cdot 1$.

Ex. 6. If $y = \log_{10} x$, find from the Tables the values of δy_1 and $\dfrac{\delta y_1}{\delta x_1}$

(i) when $x_1 = 325$ and $\delta x_1 = 2, 1, \cdot 5, \cdot 1$;

(ii) when $x_1 = 72$ and $\delta x_1 = 2, 1, \cdot 1, \cdot 01$.

§ 33. Uniform Variation. When the argument of a function takes a series of values $x_1, x_2, x_3, x_4 \ldots$ the function takes a corresponding series of values y_1, y_2, y_3, y_4. When the increment of the function is in a constant ratio to the corresponding increment of the argument the function is said to vary *uniformly* or *at a constant rate* with respect to its argument.

If the constant ratio is a, then
$$\frac{y_2 - y_1}{x_2 - x_1} = a; \quad \frac{y_4 - y_3}{x_4 - x_3} = a,$$
and
$$y_2 - y_1 = a(x_2 - x_1); \quad y_4 - y_3 = a(x_4 - x_3).$$

If the increments $(x_2 - x_1)$ and $(x_4 - x_3)$ of the argument are equal so are the corresponding increments $(y_2 - y_1)$ and $(y_4 - y_3)$ of the function. The increment $(x_2 - x_1)$ may be either positive or negative and may be of any magnitude whatever; the corresponding increment of the function is $a(x_2 - x_1)$, and always, when the argument takes two equal increments so does the function.

It follows from the definition that the uniformly varying function is a linear function of its argument. For when the argument changes from any value x_1 to any other

value x, let the function change from y_1 to y; then the increment of the argument is $(x-x_1)$, the increment of the function is $(y-y_1)$ and
$$(y-y_1):(x-x_1)=a;$$
that is, $\qquad y = ax + y_1 - ax_1.$

But x_1, y_1 are fixed values of the argument and function and the ratio a is constant, so that y is a linear function of x.

It is easy to see conversely that if y is a linear function, $ax+b$ say, of x, then y varies uniformly with respect to x.

Measure of a Uniform Rate.—The constant ratio a is taken as the measure of the rate at which the function varies with respect to its argument. Instead of saying that the ratio a measures the rate we shall generally use the briefer expression that a *is* the rate.

When a is a positive number, y increases as x increases and decreases as x decreases; when a is a negative number, y decreases as x increases and increases as x decreases. The particular case in which the function reduces to a constant, $y=b$, may be included in the general category of uniformly varying functions by saying that the function varies at the rate *zero*; $a=0$.

Since the graph of $ax+b$ is a straight line with the gradient a (§ 22) the gradient of the line measures the rate at which the function varies with respect to its argument. It should be noticed that if in plotting the graph the unit for the ordinates is not of the same length as the unit for the abscissae the tangent of the angle *shown on the diagram* will not be equal to the rate a; if the unit for abscissae is 1 inch and for ordinates, say ·1 inch, then to an increment 1 of the abscissa the diagram will show an increment, not of a but of ·1a of the ordinate, so that the real gradient or rate will be found by multiplying by 10 the tangent of the angle shown on the diagram.

§ 34. Dimensions of Magnitudes. It is customary and convenient to use such expressions as "the area of a rectangle is the product of its base and its altitude," "the speed of a body which moves uniformly is the distance

DIMENSIONS OF MAGNITUDES. 69

gone in a given time divided by the time," and these expressions are represented in the form of equations:

$$\text{area} = \text{base} \times \text{altitude}; \quad \text{speed} = \frac{\text{distance}}{\text{time}}.$$

When considered as equations in the sense commonly understood in algebra these must be interpreted as "the number of square feet (or square inches, etc.) in the area is equal to the product of the number of linear feet (or inches, etc.) in the base and in the altitude," "the number of units of speed is equal to the quotient of the number of units of length in the distance by the number of units of time."

But the equations may be interpreted in a different manner. Let capital letters denote, not numbers but magnitudes; L the straight line of unit length, T the interval of time taken as the unit. Taking as unit of area the square on the line L, and as unit of speed that of a body which moves uniformly a distance L in time T, the equations may be stated for the unit magnitudes in the form

$$\text{unit area} = L \times L; \quad \text{unit speed} = \frac{L}{T};$$

or, combining the symbols by the algebraic laws of indices,

$$\text{unit area} = L^2; \quad \text{unit speed} = LT^{-1}.$$

These equations are usually called *dimensional* equations, and the indices are said to give the *dimensions* of the magnitudes; thus the first equation states that the unit area is of 2 dimensions in L, the unit of length, and the second states that the unit of speed is of dimension 1 in L and of dimension -1 in T. Since all areas are magnitudes of the same kind as the unit area, area is said to be of 2 dimensions as to length and to have L^2 as its *dimensional formula*. Similarly, the dimensional formula of speed is LT^{-1}.

If M denote the unit mass the dimensional formula of momentum will be MLT^{-1}, because momentum is the product of mass and velocity.

It may happen that a magnitude has zero dimensions; thus angles when measured in radians have zero dimensions, because the radian is "arc divided by radius," and its dimensional formula therefore is L/L, that is L^0.

A notation that suggests the dimensional formula is sometimes used; thus an area of 10 square feet is denoted by 10 ft.^{2}, a speed of 10 feet per second by 10 ft./sec., a pressure of 14 pounds per square inch by 14 lb./in.^{2} and so on. The characteristic word for expressing a rate, namely *per*, is represented by the symbol of division.

When a function varies uniformly the number which has been defined as the rate of variation is quite independent of the magnitude of the increment which the argument takes; it is therefore possible to choose at pleasure the increment of the argument that shall be called unit increment. Thus we may speak of a speed of 30 miles per hour, although the motion may only last 5 minutes, or 1 minute or less; a rate of 30 miles per hour is the same thing as one of half a mile per minute, or of 44 feet per second. It is important to bear in mind this aspect of a rate when discussing non-uniform variation.

Again, the statement that the speed of a moving body is 30 miles per hour is equivalent to the statement that the distance travelled varies with respect to the time at the rate 30, when it is understood that the units are the mile and the hour. The latter mode of expression is more simple in many cases.

When the measure of a magnitude is interpreted as a rate the dimensional formula for the magnitude will be the quotient of the formula for the function by that for the argument. Thus force may be measured as the rate of change of momentum with respect to time; its dimensional formula is therefore MLT^{-1}/T or MLT^{-2}.

It is important to bear in mind that the measure of one magnitude can often be interpreted as the rate of change of a second magnitude with respect to a third, because it is through this connection that the calculus is applied to the investigation of the numerical relations of magnitudes, and in all such interpretations the theory of dimensions is of great service. For a full treatment of that theory the student is referred to the books named below.[1]

---

[1] Everett's *Units and Physical Constants*; Gray's *Absolute Measurements in Electricity and Magnetism*; Maclean's *Physical Units*.

AVERAGE RATE.

§ 35. Variable Rates. So far, only uniformly varying functions have been discussed. But it may happen that the increment of the function is not in a constant ratio to the corresponding increment of its argument, or in other words, that two equal increments of the argument do not always produce two equal increments of the function; in that case the function is said to vary *non-uniformly*, or *at a variable rate*, with respect to its argument.

Let $y = 3x^2$; when x varies from x_1 to $x_1 + h$ let y vary from y_1 to $y_1 + k$, and when x varies from x_2 to $x_2 + h$ let y vary from y_2 to $y_2 + k'$. Then

$$y_1 = 3x_1^2;\quad y_1 + k = 3(x_1 + h)^2;\quad k = 6x_1 h + 3h^2;$$

therefore $k/h = 6x_1 + 3h$; and in the same way we find

$$k'/h = 6x_2 + 3h.$$

The two ratios k/h, k'/h are therefore unequal, so that y varies non-uniformly with respect to x.

In this case the ratio k/h depends both on h and on x_1; the characteristic property of a uniformly varying function is that the ratio k/h depends neither on h nor on the value x_1 of x, from which the increment begins. To obtain the number which is taken as the measure of a variable rate we proceed as follows.

§ 36. Average Rate. We first define an *average rate*, thus :—The average rate at which a function varies with respect to its argument while that argument takes a given increment h is defined to be that uniform rate which would give the actual increment k taken by the function.

The average rate is thus k/h. In the example of last article the average rate at which y varies with respect to x while x varies from x_1 to $x_1 + h$ is

$$k/h = 6x_1 + 3h;$$

the average rate at which y varies while x varies from x_2 to $x_2 + h$ is
$$k'/h = 6x_2 + 3h.$$

The average rate thus depends both on x and on h.

Next, it agrees with our ordinary notions of a rate of change to suppose that the smaller h is the better will the average rate measure the rate at which the function varies as x varies from x_1 to x_1+h. But as h is taken less and less the average rate $6x_1+3h$ approximates more and more closely to the definite number $6x_1$. The average rate will never be *exactly* $6x_1$, because it would be absurd to suppose h actually zero; that would amount to supposing that x had not changed from the value x_1 at all. On the other hand, however small h may be, provided it is not zero, the quotient k/h can be calculated and the average rate for that small increment determined. We may therefore suppose h to be, not zero, but so small that the difference between $6x_1+3h$ and $6x_1$, namely $3h$, shall be less than any fraction that may be named, however small that fraction may be, provided only it is not zero; for example, the difference will be less than ·001 if h be numerically less than one third of ·001, say less than ·0003. It is natural therefore to consider $6x_1$ as measuring the rate at which y changes with respect to x as x increases or decreases from the value x_1.

We therefore *define* $6x_1$ as the rate at which the function y or $3x^2$ varies with respect to its argument x *for the value x_1 of the argument*.

In the same way $6x_2$ is, *by definition*, the rate of change for the value x_2 and in general for any value a of the argument the rate is $6a$, because the reasoning does not depend on the particular value x_1; the reasoning is the same whatever value of the argument be chosen.

When x has the values $0, \frac{1}{4}, \frac{1}{2}, \frac{3}{4}, 1, \frac{3}{2}, 2 \ldots$ the rate is equal to $0, \frac{3}{2}, 3, \frac{9}{2}, 6, 9, 12 \ldots$ respectively; thus for the value 1 of x, y is increasing twice as fast as for the value $\frac{1}{2}$, for the value $\frac{3}{2}$ thrice as fast, for the value 2 four times as fast and so on. The student should compare these statements with the information to be derived from an inspection of the graph of $3x^2$.

When x is positive the rate is positive, so that as the point x moves to the right the graphic point moves up; on the other hand, when x is negative the rate is negative, so that as the point x moves to the right the graphic point moves down.

It will be noticed that in stating a variable rate the phrase "for the value x_1 of the argument" occurs; the phrase is needed because, unlike that of a uniformly varying function, the rate is in this case itself a variable. If the number s of feet described by a moving body in t seconds be $3t^2$, the rate at which s varies with respect to t at time t_1 seconds after motion begins is $6t_1$, that is, *the speed at time t_1 is $6t_1$ feet per second.*

§ 37. Measure of a Variable Rate. The method just given of defining a variable rate is of fundamental importance, and the student should make sure that he masters the reasoning on which the definition is based. The process consists of three steps:

(i) We find the average rate k/h; the number k/h depends both on x_1 and on h.

(ii) We assume as consistent with our notions of rate of change that the smaller h is the better will the quotient k/h measure the rate at which the function changes as the argument changes from x_1 to x_1+h. It usually happens that by taking h less and less the quotient k/h gets nearer and nearer to a definite number; h is not supposed to become zero, but in general we can take h so small that the difference between k/h and a definite number will become, and for smaller values of h will remain, less than any stated, non-zero, fraction. The number will depend on x_1.

(iii) We then *define* this number as *the* rate at which the function changes with respect to the argument for the value x_1 of the argument.

The more rigorous of the older mathematicians, such as Maclaurin, starting from definitions or axioms respecting variation at a greater or less rate, *proved* that $6x_1$ is the "true measure" of the rate at which $3x^2$ varies with respect to x for the value x_1; but the reasoning on which we have based the definition seems sufficient to establish its correctness. Of course if the values considered were determined by measurement a stage of smallness for h would soon be reached at which it would become impossible to distinguish between $6x_1$ and $6x_1+3h$; the average rate determined by the smallest available value of h would therefore coincide with that determined by the process and definition we have adopted.

Ex. 1. If $s = \tfrac{1}{2}gt^2$, find the rate at which s varies with respect to t, when t has the values 0, $\tfrac{1}{2}$, 1, 2.

Ex. 2. If $p = \dfrac{a}{v}$, find the rate at which p varies with respect to v when $v = v_1$.

§ 38. Limits. It would seem at first sight as if the rate $6x_1$ could be determined from the average rate $6x_1 + 3h$ simply by putting h equal to 0. But the logic of such a step would be faulty, because the equation

$$\frac{k}{h} = 6x_1 + 3h.$$

can only be established *on the assumption that h is not zero*; in proving the laws of division in algebra the case in which the divisor is zero is expressly excluded. But further, if $h = 0$, so also is k, and the quotient k/h would appear in the form $0/0$—a symbol which has absolutely no meaning whatever. The ground in common sense for defining $6x_1$ as the rate of change for the value x_1 is that $6x_1$ is *the one definite number* towards which the average rate k/h settles down as h is taken smaller and smaller. (See the values of $\delta y_1/\delta x_1$ in examples 4, 5, 6 (§ 32) as an illustration of this settling down.)

In mathematical language we are said, in determining the number towards which the quotient k/h settles down, to find the *limit* of k/h when h tends to zero as *its limit*; in this process h is a variable number, positive or negative, and it may take any value *except zero*; zero is so to speak a boundary to which it gets nearer and nearer, but which it never actually reaches.

Before giving a formal definition of a limit we will consider a few typical cases; by carefully studying these the student will gather the necessity for the introduction of the word and will see what it really means.

§ 39. Examples of Limits. (i) Let AB (Fig. 24) be a chord of a circle whose centre is O; AT, BT the tangents at A, B. Let OT cut the chord AB at M and the arc AB at N; M and N will be the middle points of the chord and the arc respectively and OM will be perpendicular to AB.

LIMITS. EXAMPLES.

The triangles OMA, OAT are equiangular; therefore

$$\frac{MA}{AT} = \frac{OM}{OA} \quad \ldots\ldots\ldots\ldots\ldots\ldots\ldots(1)$$

Suppose now that the chord AB moves towards N, the point N remaining fixed and AB being always perpendicular to ON; let A, B always denote the ends of the chord, M its mid point and T the point where the tangents at A and B meet. So long as A and B are not coincident, that is so long as AB is really a chord, equation (1) remains true. The ratio $MA:AT$ is a function of OM, for as soon as OM is fixed every other line in the figure is fixed, and the ratio can be calculated.

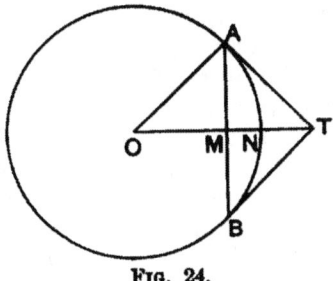

Fig. 24.

When OM is all but equal to ON both MA and AT will be all but zero; nevertheless the ratio $MA:AT$ will be all but equal to 1, because equation (1) remains true and OM is all but equal to ON which is equal to OA. Manifestly the nearer M gets to N the nearer does the ratio $MA:AT$ get to unity.

This behaviour of the ratio $MA:AT$ is expressed in the words:—as OM approaches ON as its limit the ratio $MA:AT$ approaches 1 as its limit.

Here again it has to be noted that the reasoning ceases to be just if OM becomes actually equal to ON, for the triangles will then have disappeared and the equation (1) on which the reasoning is based could not be established.

We might equally well consider the ratio as a function, not of OM but, of the angle NOA; if the angle NOA approaches zero as its limit the ratio approaches 1 as its limit.

(ii) Suppose AB (Fig. 24) to be the side of a regular polygon of n sides (regular n-gon) inscribed in the circle; then it is easy to prove that the side of the regular n-gon circumscribed about the circle is equal to $AT+BT$ or $2AT$ and that the angle NOA is $180/n$ degrees. If p, P denote the

76   AN ELEMENTARY TREATISE ON THE CALCULUS.

perimeters of the inscribed and of the circumscribed polygons respectively, then
$$p = nAB = 2nMA\ ; \quad P = 2nAT,$$
and
$$\frac{p}{P} = \frac{MA}{AT} = \frac{OM}{OA} = 1 - \frac{MN}{OA}. \quad \ldots\ldots\ldots\ldots\ldots\ldots(2)$$

Imagine a series of polygons constructed corresponding to greater and greater values of n. When n becomes very large the angle NOA will become very small; AB and MN will also become small, and therefore the ratio p/P will become nearly equal to 1. Hence when the angle NOA approaches 0 as its limit, the ratio p/P approaches 1 as its limit; or again it may be put thus, when *n becomes indefinitely large* p/P approaches 1 as its limit.

We may express the relation between p and P in a slightly different way. From equation (2) we get

$$P - p = \frac{MN}{OA} \cdot P.$$

When n is greater than 4, P will be less than the perimeter of the circumscribed square, that is less than $8OA$; hence
$$P - p < 8MN.$$

Now let ϵ be any line that is as small as we please, only not zero. By the geometry of the figure we see that we can take n so large that MN shall be less than any given line; choose n therefore so large that MN is less than $\epsilon/8$. Then for *this* and for *all greater values* of n, $8MN$ will be less than ϵ and therefore $P - p$ less than ϵ.

It is here that the limit notion comes in; no matter how large n may be P and p will never exactly coincide, but as n increases beyond all bounds the difference $P - p$ tends to zero as its limit, that is the perimeters P and p tend towards *the same limit*.

FIG. 25.

On the straight line FH (Fig. 25) mark off Fg_n, FG_n equal to the perimeters p, P respectively; then clearly for every

value of n, Fg_n is less than FG_n. But, when n has been chosen as above, $\qquad g_n G_n = P - p < \epsilon,$
and therefore n can be taken so large that $g_n G_n$ shall be less than the line ϵ. Hence the common limit of p and P is a line FG greater than every one of the lines Fg_n, but less than every one of the lines FG_n.

Since the circumference C of the circle always lies between p and P, the circumference will be equal to the line FG; the circumference may therefore be considered as the limit either of an inscribed or of a circumscribed regular polygon when the number of its sides increases indefinitely.

(iii) Show that the area of a circle may be considered as the limit either of an inscribed or of a circumscribed regular n-gon; and that an arc of a circle may be considered as the limit of the sum of n equal chords obtained by dividing the arc into n equal arcs.

The polygons have been supposed regular, but it would not be difficult to show that the theorems hold even if they be not regular, provided that as n increases beyond all bounds the length of each side of the polygons approaches zero as its limit.

(iv) Let θ be the number of radians in the angle NOA, where the angle is supposed to be acute; we have

$$\text{chord } AB < \text{arc } AB < AT + BT,$$

and therefore $\qquad MA < \text{arc } NA < AT.$

Hence $\qquad \dfrac{MA}{OA} < \dfrac{\text{arc } NA}{OA} < \dfrac{AT}{OA};$

that is, $\qquad \sin \theta < \theta < \tan \theta.$

Divide by $\sin \theta$; therefore

$$1 < \frac{\theta}{\sin \theta} < \frac{1}{\cos \theta},$$

and therefore $\qquad 1 > \dfrac{\sin \theta}{\theta} > \cos \theta.$

Thus the quotient $\sin \theta/\theta$ lies between 1 and $\cos \theta$. When θ approaches 0 as its limit $\cos \theta$ approaches 1 as its limit; therefore also $\sin \theta/\theta$ approaches 1 as its limit.

78   AN ELEMENTARY TREATISE ON THE CALCULUS.

From the nature of a limit, or from the last inequality, we see that the statement that $\sin\theta/\theta$ approaches 1 as its limit when θ approaches 0 as its limit may be put in the form: when θ is a small number $\sin\theta$ is approximately equal to θ. The student should verify this statement from the Tables; thus, for $\angle NOA = 1°$,

$$\theta = \cdot 0174533; \quad \sin\theta = \cdot 0174524;$$

for $\angle NOA = 5°$,

$$\theta = \cdot 0872665; \quad \sin\theta = \cdot 0871557.$$

(v) Show that the limit of $\tan\theta/\theta$, as θ approaches 0 as its limit, is unity.

(vi) Provided x is not equal to a,

$$(x^2 - a^2)/(x-a) = x + a.$$

The equation holds true so long as x is not equal to a; but we can take x so nearly equal to a that $x+a$ shall differ from $2a$ by as little as we please. That is, the quotient can be brought as near to $2a$ as we please simply by taking x near enough to a. Hence although the quotient has no meaning whatever, no *value*, when x is equal to a, it has a definite limit, namely $2a$, for x approaching a as its limit.

(vii) Let SPT (Fig. 26) be the tangent to a circle at P; PQ a secant and PR a given length measured along the secant. Describe a circle with centre P and radius PR, cutting PT at R'.

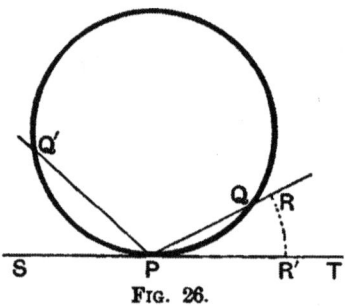

Fig. 26.

Now let Q move along the arc PQ towards P; R will therefore move along the arc RR' towards R'. The nearer Q approaches P, the nearer does R come to R', and the smaller becomes the angle TPR. If we suppose Q to approach P as its limiting position, the secant PR will approach the tangent PT as its limiting position. If we suppose the secant drawn on the other side of P, as PQ', PS will be the limiting position of the secant as Q' approaches P. Hence we may define a tangent thus:

DEFINITION.   A tangent to a curve at a point P is the

limiting position of a chord PQ as Q approaches P as its limiting position.

It is this definition of a tangent that will be subsequently used in the book.

(viii) Show from the theorem in *Exercises V*. 10, that if T is a point on the directrix KN such that the angle PST is a right angle, the line PT will be the tangent to the conic at P.

§ 40. General Explanation of a Limit. The special meaning of the word · *limit* should now be fairly clear. In each of the examples there are two variables, one being a function of the other.

One of these variables, the argument, is supposed to become all but equal to a definite number, for example to a or 0 or ON; or else it is supposed to increase beyond all bound. In the former case the definite number is called the *limit* of the argument; it is not a *value* that the argument actually takes; thus in (iv) 0 is not a value that θ assumes. In the latter case the argument is generally said to have *infinity* for limit, though this mode of expression seems rather a contradiction in terms; the argument has infinity for limit if it is supposed to become greater than any number N, no matter how great N may be.

Again, when the argument becomes nearly equal to its limit the function at the same time becomes nearly equal to a definite number; not only so, but we can make the argument differ so little from its limit that the function shall differ by as little as we please (except by the difference zero) from that definite number. This definite number therefore is called the *limit* of the function for the argument approaching its limit.

We will now give a formal definition of a limit; the first mode of statement is somewhat rough, the second is more definite, but in a first reading it may be found a little more difficult to grasp.

§ 41. Definition of a Limit. Notation. Distinction between Limit and Value.

DEFINITION 1. When it is possible to make the argument of a given function so nearly equal to a definite number a that the function will differ from another definite number A by as little as we please, that difference remaining as small as we please when the argument is taken still nearer to a, then A is called the limit of the function for the argument approaching (or converging to) a as its limit.

DEFINITION 2. Given any positive number ϵ that may be as small as we please, except that it must not be zero, given also two definite numbers a, A; if it be possible to find a positive number η such that a given function shall differ from A by less than ϵ for all values of its argument that differ from a by less than η (the value a itself being excluded), then A is called the limit of the function for the argument approaching (or converging to) a as its limit.

The modifications required when either a or A is infinite offer no difficulty. In general a variable is said to become infinite if it takes values that are numerically greater than any positive number N, no matter how large N may be; if the variable is positive it converges to $+\infty$, if negative to $-\infty$. The definite number A will be the limit of a function for its argument approaching $+\infty$ as its limit, provided that a positive number N can be found such that for every value of the argument greater than N the difference between the function and A shall be as small as we please.

The notation for a limit is the letter L or the first three letters of the word limit, namely *lim*. To state that the function $f(x)$ approaches A as its limit when x approaches a as its limit, the notation is

$$\mathrm{L}\, f(x) = A \text{ when } \mathrm{L}\, x = a,$$

or, more usually, $\mathrm{L}_{x=a} f(x) = A$;

read "limit of $f(x)$ for x equal to a is A." It must be remembered however that the more usual form is a contraction for the first, and that a, A are not values that the variables are supposed actually to take.

In this notation, if $\angle NOA = \theta$ and $OA = a$, ex. (i) of § 39 may be stated
$$\mathrm{L}_{OM=a} \frac{MA}{AT} = 1 \text{ or } \mathrm{L}_{\theta=0} \frac{MA}{AT} = 1;$$

ex. (ii): $\mathrm{L}_{n=\infty} p = \mathrm{L}_{n=\infty} P = C$;

ex. (iv), (v): $\mathrm{L}_{\theta=0} \dfrac{\sin\theta}{\theta} = 1, \quad \mathrm{L}_{\theta=0} \dfrac{\tan\theta}{\theta} = 1.$

The necessity for the introduction of the notion of a limit arose from the consideration of cases in which the function ceased to have meaning when a particular value was assigned to the argument; but the notion of a limit is not, by the definition, restricted to such cases. Whether $f(x)$ has or has not a definite value when x is equal to a the limit is found by considering the values of $f(x)$ for values of x *nearly equal to* a; the value a itself is not to be used in the process. It may of course happen that the *limit* A of the function coincides with the *value* $f(a)$; still, even when A and $f(a)$ coincide, the fact that they are determined by different processes should not be forgotten. Instances frequently occur in which the limit A and the value $f(a)$ are both definite and yet unequal.

§ 42. **Theorems on Limits.** We now state the principal rules for working with limits. In the following theorems the functions have the same argument, x say, and the limits spoken of are the limits for each function as the argument approaches a limit, say a, the limits of the functions being finite; it will be sufficient therefore to use the letter L without the subscript "$x=a$." The number of functions is supposed to be *finite;* the theorems are not necessarily true if the number be infinite.

THEOREM I. *The limit of the algebraic sum of any number of functions is equal to the like algebraic sum of the limits of the functions.*

THEOREM II. *The limit of the product of any number of functions is equal to the product of the limits of the functions.*

THEOREM III. *The limit of the quotient of two functions is equal to the quotient of the limits of the functions, provided the limit of the divisor is not zero.*

The proof of these theorems is simple; it depends on the particular cases that if the limit of each of a finite number of variables is zero, then the limit of their sum and of their product must be zero.

Let h_1, h_2, h_3, for example, be three variables the limit of each of which is zero. To prove that the limit of their sum is zero we have to show that x can be taken so near a that that sum will be numerically less than any given positive number ϵ. Now, since the limit of each is zero, we can take x so near a that each of the variables shall be numerically less than $\tfrac{1}{3}\epsilon$; hence we can take x so near a that their sum shall be less than ϵ. The same reasoning holds if there be n variables; each can be made less than ϵ/n. It does not matter whether the variables be positive or negative or whether the sum contain negative terms since it is the numerical value alone that is concerned. Manifestly the product will also have zero for limit.

Again, if C be any finite constant, the limit of Ch_1 will be zero; we need only choose x so near to a that h_1 shall be numerically less than ϵ/C.

Now, let u, v, w be functions of x whose limits are U, V, W. Then by the nature of a limit when x is nearly equal to a, u, v, w are nearly equal to U, V, W; hence we may write
$$u = U + h_1, \quad v = V + h_2, \quad w = W + h_3,$$
where h_1, h_2, h_3 are variables which have zero for limit.

Then
$$u + v - w = U + h_1 + V + h_2 - W - h_3$$
$$= U + V - W + h_1 + h_2 - h_3.$$

Hence, since the limit of each of the numbers h_1, h_2, h_3 is zero,
$$L(u + v - w) = U + V - W,$$
$$= Lu + Lv - Lw.$$

Again,
$$uv = (U + h_1)(V + h_2),$$
$$= UV + Uh_2 + Vh_1 + h_1h_2;$$
so that
$$L(uv) = UV = (Lu) \times (Lv).$$

Again,
$$L(uvw) = L(uv) \times Lw,$$
$$= Lu \times Lv \times Lw,$$
by applying twice the case for the product of two variables.

Finally,
$$\frac{u}{v} = \frac{U + h_1}{V + h_2} = \frac{U}{V} + \left(\frac{U + h_1}{V + h_2} - \frac{U}{V}\right)$$
or
$$\frac{u}{v} = \frac{U}{V} + \frac{Vh_1 - Uh_2}{V(V + h_2)}.$$

The limit of the second fraction is zero because the numerator can be made as small as we please, while the

denominator is not zero, since V is by hypothesis not zero; it follows that
$$\mathop{L}\frac{u}{v} = \frac{U}{V} = \frac{\mathop{L}u}{\mathop{L}v}.$$

We have for simplicity taken only three functions, but clearly the reasoning holds if there be more than three functions, the limits U, V, ... being of course all finite and no denominator having zero for limit.

If one or more of the functions be constant it is evident that the reasoning holds; thus u might be a constant, and then we might consider Lu as being simply u itself, without in any way violating the conditions for a limit.

§ 43. Examples. We will now give a number of examples in which the above principles come into play. In seeking the limit it is useful to bear in mind that any transformation of the function which is legitimate when the argument is not equal to its limit may be applied as a help towards the solution. Thus

$$\frac{\sqrt{(x+1)}-1}{x} = \frac{x+1-1}{x\{\sqrt{(x+1)}+1\}} = \frac{1}{\sqrt{(x+1)}+1}.$$

The division of x out of numerator and denominator is legitimate so long as x is not zero; but in finding the limit for $x=0$, x is not to become 0, and therefore the first and the last of the three fractions are equal for all values of x considered. Hence

$$\mathop{L}_{x=0}\frac{\sqrt{(x+1)}-1}{x} = \mathop{L}_{x=0}\frac{1}{\sqrt{(x+1)}+1} = \frac{1}{2}.$$

In the same way we find

$$\mathop{L}_{x=\infty}\frac{\sqrt{(x+1)}-1}{x} = \mathop{L}_{x=\infty}\frac{1}{\sqrt{(x+1)}+1} = 0.$$

We take it to be sufficiently evident that the first of these limits is $\frac{1}{2}$; by Def. 2, § 41, we should be able to find η, so that when x is numerically less than η the difference between the function and $\frac{1}{2}$ shall be less than any number we may name, say less than ·001. But the search for η is usually very troublesome, and in such simple cases as we have to deal with we shall usually dispense with that part of the investigation, as the nature of the processes involved will show that such a number can be found.

Ex. 1. $\qquad \underset{n=\infty}{L} \left\{ \dfrac{1}{n^2} + \dfrac{2}{n^2} + \dfrac{3}{n^2} + \ldots + \dfrac{n}{n^2} \right\} = \dfrac{1}{2}.$

For the sum $= \dfrac{1+2+3+\ldots+n}{n^2} = \dfrac{\frac{1}{2}n(n+1)}{n^2} = \dfrac{1}{2} + \dfrac{1}{2n}.$

This example shows that Th. I., § 42, is not necessarily true unless the number of functions is finite; for although the limit of each term in the bracket is zero, the limit of the sum is not zero.

Ex. 2. $\qquad \underset{n=\infty}{L} \dfrac{1^2 + 2^2 + 3^2 + \ldots + n^2}{n^3} = \dfrac{1}{3},$

for $\qquad 1^2 + 2^2 + 3^2 + \ldots + n^2 = \dfrac{1}{6}n(n+1)(2n+1),$

and therefore $\dfrac{1^2 + 2^2 + 3^2 + \ldots + n^2}{n^3} = \dfrac{1}{3} + \dfrac{1}{2n} + \dfrac{1}{6n^2},$

so that the limit is $\dfrac{1}{3}$.

Ex. 3. If r be a proper fraction and n a positive integer, $\underset{n=\infty}{L} r^n = 0$.

For any positive proper fraction is of the form $1/(1+a)$, where a is a positive number. Now, by the binomial theorem or otherwise we can readily show that $(1+a)^n$ is greater than $1+na$.

Hence, so far as numerical value is concerned,

$$r^n = \dfrac{1}{(1+a)^n} < \dfrac{1}{1+na},$$

and since the limit of $1/(1+na)$ for $n = \infty$ is zero, the limit of r^n is also zero.

Ex. 4. Show that if r be a proper fraction and n a positive integer
$\qquad \underset{n=\infty}{L} nr^n = 0, \quad \underset{n=\infty}{L} n^2 r^n = 0,$ etc.,

for $\qquad (1+a)^n > 1 + na + \dfrac{1}{2}n(n-1)a^2;$ so that

$$nr^n < \dfrac{1}{\dfrac{1}{n} + a + \dfrac{1}{2}(n-1)a^2}, \text{ etc.}$$

Ex. 5. $\qquad \underset{x=\infty}{L} \dfrac{2x^2 + 3x - 1}{3x^2 - 2x + 1} = \dfrac{2}{3},$

for the fraction $= \dfrac{2 + \dfrac{3}{x} - \dfrac{1}{x^2}}{3 - \dfrac{2}{x} + \dfrac{1}{x^2}},$

and the limits of numerator and denominator are 2 and 3 respectively.

EXAMPLES.

Ex. 6. $$\mathop{L}_{h=0}\frac{(x+h)^3-x^3}{h}=3x^2,$$

for $$\frac{(x+h)^3-x^3}{h}=3x^2+3xh+h^2.$$

Ex. 7. $$\mathop{L}_{x=-1}\frac{x^2+4x+3}{x^2-7x-8}=-\frac{2}{9}.$$

First remove the common factor $x+1$; it is the presence of this factor that makes the fraction take the form 0/0 when we try to calculate its value for $x=-1$.

Ex. 8. $$\mathop{L}_{x=2}\frac{x^2-4}{\sqrt{(x+2)}-\sqrt{(3x-2)}}=-8.$$

Ex. 9. $$\mathop{L}_{x=0}\frac{\sin 3x}{x}=3,$$

for $$\frac{\sin 3x}{x}=\frac{\sin 3x}{3x}\times 3\ \ \text{and}\ \ \mathop{L}_{x=0}\frac{\sin 3x}{3x}=1.$$

Ex. 10. $$\mathop{L}_{x=\frac{\pi}{2}}\frac{\cot x}{\frac{\pi}{2}-x}=1.$$

Put $x=\frac{\pi}{2}-y$; then when x approaches $\frac{\pi}{2}$ as its limit y approaches 0 as its limit. Hence

$$\mathop{L}_{x=\frac{\pi}{2}}\frac{\cot x}{\frac{\pi}{2}-x}=\mathop{L}_{y=0}\frac{\tan y}{y}=1.$$

This device of changing the variable is often useful; for example:

Ex. 11. $$\mathop{L}_{h=0}\frac{(x+h)^{\frac{3}{2}}-x^{\frac{3}{2}}}{h}=\frac{3}{2}x^{\frac{1}{2}}.$$

Put $x=y^2$ and $x+h=(y+k)^2$, so that when h approaches 0 so does k; therefore

$$\mathop{L}_{h=0}\frac{(x+h)^{\frac{3}{2}}-x^{\frac{3}{2}}}{h}=\mathop{L}_{k=0}\frac{(y+k)^3-y^3}{(y+k)^2-y^2}=\mathop{L}_{k=0}\frac{3y^2k+3yk^2+k^3}{2yk+k^2}$$

$$=\frac{3y^2}{2y}=\frac{3}{2}y=\frac{3}{2}x^{\frac{1}{2}}.$$

Ex. 12. P, P' are the perimeters of two regular n-gons circumscribed about two circles whose radii are a, a' and circumferences C, C'; show that

$$P:a=P':a'\ \ \text{and}\ \ C:a=C':a'.$$

The constant ratio of circumference to radius is denoted by 2π; π is an irrational number, approximately equal to 3·14159.

Ex. 13. Show that the area of a circle of radius a is πa^2, and that the area of a sector of the circle of angle θ radians is $\frac{1}{2}\theta a^2$.

Ex. 14. Show that the volume of a right circular cylinder, the radius of the base being a and the altitude h, is $\pi a^2 h$.

Show that the area of the curved surface is $2\pi a h$.

Ex. 15. If A is the base and h the altitude of a triangular pyramid, and if the pyramid be divided into n slices, each of height h/n, by planes parallel to the base; show that the volume of the pyramid is less than

$$\frac{h}{n}\left\{A + \frac{(n-1)^2}{n^2}A + \frac{(n-2)^2}{n^2}A + \ldots + \frac{2^2}{n^2}A + \frac{1^2}{n^2}A\right\},$$

but greater than

$$\frac{h}{n}\left\{\frac{(n-1)^2}{n^2}A + \frac{(n-2)^2}{n^2}A + \ldots + \frac{2^2}{n^2}A + \frac{1^2}{n^2}A\right\}.$$

Hence show, by ex. 2, that the volume is $\frac{1}{3}hA$. Extend the result to any pyramid.

(Let V be the vertex and DEF the base; through the line in which a plane meets the face VEF draw a plane parallel to VD to meet the two planes next above and next below the plane containing the line. Two sets of triangular prisms will be formed; the one set will lie within the pyramid, the other set will include the pyramid. The two sums are the volumes of the two sets; the highest pyramid of the upper set is got by drawing a plane through the vertex parallel to the base.)

Ex. 16. Taking a circular cone as the limit of a pyramid whose vertex is the vertex of the cone and whose base is a regular n-gon inscribed in or described about the base of the cone, deduce from ex. 15 that the volume of a cone is $\frac{1}{3}hA$, h being its altitude and A its base.

Ex. 17. Show that the volume of the frustum of a right circular cone is $\frac{1}{3}h(A+\sqrt{AB}+B)$ or $\dfrac{\pi h}{3}\left(a^2+ab+b^2\right)$, where h is the height of the frustum, A, a and B, b the areas and the radii of the circular ends.

Ex. 18. C and a are the circumference and the radius of the base, and l is the slant side of a right circular cone; show that the area of the curved surface is $\frac{1}{2}lC$ or πla.

(The curved surface may be considered as the limit of the lateral surface of either of the pyramids of ex. 16.)

Ex. 19. If the slant side of a frustum of a right circular cone is l, and if the radii of the circular ends are a, b show that the area of the curved surface is $\pi l(a+b)$; if c, c' are the circumferences of the ends, the area is $\frac{1}{2}l(c+c')$.

CHAPTER V.

CONTINUITY OF FUNCTIONS. SPECIAL LIMITS.

§ 44. Continuity of a Function. The conception of a limit enables us to put in arithmetical form the property that may be considered as most characteristic of a continuous function.

The argument will be said to vary continuously from a to b when it takes once and once only every value lying between a and b; when the argument is represented as an abscissa, the corresponding point will move along the axis from the point a to the point b as the argument varies continuously from a to b, and will coincide once and once only with every point on that segment.

In plotting the graphs of the elementary functions it was found that, except in the immediate neighbourhood of those values of the argument for which the function became infinite, a small change in the argument produced only a small change in the function. Now by the definition of a limit, when x is nearly equal to a the function, $f(x)$ say, is nearly equal to its limit A; if therefore the *limit* A be identical with the *value* $f(a)$ of the function, we see that when x either increases or decreases from the value a by a small amount the function $f(x)$ will also change by a small amount from the value $f(a)$. Hence the

DEFINITION. A function $f(x)$ is defined to be continuous for the value a of x, or more simply, continuous *at* a if

(i) $f(a)$ is a definite (finite) number, and

(ii) $\mathop{L}\limits_{x=a} f(x) = f(a)$.

For continuity therefore the *value* of $f(x)$ for $x=a$ and the *limit* of $f(x)$ for $x=a$ must coincide; since infinity is not a *value*, in the sense that is required for the application of the laws of algebra, a function ceases to be continuous, that is it becomes *discontinuous*, for those values of the argument that make it infinite.

Again it is implied in the definition that x may approach a either through values less than a or through values greater than a; that is when $f(a)$ is represented as an ordinate the point x may approach a either from the left or from the right and the limit must for both methods of approach be the same. It will sometimes happen, as for example when $f(x) = \sqrt{(a^2 - x^2)}$, that x can only approach a from one side, the function being undefined for values of x on the other side; in such cases of course the condition that the limit must be the same from whichever side x approaches a has to be modified, but the modification offers no difficulty. To express that x is to approach its limit a through values less than a the notation

$$\underset{x=a-0}{\mathrm{L}} f(x)$$

is sometimes used, and in the same way the notation $x = a+0$ implies that x is to approach a through values greater than a; but we shall as a rule use the ordinary notation and leave the student to modify it to suit special cases.

The only other type of discontinuity that needs special mention is that represented in Fig. 27. As x varies from a value a little less than a to one a little greater the function changes by the finite amount BC. Here the function $f(x)$ has not a definite *value* when $x=a$; as x approaches a from the left $f(x)$ approaches one definite *limit* AB, while as x approaches from the other side $f(x)$ approaches another definite *limit* AC. If a moving particle were at a certain instant to experience an impulse the

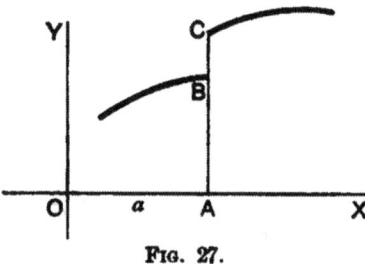

FIG. 27.

graph of its velocity would present a discontinuity of this kind for the value of the abscissa representing the instant. (See § 69, ex. 6, for an example of discontinuity.)

§ 45. Theorems on Continuous Functions. When $f(x)$ is continuous at a it is merely stating the definition of continuity in another form to say that if x be nearly equal to a, $f(x)$ is nearly equal to $f(a)$; or again we may say that $f(x) = f(a) + d$, where d is a variable which converges to zero when x converges to a.

A function is said to be continuous over the range from a to b if it is continuous for every value of its argument that lies between a and b; the range is understood, unless the contrary is stated, to include its extremities a, b. A range which includes its extremities is sometimes called a *closed* range; one which excludes its extremities an *open* range.

The following theorems are of constant application:

THEOREM I. *If $f(x)$ is continuous at a and if $f(a)$ is not zero, then for values of x near a, $f(x)$ has the same sign as $f(a)$.*

For if $f(x) = f(a) + d$, the sign of $f(x)$ will be that of the numerically greater of the two numbers $f(a)$ and d; since x may be taken so near to a that d shall be less (numerically) than any given number, and therefore less (numerically) than $f(a)$, the sign will be that of $f(a)$.

The meaning of the phrase "near a" and of similar phrases will be gathered from the proof.

THEOREM II. *If $f(x)$ be continuous over the range from a to b, and if $f(a) = A$ and $f(b) = B$, then $f(x)$ will assume once at least every value lying between A and B as x ranges continuously from a to b; in particular if A and B have opposite signs $f(x)$ will become zero for at least one value of x lying between a and b.*

A mathematical proof of this theorem lies beyond our scope; so far as a function is adequately represented by a graph the theorem is geometrically evident. It is easy also to show by use of a graph that the converse theorem is not necessarily true.

§ 46. Continuity of the Elementary Functions.

The theorems on limits stated in § 42 enable us to prove that the elementary functions of a single variable are continuous for all values of the variable except those for which a function becomes infinite.

When x varies continuously so does the product x^n and the product ax^n, n being any positive integer and a a constant. (§ 42, Th. II.). Hence by Th. I. a rational integral function is continuous for all finite values of its argument; and by Th. I. and Th. III. a rational fractional function is continuous for all finite values of its argument except such as make its denominator vanish.

From the geometrical definition or by direct application of the limit test we see that the trigonometrical functions are continuous for all values of the variable except such as make the function infinite. The sine and the cosine are continuous for all values of the argument; the tangent and the secant for all values except the odd multiples of $\pi/2$; the cotangent and the cosecant for all values except 0 and multiples of π.

A full discussion of the continuity of a^x would take us too far into abstract considerations; we will therefore assume that a^x is continuous for all finite values of x and that its inverse, $\log x$, is continuous for all finite positive values of x but discontinuous for $x=0$. When x is irrational we may in practice replace a^x by $a^{x'}$ where x' is a rational approximation to x; the simplest discussion is based on the exponential series.

Function of a Function. When y is a function of u, say $y=\phi(u)$, and u a function of x, say $u=f(x)$, then y is said to be a function of a function of x; y is thus given as a function of x *mediately*, through u. Functions of functions are of constant occurrence in the calculus, and there may be several intermediate variables such as u.

If y is a continuous function of u, and u a continuous function of x, the student will have no difficulty in showing that y is a continuous function of x; in the notation of § 42

$$\operatorname{L} \phi(u) = \phi(U) = \phi(\operatorname{L} u).$$

Again when a function is continuous so is its inverse

THE POWER LIMIT.

function. Hence x^n is continuous when n is fractional, positive or negative, except for $x=0$ when n is negative. In the same way we see that the inverse trigonometric functions are in general continuous.

§ 47. $\underset{x=a}{L}\dfrac{x^n-a^n}{x-a}$. The limits discussed in §§ 47-49 are fundamental.

$$\underset{x=a}{L}\frac{x^n-a^n}{x-a}=na^{n-1},$$

n being any rational number.

(i) Let n be a positive integer; then

$$\frac{x^n-a^n}{x-a}=x^{n-1}+x^{n-2}a+x^{n-3}a^2+\ldots+xa^{n-2}+a^{n-1};$$

$$\therefore\ \underset{x=a}{L}\left(\frac{x^n-a^n}{x-a}\right)=\underset{x=a}{L}(x^{n-1}+x^{n-2}a+x^{n-3}a^2+\ldots+xa^{n-2}+a^{n-1})$$

$$=na^{n-1},$$

since the limit of each of the n terms is a^{n-1}.

(ii) Let n be a positive proper fraction p/q, where p, q are positive integers.

Put y^q for x and b^q for a; then when $x=a$, $y=b$. Hence since $x^n=x^{p/q}=y^p$ and $a^n=b^p$,

$$\frac{x^n-a^n}{x-a}=\frac{y^p-b^p}{y^q-b^q}=\frac{y^{p-1}+y^{p-2}b+\ldots+yb^{p-2}+b^{p-1}}{y^{q-1}+y^{q-2}b+\ldots+yb^{q-2}+b^{q-1}},$$

by rejecting the common factor $y-b$;

$$\therefore\ \underset{x=a}{L}\left(\frac{x^n-a^n}{x-a}\right)=\frac{\underset{y=b}{L}(y^{p-1}+y^{p-2}b+\ldots+yb^{p-2}+b^{p-1})}{\underset{y=b}{L}(y^{q-1}+y^{q-2}b+\ldots+yb^{q-2}+b^{q-1})}$$

$$=\frac{pb^{p-1}}{qb^{q-1}}=\frac{p}{q}b^{p-q}=na^{n-1}.$$

(iii) Let n be negative, but either integral or fractional, say $n=-m$. Then

$$\frac{x^n-a^n}{x-a}=\frac{x^{-m}-a^{-m}}{x-a}=-\frac{x^m-a^m}{x-a}\times\frac{1}{x^m a^m};$$

$$\therefore \operatorname*{L}_{x=a}\left(\frac{x^n-a^n}{x-a}\right) = -\operatorname*{L}_{x=a}\left(\frac{x^m-a^m}{x-a}\right) \times \operatorname*{L}_{x=a}\frac{1}{x^m a^m}$$

$$= -ma^{m-1} \times \frac{1}{a^{2m}} = -ma^{-m-1} = na^{n-1},$$

since the limit of the first factor is ma^{m-1} by cases (i) and (ii).

The student should be able to identify the theorem in whatever notation it may be presented; thus

$$\operatorname*{L}_{h=0}\frac{(x+h)^n - x^n}{h} = nx^{n-1}; \quad \operatorname*{L}_{\delta v=0}\frac{1}{\delta v}\left(\frac{1}{v+\delta v}-\frac{1}{v}\right) = -\frac{1}{v^2}.$$

COR. If h be a small positive or negative number, $(x+h)^n$ is equal to $x^n + nhx^{n-1}$ approximately.

§ 48. $\operatorname*{L}_{m=\infty}\left(1+\dfrac{1}{m}\right)^m$. The number e.

(i) Let m be a positive integer and expand by the Binomial Theorem; then

$$\left(1+\frac{1}{m}\right)^m = 1 + \frac{m}{1}\cdot\frac{1}{m} + \frac{m(m-1)}{2!}\frac{1}{m^2}$$
$$+ \frac{m(m-1)(m-2)}{3!}\frac{1}{m^3} + \cdots$$

$$= 1 + \frac{1}{1} + \frac{1-\frac{1}{m}}{2!} + \frac{\left(1-\frac{1}{m}\right)\left(1-\frac{2}{m}\right)}{3!} + \cdots$$

In the expansion there are $(m+1)$ terms and every term after the second can be written in the form given to the 3rd and 4th terms; for example, the last or $(m+1)$th term is

$$\left(1-\frac{1}{m}\right)\left(1-\frac{2}{m}\right)\cdots\left(1-\frac{m-1}{m}\right) \div m!$$

Let n be any positive integer less than m and let n be kept fixed while m increases. Denote the first $(n+1)$ terms of the expansion by S'_{n+1} and the remaining $(m-n)$ terms by R'_{n+1}. Then

$$\left(1+\frac{1}{m}\right)^m = S'_{n+1} + R'_{n+1}.$$

THE NUMBER e.

Now,
$$S'_{n+1} = 1 + \frac{1}{1} + \frac{1 - \frac{1}{m}}{2!} + \frac{\left(1 - \frac{1}{m}\right)\left(1 - \frac{2}{m}\right)}{3!} + \cdots$$
$$+ \frac{\left(1 - \frac{1}{m}\right)\left(1 - \frac{2}{m}\right) \cdots \left(1 - \frac{n-1}{m}\right)}{n!}.$$

The limit for $m = \infty$ of each of the factors $\left(1 - \frac{1}{m}\right)$, $\left(1 - \frac{2}{m}\right) \cdots$ is 1, and since there is a finite number of factors the limit of each numerator is 1. Denote by S_{n+1} the limit for $m = \infty$ of S'_{n+1}; therefore

$$S_{n+1} = \underset{m=\infty}{\mathrm{L}} S'_{n+1} = 1 + \frac{1}{1} + \frac{1}{2!} + \frac{1}{3!} + \cdots + \frac{1}{n!}.$$

We have now to consider the limit of R'_{n+1}. The first term of R'_{n+1} is

$$\left(1 - \frac{1}{m}\right)\left(1 - \frac{2}{m}\right) \cdots \left(1 - \frac{n}{m}\right) \div (n+1)!,$$

and this term is a factor of every one that follows it.

Hence R'_{n+1} is the product of

$$\left(1 - \frac{1}{m}\right)\left(1 - \frac{2}{m}\right) \cdots \left(1 - \frac{n}{m}\right) \div (n+1)!$$

and

$$\left\{1 + \frac{1 - \frac{n+1}{m}}{n+2} + \frac{\left(1 - \frac{n+1}{m}\right)\left(1 - \frac{n+2}{m}\right)}{(n+2)(n+3)} + \cdots \text{to } (m-n) \text{ terms}\right\}.$$

Everywhere replace each of the factors $\left(1 - \frac{1}{m}\right)$, $\left(1 - \frac{2}{m}\right)$, $\cdots$ $\left(1 - \frac{m-1}{m}\right)$, which are all positive and less than 1, by the factor 1, and replace each of the factors $(n+2)$, $(n+3) \ldots m$ by $(n+1)$; by so doing we shall increase R'_{n+1}, which is therefore less than

$$\frac{1}{(n+1)!}\left\{1 + \frac{1}{n+1} + \frac{1}{(n+1)^2} + \cdots \text{ to } (m-n) \text{ terms}\right\}.$$

But the series within the bracket is a geometrical progression whose sum is

$$\frac{1-\dfrac{1}{(n+1)^{m-n}}}{1-\dfrac{1}{n+1}} = \frac{n+1}{n}\left\{1-\frac{1}{(n+1)^{m-n}}\right\};$$

and for every value of m greater than n, this sum is less than $(n+1)/n$.

Hence R'_{n+1} is a positive number which for every value of m greater than n is less than R''_{n+1}, where

$$R''_{n+1} = \frac{1}{(n+1)!} \times \frac{n+1}{n} = \frac{1}{n(n!)}.$$

But $\quad \underset{m=\infty}{\mathrm{L}}\left(1+\dfrac{1}{m}\right)^m = \underset{m=\infty}{\mathrm{L}} S'_{n+1} + \underset{m=\infty}{\mathrm{L}} R'_{n+1}.$

The first limit is S_{n+1}, and the second limit is a positive number less than R''_{n+1}; therefore, inserting the values of S_{n+1} and R''_{n+1} we get,

$$\underset{m=\infty}{\mathrm{L}}\left(1+\frac{1}{m}\right)^m > 1+1+\frac{1}{2!}+\frac{1}{3!}+\cdots+\frac{1}{n!};$$

but $\quad\quad\quad\quad < 1+1+\dfrac{1}{2!}+\dfrac{1}{3!}+\cdots+\dfrac{1}{n!}+\dfrac{1}{n(n!)},$

or $\quad \underset{m=\infty}{\mathrm{L}}\left(1+\dfrac{1}{m}\right)^m = 1+1+\dfrac{1}{2!}+\dfrac{1}{3!}+\cdots+\dfrac{1}{n!}+R_{n+1},\ \ldots\ldots$ (A)

where R_{n+1} is less than R''_{n+1} or $\dfrac{1}{n(n!)}$.

When n is even moderately large, R_{n+1} is very small; for example, when $n=12$, $1/n(n!)$ is less than 3×10^{-10}; so that the value of the limit may be obtained very approximately by calculating the series as far as $1/12!$. The calculations are very easy to effect, and the value will be found to be, for the nearest 7-figure approximation,

$$2\cdot 7182818.$$

The limit is usually denoted by e; e is really an irrational number. It is easy to see, by comparing S_{n+1} with the sum

$$1+1+\frac{1}{2}+\frac{1}{2^2}+\cdots+\frac{1}{2^{n-1}},$$

which is greater than S_{n+1} and equal to $3 - 1/2^{n-1}$, that no matter how great n may be S_{n+1} is certainly finite and less than 3. Since $e - S_{n+1}$ is equal to R_{n+1}, and since the limit for $n = \infty$ of R_{n+1} is zero, e may be considered as

$$\underset{n=\infty}{\mathrm{L}} \left(1 + 1 + \frac{1}{2!} + \frac{1}{3!} + \ldots + \frac{1}{n!}\right); \quad \ldots\ldots\ldots\ldots (\text{B})$$

or, in the usual phraseology,

$$e = 1 + 1 + \frac{1}{2!} + \frac{1}{3!} + \ldots \text{to infinity.}$$

(ii) Next suppose that m proceeds to infinity through positive fractional values; m will therefore always lie between two consecutive integers, say n and $n+1$. Hence

$$1 + \frac{1}{n} > 1 + \frac{1}{m} > 1 + \frac{1}{n+1};$$

$$\therefore \quad \left(1 + \frac{1}{n}\right)^{n+1} > \left(1 + \frac{1}{m}\right)^{m} > \left(1 + \frac{1}{n+1}\right)^{n}.$$

But $\underset{n=\infty}{\mathrm{L}} \left(1 + \frac{1}{n}\right)^{n+1} = \underset{n=\infty}{\mathrm{L}} \left(1 + \frac{1}{n}\right)^{n} \times \underset{n=\infty}{\mathrm{L}} \left(1 + \frac{1}{n}\right) = e \times 1;$

and $\underset{n=\infty}{\mathrm{L}} \left(1 + \frac{1}{n+1}\right)^{n} = \underset{n=\infty}{\mathrm{L}} \left(1 + \frac{1}{n+1}\right)^{n+1} \div \underset{n=\infty}{\mathrm{L}} \left(1 + \frac{1}{n+1}\right) = e \div 1$

by case (i).

Hence in this case also the limit is e, because as m becomes infinite so does n.

(iii) Let m be negative, $m = -n$ where n is positive but either integral or fractional. Then

$$\left(1 + \frac{1}{m}\right)^{m} = \left(1 - \frac{1}{n}\right)^{-n} = \left(\frac{n}{n-1}\right)^{n} = \left(1 + \frac{1}{n-1}\right)^{n};$$

$$\therefore \quad \underset{m=-\infty}{\mathrm{L}} \left(1 + \frac{1}{m}\right)^{m} = \underset{n=\infty}{\mathrm{L}} \left(1 + \frac{1}{n-1}\right)^{n}$$

$$= \underset{n=\infty}{\mathrm{L}} \left(1 + \frac{1}{n-1}\right)^{n-1} \times \underset{n=\infty}{\mathrm{L}} \left(1 + \frac{1}{n-1}\right)$$

$$= e \times 1$$

by cases (i) and (ii).

Hence finally $\underset{m=\pm\infty}{\mathrm{L}} \left(1 + \frac{1}{m}\right)^{m} = e,$

whether m proceeds to infinity through integral or fractional values.

Cor. $$\underset{h=0}{L}(1+h)^{\frac{1}{h}}=e.$$

§ 49. The Function e^x. If $x \neq 0$, we see, by putting $m = Mx$, that when m becomes infinite so does M; hence

$$\left(1+\frac{x}{m}\right)^m = \left(1+\frac{1}{M}\right)^{Mx} = \left\{\left(1+\frac{1}{M}\right)^M\right\}^x,$$

and $\underset{m=\infty}{L}\left(1+\frac{x}{m}\right)^m = \underset{M=\infty}{L}\left\{\left(1+\frac{1}{M}\right)^M\right\}^x = \left\{\underset{M=\infty}{L}\left(1+\frac{1}{M}\right)^M\right\}^x = e^x,$

by § 46 (Function of a Function).

Since M may be positive or negative, integral or fractional, the result holds whether x or m be positive or negative, integral or fractional.

By exactly the same method as in § 48, it may be shown that

$$e^x = \underset{n=\infty}{L}\left(1+x+\frac{x^2}{2!}+\frac{x^3}{3!}+\cdots+\frac{x^n}{n!}\right)$$

by expanding $\left(1+\frac{x}{m}\right)^m$ for positive integral values of m.

It is easy to see that this series is a finite number no matter how great n may be; for as soon as n is numerically greater than x,

$$\frac{x^{n+1}}{(n+1)!}+\frac{x^{n+2}}{(n+2)!}+\frac{x^{n+3}}{(n+3)!}+\cdots$$
$$=\frac{x^{n+1}}{(n+1)!}\left\{1+\frac{x}{n+2}+\frac{x^2}{(n+2)(n+3)}+\cdots\right\}$$

is numerically less than

$$\frac{x_1^{n+1}}{(n+1)!}\left\{1+\frac{x_1}{n+1}+\left(\frac{x_1}{n+1}\right)^2+\cdots\right\},$$

where x_1 is the numerical value of x. The series in brackets is a geometrical progression with a common ratio numerically less than 1; hence if we write

$$e^x = 1+x+\frac{x^2}{2!}+\cdots+\frac{x^n}{n!}+R_{n+1}\ldots\ldots\ldots\ldots(\text{A})$$

R_{n+1} will, for every value of n greater than x numerically, be less than

$$\frac{x_1^{n+1}}{(n+1)!} \times \frac{n+1}{n+1-x_1} = \frac{x_1^{n+1}}{(n+1-x_1)(n!)}.$$

If $x_1 = 1$, this gives the value of R_{n+1} in § 48.

COR. $$\operatorname*{L}_{x=0} \frac{e^x - 1}{x} = 1.$$

For if x be a positive proper fraction, we may put 1 for n in (A); therefore

$$e^x > 1 + x, \quad \text{but} \quad e^x < 1 + x + \frac{x^2}{2-x},$$

so that

$$\frac{e^x - 1}{x} > 1, \quad \text{but} \quad < 1 + \frac{x}{2-x},$$

from which the result follows for positive values of x.

If x be negative, $x = -h$ where h is positive, then

$$\operatorname*{L}_{x=0} \frac{e^x - 1}{x} = \operatorname*{L}_{h=0} \frac{e^{-h} - 1}{-h} = \operatorname*{L}_{h=0} \frac{e^h - 1}{h} \times \frac{1}{e^h} = 1$$

by the first case, so that the limit is the same whether x proceeds by positive or negative values towards its limit 0.

§ 50. Compound Interest Law.

When an exponential function is spoken of, the base is usually understood to be e; where the base is any other number, say a, the function a^x can be written e^{kx}, where $k = \log a$.

The rate at which ae^{kx} increases with respect to x when $x = x_1$ is kae^{kx_1}, that is, is proportional to the value of the function when $x = x_1$. For when x increases from x_1 to $x_1 + h$ the increment of the function is

$$ae^{k(x_1+h)} - ae^{kx_1} = ae^{kx_1}(e^{kh} - 1),$$

and the average rate is

$$\frac{ae^{kx_1}(e^{kh} - 1)}{h} = kae^{kx_1} \frac{e^{kh} - 1}{kh}.$$

By § 49, COR., the limit of this expression for $h = 0$ is kae^{kx_1}.

Many processes in nature follow this law; the law is sometimes quoted as *the compound interest law*, since the simplest case of it is that of compound interest. For, suppose a principal of P pounds to earn interest at the rate of p per cent. per annum; let interest be calculated at n equal intervals in each year, and let it be added to the principal as soon as it is earned, so that the interest bears

interest. It is easy to see that at the end of t years the principal will amount to
$$P\left(1+\frac{p}{100n}\right)^{nt}.$$
Let us suppose now that n increases indefinitely, that is, suppose that the interest is added on at shorter and shorter intervals; we thus approach a condition in which the interest is added on *continuously*. Put $n=\frac{pm}{100}$ so that when n becomes infinite so does m. The limit of the above expression for n increasing indefinitely is
$$A = \operatorname*{L}_{m=\infty} P\left\{\left(1+\frac{1}{m}\right)^m\right\}^{\frac{pt}{100}} = Pe^{\frac{pt}{100}}.$$
Again, we see that if t increase in any arithmetical progression, whose common difference is h, A will increase in a geometrical progression whose common ratio is $e^{\frac{ph}{100}}$; for if t become $t+h$, A will become $Pe^{\frac{p(t+h)}{100}}$, that is, $Ae^{\frac{ph}{100}}$. Hence A is a quantity which is equally multiplied in equal times.

The density of the air as we descend a hill is a quantity which is equally multiplied in equal distances of descent, for the increase in density per foot of descent is due to the weight of that layer which is itself proportional to the density. Many other instances may be found in physics.

EXERCISES VII.

1. If $f(x) = ax^n + bx^{n-1} + \ldots + kx + l$ is a rational integral function of x, show that
$$\operatorname*{L}_{x=\infty}\left(\frac{f(x)}{ax^n}\right)=1\,;\quad \operatorname*{L}_{x=-\infty}\left(\frac{f(x)}{ax^n}\right)=1\,;$$
and therefore that when x is numerically large,
$$f(x) = ax^n(1+d),$$
where d is a variable whose limit is zero for $x = \pm\infty$.

Use Th. I., § 42.

2. Show that $f(x)$ in ex. 1 has the same sign as a when x is a large positive number, but has the same sign as $(-1)^n a$ when x is numerically large but negative (that is, has the sign of $+a$ or $-a$ according as n is even or odd).

EXERCISES VII.

3. From the result of ex. 2, show, by applying Th. II. of § 45, that every equation of odd degree has at least one real root, and that if it has more than one it must have an odd number.

4. If $f(x)$ is a rational fractional function
$$f(x) = \frac{ax^m + bx^{m-1} + \ldots + kx + l}{Ax^n + Bx^{n-1} + \ldots + Kx + L},$$

prove   (i) $f(x) = \frac{a}{A} x^{m-n}(1 + d_1)$ if $m > n$;

           (ii) $f(x) = \frac{a}{A}(1 + d_2)$ if $m = n$;

           (iii) $f(x) = \frac{a}{A} \cdot \frac{1}{x^{n-m}}(1 + d_3)$ if $m < n$,

where d_1, d_2, d_3 are numerically very small when x is numerically very large.
Use Th. I. and Th. III., § 42.

5. Show that, the angle being measured in radians,
$$\operatorname*{L}_{\theta=0}\left(\frac{1 - \cos\theta}{\theta^2}\right) = \frac{1}{2}\operatorname*{L}_{\theta=0}\left(\frac{\sin\tfrac{1}{2}\theta}{\tfrac{1}{2}\theta}\right)^2 = \frac{1}{2}.$$

Hence show that when θ is small, $\cos\theta = 1 - \tfrac{1}{2}\theta^2$ approximately.

6. Prove $\quad\quad 1 > \cos\theta > 1 - \tfrac{1}{2}\theta^2$.

7. Prove   (i) $\operatorname*{L}_{\theta=0} \dfrac{\sin a\theta}{\sin b\theta} = \dfrac{a}{b}$;   (ii) $\operatorname*{L}_{\theta=0} \dfrac{\tan a\theta}{\tan b\theta} = \dfrac{a}{b}$.

8. Prove   (i) $\operatorname*{L}_{x=+\infty}(xe^{-x}) = 0$;   (ii) $\operatorname*{L}_{x=0}(x \log x) = 0$.

By § 49 (A) $\quad\quad e^x > 1 + x + \tfrac{1}{2}x^2$;

therefore $\quad xe^{-x} = \dfrac{x}{e^x} < \dfrac{x}{1 + x + \tfrac{1}{2}x^2}$, that is, $\dfrac{1}{\dfrac{1}{x} + 1 + \tfrac{1}{2}x}$,

and the limit of the fraction last written is zero.
Next put $x = e^{-y}$; then $x \log x = -ye^{-y}$ and the limit for $x = 0$ is equal to the limit for $y = +\infty$, which is zero.

9. Prove $\quad\quad \operatorname*{L}_{x=+\infty} x^n e^{-x} = 0$.

10. Prove that if n be positive $\operatorname*{L}_{x=0} x^n \log x = 0$.

For $\quad\quad x^n \log x = \dfrac{1}{n} x^n \log(x^n)$

and the limit is zero by ex. 8 since the limit of x^n is zero.

11. Prove $\quad\quad \operatorname*{L}_{x=0} \sin x \log x = 0$,

For $\quad\quad \sin x \log x = \left(\dfrac{\sin x}{x}\right)(x \log x)$.

12. If x is any finite quantity, prove
$$\operatorname*{L}_{n=\infty} \frac{x^n}{n!} = 0.$$

Suppose x equal to or less than the integer μ; then, numerically,

$$\frac{x^n}{n!} = \frac{x^\mu}{\mu!} \cdot \frac{x}{\mu+1} \cdot \frac{x}{\mu+2} \cdots \frac{x}{n} < \frac{x^\mu}{\mu!}\left(\frac{x}{\mu+1}\right)^{n-\mu}.$$

But $x/(\mu+1)$ is a proper fraction.

13. If a be a constant or a function of x which is finite for every value of x, prove

(i) $\underset{x=\infty}{\mathrm{L}}\left(\sin\frac{a}{x}:\frac{a}{x}\right)^x = 1$; (ii) $\underset{x=\infty}{\mathrm{L}}\left(\cos\frac{a}{x}\right)^x = 1$;

(iii) $\underset{x=\infty}{\mathrm{L}}\left(\tan\frac{a}{x}:\frac{a}{x}\right)^x = 1$.

14. If s_n is a variable that (i) always increases as n increases but (ii) always remains less than some definite fixed number a, show that as n tends to infinity s_n tends to a definite limit that is equal to or less than a.

Take the values $s_1, s_2, s_3 \ldots$ as the abscissae of points $A_1, A_2, A_3 \ldots$ on the x-axis and let A be the point whose abscissa is a; for every value of n, A_n will be to the *right* of A_{n-1} but to the *left* of A. As n increases the point A_n will move further and further to the right, but will not for any finite value of n coincide with A. There must therefore be some point S to the left of A or coinciding with A to which A_n may be made to approach as near as we please; if the abscissa of S is s then by the definition of a limit

$$\underset{n=\infty}{\mathrm{L}}\, s_n = s,$$

and s is less than or equal to a. (Compare § 39 (ii) and Fig. 25.)

15. If s_n is a variable that (i) always decreases as n increases but (ii) always remains greater than some definite fixed number b, show that as n tends to infinity s_n tends to a definite limit that is equal to or greater than b.

16. If
$$s_n = 1 + \frac{1}{1} + \frac{1}{2!} + \frac{1}{3!} + \cdots + \frac{1}{n!},$$

show that s_n converges, as n tends to infinity, to a number that is greater than 2 but less than 3.

17. If
$$s_n = \frac{1}{1^2} + \frac{1}{2^2} + \frac{1}{3^2} + \cdots + \frac{1}{n^2},$$

show that s_n converges to a number that lies between 1 and 2.

Let $s'_n = \frac{1}{1} + \frac{1}{1.2} + \frac{1}{2.3} + \cdots + \frac{1}{(n-1)n} = 2 - \frac{1}{n} < 2$,

then for every value of n (greater than 1) $s_n < s'_n < 2$.

18. Apply the theorems of exs. 14, 15 to establish the results of exs. (i), (ii), (iii) of § 39 when the n-gons are not regular but are such that as n increases indefinitely the length of each side diminishes indefinitely.

CHAPTER VI.

DIFFERENTIATION. ALGEBRAIC FUNCTIONS.

§ 51. Derivatives. Differentiation. The process of §§ 36, 37 can now, by making use of the notion of the limit, be stated more compactly.

The average rate at which the function $3x^2$ varies as x varies from x_1 to $x_1 + \delta x_1$, where δx_1 may be either a positive or a negative increment, is by definition

$$\frac{\delta(3x_1^2)}{\delta x_1} = \frac{3(x_1 + \delta x_1)^2 - 3x_1^2}{\delta x_1} = 6x_1 + 3\delta x_1,$$

and the number which is taken as measuring *the* rate of change when $x = x_1$ is

$$\underset{\delta x_1 = 0}{\mathrm{L}} \frac{\delta(3x_1^2)}{\delta x_1} = \underset{\delta x_1 = 0}{\mathrm{L}} (6x_1 + 3\delta x_1) = 6x_1.$$

The reasoning does not depend on the particular value x_1 of the argument, and we therefore state the result in the form, " the function $3x^2$ varies with respect to x at the rate $6x$," leaving it to be understood that when $x = x_1$ the rate is $6x_1$, when $x = x_2$ the rate is $6x_2$ and so on. It will save multiplication of symbols to use x as the symbol for the argument in general and also as the symbol for some definite value of the argument, and the student will find that, as a rule, it causes no ambiguity to do so; if he ever finds difficulty, let him choose a separate symbol as x_1 for the definite value at which the rate is measured.

Now take the general case. Let $f(x)$ be a continuous function of x; as the argument varies from x to $x + \delta x$, where δx may be either a positive or a negative increment, the function varies from $f(x)$ to $f(x+\delta x)$. The average rate of change of the function when the argument changes by δx is

$$\frac{\delta f(x)}{\delta x} = \frac{f(x+\delta x)-f(x)}{\delta x},$$

and the number which measures *the* rate of change is

$$\operatorname*{L}_{\delta x=0} \frac{\delta f(x)}{\delta x}.$$

We shall find that for all the elementary functions this limit is a definite number, except, it may be, for particular values of x. In general the limit will depend on x, and a special name is given to it, namely, "the derivative of $f(x)$ with respect to x."

Instead of "derivative," the names "differential coefficient," "derived function" are also used; in certain connections also the word "gradient," or "slope" is used. (§ 53.) The process of finding the derivative is called "differentiation"; the name "differential coefficient" was formerly more frequently used than "derivative."

Again, there are special notations for the derivative. A very convenient notation is obtained by accenting the functional letter, as $f'(x)$; another is got by prefixing the letter D, with or without the suffix x, as $D_x f(x)$ or $Df(x)$. If the function be denoted by a single letter, as y, the notation for the derivative of y with respect to the argument x is similar, as y'_x, $D_x y$ or y', Dy. As a rule the suffix is omitted when there is no ambiguity as to the argument.

Finally, to denote the value of the derivative for a special value of x, say x_1, the following notations are used:

$$f'(x_1); \quad [D_x f(x)]_{x=x_1}; \quad [y']_{x=x_1}.$$

As a matter of fact, the derivative is really formed for such a definite value, but the functional character of the derivative is more prominent when that value is denoted by the same symbol x as represents the argument in general.

To sum up then we have the defining equations:—

$$f'(x) = D_x f(x) = \operatorname*{L}_{\delta x=0} \frac{\delta f(x)}{\delta x} = \operatorname*{L}_{\delta x=0} \frac{f(x+\delta x)-f(x)}{\delta x}$$

$$y' = D_x y = \operatorname*{L}_{\delta x=0} \frac{\delta y}{\delta x}.$$

DERIVATIVES. NOTATION. 103

The function thus determined is called *the derivative*, or *the differential coefficient*, or *the derived function* of $f(x)$ with respect to x and it measures, or briefly it is, *the* rate at which the function varies with respect to its argument for the particular value x.

Of course other letters than x, f, y, may be used; thus

$$\phi'(t) = D_t\phi(t) = \operatorname*{L}_{\delta t = 0} \frac{\delta \phi(t)}{\delta t},$$

and $\phi'(t)$ is the derivative of $\phi(t)$ with respect to t.

It will be convenient often to use such expressions as *the x-derivative of $f(x)$*, or *the time-rate of change of a function*, instead of *the derivative with respect to x*, or *the rate of change with respect to the time*.

Ex. 1. $D_x(3x^2 - 4x + 3) = 6x - 4$,

for $D_x(3x^2 - 4x + 3) = \operatorname*{L}_{\delta x = 0} \frac{\delta(3x^2 - 4x + 3)}{\delta x}$.

Now $\delta(3x^2 - 4x + 3) = 3(x + \delta x)^2 - 4(x + \delta x) + 3 - (3x^2 - 4x + 3)$
$$= 6x\delta x + 3(\delta x)^2 - 4\delta x;$$

$\therefore \operatorname*{L}_{\delta x = 0} \frac{\delta(3x^2 - 4x + 3)}{\delta x} = \operatorname*{L}_{\delta x = 0} (6x + 3\delta x - 4) = 6x - 4.$

Ex. 2. $D_v\left(\frac{c}{v}\right) = -\frac{c}{v^2}$ (c constant),

for $\delta\left(\frac{c}{v}\right) = \frac{c}{v + \delta v} - \frac{c}{v} = \frac{-c\delta v}{v^2 + v\delta v}$;

$\therefore D_v\left(\frac{c}{v}\right) = \operatorname*{L}_{\delta v = 0} \frac{\delta\left(\frac{c}{v}\right)}{\delta v} = -\frac{c}{v^2}.$

If $v = 0$, the above process cannot be carried out.

§ 52. Increasing and Decreasing Functions. By definition of a limit and of a derivative

$$\frac{\delta f(x)}{\delta x} = f'(x) + a,$$

where a is a variable which is very small when δx is very small and converges to 0 when δx converges to 0.

For an illustration of the difference between $\delta f(x)/\delta x$ and $f'(x)$, see the results of examples 4, 5, 6, § 32.

Hence if $f'(x)$ is not zero the sign of $f'(x) + a$ and therefore of $\delta f(x)/\delta x$ will be, for sufficiently small values of δx, the same as that of $f'(x)$ (compare § 45 Th. I.); therefore the sign of $\delta f(x)$ will be that of $f'(x)\delta x$.

Now suppose δx a positive increment; then $\delta f(x)$ will be positive or negative according as $f'(x)$ is positive or negative. But
$$\delta f(x) = f(x+\delta x) - f(x);$$
hence $f(x+\delta x)$ is algebraically greater or less than $f(x)$ according as $f'(x)$ is positive or negative. In other words, $f(x)$ increases as x increases so long as $f'(x)$ is positive, but $f(x)$ decreases as x increases so long as $f'(x)$ is negative, increase and decrease being *algebraical* and not *numerical* increase or decrease.

If we suppose δx a negative increment then $\delta f(x)$ will be negative or positive according as $f'(x)$ is positive or negative; $f(x)$ will decrease as x decreases so long as $f'(x)$ is positive but will increase as x decreases so long as $f'(x)$ is negative.

Hence *the mere sign* of $f'(x)$ tells *how* the function changes as x changes; if $f(x) = ax+b$, $f'(x) = a$ and the conclusions agree with the statements of § 33 for the uniformly varying function.

DEFINITION. A function which increases as its argument increases and decreases as its argument decreases is called an *increasing function*; one which decreases as its argument increases and increases as its argument decreases is called a *decreasing function*.

Thus since $D(3x^2) = 6x$, $3x^2$ is a decreasing function for all negative values of x and an increasing function for all positive values of x. The function ceases to decrease and begins to increase as x passes through the value 0; hence when $x=0$ the function is a *minimum* (§ 17, iv), and its value is then 0. It will be noticed that when $x=0$ the derivative is 0; *the rate* of change is therefore zero for the minimum value.

The derivative of $3x^2 - 4x + 3$ is $6x - 4$; hence so long as $6x - 4$ is positive, that is, so long as $6x$ is greater than 4, that is, so long as x is greater than $\frac{2}{3}$, the function is an increasing one; on the other hand so long as x is (algebraically) less than $\frac{2}{3}$ it is a decreasing function. When $x = \frac{2}{3}$ the function is a minimum, the minimum value being $\frac{5}{3}$. Here again when $x = \frac{2}{3}$ the derivative is zero, that is the rate of change of the function is zero.

INCREASING AND DECREASING FUNCTIONS. 105

Stationary Values. The conclusions about increasing and decreasing cease to hold for those values of x for which $f'(x)$ is zero. Since $f'(x)$ measures the rate of change of the function it is usual to class those values of the function for which $f'(x)$ is zero as **stationary** values.

Ex. Show that the function x^3+1 has a stationary value when $x=0$, and that for all other finite values of x it is an increasing function.

§ 53. Geometrical Interpretation of a Derivative. A specially useful interpretation of a derivative is obtained from the graphic representation of a function.

Let ABP (Fig. 28 a, b) be the graph of $f(x)$. Take a point P on the graph; $OM=x$, $MP=y=f(x)$.

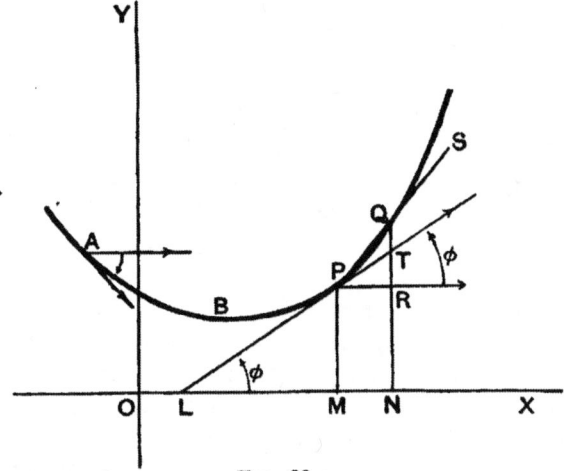

FIG. 28 a.

Let $MN=\delta x$; then $ON=x+\delta x$, $NQ=y+\delta y=f(x+\delta x)$. From P draw PR parallel to the x-axis to meet NQ (or NQ produced) at R; then, both in sign and in magnitude,

$$RQ = NQ - NR = NQ - MP = f(x+\delta x) - f(x) = \delta f(x),$$

and $$\tan RPQ = \frac{RQ}{PR} = \frac{RQ}{MN} = \frac{\delta f(x)}{\delta x}.$$

When δx converges to 0 as its limit, the quotient $\delta f(x)/\delta x$ converges to $f'(x)$ as its limit. But as δx converges to 0, N tends towards coincidence with M and Q tends towards coincidence with P. Hence since $\tan RPQ$ converges to a definite value, namely $f'(x)$, the angle RPQ converges to a definite angle and therefore the secant PQS tends to a definite limiting position PT. The line PT is, by definition (§ 39, ex. vii.) the tangent to the curve at P.

Hence $f'(x)$ is the trigonometrical tangent of the angle that the tangent to the graph at P, the point $(x, f(x))$, makes with the x-axis. From this property of the derivative, the name *gradient* is used (see § 22).

In Fig. 28a, the tangent PT makes with the x-axis the positive angle RPT or XLP; in Fig. 28b it makes the negative angle RPT or XLP'. We will usually denote the angle by ϕ, so that $\tan \phi = f'(x)$.

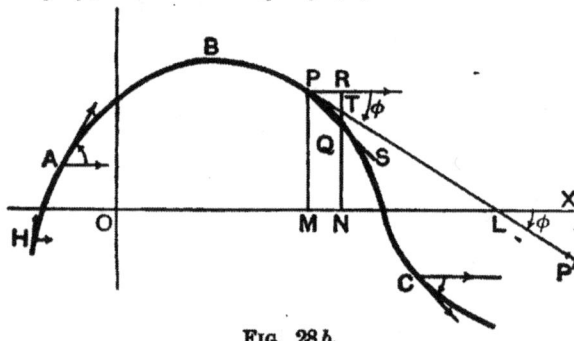

Fig. 28 b.

In the diagrams δx is positive, but it is evident that the same conclusions can be drawn if δx is negative, that is if Q is on the opposite side of P. In particular cases it may happen that P can only be approached from one side. Thus if $f(x) = \sqrt{x^3}$, x cannot take negative values; in finding $f'(0)$ therefore δx must be positive. Here

$$f'(0) = \underset{\delta x = 0}{L} \frac{f(\delta x) - f(0)}{\delta x} = \underset{\delta x = 0}{L} \frac{\sqrt{(\delta x)^3}}{\delta x} = \underset{\delta x = 0}{L} \sqrt{(\delta x)} = 0,$$

and the tangent makes a zero angle with the x-axis; since $f(0) = 0$, the x-axis is itself the tangent at the origin.

Ex. Find the gradient of the graph of $3x^2 - 4x + 3$ at the points whose abscissae are $-1, 0, \frac{2}{3}, 1, 2$.

§ 54. **Derivative as an Aid in Graphing a Function.** The conclusions drawn in § 52 from the sign of the derivative are valuable as an aid to a mental representation of the variation of a function; those of § 53 are equally valuable in helping us to graph the function.

The diagrams of § 53 may be considered the standard ones. We see that when the gradient $f'(x)$ is positive the graphic point moves upward as the point x moves to the right—as along BPQ Fig. 28a, along HAB Fig. 28b; when the gradient is negative the graphic point moves down as the point x moves to the right—as along AB Fig. 28a, along BPQ Fig. 28b. At B the gradient is 0, and the tangent is parallel to the x-axis; the graphic point is for the moment stationary.

The student must not confuse moving upwards with motion away from the x-axis; thus near H (Fig. 28b) the graphic point in moving up gets nearer the axis. The graphic point moves up or down when the point x moves to the right according as NQ is *algebraically* greater or less than MP; for $NQ - MP = f'(x)\delta x$ approximately, and when $f'(x)$ is positive, δx being supposed also positive, NQ is algebraically greater than MP. If MP and NQ are both negative this implies that NQ is numerically less than MP.

As an exercise, trace the graph of $f(x) = x^3 - 3x + 1$, already shown in § 23. Here it is easily found that

$$f'(x) = 3x^2 - 3 = 3(x+1)(x-1).$$

So long as x is less than -1, that is, so long as the point x is to the left of the point -1, both $x+1$ and $x-1$ are negative, and therefore $f'(x)$ is positive. Hence, as the point x moves from the extreme left of the x-axis to the point -1, the graphic point moves steadily upwards.

So long as x is greater than -1, but less than 1, $x+1$ is positive and $x-1$ negative, and therefore $f'(x)$ is negative. Hence, as the point x moves from the point -1 to the point 1, the graphic point moves downwards.

If x be greater than 1, $f'(x)$ is positive. Hence, as the point x moves from the point 1 to the extreme right, the graphic point moves steadily up.

The turning points of the graph are found where $x = -1$ and where $x = +1$; when $x = -1$, the function has a *maximum* value 3, and when $x = +1$, it has a *minimum* value -1.

§ 55. Derivative not definite. It may happen that the limit of $\delta f(x)/\delta x$ is not a definite finite number. There are two chief cases.

108   AN ELEMENTARY TREATISE ON THE CALCULUS.

(i) $f'(x)$ may be infinite for particular values of x. Thus if $f(x) = \sqrt{x}$, then

$$f'(0) = \operatorname*{L}_{\delta x = 0} \frac{\sqrt{(\delta x)} - \sqrt{0}}{\delta x} = \operatorname*{L}_{\delta x = 0} \frac{1}{\sqrt{\delta x}} = \infty,$$

but, for all other positive finite values of x,

$$f'(x) = \operatorname*{L}_{\delta x = 0} \frac{\sqrt{(x + \delta x)} - \sqrt{x}}{\delta x} = \operatorname*{L}_{\delta x = 0} \frac{1}{\sqrt{(x + \delta x)} + \sqrt{x}} = \frac{1}{2\sqrt{x}}.$$

We see that as x approaches the origin the gradient gets greater and greater, and when x coincides with the origin the tangent to the graph is perpendicular to the x-axis. In general the tangent at a point on the graph at which $f'(x)$ is infinite will be perpendicular to the x-axis. When $f(x)$ is infinite for a finite value of x as in the case of $1/x$ for $x=0$, it will usually be found that as x tends towards that value $f'(x)$ tends towards infinity; we may say, therefore, that the tangent which meets the graph at the infinitely distant point is perpendicular to the x-axis. Such a tangent is an asymptote. (See the graphs of § 24.)

(ii.) It may happen that at particular points of the graph there are two tangents, as at A, Fig. 29. Although the function is continuous when $x = a$, the gradient $f'(x)$ is not. There is one gradient as we approach A from the left, another as we approach A from the right; as x increases through the value a, $f'(x)$ changes suddenly from $\tan XBA$ to $\tan XCD$.

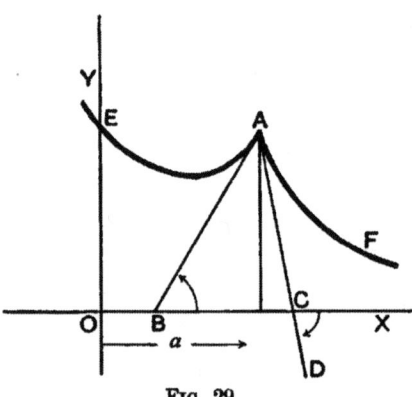

FIG. 29.

It will be found that for all the ordinary functions the derivative $f'(x)$ is, except for particular values of x, a continuous function and therefore these functions can be appropriately discussed by means of their graph.

§ **56. Fluxions.** Newton founded his treatment of the calculus on the conception of the growth of mathematical quantities by a continuous motion; he called the *time-rate* of change of a variable its *fluxion*, the variable itself being called the *fluent*. He laid little stress on notations but sometimes denoted the fluxion of a variable, say x, by the symbol $\dot{x}$, and this notation is still often used in works on mechanics to denote a time-rate of change.

We may take one illustration of a time-rate of change. Suppose a particle to move along the path APQ (Fig. 28) and at time t seconds from a chosen instant let it be at P, the point (x, y), where $y = f(x)$ is the equation to the path. Let s be the length, in feet say, of the arc ABP measured from some fixed point A on the path. x, y, s are then all functions of t.

Suppose that when the time increases from t to $t + \delta t$ the particle comes to Q (Fig. 28 a, b) and denote the increments MN, RQ, arc PQ of x, y, s by $\delta x, \delta y, \delta s$. By the usual definitions, the chord PQ is the displacement of the particle in time δt and the quotient of the displacement by δt is the average velocity of the particle during the interval, the direction of this velocity being given by the angle RPQ. To get *the* velocity at time t, find the limit of the average velocity for δt approaching 0.

Now the limit of the angle RPQ is RPT, so that the direction of the velocity at time t will be along the tangent PT.

Again, to find the magnitude of the velocity, or the *speed*, as the magnitude is now usually called, we have to find
$$\underset{\delta t = 0}{L} \frac{\text{chord } PQ}{\delta t}.$$

We will assume as an axiom that when the chord PQ is very small, the arc and the chord are nearly equal; or, in the more definite language of limits, we will assume
$$\underset{\text{chord } PQ = 0}{L} \left(\frac{\text{chord } PQ}{\text{arc } PQ} \right) = 1.$$

Now,
$$\frac{\text{chord } PQ}{\delta t} = \frac{\text{chord } PQ}{\text{arc } PQ} \cdot \frac{\text{arc } PQ}{\delta t} = \frac{\text{chord } PQ}{\text{arc } PQ} \cdot \frac{\delta s}{\delta t}.$$

Hence, since $\delta s = 0$ when $\delta t = 0$, we have

$$\text{speed} = \underset{\delta t=0}{\text{L}} \frac{\text{chord } PQ}{\delta t} = \underset{PQ=0}{\text{L}} \left(\frac{\text{chord } PQ}{\text{arc } PQ}\right) \times \underset{\delta t=0}{\text{L}} \frac{\delta s}{\delta t} = \dot{s}.$$

This equation of course simply states that the speed is the time-rate of change of s, and may be considered merely as the symbolic statement of the definition of speed; but, however simple the conception of a rate is at bottom, it will be well for the student to recur again and again to the process by which the number is determined.

Again, $\dot{x}$ is the rate of change of x, that is, $\dot{x}$ is the rate at which the point moves to the right, and in the same way $\dot{y}$ is the rate at which the point moves upward. These two rates are called *the components of the velocity* parallel to the coordinate axes.

From the diagram we see that

$$(\delta x)^2 + (\delta y)^2 = (\text{chord } PQ)^2 = \left(\frac{\text{chord } PQ}{\text{arc } PQ}\right)^2 \cdot (\delta s)^2,$$

and therefore

$$\left(\frac{\delta x}{\delta t}\right)^2 + \left(\frac{\delta y}{\delta t}\right)^2 = \left(\frac{\text{chord } PQ}{\text{arc } PQ}\right)^2 \cdot \left(\frac{\delta s}{\delta t}\right)^2.$$

Hence, taking the limit for $\delta t = 0$, we get

$$(\dot{x})^2 + (\dot{y})^2 = (\dot{s})^2,$$

a result that expresses the usual rule for determining the resultant velocity $\dot{s}$, when the component velocities $\dot{x}$, $\dot{y}$ are given.

Ex. Suppose $x = t$, $y = t^2$. For every value of t, $y = x^2$; that is, the point P lies on the parabola whose equation is $y = x^2$. The component velocities are $\dot{x} = 1$, $\dot{y} = 2t$, and the magnitude of the resultant velocity $\dot{s}$ is $\sqrt{(\dot{x}^2 + \dot{y}^2)} = \sqrt{(1 + 4t^2)}$. The direction of the velocity is given by $\tan \phi = D_x y = 2x = 2t$.

It will be observed that the path of the point is given by stating where at *each* instant the point is, because whenever the instant is named, that is, whenever the value of t is given, the coordinates x, y can at once be calculated. By eliminating t between the equations determining x and y, we find a relation that holds between the coordinates of *every* point on the path, that is, we find the equation of the path in the usual form. (See Exercises IV. 10, VI. 4, 6, 10, 11.)

We will now show how to find the derivatives of the ordinary functions; in the exercises examples will be found illustrating the geometrical and the mechanical applications of the derivative. After the student has gained some facility in differentiating, other examples will be considered.

§ 57. Derivative of a Power. By definition
$$D_x(x^n) = \operatorname*{L}_{\delta x=0} \frac{(x+\delta x)^n - x^n}{\delta x},$$
and, by § 47, this limit is nx^{n-1}; that is,
$$D(x^n) = nx^{n-1}.$$

Hence the derivative of a power with respect to its base is got by multiplying by the index and then diminishing the index by 1.

It is obvious that the derivative is a continuous function for all finite values of x, except for $x=0$, and it is then discontinuous only when $n-1$ is negative; that is, when n is less than 1 algebraically.

COR. If a be a constant, $D_x(ax^n) = nax^{n-1}.$

Ex. 1. $\quad D(x^5) = 5x^4; \quad D(\sqrt{x^5}) = D\left(x^{\frac{5}{2}}\right) = \frac{5}{2}x^{\frac{3}{2}};$

$$D\left(\frac{5}{\sqrt{x}}\right) = D\left(5x^{-\frac{1}{2}}\right) = -\frac{1}{2} \cdot 5x^{-\frac{3}{2}} = -\frac{5}{2\sqrt{x^3}}.$$

Ex. 2. Write down the derivatives with respect to t of
$$t^5, \quad \frac{2}{t^2}, \quad 3\sqrt{t}, \quad \frac{4}{\sqrt{t}}.$$

Ex. 3. Write down a function of x which has x^2 as its derivative.

Reverse the process for obtaining a derivative, that is, increase the index by 1, and then divide the result by the new index. Thus, one function whose x-derivative is x^2 is $\frac{1}{3}x^3$, as may at once be tested by differentiation.

Ex. 4. Write down for each of the following functions a function of which it is the derivative,
$$5x^4, \quad \sqrt{x}, \quad \frac{2}{\sqrt{x}}, \quad \frac{1}{x^3}, \quad \frac{3}{x^5}.$$

§ 58. General Theorems. The following theorems are of constant application. We suppose x to be the independent variable, so that the suffix may be omitted in indicating derivatives.

THEOREM I. *An additive constant disappears in differentiation; or, two functions which differ only by a constant have the same derivative.*

For let $f(x) = \phi(x) + C$, where C is a constant, that is, does not change as x changes; $f(x)$ and $\phi(x)$ therefore differ only by the constant C,

$$\frac{f(x+\delta x)-f(x)}{\delta x} = \frac{[\phi(x+\delta x)+C]-[\phi(x)+C]}{\delta x}$$

$$= \frac{\phi(x+\delta x)-\phi(x)}{\delta x}.$$

Take the limit of these equal quantities for δx converging to 0 and we find $f'(x) = \phi'(x)$.

Ex. $D(x^3 - 4) = 3x^2$.

THEOREM II. *A constant factor remains as a constant factor in the derivative.*

For $D[Cf(x)] = L\dfrac{Cf(x+\delta x) - Cf(x)}{\delta x} = C\, L\dfrac{f(x+\delta x) - f(x)}{\delta x}$

therefore $D[Cf(x)] = CDf(x)$.

THEOREM III. *The derivative of an algebraic sum of a finite number of functions is equal to the like algebraic sum of the derivatives of the functions.*

Let $f(x)$, $F(x)$, $\phi(x)$ be three functions of x; then it is easy to see that

$$\delta[f(x) + F(x) - \phi(x)] = \delta f(x) + \delta F(x) - \delta \phi(x).$$

Therefore, dividing by δx and taking the limit, we get

$$D[f(x) + F(x) - \phi(x)] = Df(x) + DF(x) - D\phi(x).$$

The same proof holds for more than three functions; the number of them, however, must be finite, for if there be an infinite number the theorem is not necessarily true, just as in the case of the corresponding theorem in limits (§ 42, Th. I.).

Ex. $D(3x^2 - 5x + 1) = D(3x^2) - D(5x)$ Th. III. and I.
                 $= 3D(x^2) - 5D(x)$ Th. II.
                 $= 6x - 5$.

THEOREMS ON DIFFERENTIATION.

THEOREM IV. $D(uv) = vDu + uDv$, where u, v are functions of x.

When x takes the increment δx, let u, v take the increments $\delta u, \delta v$ respectively, then

$$\delta(uv) = (u + \delta u)(v + \delta v) - uv$$
$$= v\delta u + u\delta v + \delta u \delta v;$$
$$\therefore \frac{\delta(uv)}{\delta x} = v\frac{\delta u}{\delta x} + u\frac{\delta v}{\delta x} + \frac{\delta u}{\delta x} \cdot \delta v.$$

When δx converges to 0 so does δv; the limit of the last term is therefore 0, and we get

$$D(uv) = vDu + uDv.$$

If there be more than two factors, say u, v, w, we may extend the theorem by applying it twice; thus, first consider vw as forming one factor, we get

$$D(uvw) = D(u \cdot vw) = vwDu + uD(vw).$$

But $\qquad D(vw) = wDv + vDw;$

$\therefore \qquad D(uvw) = vwDu + uwDv + uvDw.$

If we divide both sides by uvw we get

$$\frac{D(uvw)}{uvw} = \frac{Du}{u} + \frac{Dv}{v} + \frac{Dw}{w}.$$

More generally, if there be n factors, $u_1, u_2, \ldots u_n$, we have

$$\frac{D(u_1 u_2 \ldots u_n)}{u_1 u_2 \ldots u_n} = \frac{Du_1}{u_1} + \frac{Du_2}{u_2} + \ldots + \frac{Du_n}{u_n}.$$

Logarithmic Differentiation. When the differentiation is carried out in the form last written, the process is usually called *logarithmic differentiation*. (See § 65, ex. 3.)

The student must particularly notice that the derivative of a product is *not* the product of the derivatives of its factors.

Ex. $\quad D[(5x+2)(3x-7)] = (3x-7)D(5x+2) + (5x+2)D(3x-7)$
$\qquad\qquad\qquad\qquad = (3x-7) \cdot 5 + (5x+2) \cdot 3$
$\qquad\qquad\qquad\qquad = 30x - 29.$

The result may be verified by first distributing the product and then differentiating.

Theorem V. $D\left(\dfrac{u}{v}\right) = \dfrac{vDu - uDv}{v^2}$ provided v is not zero for the values of x considered.

For
$$\delta\left(\frac{u}{v}\right) = \frac{u+\delta u}{v+\delta v} - \frac{u}{v} = \frac{v\delta u - u\delta v}{v^2 + v\delta v};$$

$$\therefore \frac{\delta\left(\dfrac{u}{v}\right)}{\delta x} = \frac{v\dfrac{\delta u}{\delta x} - u\dfrac{\delta v}{\delta x}}{v^2 + v\delta v}$$

Since the limit for $\delta x = 0$ of the denominator is v^2 and v^2 is not zero, we can apply the theorem that the limit of a quotient is the quotient of the limits of numerator and denominator. Hence the theorem.

If we divide by $\dfrac{u}{v}$ we get

$$\frac{D\left(\dfrac{u}{v}\right)}{\dfrac{u}{v}} = \frac{Du}{u} - \frac{Dv}{v}.$$

Ex. $\quad D\left(\dfrac{x^2-1}{x^2+1}\right) = \dfrac{(x^2+1)D(x^2-1) - (x^2-1)D(x^2+1)}{(x^2+1)^2}$

$$= \frac{(x^2+1).2x - (x^2-1).2x}{(x^2+1)^2}$$

$$= \frac{4x}{(x^2+1)^2}.$$

Theorem VI. *If the derivatives of two functions are equal for every value of the argument, the functions can only differ, if at all, by a constant.*

This theorem is the converse of Theorem I. and seems hardly to require proof for the ordinary functions. For if $f'(x) = \phi'(x)$ for every value of x, then putting y equal to $f(x) - \phi(x)$ we have

$$D_x y = D_x[f(x) - \phi(x)] = f'(x) - \phi'(x) = 0.$$

Hence the gradient of the graph of y is zero for every value of x; the graph must therefore be either the x-axis or a straight line parallel to that axis. But the equation of every line parallel to the x-axis is $y = \text{const.} = C$; the equation will represent the axis itself if $C = 0$.

Therefore $f(x) - \phi(x) = C$ or $f(x) = \phi(x) + C$.

Ex. If $D_x y = x^2 - 1$, determine the general value of y.

The derivative of $\tfrac{1}{3}x^3 - x$ is $x^2 - 1$, as may be tested by differentiation; therefore the derivatives of y and of $\tfrac{1}{3}x^3 - x$ are the same for every value of x. Hence y and $\tfrac{1}{3}x^3 - x$ can only differ by a constant, that is, $y = \tfrac{1}{3}x^3 - x + C$. This value is called the *general* value, because *every* function which has the same derivative as y will at most differ from $\tfrac{1}{3}x^3 - x$ by a constant, and C may be *any* constant.

The particular function which has the value 2, say, when x has the value 1, will require a particular value of the constant C. But *always*

$$y = \tfrac{1}{3}x^3 - x + C;$$

therefore $\quad 2 = \tfrac{1}{3} - 1 + C, \;\; \therefore \; C = \tfrac{8}{3};$

and $\quad y = \tfrac{1}{3}x^3 - x + \tfrac{8}{3}.$

It is to be observed that the derivatives must be equal for *every* value of the argument; thus $x^2 - 1$ and $x^3 - 1$ are equal when x is 0 or 1, yet the functions $\tfrac{1}{3}x^3 - x + C$ and $\tfrac{1}{4}x^4 - x + C$, of which they are the derivatives, do not differ merely by a constant: they are different functions.

EXERCISES VIII.

Differentiate with respect to x, examples 1-10:

1. $7x^3 + 5x^2 + 4x - 2$.
2. $(7x - 3)(8x + 2)$.
3. $(x - 1)(x + 2)(x - 3)$.
4. $(3x - 7)/(5 - 2x)$.
5. $\sqrt{x} + \dfrac{1}{\sqrt{x}}$.
6. $\left(\sqrt{x} - \dfrac{1}{\sqrt{x}}\right)^2$.
7. $x^n + \dfrac{1}{x^n}$.
8. $ax^m + \dfrac{b}{x^m}$.
9. $4x^{\frac{5}{4}} + 2x^{\frac{1}{4}} - 2x^{-\frac{1}{4}} + x^{-\frac{5}{4}}$.
10. $\dfrac{x^3 - 2 + x^{-3}}{x - 2 + x^{-1}}$.

Differentiate with respect to t, examples 11-14:

11. $(at + b)/(ct + d)$.
12. $a/(b + ct)$.
13. $(at^2 + 2bt + c)/(At^2 + 2Bt + C)$.
14. $\dfrac{(t - 1)(t - 2)}{(t + 1)(t + 2)}$.

15. Give a geometrical interpretation of Th. I. § 58.

Deduce Th. V. from Th. IV. by putting $\dfrac{u}{v} = w$, so that $Du = D(vw)$.

16. If at time t the adjacent sides of a rectangle are u and v feet respectively, where u, v are both functions of t, show that at time t the area is growing at the rate $v\dot{u} + u\dot{v}$.

If at time t the three edges of a rectangular parallelepiped which meet at one corner are u, v, w feet respectively, find the rate at which the volume is increasing.

Show that these results give a geometrical interpretation of Th. IV., § 58.

116   AN ELEMENTARY TREATISE ON THE CALCULUS.

17. Find the values of x for which the following functions are (i) increasing, (ii) decreasing, (iii) stationary. Apply the results to the graphing of the functions, and state the turning points.

(a) $3-x+x^2$; (b) x^3-3x+2; (c) x^4-2x^2-1.

18. State the most general function which has as its derivative

(i) $2x-1$; (ii) $3x-\dfrac{1}{x^2}$; (iii) ax^2+bx+c.

19. The gradient of a curve is x^2-x+1, and the curve passes through the point $(1, \frac{5}{6})$; find the equation of the curve.

20. If $pv = p_0 v_0$ where p_0, v_0 are constants, show that
$$-vD_v p = p.$$

21. The speed of a particle at time t seconds from the beginning of its motion is $V-gt$ feet per second; find how far it has moved in t seconds.

§ 59. Derivative of a Function of a Function and of Inverse Functions.

The derivative of such a function as $(x^2-x+1)^{\frac{1}{2}}$ cannot be found by immediate application of the rule for the derivative of a power. In a case like this we may proceed as follows:—Denote $(x^2-x+1)^{\frac{1}{2}}$ by y; now put $x^2-x+1=u$. Then $y = u^{\frac{1}{2}}$ where $u = x^2-x+1$; that is, y is a function of u where u is a function of x. In other words, y is a function of a function of x (§ 46).

When x takes the increment δx let u take the increment δu; when u takes the increment δu let y take the increment δy. Hence when x takes the increment δx, y takes the increment δy, and when δx converges to zero so do δu and δy. Now
$$\frac{\delta y}{\delta x} = \frac{\delta y}{\delta u} \cdot \frac{\delta u}{\delta x};$$

therefore
$$\mathop{L}_{\delta x=0} \frac{\delta y}{\delta x} = \mathop{L}_{\delta u=0} \frac{\delta y}{\delta u} \cdot \mathop{L}_{\delta x=0} \frac{\delta u}{\delta x};$$

that is,
$$D_x y = D_u y \times D_x u.$$

In the derivative $D_x y$, y is supposed to be expressed as a function of x, while in the derivative $D_u y$, y is supposed to be expressed as a function of u. That is,
$$D_x(x^2-x+1)^{\frac{1}{2}} = D_u u^{\frac{1}{2}} \times D_x(x^2-x+1)$$
$$= \tfrac{1}{2}u^{-\frac{1}{2}} \times (2x-1) = \frac{2x-1}{2(x^2-x+1)^{\frac{1}{2}}}.$$

DERIVATIVE OF FUNCTION OF FUNCTION. 117

where, after the differentiation is effected, u is replaced by its value in terms of x, namely x^2-x+1.

The reasoning is perfectly general, so that we have the theorem:—If $y=f(u)$ and $u=\phi(x)$, then y is a function of a function of x, and

$$D_x y = D_u f(u) \times D_x \phi(x) \text{ or } D_x y = D_u y \times D_x u.$$

If we had $y=f(u)$, $u=\phi(v)$, $v=\psi(x)$, we should get in exactly the same way

$$D_x y = D_u f(u) \times D_v \phi(v) \times D_x \psi(x)$$

or

$$D_x y = D_u y \times D_v u \times D_x v.$$

The same method shows how to obtain the derivative of an inverse function. Let $y=f(x)$ so that x is the independent variable. The inverse function is $x=f^{-1}(y)$ and y is now considered to be the independent variable.

Let δx and δy be two corresponding increments of x and y, so that δx and δy vanish together. Then

$$\frac{\delta y}{\delta x} \times \frac{\delta x}{\delta y} = 1;$$

therefore $\underset{\delta x=0}{\mathrm{L}}\frac{\delta y}{\delta x} \times \underset{\delta y=0}{\mathrm{L}}\frac{\delta x}{\delta y} = 1$; that is, $D_x y \times D_y x = 1$.

The result is evident geometrically. In Fig. 28 (§ 53) $D_x y$ is the tangent of the angle that PT makes with OX, $D_y x$ is the tangent of the angle that PT makes with OY, and since these two angles are complementary the product of their tangents is 1.

This theorem is of great use in finding the derivatives of inverse functions (§§ 64, 65); meanwhile we note that

$$D_y x = \frac{1}{D_x y},$$

and the theorem remains true even if one of the derivatives is zero.

The student should carefully note the following examples; at all stages the rule for differentiating a function of a function has constantly to be used.

Ex. 1. $D_x(ax+b)^n = na(ax+b)^{n-1}$.
Put $ax+b=u$,
then $D_x(ax+b)^n = D_u u^n \times D_x u = nu^{n-1} \times a = na(ax+b)^{n-1}$.

With a little practice, the student will be able to dispense with the actual substitution of u. Thus he will write

$$D_x(ax+b)^n = n(ax+b)^{n-1} \times a = na(ax+b)^{n-1};$$

$$D_x(3x-2)^{\frac{1}{2}} = \tfrac{1}{2}(3x-2)^{-\frac{1}{2}} \times 3 = \frac{3}{2(3x-2)^{\frac{1}{2}}}.$$

Ex. 2. $\quad D_x(x^2-a^2)^{\frac{1}{2}} = \tfrac{1}{2}(x^2-a^2)^{-\frac{1}{2}} \times 2x = \dfrac{x}{\sqrt{(x^2-a^2)}}.$

Ex. 3. If $D_x y = x/(x^2-a^2)$ and $u = x^2-a^2$, find $D_u y$.

$$D_u y = D_x y \times D_u x = D_x y / D_x u = x/(x^2-a^2)/2x;$$

therefore $\quad D_u y = \tfrac{1}{2}/(x^2-a^2) = \tfrac{1}{2}/u.$

Ex. 4. If y is a function of x, so is y^2, y^3, ... xy, x^2y ..., and

$$D_x(y^2) = D_y(y^2) \times D_x y = 2y D_x y,$$

$$D_x(xy) = x D_x y + y D_x x = x D_x y + y;$$

and generally, using y' for $D_x y$,

$$D_x(y^n) = n y^{n-1} y',$$

$$D_x(x^m y^n) = x^m D_x(y^n) + y^n D_x(x^m)$$
$$= x^m n y^{n-1} y' + y^n m x^{m-1}$$
$$= x^{m-1} y^{n-1}(nxy' + my).$$

Conversely, $\quad yy' = D_x(\tfrac{1}{2}y^2), \quad y^{n-1}y' = D_x\left(\dfrac{1}{n}y^n\right).$

This transformation is specially useful in mechanical problems. Thus, t being the argument,

$$x\dot{x} = D_t(\tfrac{1}{2}x^2); \quad y\dot{y} = D_t(\tfrac{1}{2}y^2).$$

Ex. 5. If $v = \dot{s}$, prove $\dot{v} = D_s(\tfrac{1}{2}v^2)$.

$$\dot{v} = D_t v = D_s v \times D_t s = \dot{s} D_s v = v D_s v = D_s(\tfrac{1}{2}v^2),$$

or, in words, the time-rate of change of v is equal to the space-rate of change of $\tfrac{1}{2}v^2$ (see § 69).

Ex. 6. If the coordinates of a point on a curve are given in the form $x = f(t)$, $y = \phi(t)$, where, for example, t may denote the time, find $D_x y$.

y is a function of t, and t may be supposed to be determined as a function of x by the first equation. Hence

$$D_x y = D_t y \times D_x t.$$

But $D_x t = 1/D_t x$ by the rule for inverse functions; therefore

$$D_x y = \frac{D_t y}{D_t x} = \frac{\dot{y}}{\dot{x}}.$$

Thus, if $x = at^2$, $y = 2at$,

$$D_x y = \frac{\dot{y}}{\dot{x}} = \frac{2a}{2at} = \frac{1}{t} = \frac{2a}{y}.$$

Ex. 7. When y is given as an implicit function of x by an equation of the form
$$Ax^my^n + Bx^py^q + \ldots + Kx + Ly + M = 0, \ldots\ldots\ldots\ldots(\alpha)$$
we can find y' by the method of ex. 4. For in whatever way x changes y must change so that the equation (α) always remains true; therefore *the rate* at which the expression on the left side of (α) changes as x changes must always be zero; that is,
$$D_x(Ax^my^n + Bx^py^q + \ldots + Kx + Ly + M) = 0 ;$$
that is, $\quad AD_x(x^my^n) + BD_x(x^py^q) + \ldots + K + LD_xy = 0.$

Each term may now be differentiated and the equation solved for D_xy or y'. For example, given
$$x^2 + xy + y^2 - 1 = 0 ; \ldots\ldots\ldots\ldots(\beta)$$
then $\qquad D_x(x^2 + xy + y^2 - 1) = 0 ;$
that is, $\qquad 2x + xy' + y + 2yy' = 0 ;$
and therefore $\qquad y' = -\dfrac{2x+y}{x+2y}\ldots\ldots\ldots\ldots(\gamma)$

To find the gradient of the ellipse represented by (β) at particular points, we proceed as follows:

When $x=1$, $y^2 + y = 0$; that is, $y = 0$ or -1;

at the point $(1, 0)$, $\qquad y' = -\dfrac{2+0}{1+0} = -2;$

at the point $(1, -1)$, $\qquad y' = -\dfrac{2-1}{1-2} = 1.$

To find where the tangent is parallel to the x-axis, we have to solve (β) and the equation $y' = 0$, taking care that the values which make the numerator of y' vanish do not also *make the denominator* vanish. If this were to happen, then y' would for such values take the form $0/0$, and y' might or might not be zero. In the above case we have to solve (β) and $2x+y=0$. The values are $x = \dfrac{1}{\sqrt{3}}$, $y = -\dfrac{2}{\sqrt{3}}$, and $x = -\dfrac{1}{\sqrt{3}}$, $y = \dfrac{2}{\sqrt{3}}$; at these points the tangent is parallel to the x-axis.

In the same way we find where it is perpendicular to the x-axis by solving (β) and $x + 2y = 0$, which makes y' infinite. The points are $\left(\dfrac{2}{\sqrt{3}}, -\dfrac{1}{\sqrt{3}}\right)$, $\left(-\dfrac{2}{\sqrt{3}}, \dfrac{1}{\sqrt{3}}\right)$.

EXERCISES IX.

Differentiate as to x, examples 1-8:

1. $\sqrt{(1-x)}$.
2. $x/\sqrt{(1-x)}$.
3. $\sqrt{(2x+1)(x-2)}$.
4. $x/\sqrt{(a^2-x^2)}$.
5. $x/\sqrt{(a^2+x^2)}$.
6. $\sqrt{(ax^2+bx+c)}$.
7. $\sqrt{(x^2+1)}/\sqrt{(x^2-1)}$.
8. $\sqrt{(ax^2+2bx+c)}/\sqrt{(Ax^2+2Bx+C)}$.

In differentiating a quotient of the form $(x+a)^m/(x+b)^n$, it is often advisable to write the quotient as a product in the form $(x+a)^m(x+b)^{-n}$; when simplified, the result will appear in its lowest terms. Differentiate in this way:

9. $(x+1)^3/(x-1)^4$.   10. $(x+a)^m/(x+b)^n$.   11. $1/x^2(x-1)^3$.

12. State in words the equation
$$D_t y = D_x y \times D_t x.$$

13. If $v^2 = 2k\left(\dfrac{1}{s} - \dfrac{1}{a}\right)$, show that $\dot{v} = -\dfrac{k}{s^2}$, s and v being functions of the time t.

14. If $2x^2 + 3y^2 = 5$, find y'. Then find the gradient at the points:
 (i) (1, 1),   (ii) (−1, 1),   (iii) (−1, −1),   (iv) (1, −1).

15. If $(x+y)^2 - 5x + y = 1$, find y'. Find the gradient at the point or points where the line whose equation is $x+y=1$ cuts the graph.

16. If $x = at$, $y = bt - \tfrac{1}{2}ct^2$, find the components parallel to the axes of the velocity of the point (x, y), and find the direction in which the point is moving at time t. (Compare Ex. VI. 4.)

17. Find $D_x y$ in the following cases:
 (i) $(x-a)^2 + (y-b)^2 = c^2$.   (ii) $y^2 = Ax + Bx^3$.
 (iii) $xy = c^2$.   (iv) $x^m y^n = c^{m+n}$.

18. If $D_x y = x^2 \sqrt{(ax^3 + b)}$ and $u = ax^3 + b$, find $D_u y$.

19. If $D_x y = (x+a)(x^2 + 2ax + b)^n$ and $u = x^2 + 2ax + b$, find $D_u y$.

20. If $D_x y = f(ax+b)$ and $u = ax+b$, find $D_u y$.

§ 60. Differentials. In Fig. 28 a, b, § 53, the value of $f'(x)$ or $D_x y$ is $\tan RPT$, and
$$f'(x) = \frac{RT}{PR} = \frac{RT}{MN}.$$

Now, suppose that as x increases from OM to ON the ordinate y or $f(x)$ increases *uniformly* at the rate $f'(x)$ or $\tan RPT$; then the point P will move, not along the arc PQ but along the tangent PT, and the increment that y on this supposition will take will be, not RQ but RT.

This *hypothetical* increment of y is called the *differential* of the function y or $f(x)$ and is denoted by dy or $df(x)$. The *actual* increment of y, denoted by δy or $\delta f(x)$, is not RT but RQ. Writing as usual δx for the increment MN of x we have
$$dy = RT = f'(x)\delta x; \quad \delta y = RQ = (f'(x) + a)\delta x,$$
where a is used in the same meaning as in § 52.

DIFFERENTIALS.

If $f(x)$ is the function x, then $f'(x)=1$, and we have
$$df(x)=dx=1.\,\delta x,$$
so that for the *independent* variable δx and dx may be considered to be the same thing. We may therefore write
$$dy=RT=f'(x)dx;\quad \delta y=RQ=(f'(x)+a)dx.$$
The first of these equations gives a new notation for the derivative, namely
$$f'(x)=\frac{dy}{dx}=\frac{df(x)}{dx}.$$
This notation, which is perhaps the most common, has the advantage that its form recalls the process by which the derivative is obtained. Again, we have another advantage. For
$$\delta y-dy=(f'(x)+a)dx-f'(x)dx=adx,$$
and (see § 52) when dx or MN is very small a is also very small, and therefore δy is very approximately equal to dy.

The notation of differentials is due to Leibniz; the above mode of defining a differential is usually attributed to Cauchy, but the differential is equivalent to Newton's "moment," which is explained in exactly the same way by Benjamin Robins (see his *Mathematical Tracts*, London, 1761). A reading of Robins' *Tracts* would well repay the student who is fortunate enough to get hold of a copy; the book is now somewhat rare.

The notation of differentials is practically a necessity in the integral calculus, and the student should accustom himself to it. In practical work dx and therefore dy are usually supposed to be very small quantities; but it is only *their ratio* that is of importance.

The symbol $\frac{dy}{dx}$ is often written as $\frac{d}{dx}y$; but when used in this way the symbol $\frac{d}{dx}$ is to be taken *as a whole* and as meaning exactly the same thing as D_x.

Since $du=D_x u\,dx$, $dv=D_x v\,dx$, etc., when the independent variable is x, we have
$$d(u+v-w)=du+dv-dw,$$
$$d(uv)=vdu+udv,$$
and so on. We may, in fact, replace D in the theorems of § 58 by d.

122   AN ELEMENTARY TREATISE ON THE CALCULUS.

Again, since $\dfrac{du}{dx}$ means $D_x u$ we have

$$\frac{d(u+v-w)}{dx} \text{ or } \frac{d}{dx}(u+v-w) = \frac{du}{dx} + \frac{dv}{dx} - \frac{dw}{dx},$$

$$\frac{d(uv)}{dx} = v\frac{du}{dx} + u\frac{dv}{dx},$$

$$\frac{dy}{dx} = \frac{dy}{du}\frac{du}{dx}, \quad \frac{dy}{dx} = \frac{1}{\frac{dx}{dy}} \quad (\S\ 59),$$

and so on.

Ex. 1. $d(3x^2 - x + 1) = D_x(3x^2 - x + 1).\,dx = (6x - 1)dx$.

Ex. 2. $d\,.\,\sqrt{(x^2 - a^2)} = \tfrac{1}{2}(x^2 - a^2)^{-\tfrac{1}{2}}\,.\,2x\,dx = \dfrac{x\,dx}{\sqrt{(x^2 - a^2)}}$.

Ex. 3. $x\,dx = d(\tfrac{1}{2}x^2)$; $(x^2 - 1)dx = d(\tfrac{1}{3}x^3 - x)$.

Ex. 4.   State in the form of differentials Ex. IX. 1-6.

§ 61. Geometrical Applications. Let OM be the abscissa and MP the ordinate of the point P on the curve whose equation is $y = f(x)$; and let the tangent at P meet the axes at L, K (Fig. 30).

The line CPG drawn through P perpendicular to the tangent is called the *normal* to the curve at P.

When the tangent and the normal are spoken of as finite segments the portions LP, GP, intercepted between P and the x-axis, are the segments referred to.

In the same way the projections of these segments on the x-axis, namely LM and MG, are called the *subtangent* and the *subnormal* respectively.

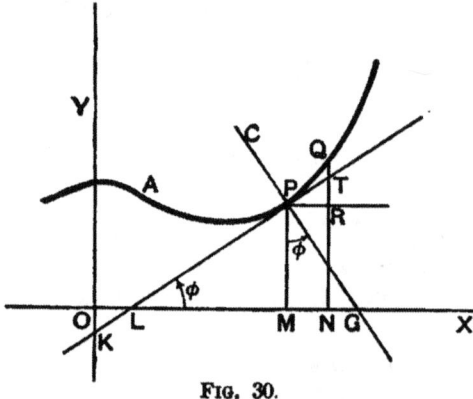

Fig. 30.

These segments can be expressed in terms of the values of x, y, y' at P.

$$\text{Subtangent} = LM = \frac{y}{\tan \phi} = \frac{y}{y'};$$

$$\text{Subnormal} = MG = y \tan \phi = yy';$$

$$\text{Tangent} = LP = y \operatorname{cosec} \phi = \frac{y\sqrt{(1+y'^2)}}{y'};$$

$$\text{Normal} = GP = y \sec \phi = y\sqrt{(1+y'^2)};$$

$$OL = OM - LM = x - \frac{y}{y'} = \frac{xy' - y}{y'};$$

$$OK = -OL \tan \phi = -\frac{xy'-y}{y'} y' = y - xy'.$$

These expressions are true for all positions of P, provided the *signs* of the segments be attended to. Thus, if LM is expressed by a negative number, L will be to the *right* of M, since, in the above diagram which is taken as the standard, LM is positive when L is to the left of M. There is no need to commit these formulae to memory; the values can at once be obtained in any given case by drawing a diagram.

We may also find the *equations* of the tangent and normal. For this purpose let the values of x, y, y' at P be denoted by x_1, y_1, y_1' in order to distinguish them from the coördinates (x, y) of a point on the tangent LP or the normal GP.

The equation of the tangent is

$$y - y_1 = (x - x_1) \tan \phi \text{ or } y - y_1 = y_1'(x - x_1),$$

since it is a straight line passing through (x_1, y_1) and making an angle ϕ with the x-axis.

The *acute* angle that the normal makes with the x-axis is $\phi - \frac{\pi}{2}$ and $\tan\left(\phi - \frac{\pi}{2}\right) = -\cot \phi = -1/y_1'$; hence the equation of the normal is

$$y - y_1 = -\frac{1}{y_1'}(x - x_1).$$

Ex. 1. Find the subtangent and the subnormal in the ellipse given by $\dfrac{x^2}{a^2}+\dfrac{y^2}{b^2}=1$.

If we suppose y to be positive, then

$$y = +\frac{b}{a}\sqrt{(a^2-x^2)}\;;\quad y' = -\frac{b}{a}\cdot\frac{x}{\sqrt{(a^2-x^2)}}\;;$$

$$\text{subtangent} = \frac{y}{y'} = -\frac{a^2-x^2}{x}\;;$$

$$\text{subnormal} = yy' = -\frac{b^2 x}{a^2}.$$

When x is *positive*, both these numbers are negative; L therefore lies to the right of M and G to the left of M; when x is *negative*, the positions are reversed.

$$OL = x - \text{subtangent} = \frac{a^2}{x}\;;$$

$$\therefore\; OL\cdot OM = \frac{a^2}{x}\cdot x = a^2,$$

a well-known property of the ellipse. O is of course the centre of the ellipse, denoted in § 26 by C.

Ex. 2. If the equation of the curve is $xy = c^2$, find the ratio of KP to LP.

Here
$$\frac{KP}{LP} = \frac{OM}{LM} = \frac{xy'}{y} = -\frac{c^2}{x}\div\frac{c^2}{x} = -1.$$

The ratio is given both in sign and in magnitude; hence P lies *between* K and L, and KL is bisected at P. The curve is a hyperbola (§ 27, II., ex.), and this is a well-known property.

§ 62. Derivative of the Arc. Let s be the length of the arc AP measured from a fixed point A on the curve (Fig. 30); to find $D_x s$, $D_y s$.

Proceeding exactly as in § 56 we get the equation

$$(\delta x)^2 + (\delta y)^2 = \left(\frac{\text{chord } PQ}{\text{arc } PQ}\right)^2\cdot(\delta s)^2,\dots\dots\dots\dots(\text{A})$$

where δs and δy are the increments of the arc s and of the ordinate y due to the increment δx of x; $\delta s = \text{arc } PQ$.

The average rate of change of s with respect to x, namely $\delta s/\delta x$, is determined by the equation

$$1 + \left(\frac{\delta y}{\delta x}\right)^2 = \left(\frac{\text{chord } PQ}{\text{arc } PQ}\right)^2\left(\frac{\delta s}{\delta x}\right)^2.$$

DERIVATIVE OF ARC. VELOCITIES.

Since the limit for $\delta x = 0$ of the first factor on the right is 1 we get
$$1 + (D_x y)^2 = (D_x s)^2,$$
or
$$D_x s = \sqrt{\{1 + (D_x y)^2\}}.$$

In exactly the same way we obtain
$$D_y s = \sqrt{\{1 + (D_y x)^2\}}.$$

Again,
$$\cos \phi = \underset{PQ=0}{L} \frac{\delta x}{PQ} = \underset{PQ=0}{L} \left(\frac{\delta x}{\delta s} \cdot \frac{\text{arc } PQ}{\text{chord } PQ} \right) = D_s x = \frac{dx}{ds}$$

$$\sin \phi = \underset{PQ=0}{L} \frac{\delta y}{PQ} = \underset{PQ=0}{L} \left(\frac{\delta y}{\delta s} \cdot \frac{\text{arc } PQ}{\text{chord } PQ} \right) = D_s y = \frac{dy}{ds}.$$

Using the notation of differentials we take $PR = dx$; then $RT = dy$ and $PT = ds$. The equation in differentials is
$$(ds)^2 = (dx)^2 + (dy)^2 \quad \ldots\ldots\ldots\ldots\ldots\ldots (A')$$
and division by $(dx)^2$ or $(dy)^2$ at once gives the derivative of s with respect to x or y.

If t be the independent variable and dt its differential, then, since x, y, s are all functions of t, we shall have
$$dx = \dot{x}\, dt, \quad dy = \dot{y}\, dt, \quad ds = \dot{s}\, dt,$$
and the substitution of these values in (A') gives, as in § 56, $\dot{x}^2 + \dot{y}^2 = \dot{s}^2$.

We also have
$$\dot{x} = \frac{dx}{dt} = \frac{dx}{ds}\frac{ds}{dt} = \dot{s} \cos \phi; \quad \dot{y} = \frac{dy}{dt} = \frac{dy}{ds}\frac{ds}{dt} = \dot{s} \sin \phi.$$

EXERCISES X.

1. Show that in the parabola* $y^2 = 4ax$ the subnormal is constant.

2. If the subnormal is constant ($2a$) show that the curve is a parabola $y^2 = 4ax + C$.

3. Find the equation of the tangent and of the normal at the point (x_1, y_1) of the parabola $y^2 = 4ax$.
Show that the subtangent is bisected at the vertex.

---

* It is customary to abbreviate the phrase "the curve of which the equation is $y = f(x)$" to "the curve $y = f(x)$."

4. If the tangent at P in the ellipse (Fig. 20, § 26) meet the major axis at T and the minor axis at t, prove that
$$CM \cdot CT = CA^2, \quad Cm \cdot Ct = CB^2,$$
where m is the projection of P on $B'B$.

5. Show that the equation of the tangent at (x_1, y_1) to the hyperbola $x^2/a^2 - y^2/b^2 = 1$ is
$$x_1 x/a^2 - y_1 y/b^2 = 1.$$
With the same notation as in ex. 4 show that
$$CM \cdot CT = CA^2, \quad Cm \cdot Ct = -CB^2,$$
and explain the meaning of the minus sign.

6. The equation of the normal to the ellipse at (x_1, y_1) is
$$(x - x_1) a^2/x_1 = (y - y_1) b^2/y_1.$$

7. If the normal at P to the ellipse (Fig. 20) meets the major axis at G show that $CG = e^2 CM$ in magnitude and in sign.

Prove also that
$$SG = e(AC + eCM) = eSP; \quad GS' = eS'P;$$
$$SG : GS' = SP : S'P.$$

The last equation shows (*Euc.* vi. 3) that the normal at P bisects the interior angle and that the tangent at P bisects the exterior angle between the focal distances of P.

8. State and prove for the hyperbola the results corresponding to those of ex. 7 for the ellipse.

9. If SZ, $S'Z'$ are the perpendiculars from the foci S, S' on the tangent at P to a central conic (Figs. 20, 21) show that
$$SZ \cdot S'Z' = CB^2.$$

For the ellipse $\quad SZ \cdot S'Z' = \dfrac{\dfrac{-ex_1}{a} - 1}{D} \cdot \dfrac{\dfrac{ex_1}{a} - 1}{D} = \dfrac{1 - \dfrac{e^2 x_1^2}{a^2}}{D^2},$

where $\quad D^2 = \dfrac{x_1^2}{a^4} + \dfrac{y_1^2}{b^4} = \dfrac{x_1^2}{a^4} + \dfrac{1}{b^2}\left(1 - \dfrac{x_1^2}{a^2}\right),$

since (x_1, y_1) is on the ellipse. A little reduction shows that
$$D^2 = \frac{1}{b^2}\left(1 - \frac{e^2 x_1^2}{a^2}\right), \text{ since } e^2 = \frac{a^2 - b^2}{a^2}.$$
(See Exercises VI. 18.)

10. If P is the point $(a \cos \theta, b \sin \theta)$ show that the equations of the tangent and normal at P are (see Exercises V. 5)
$$\frac{x}{a} \cos \theta + \frac{y}{b} \sin \theta = 1; \quad \frac{a}{\cos \theta} x - \frac{b}{\sin \theta} y = a^2 - b^2.$$

11. If P is the point $(at^2, 2at)$ on the parabola $y^2 = 4ax$ show that the equations of the tangent and normal at P are (see Exercises V. 6)

$$y = \frac{x}{t} + at; \quad y = -tx + 2at + at^3.$$

12. From the result of ex. 3 or otherwise show that if the tangent at P to the parabola (Fig. 19) meets the axis at T

$$TS = AS + AM = SP.$$

If NP is produced to Q show that TP bisects the angle SPN and PG bisects the angle SPQ. Also that, if SN cuts the tangent at the vertex at Z, SZ is perpendicular to and bisects TP and $SZ^2 = AS \cdot SP$.

13. In the notation of § 61 show that for the curves $x^m y^n = c^{m+n}$

$$KP : LP = -m : n.$$

Sketch the curve (i) if $m=7$, $n=5$; (ii) if $m=10$, $n=9$. These are *Adiabatic Curves*.

14. Show that for the parabola $y^2 = 4ax$

$$\frac{ds}{dx} = \sqrt{\left(1 + \frac{a}{x}\right)}; \quad \frac{ds}{dy} = \sqrt{\left(1 + \frac{y^2}{4a^2}\right)}.$$

15. In the *semi-cubical parabola* $ay^2 = x^3$ show that

$$LM = \frac{2}{3}x; \quad MG = \frac{3}{2a}x^2; \quad MG = \frac{27}{8a}LM^2.$$

Show also that

$$\frac{ds}{dx} = \sqrt{\left(1 + \frac{9x}{4a}\right)};$$

and verify that if the arc s is measured from the origin

$$s = \frac{8a}{27}\left(1 + \frac{9x}{4a}\right)^{\frac{3}{2}} - \frac{8a}{27}.$$

16. Show that the tangents at the points where the straight line $ax + hy = 0$ meets the ellipse

$$ax^2 + 2hxy + by^2 = 1$$

are parallel to the x-axis, and that the tangents at the points where the straight line $hx + by = 0$ meets the ellipse are parallel to the y-axis.

17. Show that the tangents at the points where the parabola $ay = x^2$ meets the folium of Descartes, whose equation is (compare Exercises VI., 13)

$$x^3 + y^3 = 3axy,$$

are parallel to the x-axis, and that the tangents at the points where the parabola $y^2 = ax$ meets the folium are parallel to the y-axis.

The origin $(0, 0)$ is one of the points, and the coordinate axes are tangents, though this is one of the exceptional cases referred to in § 59, ex. 7. The other points are $(a\sqrt[3]{2}, a\sqrt[3]{4})$, $(a\sqrt[3]{4}, a\sqrt[3]{2})$.

18. Show that for the ellipse $x^2/a^2 + y^2/b^2 = 1$

$$\frac{ds}{dx} = \sqrt{\left(\frac{a^2 - e^2x^2}{a^2 - x^2}\right)},$$

and that for the hyperbola $x^2/a^2 - y^2/b^2 = 1$

$$\frac{ds}{dx} = \sqrt{\left(\frac{e^2x^2 - a^2}{x^2 - a^2}\right)}.$$

19. Show that for the curve $y = cx^m$

$$\frac{ds}{dx} = \sqrt{(1 + m^2c^2x^{2m-2})}.$$

20. Show that for the curve $x^{\frac{2}{3}} + y^{\frac{2}{3}} = a^{\frac{2}{3}}$

$$\frac{ds}{dx} = \left(\frac{a}{x}\right)^{\frac{1}{3}}, \quad s = \frac{3}{2}a^{\frac{1}{3}}x^{\frac{2}{3}},$$

the arc being measured from the point (o, a).

CHAPTER VII.

DIFFERENTIATION (*continued*). TRANSCENDENTAL FUNCTIONS. HIGHER DERIVATIVES.

§ 63. Derivatives of the Trigonometric Functions. The fundamental limit is that proved in § 39 (iv.), the angle being measured in radians, namely

$$\operatorname*{L}_{\theta=0} \frac{\sin \theta}{\theta} = 1.$$

(i) $D_x \sin x = \cos x$.

For
$$D_x \sin x = \operatorname*{L}_{\delta x = 0} \frac{\sin(x+\delta x) - \sin x}{\delta x}.$$

Now,
$$\frac{\sin(x+\delta x) - \sin x}{\delta x} = \frac{2 \sin \frac{\delta x}{2} \cos\left(x + \frac{\delta x}{2}\right)}{\delta x}$$

$$= \left(\frac{\sin \frac{\delta x}{2}}{\frac{\delta x}{2}}\right) \cos\left(x + \frac{\delta x}{2}\right).$$

The limit for $\delta x = 0$ of the first factor is 1, and of the second factor is $\cos x$. Hence

$$D_x \sin x = \cos x.$$

(ii) $D_x \cos x = -\sin x$.

For
$$D_x \cos x = \operatorname*{L}_{\delta x = 0} \frac{\cos(x+\delta x) - \cos x}{\delta x},$$

and
$$\cos(x+\delta x) - \cos x = -2 \sin \frac{\delta x}{2} \sin\left(x + \frac{\delta x}{2}\right).$$

The rest of the work is the same as in (i).

(iii) $D_x \tan x = \dfrac{1}{\cos^2 x} = \sec^2 x.$

For $\quad D_x \tan x = \underset{\delta x = 0}{L} \dfrac{\tan(x + \delta x) - \tan x}{\delta x}$

$\qquad = \underset{\delta x = 0}{L} \dfrac{1}{\cos(x + \delta x)\cos x} \cdot \dfrac{\sin \delta x}{\delta x}$

$\qquad = \dfrac{1}{\cos^2 x} = \sec^2 x.$

The result may, of course, be obtained by writing $\tan x$ in the form $\sin x / \cos x$, and applying the rule for differentiating a quotient.

Directly from the definition or by applying the rule for differentiating a quotient we obtain

$\quad$ (iv) $\quad D_x \operatorname{cosec} x = -\operatorname{cosec} x \cot x;$

$\quad$ (v) $\quad D_x \sec x = \sec x \tan x;$

$\quad$ (vi) $\quad D_x \cot x = -\dfrac{1}{\sin^2 x} = -\operatorname{cosec}^2 x.$

The knowledge of the derivatives makes it easier to graph the functions, and the student should test such graphs as he has already drawn by examining the gradient in the light of the derivative.

The derivatives of the sine and cosine are continuous for all values of the argument. The derivatives of the other functions become discontinuous for the values for which the functions become discontinuous.

The rule for differentiating a function of a function has often to be applied, for it is very seldom that the argument is x simply. The most important case is that in which the argument is a linear function $ax + b$.

Put $ax + b = u$, and we have

$\qquad D_x \sin(ax + b) = D_u \sin u \times D_x(ax + b)$

$\qquad\qquad = \cos u \times a = a \cos(ax + b).$

In the same way we find

$\quad D_x \cos(ax + b) = -a \sin(ax + b); \quad D_x \tan(ax + b) = a \sec^2(ax + b),$

and so on. In fact the student should from the first accustom himself to these forms.

Again, to find the derivative of $\sin^2(ax + b)$, let $\sin(ax + b)$ be denoted by u; then

DIRECT TRIGONOMETRIC FUNCTIONS.

$$D_x[\sin^2(ax+b)] = D_u u^2 \times D_x \sin(ax+b)$$
$$= 2u \times a \cos(ax+b)$$
$$= 2a \sin(ax+b)\cos(ax+b).$$

With a little practice, and the application of common sense, even this substitution will not be necessary.

NOTE.—If the angle is measured in degrees, then $D_x \sin x$ is not $\cos x$ but $\dfrac{\pi}{180}\cos x$, because x degrees make $\pi x/180$ radians, and

$$\sin(x \text{ deg.}) = \sin\left(\frac{\pi x}{180} \text{ rad.}\right);$$

$$D_x \sin(x \text{ deg.}) = D_x \sin\left(\frac{\pi x}{180} \text{ rad.}\right)$$

$$= \frac{\pi}{180}\cos\left(\frac{\pi x}{180} \text{ rad.}\right) = \frac{\pi}{180}\cos(x \text{ deg.}).$$

EXERCISES XI.

Differentiate with respect to x, ex. 1-9:

1. $\sin 3x + \cos 3x$.   2. $\sin \dfrac{2\pi}{a}(x+b)$.   3. $\sin mx \cos nx$.
4. $x \sin x + \cos x$.   5. $\sin x - x \cos x$.   6. $\tfrac{1}{2}x - \tfrac{1}{4}\sin 2x$.
7. $\tfrac{1}{2}x + \tfrac{1}{4}\sin 2x$.   8. $\tfrac{3}{4}\sin x + \tfrac{1}{12}\sin 3x$.   9. $-\tfrac{3}{4}\cos x + \tfrac{1}{12}\cos 3x$.

Write down for each of the functions 10-15 a function of which it is the x-derivative.

10. $\cos 3x - \sin 3x$.   11. $\cos(ax+b)$.   12. $\sec^2(ax+b)$.
13. $\cos^2 x$.   14. $\sin^2 x$.   15. $\sin 4x \cos 2x$.

Differentiate with respect to x, ex. 16-22.

16. $\cos^2(ax+b)$.   17. $\tan^2(\tfrac{1}{2}x+1)$.   18. $\sqrt{\sin 2x}$.
19. $\sin^2 x/\cos^3 x$.   20. $\dfrac{1}{1+\cos x}$.   21. $\dfrac{1-\cos x}{1+\cos x}$.
22. $\dfrac{\sin x}{1+\tan x}$.

23. Show that $D_x[\tan \tfrac{1}{2}x] = \dfrac{1}{1+\cos x}$,

and that $D_x[x \tan \tfrac{1}{2}x] = \dfrac{x+\sin x}{1+\cos x}$.

24. Show that $\dfrac{\sin x}{x}$ steadily decreases as x increases from 0 to $\dfrac{\pi}{2}$; graph the function from $x=0$ to $x=\pi$ (see also ex. 34).

To prove the theorem, show that the derivative of $\sin x/x$ is negative, and therefore $\sin x/x$ a decreasing function. Since $\sin x/x = 2/\pi$ when $x=\pi/2$ and $\sin x < x$ we get the inequalities
$$\frac{2}{\pi}x < \sin x < x,$$
which hold for the range 0 to $\pi/2$.

25. A point moves on a straight line and at time t its distance s from a fixed point on the line is given by the equation $s = a\cos(nt - e)$. Find for what values of t its velocity is greatest and state where the point then is. For what values of t is its velocity zero and where is the point at these instants?

26. The coordinates x, y of a point at time t are given by the equations $x=a\cos t$, $y=b\sin t$. Show that as t varies from 0 to 2π (or from t_1 to $t_1 + 2\pi$) the point describes an ellipse, and find the components of the velocity and the direction of motion at time t.

27. The coordinates of a point are given by
$$x = a(\theta - \sin\theta), \quad y = a(1 - \cos\theta),$$
where $0 \leqq \theta \leqq 2\pi$. Show that the tangent to the locus of the point makes with the x-axis the angle $\dfrac{\pi}{2} - \dfrac{\theta}{2}$ and that if the arc s is measured from the origin, $s = 4a\left(1 - \cos\dfrac{\theta}{2}\right)$. The locus of the point is called a *cycloid* (§ 146).

28. Find the subtangent and the subnormal of the curve of sines whose equation is $y = a\sin(x/b)$.

29. If $D_x y = \sqrt{(a^2 - x^2)}$ and $x = a\sin\theta$, show that $D_\theta y = a^2\cos^2\theta$.
In the notation of differentials, we may write
$$dy = \sqrt{(a^2 - x^2)}dx;\quad dx = a\cos\theta\, d\theta;\quad dy = a^2\cos^2\theta\, d\theta.$$

30. If $dy = \sqrt{(x^2 + a^2)}dx$ and $x = a\tan\theta$, show that
$$dy = a^2\sec^3\theta\, d\theta.$$

31. If $dy = \dfrac{dx}{\sqrt{(a^2 - x^2)}}$ and $x = a\sin\theta$, show that $dy = d\theta$.

32. If $dy = \dfrac{dx}{\sqrt{(2ax - x^2)}}$ and $x = a(1 + \sin\theta)$, show that $dy = d\theta$.

33. If $f(x) = 1 - \tfrac{1}{2}x^2 - \cos x$, show that when x is positive $f'(x)$ is negative. Hence show that for positive values of x
$$1 - \tfrac{1}{2}x^2 < \cos x < 1.$$

$f(x)$ is a decreasing function. Since $f(x) = 0$ when $x = 0$, it must therefore be negative for every positive value of x.

INVERSE TRIGONOMETRIC FUNCTIONS.

34. Show that when x is positive
$$x - \tfrac{1}{6}x^3 < \sin x < x.$$
Take $\phi(x) = x - \tfrac{1}{6}x^3 - \sin x$; then (by ex. 33) $\phi'(x)$ is negative, since $\phi'(x) = f(x)$.

35. Prove in the same way that when x is positive
$$1 - \tfrac{1}{2}x^2 < \cos x < 1 - \tfrac{1}{2}x^2 + \tfrac{1}{4!}x^4\,;$$
$$x - \tfrac{1}{3!}x^3 < \sin x < x - \tfrac{1}{3!}x^3 + \tfrac{1}{5!}x^5.$$
These inequalities may be carried out to any number of terms.

36. How should the inequalities of examples 33, 34, 35 be stated for negative values of x?

37. Show that if x is positive and less than $\pi/2$
$$x < \tfrac{1}{3}\tan x + \tfrac{2}{3}\sin x.$$

§ 64. Inverse Trigonometric Functions. The direct trigonometric functions are single-valued but the angle has to be restricted to a certain range in order that the inverse functions may be single-valued (see § 28). The range is from $-\pi/2$ to $\pi/2$ for the functions inverse to the sine, the cosecant, the tangent, and the cotangent, but from 0 to π for those inverse to the cosine and the secant.

In finding the derivatives the theorem expressed in the equation $D_x y = 1/D_y x$ is used (§ 59).

(i) $D_x \sin^{-1} x = + \dfrac{1}{\sqrt{(1-x^2)}}.$

Let $y = \sin^{-1} x$; then $x = \sin y$ and
$$D_y x = \cos y = +\sqrt{(1-x^2)},$$
because $\cos y$ is positive, y lying between $-\pi/2$ and $\pi/2$.
Hence $\qquad D_x \sin^{-1} x = D_x y = \dfrac{1}{D_y x} = \dfrac{1}{\sqrt{(1-x^2)}}.$

(ii) $D_x \cos^{-1} x = - \dfrac{1}{\sqrt{(1-x^2)}}.$

Let $y = \cos^{-1} x$; then $x = \cos y$, and
$$D_y x = -\sin y = -[+\sqrt{(1-x^2)}],$$
because $\sin y$ is positive, y lying between 0 and π. Hence
$$D_x \cos^{-1} x = D_x y = \dfrac{1}{D_y x} = -\dfrac{1}{\sqrt{(1-x^2)}}.$$

This result may also be obtained from the equation

$$\cos^{-1}x = \frac{\pi}{2} - \sin^{-1}x.$$

In the same way the following results are established:

(iii) $D_x \tan^{-1}x = \dfrac{1}{1+x^2}$;   (iv) $D_x \cot^{-1}x = -\dfrac{1}{1+x^2}$;

(v) $D_x \operatorname{cosec}^{-1}x = -\dfrac{1}{x\sqrt{(x^2-1)}}$; (vi) $D_x \sec^{-1}x = \dfrac{1}{x\sqrt{(x^2-1)}}.$

Of the above results (i), (iii) are the most important. The root is a *positive* number, so that, for example, $\sqrt{(x^2)}$ means $+x$ when x is positive, but $-x$ when x is negative. The results (v), (vi) hold so long as x is positive; when x is negative the sign of each must be changed.

It is worth noting that the derivatives of the inverse trigonometric functions are *not transcendental* but are *algebraic* functions.

The derivatives (i), (ii), (v), (vi) become discontinuous for $x = \pm 1$; (iii), (iv) are continuous for every finite value of x.

In the case of the inverse functions also the student should accustom himself to the form in which the argument is not x but a linear function of x, specially x/a or $x/\sqrt{k}$. Thus, if $x/a = u$

$$D_x \sin^{-1}\!\left(\frac{x}{a}\right) = D_u \sin^{-1}u \times D_x\!\left(\frac{x}{a}\right) = \frac{1}{\sqrt{(1-u^2)}} \cdot \frac{1}{a} = \frac{1}{\sqrt{(a^2-x^2)}};$$

$$D_x \tan^{-1}\!\left(\frac{x}{a}\right) = D_u \tan^{-1}u \times D_x\!\left(\frac{x}{a}\right) = \frac{1}{1+u^2} \cdot \frac{1}{a} = \frac{a}{a^2+x^2};$$

$$D_x \sin^{-1}\!\left(\frac{x}{\sqrt{k}}\right) = \frac{1}{\sqrt{(k-x^2)}}; \quad D_x \tan^{-1}\frac{x}{\sqrt{k}} = \frac{\sqrt{k}}{k+x^2}.$$

EXERCISES XII.

Differentiate with respect to x, ex. 1-6:

1. $\sin^{-1}3x.$ 2. $\sin^{-1}\!\left(\dfrac{2x-1}{3}\right).$ 3. $\tan^{-1}\!\left(\dfrac{2x-1}{3}\right).$

4. $\sin^{-1}(1-x).$ 5. $x\sin^{-1}x.$ 6. $x\tan^{-1}x.$

EXPONENTIAL FUNCTION.

Write down for each of the functions 7-9 a function of which it is the x-derivative:

7. $\dfrac{1}{\sqrt{(3-x^2)}}.$ **8.** $\dfrac{1}{3+x^2}.$ **9.** $\dfrac{1}{\sqrt{\{4-(x-1)^2\}}}.$

10. Prove that
$$D_x\left\{\tfrac{1}{2}x\sqrt{(a^2-x^2)}+\tfrac{1}{2}a^2\sin^{-1}\tfrac{x}{a}\right\}=\sqrt{(a^2-x^2)}.$$

11. Show that*
$$D_x\cos^{-1}\left(\dfrac{b+a\cos x}{a+b\cos x}\right)=\dfrac{\sqrt{(a^2-b^2)}}{a+b\cos x}.$$
If a^2 is less than b^2 the derivative is imaginary; explain this.

12. Show that
$$D_x\left\{\dfrac{a\sin x}{a+b\cos x}-\dfrac{b}{\sqrt{(a^2-b^2)}}\cos^{-1}\left(\dfrac{b+a\cos x}{a+b\cos x}\right)\right\}=\dfrac{(a^2-b^2)\cos x}{(a+b\cos x)^2}.$$

13. Show that
$$D_x\tan^{-1}(\sqrt{1-k^2}\tan x)=\dfrac{\sqrt{(1-k^2)}}{1-k^2\sin^2 x}.$$

14. Show that
$$D_x\sin^{-1}\left(\dfrac{b+a\sin x}{a+b\sin x}\right)=\dfrac{\sqrt{(a^2-b^2)}}{a+b\sin x}.$$

15. If $x=r\cos\theta$, $y=r\sin\theta$, and x, y, r, θ are all functions of t, prove
  (i) $\dot{x}=\dot{r}\cos\theta-r\sin\theta\,\dot{\theta}.$ (ii) $\dot{y}=\dot{r}\sin\theta+r\cos\theta\,\dot{\theta}.$
  (iii) $x\dot{y}-y\dot{x}=r^2\dot{\theta}.$

§ 65. Exponential and Logarithmic Functions. The fundamental limit is now that discussed in § 48, namely,
$$\underset{m=\infty}{\mathrm{L}}\left(1+\dfrac{1}{m}\right)^m=e$$
and that stated in the corollary to § 49, which may be put in the form
$$\underset{\delta x=0}{\mathrm{L}}\dfrac{e^{\delta x}-1}{\delta x}=1.$$

---

*The value of the derivative given in ex. 11 is only true if a is positive and x lies in the first or second positive quadrant. If a is negative, or if x lies in the first or second negative quadrant, the sign of the result must be changed. A similar remark applies to ex. 12. In ex. 14 the result holds if a is positive and if x lies in the first positive or in the first negative quadrant.

I. $D_x e^x = e^x$.

For $\quad D_x e^x = \underset{\delta x = 0}{L} \dfrac{e^{x+\delta x} - e^x}{\delta x} = e^x \underset{\delta x = 0}{L} \dfrac{e^{\delta x} - 1}{\delta x} = e^x$.

COR. $\qquad D_x a^x = \log a \cdot a^x$.

For if $k = \log a$, $a^x = e^{kx}$, so that putting $kx = u$

$$D_x a^x = D_u e^u \times D_x(kx) = e^{kx} \times k = \log a \times a^x.$$

II. $D_x \log x = \dfrac{1}{x}$.

For $D_x \log x = \underset{\delta x = 0}{L} \dfrac{\log(x+\delta x) - \log x}{\delta x} = \underset{\delta x = 0}{L} \dfrac{1}{\delta x} \log\left(1 + \dfrac{\delta x}{x}\right)$.

Put $\dfrac{\delta x}{x} = \dfrac{1}{m}$, so that if $x \neq 0$, as δx converges to 0, m converges to ∞. Now

$$\dfrac{1}{\delta x} \log\left(1 + \dfrac{\delta x}{x}\right) = \dfrac{m}{x} \log\left(1 + \dfrac{1}{m}\right) = \dfrac{1}{x} \log\left[\left(1 + \dfrac{1}{m}\right)^m\right];$$

and $\quad \underset{\delta x = 0}{L} \dfrac{1}{\delta x} \log\left(1 + \dfrac{\delta x}{x}\right) = \dfrac{1}{x} \underset{m = \infty}{L} \log\left[\left(1 + \dfrac{1}{m}\right)^m\right]$

$$= \dfrac{1}{x} \log\left[\underset{m = \infty}{L} \left(1 + \dfrac{1}{m}\right)^m\right]$$

$$= \dfrac{1}{x} \log e.$$

Since the base of the logarithms is supposed to be e the result is established.

COR. $\qquad D_x \log_{10} x = \dfrac{1}{x} \log_{10} e$.

Assuming the derivative of $\log x$ the derivative of e^x may be obtained by the rule for the derivative of inverse functions; and conversely that of $\log x$ may be obtained from that of e^x. Thus,

Let $\qquad y = e^x$; then $x = \log y$, and $D_y x = 1/y$,

$$D_x e^x = D_x y = \dfrac{1}{D_y x} = y = e^x.$$

Again, $\qquad D_x \log(ax+b) = \dfrac{a}{ax+b}$.

LOGARITHMIC FUNCTION. 137

For, put $ax+b=u$ and we get
$$D_x\log(ax+b)=D_u\log u \times D_x(ax+b)=\frac{1}{u}\times a=\frac{a}{ax+b}.$$

Since $\log x$ is a real number only when x is positive, $\log(-x)$ will be real only if x is negative. The x-derivative of $\log(-x)$ is however $1/x$, as may be seen by putting $a=-1, b=0$. Hence the function whose x-derivative is $1/x$ is $\log x$ or $\log(-x)$, according as x is positive or negative.

It will be noticed that the derivative of $\log x$ is an algebraic function, discontinuous for $x=0$ like the function itself.

Ex. 1. $\quad D_x\log(x+\sqrt{x^2+k})=\dfrac{1}{\sqrt{(x^2+k)}}.$

Let $u=x+\sqrt{(x^2+k)}$; then
$$D_x\log(x+\sqrt{x^2+k})=D_u\log u \times D_x u=\frac{1}{u}\times D_x u$$
and $\quad D_x u=1+\tfrac{1}{2}(x^2+k)^{-\frac{1}{2}}\cdot 2x=\dfrac{\sqrt{(x^2+k)}+x}{\sqrt{(x^2+k)}}=\dfrac{u}{\sqrt{(x^2+k)}},$
and the result follows at once.

The student should note that
$$\frac{1}{\sqrt{(k+x^2)}}=x\text{-derivative of }\log(x+\sqrt{x^2+k}),$$
but $\quad \dfrac{1}{\sqrt{(k-x^2)}}= \ldots\ldots\ldots\ldots$ of $\sin^{-1}\left(\dfrac{x}{\sqrt{k}}\right)$ or of $-\cos^{-1}\left(\dfrac{x}{\sqrt{k}}\right)$.

These results are frequently required in the Integral Calculus.

Ex. 2. Find the derivative of $e^{ax}\sin(bx+c)$ and of $e^{ax}\cos(bx+c)$.

These functions are of very frequent occurrence in certain branches of physics.
$$D_x\{e^{ax}\sin(bx+c)\}=ae^{ax}\sin(bx+c)+be^{ax}\cos(bx+c)$$
$$=e^{ax}\{a\sin(bx+c)+b\cos(bx+c)\}.$$

This result can be put into a form that is very convenient. Whatever be the values of a and b, it is always possible to find R and θ, so that $\quad R\cos\theta=a, \quad R\sin\theta=b$;
for these equations give
$$R=\sqrt{(a^2+b^2)}, \quad \tan\theta=\frac{b}{a}.$$

Replacing a and b by $R\cos\theta$ and $R\sin\theta$, we get
$$D_x\{e^{ax}\sin(bx+c)\}=Re^{ax}\{\cos\theta\sin(bx+c)+\sin\theta\cos(bx+c)\}$$
$$=Re^{ax}\sin(bx+c+\theta).$$

In the same way we find
$$D_x\{e^{ax}\cos(bx+c)\}=Re^{ax}\cos(bx+c+\theta),$$
where R and θ have the same meaning as before.

138 AN ELEMENTARY TREATISE ON THE CALCULUS.

Some care is necessary however in making the transformation, because θ is not uniquely determined by its tangent; the quadrant in which θ lies is determined by the signs of a and b. Thus, R being taken positive, if a and b are both positive, $\tan\theta$ is positive and θ is in the first quadrant; but if a and b are both negative, $\tan\theta$ is also positive, but θ is now in the third quadrant. Similar observations hold when a and b have opposite signs.

In practice it is usually simplest to choose R positive when a is positive, but negative when a is negative; then to choose θ as a positive or negative acute angle. When numbers are given it is best to work the example without reference to the general formula. Thus,

$$D_x\{e^{-3x}\cos(4x+1)\} = -e^{-3x}\{3\cos(4x+1)+4\sin(4x+1)\}.$$

Choose $R\cos\theta=3$, $R\sin\theta=4$, and therefore $R=5$, $\tan\theta=\tfrac{4}{3}$. Now $\tfrac{4}{3}=\tan 53°\,8'$ and $53°\,8'=\cdot 9274$ radian, so that

$$D_x\{e^{-3x}\cos(4x+1)\} = -5e^{-3x}\{\cos\theta\cos(4x+1)+\sin\theta\sin(4x+1)\}$$
$$= -5e^{-3x}\cos(4x+1-\theta)$$
$$= -5e^{-3x}\cos(4x+\cdot 0726).$$

Ex. 3. Find the x-derivative of $\sqrt{(x-1)(x-2)}/\sqrt{(x-3)(x-4)}$.

In this and in similar cases where the function is a product, it is often simplest first to take the logarithm of the function and then differentiate. Denote the function by y; then

$$\log y = \tfrac{1}{2}\log(x-1) + \tfrac{1}{2}\log(x-2) - \tfrac{1}{2}\log(x-3) - \tfrac{1}{2}\log(x-4).$$

Now $\quad D_x\log y = D_y\log y \times D_x y = \dfrac{1}{y}D_x y$;

$$\therefore \frac{1}{y}D_x y = \tfrac{1}{2}\cdot\frac{1}{x-1} + \tfrac{1}{2}\frac{1}{x-2} - \tfrac{1}{2}\frac{1}{x-3} - \tfrac{1}{2}\frac{1}{x-4}$$
$$= -\frac{2x^2-10x+11}{(x-1)(x-2)(x-3)(x-4)};$$

$$\therefore D_x y = -\frac{2x^2-10x+11}{(x-1)^{\frac{1}{2}}(x-2)^{\frac{1}{2}}(x-3)^{\frac{3}{2}}(x-4)^{\frac{3}{2}}}.$$

In the same way, if the function be uvw/UVW, where $u\ldots W$ are all functions of x, we should get, denoting the function by y and taking logarithms (see §58 Th. IV.),

$$\frac{Dy}{y} = \frac{Du}{u} + \frac{Dv}{v} + \frac{Dw}{w} - \frac{DU}{U} - \frac{DV}{V} - \frac{DW}{W}.$$

Ex. 4. If u, v are both functions of x, we may find the derivative of u^v as follows: Put $y=u^v$ and take logarithms; then

$$\log y = v\log u,$$

$$\frac{Dy}{y} = Dv \times \log u + v\cdot\frac{Du}{u};\quad Du^v = u^v\Big\{Dv\times\log u + \frac{v}{u}Du\Big\}.$$

For example, $\quad Dx^x = x^x(\log x + 1).$

EXERCISES XIII.

Differentiate with respect to x, examples 1-13:

1. $x \log x$. 2. $x^n \log x$. 3. $\log \sin x$. 4. $\log \cos x$.
5. $\log \tan \tfrac{1}{2}x$. 6. $\log\left(\dfrac{1+\sin x}{1-\sin x}\right)$. 7. $\log\left(\dfrac{1-\cos x}{1+\cos x}\right)$.
8. $\log \dfrac{a+\sqrt{x}}{a-\sqrt{x}}$. 9. $\log\{\sqrt{(x+a)}+\sqrt{(x-a)}\}$. 10. xe^x.
11. $x^n e^x$. 12. $e^{-x}(\sin x + \cos x)$. 13. $\dfrac{e^x}{1+x}$.

Write down for each of the functions 14-18 a function of which it is the x-derivative:

14. $\dfrac{1}{3x+4}$. 15. $\dfrac{1}{x^2-a^2}\left[=\dfrac{1}{2a}\left(\dfrac{1}{x-a}-\dfrac{1}{x+a}\right)\right]$.
16. $\dfrac{1}{4x^2-9}$. 17. $\dfrac{1}{\sqrt{(x^2+1)}}$. 18. e^{ax}.

19. If $\quad y=\tfrac{1}{2}x\sqrt{(x^2+k)}+\tfrac{1}{2}k\log\{x+\sqrt{x^2+k}\},$
show that $\quad D_x y = \sqrt{(x^2+k)}.$
Compare Exercises XII. 10.

20. If $\quad y=\sqrt{(x^2+k)}-\sqrt{k}\log\left\{\dfrac{\sqrt{(x^2+k)}+\sqrt{k}}{x}\right\},$
show that $\quad D_x y = \sqrt{(x^2+k)}/x.$

21. If $\quad y=\log\dfrac{b+a\cos x+\sqrt{(b^2-a^2)}\sin x}{a+b\cos x},$
show that $\quad D_x y = \dfrac{\sqrt{(b^2-a^2)}}{a+b\cos x}.$
Compare Exercises XII. 11.

22. In the exponential curve, the equation being $y=ce^{\frac{x}{a}}$, find the subtangent and the subnormal.

23. The curve whose equation is $y=\tfrac{1}{2}a(e^{\frac{x}{a}}+e^{-\frac{x}{a}})$ is called a "catenary"; find the subtangent, the subnormal, and the normal. Show that the perpendicular drawn from the foot of the ordinate at any point to the tangent at that point is of constant length. Graph the curve.

24. In the catenary, show that, the arc s being measured from $x=0$,
$$\dfrac{ds}{dx}=\tfrac{1}{2}(e^{\frac{x}{a}}+e^{-\frac{x}{a}}) \text{ and } s=\dfrac{a}{2}(e^{\frac{x}{a}}-e^{-\frac{x}{a}}).$$

§ 66. Hyperbolic Functions. In recent years certain functions called *Hyperbolic Functions* have been introduced; these have many analogies with the trigonometric or circular functions, and in some respects have the same relation to the rectangular hyperbola as the trigonometric functions to the circle. We shall not make much use of them, but it seems proper to define them, so that the student may not be altogether at a loss should he fall in with them in his reading. They are called the hyperbolic sine, cosine, etc., and are defined as follows, the symbol *sinh* meaning *hyperbolic sine of*; *cosh*, *hyperbolic cosine of*, and so on.

$$\sinh x = \tfrac{1}{2}(e^x - e^{-x}); \quad \cosh x = \tfrac{1}{2}(e^x + e^{-x});$$

$$\tanh x = \frac{\sinh x}{\cosh x} = \frac{e^x - e^{-x}}{e^x + e^{-x}}; \quad \coth x = \frac{\cosh x}{\sinh x};$$

$$\operatorname{cosech} x = \frac{1}{\sinh x}; \quad \operatorname{sech} x = \frac{1}{\cosh x}.$$

Identities. The following identities, similar to those for the trigonometric functions, are readily established by substituting the values of the functions in terms of x.

(i) $\cosh^2 x - \sinh^2 x = 1$; (ii) $1 - \tanh^2 x = \operatorname{sech}^2 x$;

(iii) $\coth^2 x - 1 = \operatorname{cosech}^2 x$,

where $\cosh^2 x$ means $(\cosh x)^2$, etc.

Addition Theorem. Again, corresponding to the addition theorem in trigonometry, we have

(iv) $\sinh(x \pm y) = \sinh x \cosh y \pm \cosh x \sinh y$;

(v) $\cosh(x \pm y) = \cosh x \cosh y \pm \sinh x \sinh y$;

By putting $y = x$ we get

(vi) $\sinh 2x = 2 \sinh x \cosh x$;

(vii) $\cosh 2x = \cosh^2 x + \sinh^2 x$;

$$= 2\cosh^2 x - 1 = 1 + 2\sinh^2 x.$$

In drawing the graphs of these functions it should be noted that the sine, the tangent, and their reciprocals are odd functions, but that the cosine and its reciprocal are even functions. The sine may take any value from $-\infty$ to $+\infty$; the cosine is never less than 1 and is always positive; the tangent may take any value between -1 and 1, and the

HYPERBOLIC FUNCTIONS.

lines whose equations are $y = \pm 1$ are asymptotes to the graph of $\tanh x$.

Derivatives. The derivatives are readily found:

$$D_x \sinh x = \cosh x; \quad D_x \cosh x = \sinh x;$$
$$D_x \tanh x = \operatorname{sech}^2 x; \quad D_x \coth x = -\operatorname{cosech}^2 x;$$
$$D_x \operatorname{cosech} x = -\operatorname{cosech} x \coth x;$$
$$D_x \operatorname{sech} x = -\operatorname{sech} x \tanh x.$$

Inverse Functions. The inverse functions can be expressed by means of the logarithm.

If $y = \sinh^{-1} x$, then $x = \sinh y$, just as when $y = \sin^{-1} x$, $x = \sin y$. To find the logarithmic form of y we have to solve the equation

$$x = \tfrac{1}{2}(e^y - e^{-y}) \text{ or } e^{2y} - 2xe^y - 1 = 0,$$

which gives $\quad e^y = x \pm \sqrt{(x^2 + 1)}.$

Since e^y is always positive the $+$ sign can alone be taken; therefore

$$e^y = x + \sqrt{(x^2 + 1)}, \text{ and } \sinh^{-1} x = y = \log(x + \sqrt{x^2 + 1}).$$

In the same way we find

$$\cosh^{-1} x = \log(x \pm \sqrt{x^2 - 1}).$$

Since $\quad (x - \sqrt{x^2 - 1}) = 1/(x + \sqrt{x^2 - 1})$ we have

$$\log(x - \sqrt{x^2 - 1}) = -\log(x + \sqrt{x^2 - 1}).$$

In this case the inverse function is not single-valued; to each value of x greater than 1 there are two values of $\cosh^{-1} x$, equal numerically but of opposite sign. The graph of $\cosh x$ is in general appearance like that of $1 + x^2$; by rotating the graph of $1 + x^2$ about the bisector of the angle XOY we should get a curve resembling that of $\cosh^{-1} x$, and the curve would be symmetrical about the x-axis as the graph of $\cosh x$ is symmetrical about the y-axis.

If $\qquad x^2 < 1, \; \tanh^{-1} x = \tfrac{1}{2} \log \dfrac{1+x}{1-x};$

if $\qquad x^2 > 1, \; \coth^{-1} x = \tfrac{1}{2} \log \dfrac{x+1}{x-1}.$

Derivatives of Inverse Functions. The derivatives of the inverse functions, taking for greater convenience x/a instead of x, are

$$D_x \sinh^{-1} \frac{x}{a} = \frac{1}{\sqrt{(x^2 + a^2)}};$$

$$D_x \cosh^{-1} \frac{x}{a} = \pm \frac{1}{\sqrt{(x^2 - a^2)}};$$

$$D_x \tanh^{-1} \frac{x}{a} = \frac{a}{a^2 - x^2} \quad (x^2 < a^2);$$

$$D_x \coth^{-1} \frac{x}{a} = \frac{-a}{x^2 - a^2} \quad (x^2 > a^2).$$

For the positive ordinate of $\cosh^{-1} \frac{x}{a}$ the $+$ sign must be taken.

It should be noticed that

$$\sinh^{-1} \frac{x}{a} = \log \frac{x + \sqrt{x^2 + a^2}}{a} = \log(x + \sqrt{x^2 + a^2}) - \log a,$$

so that the derivative of $\sinh^{-1} \frac{x}{a}$ is the same as that of $\log(x + \sqrt{x^2 + a^2})$, the constant $\log a$ disappearing in the differentiation. The occurrence of the divisor a in the logarithmic form of $\sinh^{-1} \frac{x}{a}$ has to be borne in mind when comparing the same result expressed in logarithms and in inverse hyperbolic sines (or cosines).

§ 67. Higher Derivatives. The derivative of $f(x)$ is usually itself a function of x and may therefore be differentiated with respect to x. Thus the derivative of x^3 is $3x^2$ and the derivative of $3x^2$ is $6x$. $6x$ is therefore called the second derivative of x^3, while $3x^2$, which has hitherto been called simply the derivative of x^3, may be called for distinction the first derivative of x^3.

The notation for derivatives higher than the first is modelled on the analogy of indices. Thus

the first x-derivative of y is $D_x y$,
the second „   „ $D_x(D_x y)$ written $D_x^2 y$,
the third   „   „ $D_x(D_x^2 y)$ „ $D_x^3 y$,
the nth „   „ $D_x(D_x^{n-1} y)$ „ $D_x^n y$.

HIGHER DERIVATIVES.

When the derivative is written in the form $\dfrac{dy}{dx}$ the higher derivatives are written by considering $\dfrac{d}{dx}$ as the equivalent of D_x and $D_x^2 y$ becomes $\dfrac{d^2}{dx^2} \cdot y$, usually written $\dfrac{d^2 y}{dx^2}$; generally $D_x^n y$ becomes $\dfrac{d^n}{dx^n} \cdot y$ or $\dfrac{d^n y}{dx^n}$.

The accent notation is also used; thus $f''(x)$, $f'''(x)$, $f^{\text{iv}}(x) \ldots f^{(n)}(x)$ mean the 2nd, 3rd, 4th ... nth derivatives of $f(x)$, n being enclosed in brackets to distinguish the nth derivative from the nth power. In the same way y'', $y''' \ldots \ddot{x}, \dddot{x} \ldots$ are used, but the notation is rather cumbrous when more than two accents are used.

Ex. 1. If $f(x) = ax^4 + bx^3 + cx^2 + dx + e$, find $f''(x)$, $f'''(x)$, $f^{\text{iv}}(x)$.

$$f'(x) = 4ax^3 + 3bx^2 + 2cx + d$$
$$f''(x) = 12ax^2 + 6bx + 2c$$
$$f'''(x) = 24ax + 6b$$
$$f^{\text{iv}}(x) = 24a.$$

Since $f^{\text{iv}}(x)$ is a constant, the fifth and all higher derivatives will be zero.

It will be readily seen that the nth derivative of x^n is $n!$ and all derivatives of higher order than n are zero.

Ex. 2. If $x = a \cos nt$, find $\ddot{x}$.

$$\dot{x} = -na \sin nt \, ; \quad \ddot{x} = -n^2 a \cos nt = -n^2 x.$$

Ex. 3. If $y = e^{ax}$, prove $D^n y = a^n e^{ax} = a^n y$.

$$Dy = ae^{ax}\,; \quad D^2 y = aDe^{ax} = a^2 e^{ax}, \text{ etc.}$$

Each differentiation in this case is equivalent to the multiplication of the function by a.

Ex. 4. If $y = e^{ax} \sin(bx + c)$ find $D^2 y$ and $D^n y$.

$$Dy = Re^{ax} \sin(bx + c + \theta) \quad (\S\ 65,\ \text{ex. 2}),$$
$$D^2 y = RRe^{ax} \sin(bx + c + \theta + \theta)$$
$$= R^2 e^{ax} \sin(bx + c + 2\theta).$$

It is easy now to see, and the result may be strictly proved by the method of induction, that

$$D^n y = R^n e^{ax} \sin(bx + c + n\theta).$$

Ex. 5. Prove $D^n \sin(ax+b) = a^n \sin\left(ax+b+\dfrac{n\pi}{2}\right)$,

$$D \sin(ax+b) = a\cos(ax+b) = a\sin\left(ax+b+\dfrac{\pi}{2}\right), \text{ etc.};$$

and in the same way it may be shown that

$$D^n \cos(ax+b) = a^n \cos\left(ax+b+\dfrac{n\pi}{2}\right).$$

Ex. 6. Prove $D^n \log x = (-1)^{n-1}(n-1)!\, x^{-n}$.

§ 68. Leibniz's Theorem. Examples. The calculation of higher derivatives is, as a rule, a laborious process, and there are only a few functions such as x^n or e^{ax} of which the nth derivative can be stated explicitly. The following theorem, named after its discoverer, is useful in finding the nth derivative of a product.

LEIBNIZ'S THEOREM. *If y is the product of two functions, u and v, of x then*

$$D^n y = vD^n u + {}_nC_1 Dv D^{n-1}u + {}_nC_2 D^2v D^{n-2}u + \ldots$$
$$+ {}_nC_2 D^{n-2}v D^2 u + {}_nC_1 D^{n-1}v Du + D^n v \cdot u \ldots \text{(i)}$$

where ${}_nC_1, {}_nC_2, \ldots$ are the binomial coefficients.

The proof is obtained by repeated application of Th. IV., § 58. Using the accent notation we have, since $y = uv$,

$$y' = vu' + v'u,$$
$$y'' = vu'' + v'u' + v'u' + v''u = vu'' + 2v'u' + v''u,$$
$$y''' = vu''' + v'u'' + 2v'u'' + 2v''u' + v''u' + v'''u$$
$$= vu''' + 3v'u'' + 3v''u' + v'''u.$$

These expressions for y'', y''' clearly obey the law given by (i). The general theorem may now be proved by induction. The rth and the $(r+1)^\text{th}$ terms in (i) are

$${}_nC_{r-1} v^{(r-1)} u^{(n-r+1)} + {}_nC_r v^{(r)} u^{(n-r)},$$

and if (i) is differentiated the coefficient of $v^{(r)} u^{(n-r+1)}$ in the expression thus obtained for $D^{n+1}y$ will be

$${}_nC_{r-1} + {}_nC_r, \text{ that is, } {}_{n+1}C_r.$$

Hence $\quad D^{n+1}y = vu^{(n+1)} + {}_{n+1}C_1 v'u^{(n)} + {}_{n+1}C_2 v''u^{(n-1)} + \ldots,$

so that the expression for $D^{n+1}y$ obeys the law given by (i). But the theorem has been proved to be true when $n=2$ or 3; therefore it is true when n is any positive integer.

The theorem will be found very useful at a later stage when the expansion of functions in series is taken up (Chapter XVIII.); meantime the student will find among the exercises a number of examples to which the theorem may be applied.

Geometrical and physical interpretations of the higher derivatives will be given in the next and following chapters. The student may however try to interpret the geometrical signification of the second derivative $f''(x)$ as measuring the rate of change of the gradient $f'(x)$; for example, if $f''(x)$ is positive how will the tangent at the point $P(x, f(x))$ turn about its point of contact as x moves to the right?

We will conclude the chapter with one or two examples.

Ex. 1. Find the derivative of $(Dy)^2$, the argument being x.

$(Dy)^2$ means the square of the derivative of y; D^2y means the second derivative of y. The derivative of y^2 should be written $D(y^2)$ or $D \cdot y^2$. These three forms $(Dy)^2$, D^2y, $D(y^2)$ mean quite different things, and must be carefully distinguished; $\left(\dfrac{dy}{dx}\right)^2$, $\dfrac{d^2y}{dx^2}$, $\dfrac{d(y^2)}{dx}$ or $\dfrac{d \cdot y^2}{dx}$ mean respectively $(Dy)^2$, D^2y, $D(y^2)$ or $D \cdot y^2$.

Now put u for Dy; then

$$D \cdot (Dy)^2 = D_x(u^2) = D_u(u^2) \times D_x u = 2u D_x u.$$

But $D_x u = D \cdot Dy = D^2y$, and therefore

$$D \cdot (Dy)^2 = 2Dy \, D^2y.$$

This equation may also be written in such forms as

$$D_x \cdot (y')^2 = 2y'y''\, ; \quad \frac{d}{dx}\left(\frac{dy}{dx}\right)^2 = 2\frac{dy}{dx}\frac{d^2y}{dx^2}.$$

In the same way it may be shown that

$$D_x \cdot (y')^3 = 3(y')^2 y''\, ; \quad \frac{d}{dx}\left(\frac{dy}{dx}\right)^n = n\left(\frac{dy}{dx}\right)^{n-1}\frac{d^2y}{dx^2}.$$

Ex. 2. If x and y are functions of t, find $D_x^2 y$ in terms of derivatives with respect to t.

Here $\qquad D_x y = D_t y / D_t x = \dot{y}/\dot{x}$;

therefore $\qquad D_x^2 y = D_x(\dot{y}/\dot{x}) = D_t(\dot{y}/\dot{x})/D_t x.$

But $\qquad D_t\left(\dfrac{\dot{y}}{\dot{x}}\right) = \dfrac{\dot{x}D_t\dot{y} - \dot{y}D_t\dot{x}}{(\dot{x})^2} = \dfrac{\dot{x}\ddot{y} - \dot{y}\ddot{x}}{(\dot{x})^2}$;

therefore $\qquad D_x^2 y = (\dot{x}\ddot{y} - \dot{y}\ddot{x})/\dot{x}^3,$

where $\dot{x}^3$ means $(\dot{x})^3$.

Ex. 3. If $y = Ax^2 + B/x$, prove $x^2y'' = 2y$.
We have
$$y = Ax^2 + B/x\,; \quad y' = 2Ax - B/x^2\,; \quad y'' = 2A + 2B/x^3.$$

Eliminate A and B between the three equations; in this case the second equation is not really needed because if we multiply the third equation by x^2 we get
$$x^2y'' = 2A.x^2 + 2B/x = 2y.$$

In general, however, all three equations would be required for the elimination of the two constants A, B. The equation obtained is called a *differential equation*.

Ex. 4. If $y = x^2u$ where u is a function of x, find $D^n y$.
By Leibniz's theorem
$$D^n y = x^2 D^n u + n(2x) D^{n-1} u + \frac{n(n-1)}{2}(2) D^{n-2} u,$$
since every derivative of x^2 above the second is zero. Thus
$$D^n y = x^2 D^n u + 2nx D^{n-1} u + n(n-1) D^{n-2} u.$$

EXERCISES XIV.

1. If $y = 7x^4 - 2x^3 + 4$, find y', y'', y''', y^{iv}.
2. If $y = \sqrt{(x^2+1)}$, find y''.
3. If $y = x^2(a-x)^2$, find y'' and y'''.
4. If $y = \dfrac{x^2 + 4x + 1}{x^3 + 2x^2 - x - 2}$, show that $y = \dfrac{1}{x-1} + \dfrac{1}{x+1} - \dfrac{1}{x+2}$, and then find y', y'', $y^{(n)}$.
5. If $y = \sin^2 x$, find y'', and $y^{(n)}$ $\left[\sin^2 x = \dfrac{1}{2} - \dfrac{1}{2}\cos 2x\right]$.
6. If $y = x^2 \cos x$, find y'' and $y^{(n)}$.
7. If $y = \sin x \cos^3 x$, find y'' and $y^{(n)}$.
8. If $y = x \log x$, find y'' and $y^{(n)}$.
9. If $y = xe^x$, find $y^{(n)}$. 10. If $y = x^2 e^x$, find $y^{(n)}$.
11. If y is a rational integral function of x of degree n, say
$$y = ax^n + bx^{n-1} + \ldots + kx + l,$$
prove $y^{(n)} = n!\,a$, $y^{(n+1)} = 0$, $y^{(n+2)} = 0\ldots$.
12. Find the turning values of the functions in examples 1 and 2, and graph the functions.
13. If $y = \dfrac{1}{1+x^2}$, find the turning points of the graph. Find also where y'' is zero, and show that at these points the tangent changes its direction of rotation.

14. If $y = ax^{n+1} + bx^{-n}$, prove that $x^2 y'' = n(n+1)y$.

15. If $y = ae^{nx} + be^{-nx}$, prove that $y'' - n^2 y = 0$.

16. If $y = a \cos nx + b \sin nx$, prove that $y'' + n^2 y = 0$.

17. If $y = e^{-\frac{1}{2}kx}(a \cos nx + b \sin nx)$, prove that
$$y'' + ky' + (n^2 + \tfrac{1}{4}k^2)y = 0.$$

18. If $f(x) = (x-a)^2 \phi(x)$, where $\phi(x)$ is a rational integral function that does not vanish when $x=a$, show that
$$f(a) = 0, \quad f'(a) = 0, \quad f''(a) = 2\phi(a).$$

19. If $f(x) = (x-a)^r \phi(x)$ where r is a positive integer and $\phi(x)$ as in Ex. 18, show that
$$f(a)=0, \quad f'(a)=0, \ldots \ f^{(r-1)}(a)=0, \quad f^{(r)}(a) = r!\,\phi(a).$$

20. If x is positive, show that
$$x - \tfrac{1}{2}x^2 < \log(1+x) < x.$$

Take $f(x) = x - \tfrac{1}{2}x^2 - \log(1+x)$, $\phi(x) = x - \log(1+x)$; then see Exercises XI. 33.

21. If x is positive and less than 1, show that
$$-\log(1-x) > x.$$

22. Show that the limit for $n = \infty$ of $s_n - \log n$, where
$$s_n = 1 + \tfrac{1}{2} + \tfrac{1}{3} + \ldots + \frac{1}{n}$$
is a finite quantity (called Euler's Constant) lying between 0 and 1.

From the inequalities of examples 20, 21,
$$-\log\left(1 - \frac{1}{n}\right) > \frac{1}{n} > \log\left(1 + \frac{1}{n}\right),$$
or $\quad \log\{n/(n-1)\} > \dfrac{1}{n} > \log\{(n+1)/n\}.$

Hence $\quad \log\{(n-1)/(n-2)\} > \dfrac{1}{n-1} > \log\{n/(n-1)\},$

....
$$\log\{2/1\} > \tfrac{1}{2} > \log\{3/2\},$$
$$1 = 1 > \log\{2/1\}.$$

By addition, $\quad 1 + \log n > s_n > \log(n+1),$

therefore $\quad 1 > s_n - \log n > \log\left\{1 + \dfrac{1}{n}\right\},$

from which the result follows at once. The value of the constant is
·577 215 664 90.......

23. If $x = at^2$, $y = 2at$, find $D_x^2 y$ in terms of t.

24. If $x = a\cos t$, $y = b\sin t$, find $D_x^2 y$ in terms of t.

25. If $v^2 = \dot{x}^2 + \dot{y}^2$, show that in the notation of § 62
$$\dot{v} = \ddot{x}\cos\phi + \ddot{y}\sin\phi.$$

26. If $ax^2 + 2hxy + by^2 = 1$, show that
$$D^2 y = (h^2 - ab)/(hx + by)^3.$$

27. If $ax^2 + 2hxy + by^2 + 2gx + 2fy + c = 0$, show that
$$D^2 y = \triangle/(hx + by + f)^2$$
where
$$\triangle = abc + 2fgh - af^2 - bg^2 - ch^2.$$

28. If $x^3 + y^3 - 3axy = 0$, show that
$$D^2 y = 2a^3 xy/(ax - y^2)^3.$$

29. If u is a function of x, show that
$$D^n(e^{ax} u) = e^{ax}(a^n u + {}_n c_1 a^{n-1} Du + {}_n c_2 a^{n-2} D^2 u + \ldots + D^n u).$$

30. If $y = \tan^{-1} x$, show that

(i) $Dy = \cos^2 y$; (ii) $D^2 y = \cos\left(2y + \dfrac{\pi}{2}\right)\cos^2 y$;

(iii) $D^3 y = 2\cos\left(3y + 2\dfrac{\pi}{2}\right)\cos^3 y$;

(iv) $D^n y = (n-1)!\cos\left(ny + \dfrac{(n-1)\pi}{2}\right)\cos^n y.$

CHAPTER VIII.

PHYSICAL APPLICATIONS.

§ 69. Applications of Derivatives in Dynamics. We give in this chapter a few simple examples of the use of derivatives in physical problems.

Take first the case of the rectilinear motion of a particle and let the units of time, length, and mass be the second, the foot, and the pound respectively, and the units of force and work the poundal and the foot-poundal.

At time t, that is, t seconds from some chosen instant, let the particle be at P, distant x feet from a fixed point O on the line of motion and let the mass of the particle be m pounds. Denote the velocity at time t by v, the acceleration by a, the momentum by M, the force by F, the kinetic energy by E; these quantities may be expressed in terms of t, x, m.

X'   O   P Q   X
Fig. 31.

When t increases by δt let x increase by $\delta x = PQ$; then the average velocity during the interval δt in the direction in which x increases, namely, in the direction OX, is $\delta x/\delta t$, and the velocity at time t is the limit of this quotient for $\delta t = 0$. Therefore

$$v = \mathop{\mathrm{L}}_{\delta t = 0} \frac{\delta x}{\delta t} = \frac{dx}{dt} = \dot{x}.$$

v is in general a function of t. The average acceleration during an interval δt in the direction in which x increases is $\delta v/\delta t$, where δv is the increment of v in time δt; the

acceleration at time t is the limit of this quotient for $\delta t = 0$. Hence
$$a = \underset{\delta t = 0}{\mathrm{L}} \frac{\delta v}{\delta t} = \frac{dv}{dt} = \frac{d}{dt} \cdot \frac{dx}{dt} = \frac{d^2x}{dt^2} = \ddot{x}.$$

The momentum in the direction in which x increases is
$$M = mv = m\dot{x}.$$

By the second law of motion the force F in the direction in which x increases is the time-rate of change of the momentum in that direction. Hence
$$F = \frac{dM}{dt} = m\dot{v} = m\ddot{x}.$$

We may express F in another form, by considering v as a function of x, and x as a function of t, so that (see § 59, Ex. 5)
$$\frac{dv}{dt} = \frac{dv}{dx}\frac{dx}{dt} = \frac{dv}{dx}v = \frac{d}{dx}(\tfrac{1}{2}v^2).$$

Now $E = \tfrac{1}{2}mv^2$ and therefore
$$F = m\frac{dv}{dt} = \frac{d}{dx}(\tfrac{1}{2}mv^2) = \frac{dE}{dx}.$$

Hence the force may be defined either as the time-rate of change of momentum dM/dt or as the space-rate of change of kinetic energy dE/dx.

Let W denote the work done on the particle by the force F in moving it from some standard position, say from the position at which $x = a$, to the position P; δW the work done in moving it from P to Q. At Q the force is $F + \delta F$; hence when δx is small the work done will lie between $F\delta x$ and $(F + \delta F)\delta x$. For $F\delta x$ is the work done on the supposition that the force is constant over PQ and equal to its value at P, while $(F + \delta F)\delta x$ is the work done on the supposition that the force is constant over PQ and equal to its value at Q; evidently the work will lie between these two values. Hence $\delta W/\delta x$ lies between F and $F + \delta F$ and therefore
$$\frac{dW}{dx} = F.$$

Since dE/dx is also equal to F, E and W differ only by a constant.

DYNAMICAL FORMULAE.

Again, the time-rate at which the force works is dW/dt, and W may be considered as a function of x and x, a function of t.

Therefore
$$\frac{dW}{dt} = \frac{dW}{dx}\frac{dx}{dt} = Fv.$$

The student should note the dimensional formula for these magnitudes (§ 34). If x is the measure of a length so is dx, and the dimensional formula for v or dx/dt is LT^{-1}; similar observations hold for the other quantities.

Ex. 1. Suppose F constant; then the acceleration will be constant, equal to f say. Hence $\dot{v} = f$, and therefore
$$v = ft + \text{const.}$$

Let the motion be such that when $t=0$, $v=V$ and $x=a$; these are called the initial conditions. The constant in the value of v is therefore V. We can now find x; for
$$\dot{x} = v = ft + V; \quad x = \tfrac{1}{2}ft^2 + Vt + \text{const.}$$
as may be tested by differentiation. The constant is a, since when $t=0$, $x=a$, so that finally
$$x = \tfrac{1}{2}ft^2 + Vt + a.$$

To get E in terms of t we have $E = \tfrac{1}{2}mv^2 = \tfrac{1}{2}m(ft+V)^2$. Using the value found for x and putting E_0 for $\tfrac{1}{2}mV^2$ we get
$$E - E_0 = mf(x-a) = F(x-a).$$

This form may be obtained at once from the energy equation $dE/dx = F$.

Finally since $dW/dx = F$ we have $W = F(x-a)$, W being zero when $x = a$. Hence $E - E_0 = W$; that is, the gain in kinetic energy is equal to the work done by the force.

Ex. 2. Suppose F to be an *attraction* proportional to the distance of the particle from O.

Let the intensity of the attraction, that is, the force on unit mass at unit distance from O, be μ. If x be positive, that is, if the particle be to the right of O, the force towards O is μmx; if x be negative, that is, if the particle be to the left of O, the force towards O is $m\mu(-x)$. In both cases therefore the force in the direction in which x increases is $-\mu mx$. But the force in the direction in which x increases is always $m\ddot{x}$. Hence
$$m\ddot{x} = -\mu mx, \text{ or } \ddot{x} + \mu x = 0.$$

This equation is called *the differential equation* of the motion of the particle, the word "differential" being used because the equation contains the differential coefficient $\ddot{x}$.

152   AN ELEMENTARY TREATISE ON THE CALCULUS.

The student will easily verify that the equation will be satisfied (see Exercises XIV., 16) by
$$x = A\cos\sqrt{\mu}\,t + B\sin\sqrt{\mu}\,t,$$
where A, B are any constants whatever. The motion becomes definite when in addition to the law of force we are told the position and the velocity of the particle at any one instant. Suppose for example that when $t=0$, $x=a$, $v=0$. Putting $t=0$ and $x=a$ in the equation for x we find $A=a$.

Again v is found by differentiating x with respect to t; therefore
$$v = \dot{x} = -\sqrt{\mu}\,A\sin\sqrt{\mu}\,t + \sqrt{\mu}\,B\cos\sqrt{\mu}\,t.$$
But when $t=0$, $v=0$; therefore we get $0=\sqrt{\mu}B$, that is, $B=0$, and we find that $x=a\cos\sqrt{\mu}\,t$.

Simple Harmonic Motion. When the law of force is that stated in the example the motion is called simple harmonic motion, and the form $x=a\cos\sqrt{\mu}\,t$ is the simplest way of stating the relation between x and t. Obviously the motion is periodic, the *period* being $2\pi/\sqrt{\mu}$ because while t increases from a value t_1 to the value $t_1+2\pi/\sqrt{\mu}$ both x and $\dot{x}$ go through their complete range of values. a is called the *amplitude* of the motion.

The student may show that if $x=c$, $v=V$ when $t=0$, then
$$x = c\cos\sqrt{\mu}\,t + \frac{V}{\sqrt{\mu}}\sin\sqrt{\mu}\,t = a\cos(\sqrt{\mu}\,t - \theta),$$
where
$$a = \sqrt{\left(c^2 + \frac{V^2}{\mu}\right)}, \quad a\cos\theta = c, \quad a\sin\theta = \frac{V}{\sqrt{\mu}}.$$
a is again the amplitude and $2\pi/\sqrt{\mu}$ the period.

Ex. 3. A rod is stretched from its natural length a to the length $a+x$: assuming Hooke's Law to hold, find the work done.

The ratio x/a is called *the extension*, and by Hooke's Law the force required to produce that extension is proportional to it. Denoting this force by F, we have $F = Ex/a$, where E is a constant. When the extension is $(x+\delta x)/a$, the force will be $F+\delta F = E(x+\delta x)/a$. If the work done in producing the extension x/a is W, and if δW is the work done in producing the further extension, then δW will lie between $F\delta x$ and $(F+\delta F)\delta x$, so that $\delta W/\delta x$ will lie between F and $F+\delta F$. Taking the limit for δx converging to zero, we get
$$\frac{dW}{dx} = F = E\frac{x}{a}.$$
Hence
$$W = \tfrac{1}{2}E\frac{x^2}{a} + \text{const.}$$
Since $W=0$ when $x=0$, the constant is zero, so that
$$W = \tfrac{1}{2}E\frac{x^2}{a} = \tfrac{1}{2}E\frac{x}{a}\cdot x = \tfrac{1}{2}Fx.$$

EXAMPLES FROM DYNAMICS.

Ex. 4. A fluid is in communication with a cylinder in which a piston is free to slide, the cross section of the cylinder being S, a constant. Let W be the work done by the fluid in pushing out the piston a distance x, and let the intensity of pressure on the piston be p. Show that $dW/dx = pS$.

The force on the piston due to pressure is pS; when the piston is pushed out the further distance δx, let the intensity of pressure be $p+\delta p$ so that the force on the piston is $(p+\delta p)S$. The work δW done in pushing out the piston through the distance δx will lie between $pS\delta x$ and $(p+\delta p)S\delta x$, and therefore $\delta W/\delta x$ will lie between pS and $pS+\delta pS$. Hence $dW/dx=pS$.

The result may be put in another form. If v be the volume of the fluid, then $S\delta x$ is the increment of volume which may be called δv. Hence $\delta W/\delta v$ lies between p and $p+\delta p$, and we get

$$\frac{dW}{dv}=p.$$

Ex. 5. A body is rotating about an axis; a line fixed in the body and perpendicular to the axis makes at time t an angle θ with another line fixed in space and perpendicular to the axis. What do $\dot\theta$ and $\ddot\theta$ measure?

$\dot\theta$ is the time-rate of increase of θ, that is, $\dot\theta$ is the angular velocity of the body about the axis. In the same way we see that $\ddot\theta$ is the angular acceleration.

If a point P is moving in a plane, and if θ is the angle which the line joining the point P to a fixed point O in the plane makes with a fixed line through O, $\dot\theta$ and $\ddot\theta$ are sometimes called *the angular velocity* and *the angular acceleration* of the *point P* about O.

Ex. 6. A positive charge m of electricity is concentrated at a point O; the repulsion on unit charge at P (Fig. 31) is m/x^2 where $x=OP$. Find the work done as unit charge moves from A to B where $OA=a$, $OB=b$.

Let W be the work done from A to P; then

$$\frac{dW}{dx}=\frac{m}{x^2} \quad \text{and} \quad W=-\frac{m}{x}+\text{const.}$$

When $x=a$, $W=0$, and the constant is therefore m/a. Hence at P the work is

$$W=\frac{m}{a}-\frac{m}{x}.$$

The work in moving from A to B is therefore

$$W_1=\frac{m}{a}-\frac{m}{b}.$$

Potential. When B is so far off that m/b is negligible in comparison with m/a then $W_1=m/a$. Hence in this case the work done as unit charge moves from A out of the field is m/OA. This function m/OA is called the *potential* of the charge m at A.

At P the potential is m/OP. Denoting it by V we have
$$V = \frac{m}{x}; \qquad -\frac{dV}{dx} = \frac{m}{x^2},$$
so that the force at P is the space-rate of *diminution* of the potential V at P, and the direction of the force is from that of higher to that of lower potential.

For *gravitational* forces the *attraction* between two particles, m, m' (grammes) at a distance from each other of x cm. is kmm'/x^2 dynes where k is the constant of gravitation (equal to 6.6×10^{-8}). See Gray's *Treatise on Physics*, § 195. [London: J. & A. Churchill.] The potential V of m at the point x is km/x and the attraction towards m is $-D_x V$; the force outwards from m is $+D_x V$.

It is proved in works on Dynamics (*e.g.* Gray, § 484; see also Exercises XXX., 24) that the potential at the point x of a sphere of radius a and uniform density ρ is

$$V = 2\pi k \rho (a^2 - \tfrac{1}{3} x^2) \text{ for an internal point } (x < a) \ldots \ldots (i)$$

$$V = \frac{4\pi k\rho}{3} \frac{a^3}{x} \text{ for an external point } (x > a) \ldots \ldots \ldots (ii)$$

Since the field is symmetrical the force is radial at every point and the attraction at the point x is therefore

$$-D_x V = \frac{4\pi k\rho}{3} x, \quad (x < a); \quad -D_x V = \frac{4\pi k\rho}{3} \frac{a^3}{x^2} (x > a).$$

The functions V and $D_x V$ have each different analytical expressions according as x is less or greater than a, but they are each continuous functions near $x = a$; for we see from (i) and (ii) that whether x tends to a through values less or through values greater than a, V tends to $4\pi k\rho a^2/3$ and $D_x V$ to $-4\pi k\rho a/3$, and these are the *values* of V and $D_x V$ when $x = a$.

On the other hand the function $D_x^2 V$ is *discontinuous* at a; for when x tends to a through values less than a we find from (i) that $D_x^2 V$ tends to $-4\pi k\rho/3$ and when x tends to a through values greater than a we find from (ii) that $D_x^2 V$ tends to $+8\pi k\rho/3$. The function $D_x^2 V$ has therefore no *value* when $x=a$, but has one definite limit for x approaching a from one side, and another definite limit for x approaching a from the other side (see § 44).

To graph the functions V, D_xV, D_x^2V suppose for simplicity $a=1$, $4\pi k\rho/3=1$; the graphs for other values can be derived in the usual way (Fig. 32).

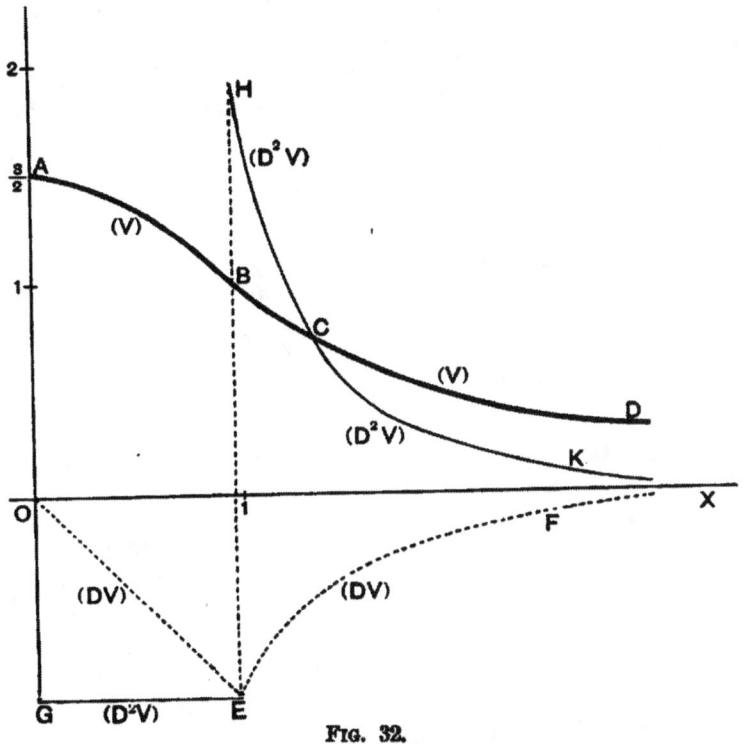

FIG. 32.

$ABCD$ is the graph of V; the part AB is a parabola, the part BCD a rectangular hyperbola.

The dotted curve OEF is the graph of D_xV; the part OE is straight.

The graph of D_x^2V is the straight line GE parallel to OX and the curve HCK.

The parts to the left of the vertical dotted line represent the functions for $x<a$, the parts to the right for $x>a$.

§ **70. Coefficients of Elasticity and Expansion.** Let p be the intensity of pressure and v the volume of unit mass of fluid, p being a definite function of v. When p increases by δp let v increase by δv; if we suppose δp positive then δv will be negative. The quotient $-\delta v/v$, that is, the ratio of the diminution of the volume to the volume at pressure p, is called the *compression* or *the mean compression*, and the limit of the increment of pressure, δp, to the compression produced, $-\delta v/v$, is called *the coefficient of the elasticity of volume*, or simply *the elasticity of volume*, or sometimes *the coefficient of the resilience of volume*. Hence the elasticity of volume is

$$\underset{\delta v=0}{L} -\frac{\delta p}{\dfrac{\delta v}{v}} = -v\frac{dp}{dv}$$

For a gas expanding at constant temperature $pv=k$, a constant, so that the elasticity of volume is

$$-v\frac{d(k/v)}{dv} = -v\cdot\frac{-k}{v^2} = p.$$

For a gas expanding adiabatically $pv^\gamma = c$, a constant, and in this case the elasticity is γp.

A rod whose length at a standard temperature, say at 0°C., is the unit of length expands when heated to a temperature θ so that its length becomes $1+f(\theta)$; denote $1+f(\theta)$ by x, and when the temperature becomes $\theta+\delta\theta$ let the length become $x+\delta x$. The quotient $\delta x/\delta\theta$ is called *the mean coefficient of linear expansion* as the temperature increases from θ to $\theta+\delta\theta$, and $dx/d\theta$ is called *the coefficient of linear expansion at the temperature θ*.

Usually $f(\theta)$ is of the form $a\theta$ or $a\theta+b\theta^2$ where a, b are very small constants. When $x=1+a\theta$, the coefficient $dx/d\theta$ is a and is independent of θ; if $f(\theta)=a\theta+b\theta^2$ and $x=1+a\theta+b\theta^2$, the coefficient is $a+2b\theta$ and depends on θ.

If a solid expand equally in all directions, the area and the volume which are unity at 0°C. would become $y=(1+f(\theta))^2$ and $z=(1+f(\theta))^3$ at temperature θ. The numbers $dy/d\theta$, $dz/d\theta$ are called the coefficients of superficial and of cubical expansion respectively at temperature θ. If $f(\theta) = a\theta$, then

$$y = (1+a\theta)^2; \quad \frac{dy}{d\theta} = 2a + 2a^2\theta;$$

$$z = (1+a\theta)^3; \quad \frac{dz}{d\theta} = 3a + 6a^2\theta + 3a^3\theta^2.$$

Since a is very small a^2 and a^3 will be much smaller and the coefficients will be very approximately $2a$ and $3a$.

Ex. The volume at temperature θ of the water which occupies unit volume at 4° is approximately $1 + a(\theta - 4)^2$ where $a = 8\cdot38 \times 10^{-6}$; find the coefficients of cubical expansion at temperatures 0° and 10°.

§ 71. Conduction of Heat. A slab of thickness d whose opposite faces are parallel planes has one face maintained at constant temperature v and the opposite face at constant temperature v_1 ($v > v_1$); the quantity Q of heat which in time t crosses an area A forming a part of a section parallel to the faces and lying between them is

$$Q = kA(v - v_1)t/d,$$

where k is a constant, called the conductivity, depending on the material of the slab. This equation expresses the law of *steady flow* of heat in a conducting solid and is a result of experiment.

If the temperature v of a solid vary from point to point of the body at the same instant, and from one instant to another at the same point in the body, v will be a function of more than one variable, namely of t and of the coordinates of the point.

At a given point in the solid the time-rate of change of v is

$$\mathop{\mathrm{L}}_{\delta t = 0} \frac{\delta v}{\delta t} = D_t v.$$

In forming this derivative the coordinates of the point do not change; v changes through lapse of time at a given point.

On the other hand, let P be a point in the body whose distance from a fixed plane is $MP = s$, and R a point in MP produced such that $PR = \delta s$; then *at the same instant* the temperature v at P will be different from that at R, which may be denoted by $v+\delta v$. At the time t the space-rate of variation of v at the point P in the direction PR will be

$$\underset{\delta s=0}{\mathrm{L}} \frac{\delta v}{\delta s} = D_s v.$$

Let us assume that at any given time t the temperature is the same at every point in any plane perpendicular to MP though different for different planes. We may assume therefore that the heat flows in straight lines parallel to MP; let v be the temperature at P, $v+\delta v$ the temperature at R where $PR=\delta s$, and let δQ be the amount of heat which in time δt crosses unit area of a plane perpendicular to PR and lying between P and R. The formula given above for Q is assumed to give the *average* value of the amount of heat crossing a section when the flow is not steady, δt and δs being small. In that formula, therefore, put δQ for Q, 1 for A, $v+\delta v$ for v_1, δt for t, δs for d, and we get

$$\delta Q = k\{v - (v+\delta v)\}\delta t/\delta s,$$

and
$$\frac{\delta Q}{\delta t} = -k\frac{\delta v}{\delta s}.$$

Take the limit for δt and δs converging to zero, and we get

$$D_t Q = -k D_s v;$$

in words, the time-rate at which heat crosses the section of unit area at P is k times the space-rate of diminution of temperature in the direction perpendicular to the area.

$D_t Q$ or its equal $-kD_s v$ is called *the flux* in the direction in which s increases; obviously the flux is from places of higher to places of lower temperature, and this is shown by the form $-kD_s v$ since if v decreases as s increases $D_s v$ is negative and $-D_s v$ is positive.

Ex. $v = Ve^{-\frac{kt}{c}} \sin x$ where V, c are constants.

$$D_t Q = -k D_x v = -k V e^{-\frac{kt}{c}} \cos x.$$

When $x=\pi/2$, $D_t Q = 0$ whatever t may be; that is, there is no flow of heat across this plane; when $x<\pi/2$, the flow is towards the left, when $x>\pi/2$ it is towards the right, the positive direction of x being towards the right.

This problem gives an example of a function of more than one variable; such functions will be taken up later.

EXERCISES XV.

1. A point P moves with uniform velocity V along a straight line AB; OA is perpendicular to AB and equal to a. Find the angular velocity of P about O.

2. A point P moves with uniform velocity u along a straight line AB, and another point Q with uniform velocity v along an intersecting straight line AC. Find the rate at which the distance between P and Q increases.

3. If ρ is the density and p the intensity of pressure of the atmosphere at a height of x feet above sea-level, express in symbols the statement that the rate of increase of pressure per unit of length downwards is equal to the density multiplied by the acceleration due to gravity. Assuming that $p = k\rho$ where k is a constant, and that at sea-level $p = p_0$, show that

$$p = p_0 e^{-\frac{gx}{k}}.$$

4. If N be the number of lines of force passing through a circuit, state in words the meaning of $-dN/dt$.

5. Express in symbols the statement that the electromotive force E is the sum of two terms of which the first is the product of the resistance R and the current C, and the second is the product of the self-inductance L and the time-rate of increase of C.

6. Express in symbols the statement that the force X acting on a magnetic shell in the direction x is equal to the space-rate of diminution in that direction of the energy E.

7. If in ex. 4, § 69, W_1 is the work done as the fluid expands from volume v_1 to volume v_2, find W_1 (i) if $pv = k$, (ii) if $pv^\gamma = k$, k being constant.

8. The potential of a long uniform rod of linear density σ at a point P whose distance PC from the rod is x is

$$V = 2k\sigma \log(c/x).$$

Show that the attraction of the rod on a unit particle at P is towards C and equal to $2k\sigma/x$.

9. The potential of a thin circular disc, of surface density σ, at a point P on the normal to the disc through its centre O is

$$V = 2\pi k\sigma \{\sqrt{(a^2 + x^2)} - x\}$$

where a is the radius of the disc and $OP = x$. Show that the attraction on unit mass at P is

$$2\pi k\sigma \left\{1 - \frac{x}{\sqrt{(a^2 + x^2)}}\right\}.$$

Show that if x is small compared with a, the attraction is $2\pi k\sigma$ approximately.

10. The coórdinates of a point at time t are given by
$$x = a\cos(2nt - a), \quad y = b\cos nt.$$
Show that the equation of the path of the point is
$$x = a\left\{\left(\frac{2y^2}{b^2} - 1\right)\cos a + \frac{2y}{b}\sqrt{\left(1 - \frac{y^2}{b^2}\right)}\sin a\right\}.$$

The x-coordinate is a simple harmonic function of amplitude a and period π/n, while the y-coordinate is a simple harmonic function of amplitude b and period $2\pi/n$, double that of the x-coordinate. The motion is therefore said to be compounded of two simple harmonic motions in rectangular directions and of periods in the ratio 1 : 2.

When $a = 0$, the path is a parabola. Figures of the curves for different values of a will be found in Gray's *Physics*, Vol. I., p. 70, and in various other books.

11. Show that two simple harmonic motions of the same period and in the same straight line compound into a simple harmonic motion of the same period and in the same straight line.

12. If in ex. 11 the motions are in rectangular directions, show that the curve compounded of the motions will be an ellipse.

CHAPTER IX.

MEAN VALUE THEOREMS. MAXIMA AND MINIMA. POINTS OF INFLEXION.

§ 72. Rolle's Theorem and the Theorems of Mean Value.
The following theorems are of constant application.

THEOREM I. *If $F(x)$ and $F'(x)$ are continuous as x varies from a to b, and if $F(x)$ is zero when $x=a$ and when $x=b$, then $F'(x)$ will be zero for at least one value of x between a and b.* (Rolle's Theorem.)

In geometrical language, the theorem simply states that at one point at least on the graph of $F(x)$ the tangent is parallel to the x-axis. There may be more points than one; if there are more than one there must be an *odd*

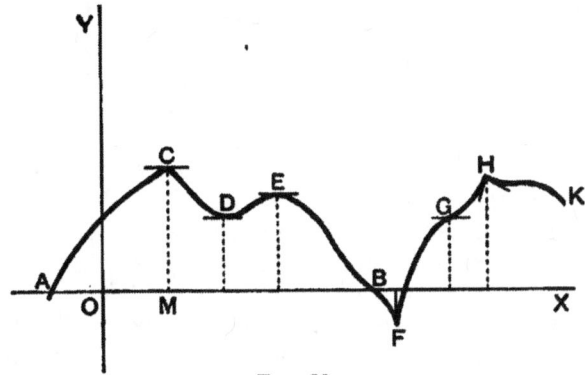

Fig. 33.

number of such points, as C, D, E (Fig. 33). The student should show by a graph that the theorem is not necessarily true if either $F(x)$ or $F'(x)$ becomes discontinuous at a point in the range from a to b.

The theorem is otherwise obvious, because $F(x)$ cannot either always increase or always decrease as x increases from a to b, since $F(a)=0$ and $F(b)=0$. Hence for at least one value of x between a and b, $F(x)$ must cease increasing and begin to decrease, or else cease decreasing and begin to increase; for that value of x, $F'(x)$ will be zero. Obviously a may be either less or greater than b.

THEOREM II. *If $f(x)$ and $f'(x)$ are continuous as x varies from a to b, then there is at least one value of x, x_1 say, between a and b such that*

$$\frac{f(b)-f(a)}{b-a}=f'(x_1) \text{ or } f(b)=f(a)+(b-a)f'(x_1)\ldots\ldots(1)$$

(Theorem of Mean Value).

In Fig. 34 let A be the point $(a, f(a))$, B the point $(b, f(b))$; the gradient of the chord AB is

$$\frac{f(b)-f(a)}{b-a},$$

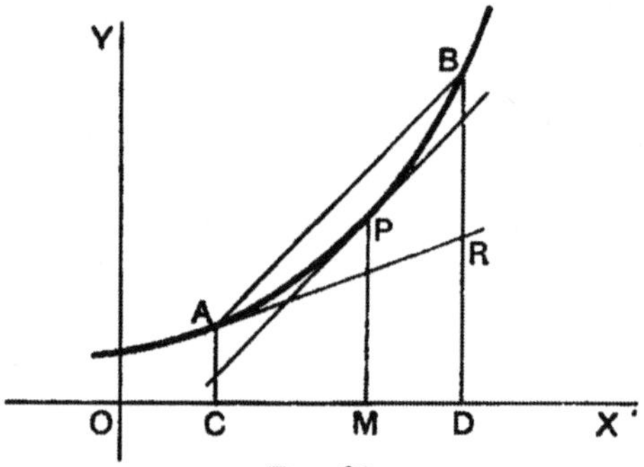

and the theorem simply asserts that there is at least one point, as P, on the graph between A and B such that the tangent at P is parallel to the chord AB. If the abscissa of P is x_1 the gradient at P is $f'(x_1)$ and the equation is established. The student should draw graphs to show that there may be more than one point such as P, and that on the other hand the theorem may not be true if either $f(x)$ or $f'(x)$ becomes discontinuous for a value of x between a and b.

FIG. 34.

The theorem may however be deduced from Th. I., and the method of deduction is important as it leads to the theorem known as Taylor's Theorem, one of the most far reaching in the *Calculus*; indeed the present theorem is only a special case of Taylor's.

Consider the quantity Q defined by the equation

$$\frac{f(b)-f(a)}{b-a}=Q \text{ or } f(b)-f(a)-(b-a)Q=0\ldots\ldots\ldots(2)$$

Let $F(x)$ denote the function
$$f(x)-f(a)-(x-a)Q$$
formed by replacing b by x in the expression
$$f(b)-f(a)-(b-a)Q.$$
By (2) $F(b)$ is zero; also $F(a)$ is zero. Hence the conditions of Th. I. hold for $F(x)$ since $F(x)$, $F'(x)$ are continuous. Therefore $F'(x)$ will be zero for at least one value of x, x_1 say, between a and b. But
$$F'(x)=f'(x)-Q;$$
and therefore $f'(x_1)-Q=0$ or $Q=f'(x_1)$
so that the theorem is established.

THEOREM III. *If $f(x)$, $f'(x)$, $f''(x)$ are continuous as x varies from a to b, then there is at least one value of x, x_2 say, between a and b such that*
$$f(b)=f(a)+(b-a)f'(a)+\tfrac{1}{2}(b-a)^2 f''(x_2).$$

This theorem is an extension of Th. II. To prove it consider the quantity R defined by the equation
$$f(b)-f(a)-(b-a)f'(a)-\tfrac{1}{2}(b-a)^2 R=0. \ldots\ldots\ldots(3)$$
As before, take the function $F(x)$, such that
$$F(x)=f(x)-f(a)-(x-a)f'(a)-\tfrac{1}{2}(x-a)^2 R.$$
Here, $F(a)=0$, $F(b)=0$ (by (3)), and $F(x)$ satisfies the conditions of Th. I. Now,
$$F'(x)=f'(x)-f'(a)-(x-a)R,$$
and therefore for at least one value of x, x_1 say, between a and b $\quad F'(x_1)=f'(x_1)-f'(a)-(x_1-a)R=0.$
Hence $F'(x)$ vanishes when $x=x_1$; obviously it also vanishes when $x=a$; the conditions of Th. I. apply therefore to $F'(x)$ so that *its* derivative must vanish for at least one value of x, x_2 say, between a and x_1, and therefore between a and b. But the derivative of $F'(x)$ is $F''(x)$ and
$$F''(x)=f''(x)-R,$$
and therefore
$$F''(x_2)=f''(x_2)-R=0;\text{ or }R=f''(x_2)$$
and we get
$$f(b)-f(a)-(b-a)f'(a)-\tfrac{1}{2}(b-a)^2 f''(x_2)=0$$
which establishes the theorem.

The theorem has the following geometrical interpretation. If the tangent at A (Fig. 34) meet DB at R, then
$$DR = f(a) + (b-a)f'(a); \quad DB = f(b),$$
and therefore, both in sign and in magnitude,
$$RB = DB - DR = \tfrac{1}{2}(b-a)^2 f''(x_2).$$
Hence the deviation of the curve at B from the tangent at A, that deviation being measured along the ordinate at B, is equal to $\tfrac{1}{2}(b-a)^2 f''(x_2)$.

§ 73. Other Forms of the Theorems of Mean Value. The following forms may be given to Theorems II., III.

If x be any number lying between a and b, then $x-a$ and $b-a$ are of the same sign whether a is less or greater than b; therefore $(x-a)/(b-a)$ is a positive proper fraction, θ say, and we can write $x = a + \theta(b-a)$, so that any number between a and b is of the form $a + \theta(b-a)$ where θ is a positive proper fraction.

Now let $b = a+h$, $b-a = h$; Th. II. will become
$$f(a+h) = f(a) + h f'(a+\theta h) \quad\quad\quad\quad\text{(II.}a\text{)}$$
and Th. III. will become
$$f(a+h) = f(a) + h f'(a) + \tfrac{1}{2}h^2 f''(a+\theta_1 h)\quad\quad\text{(III.}a\text{)}$$
The θ of Th. III. is not necessarily the same as the θ of Th. II. and θ_1 is used for distinction. All that is known of θ is that it is a positive proper fraction; it depends in general both on a and h. In special cases its value may be found. Thus, if $f(x) = x^2$
$$f'(x) = 2x; \quad f'(a+\theta h) = 2(a+\theta h).$$
But $\quad (a+h)^2 = a^2 + 2ah + h^2 = a^2 + h \cdot 2(a + \tfrac{1}{2}h),$

and $\quad (a+h)^2 = f(a) + hf'(a+\theta h) = a^2 + h \cdot 2(a+\theta h),$

so that in this case $\theta = \tfrac{1}{2}$. In Fig. 34 if APB is an arc of a parabola, M is the mid point of CD, and MP bisects the chord AB.

If we replace a by x the above forms become
$$f(x+h) = f(x) + h f'(x+\theta h)\quad\quad\quad\quad\text{(II.}b\text{)}$$
$$f(x+h) = f(x) + h f'(x) + \tfrac{1}{2}h^2 f''(x+\theta_1 h)\quad\quad\text{(III.}b\text{)}$$

OTHER FORMS OF THEOREMS.

If we make a zero and then put x for h we get
$$f(x)=f(0)+x\,f'(\theta x)\ldots\ldots\ldots\ldots\ldots\ldots(\text{II}.c)$$
$$f(x)=f(0)+x\,f'(0)+\tfrac{1}{2}x^2 f''(\theta_1 x)\ldots\ldots(\text{III}.c)$$

Theorem II. affords another proof, though really at bottom it is not different, of Theorem VI. § 58. For if $f'(x)$ is zero for every value of x, then $f'(x_1)$ is zero, and we get $f(b)=f(a)$, that is, any two values $f(a), f(b)$ of $f(x)$ are equal; in other words $f(x)$ is a constant. Hence if $\phi'(x)-F'(x)$ is zero, the function $\phi(x)-F(x)$ is a constant.

Ex. 1. If x is positive, show that $\log(1+x)$ is less than x but greater than $x-\tfrac{1}{2}x^2$.

$$f(x)=\log(1+x);\quad f'(x)=\frac{1}{1+x},\quad f''(x)=-\frac{1}{(1+x)^2};$$

$$f(0)=\log 1 = 0;\quad f'(0)=1,\quad f'(\theta x)=\frac{1}{1+\theta x};$$

$$f''(\theta_1 x)=-\frac{1}{(1+\theta_1 x)^2}$$

By Th. II.c, $\quad \log(1+x)=f(0)+xf'(\theta x)=\dfrac{x}{1+\theta x}<x$.

By Th. III.c, $\quad \log(1+x)=f(0)+xf'(0)+\tfrac{1}{2}x^2 f''(\theta_1 x)$
$$=x-\tfrac{1}{2}\frac{x^2}{(1+\theta_1 x)^2}>x-\tfrac{1}{2}x^2.$$

Ex. 2. Show that $\cos x$ is greater than $1-\tfrac{1}{2}x^2$.

$$f(x)=\cos x;\quad f'(x)=-\sin x;\quad f''(x)=-\cos x;$$
$$f(0)=1;\quad f'(0)=0;\quad f''(\theta_1 x)=-\cos(\theta_1 x).$$

By Th. III.c, $\quad \cos x = 1-\tfrac{1}{2}x^2 \cos(\theta_1 x) > 1-\tfrac{1}{2}x^2$,

since $\cos(\theta_1 x)$ is numerically less than unity. It is easy to deduce that $\cos x = 1-\theta x^2$ where θ is a positive proper fraction less than $\tfrac{1}{2}$.

Ex. 3. The student may try to prove by assuming
$$f(b)-f(a)-(b-a)f'(a)-\tfrac{1}{2}(b-a)^2 f''(a)-\tfrac{1}{6}(b-a)^3 S=0$$
that if $f(x)$ and its first three derivatives are continuous, S will be equal to $f'''(x_3)$, where x_3 lies between a and b. By putting 0 for a and x for b we should get
$$f(x)=f(0)+xf'(0)+\tfrac{1}{2}x^2 f''(0)+\tfrac{1}{6}x^3 f'''(\theta_3 x),$$
where θ_3 is a positive proper fraction.

Ex. 4. By using the theorem of ex. 3, show that if x lies between 0 and $\pi/2$ $\quad x>\sin x>x-\tfrac{1}{6}x^3;\quad \tan x > x+\tfrac{1}{3}x^3$.
How would the inequalities be stated if x lay between $-\pi/2$ and 0?

Ex. 5. If $f(x)=(x-1)^{\frac{2}{3}}-1$, $f(x)$ is zero when $x=0$ and when $x=2$. Does $f'(x)$ vanish for any value of x between 0 and 2?

§ 74. Maxima and Minima. In §§ 17, 52 attention has been called to the turning values of a function; a turning value may be either a maximum or a minimum value of the function. A formal definition of such values may be given.

DEFINITION. $f(a)$ is defined to be a maximum value of $f(x)$ if $f(a)$ is (algebraically) greater both than $f(a-h)$ and than $f(a+h)$ for every positive value of h less than a small but finite positive quantity η. $f(a)$ is defined to be a minimum value of $f(x)$ if $f(a)$ is (algebraically) less both than $f(a-h)$ and than $f(a+h)$ for every positive value of h less than η.

It is to be noticed that a maximum value is not necessarily the *greatest* value the function can have nor a minimum the *least*; $f(a)$ is a maximum if it be greater than any other value of $f(x)$ near $f(a)$ and on either side of it.

The condition for a maximum or a minimum value is easily obtained. If $f(a)$ is a maximum value of $f(x)$, then as x increases from $a-h$ to a, $f(x)$ is increasing, and therefore $f'(x)$ is positive (§ 52); on the other hand as x increases from a to $a+h$, $f(x)$ is decreasing, and therefore $f'(x)$ is negative. Hence as x increases through a, $f'(x)$ must change from a positive to a negative value. Conversely, if as x increases through a, $f'(x)$ changes from a positive to a negative value, $f(a)$ will be a maximum value of $f(x)$.

Hence $f(a)$ will be a maximum value of $f(x)$ if and only if $f'(x)$ changes from a positive to a negative value as x increases through a.

In the same way it will be seen that $f(a)$ will be a minimum value of $f(x)$ if and only if $f'(x)$ changes from a negative to a positive value as x increases through a.

This condition may be called the *fundamental* condition or test.

For ordinary cases a simpler form may be given to the condition. Usually $f'(x)$ will be continuous; now a continuous function can only change sign by passing through the value zero (§ 45, Th. II.). Therefore, if $f(a)$ is a turning value of $f(x)$, $f'(a)$ will be zero.

Again, if $f(a)$ is a maximum value of $f(x)$, $f'(x)$ changes from a positive to a negative value as x increases through a; therefore near a, $f'(x)$ is a decreasing function, and therefore *its* derivative, namely $f''(x)$, must be negative near a. But if $f''(a)$ is not zero, then near a the sign of $f''(x)$ is that of $f''(a)$. Hence $f''(a)$, if it is not zero, will be negative when $f(a)$ is a maximum value of $f(x)$. In the same way we see that $f''(a)$, if it is not zero, will be positive when $f(a)$ is a minimum value of $f(x)$. Conversely, $f(a)$ will be a maximum or a minimum value of $f(x)$ according as $f''(a)$ is negative or positive.

Hence the rule for determining the maxima and minima values of $f(x)$ when $f(x)$, $f'(x)$ are continuous:

The roots of the equation $f'(\mathrm{x})=0$ are, in general, the values of x which make $f(\mathrm{x})$ a maximum or a minimum. Let a be a root of $f'(\mathrm{x})=0$; then $f(\mathrm{a})$ will be a maximum value of $f(\mathrm{x})$ if $f''(\mathrm{a})$ is negative but a minimum if $f''(\mathrm{a})$ is positive.

When $f''(a)$ is zero this rule for testing whether $f(a)$ is a turning value fails; $f'(a)$ may be zero and yet $f(a)$ neither a maximum nor a minimum. When $f'(a)=0$ and also $f''(a)=0$, recourse may be had to the fundamental test that $f'(x)$ must change sign. It will be seen in § 78 that, in general, the point on the graph of $f(x)$ for which both $f'(x)$ and $f''(x)$ are zero is a point of inflexion.

We leave it as an exercise to the student to show that maxima and minima values occur alternately. Thus in Fig. 33, § 72, which is the graph of $F(x)$, the function is a maximum at C, then a minimum at D, then a maximum at E. At F and H on that graph the function turns though $F'(x)$ is not zero at these points; however $F'(x)$ has opposite signs on opposite sides of F and H. Again at G, $F'(x)$ is zero, yet the graph has no turning point there; $F'(x)$ has the *same* sign on opposite sides of G, and G is a point of inflexion.

The above conclusions, when $f(x)$ and its derivatives are continuous at a, may also be deduced from the Theorem of Mean Value. For if $f(a)$ is a turning value of $f(x)$ the differences
$$D_1 = f(a+h) - f(a), \quad D_2 = f(a-h) - f(a),$$
must have the same sign for small values of h: the negative sign

when $f(a)$ is a maximum, but the positive sign when $f(a)$ is a minimum.

Now, by § 73 (III a),
$$D_1 = hf'(a) + \tfrac{1}{2}h^2 f''(a+\theta h) = h\{f'(a) + \tfrac{1}{2}hf''(a+\theta h)\},$$
$$D_2 = -hf'(a) + \tfrac{1}{2}h^2 f''(a-\theta' h) = h\{-f'(a) + \tfrac{1}{2}hf''(a-\theta' h)\}.$$

When h is a very small positive number the signs of D_1 and D_2 will, if $f'(a)$ is not zero, be the same as the signs of $f'(a)$ and $-f'(a)$ respectively (*compare* § 45, Th. I.); therefore D_1 and D_2 cannot have the same sign, and therefore $f(a)$ cannot be a turning value unless $f'(a) = 0$.

Again, if $f''(a)$ is not zero the sign of $f''(a+\theta h)$ and of $f''(a-\theta' h)$ is the same as that of $f''(a)$; therefore if $f'(a) = 0$ both D_1 and D_2 will be negative when $f''(a)$ is negative, but positive when $f''(a)$ is positive. We thus get the same rule as before.

By Taylor's Theorem (chapter XVIII.) D_1 and D_2 can be expressed in a series of ascending powers of h; the same line of argument as that just followed leads to the conclusion that if $f'(a), f''(a), \ldots, f^{(n-1)}(a)$ all vanish, but $f^{(n)}(a)$ does not vanish, then $f(a)$ will be a turning value of $f(x)$, provided that n (the order of the first of the derivatives that is not zero) is an *even* integer, but not a turning value when n is an *odd* integer: the turning value will be a maximum or a minimum according as $f^{(n)}(a)$ is negative or positive. It will be a good exercise to deduce this conclusion by examining the signs of the derivatives near a; for example, show that if $f'(a)$ and $f''(a)$ are zero but $f'''(a)$ not zero, $f''(x)$ *changes sign*, and therefore $f'(x)$ *does not change sign* as x increases through a, but that if $f'''(a)$ is zero and $f^{iv}(a)$ not zero $f'''(x)$ does, $f''(x)$ does not and $f'(x)$ does change sign as x increases through a.

§ 75. Examples.

Ex. 1. Find the turning values of $3x^4 - 4x^3 + 1$.

Denote the function by $f(x)$; then
$$f'(x) = 12x^3 - 12x^2; \quad f''(x) = 36x^2 - 24x.$$

Now $f'(x) = 12x^2(x-1)$, and is therefore zero if $x = 0$ or 1.
$$f''(1) = 36 - 24 = 12 = \text{positive number.}$$

Since $f''(1)$ is positive, $f(1) = 0$ is a minimum value of $f(x)$.

Again $f''(0) = 0$; in this case consider the sign of $f'(x)$ near 0. Let h be a small positive number; then
$$f'(-h) = 12(-h)^2(-h-1) = (+)(-) = -$$
$$f'(+h) = 12(h)^2(h-1) = (+)(-) = -,$$

where only the *sign* of each factor and of the product needs to be written. $f'(x)$ is therefore negative both when x is a little less and when x is a little greater than 0; that is, $f(x)$ decreases as x increases from $-h$ to 0 and continues to decrease as x increases from 0 to h.

Hence $f(0)$ is not a turning value of $f(x)$; on the graph of $f(x)$ there is a point of inflexion where $x=0$.

We may prove otherwise that $f(0)$ is not a turning value; for $f'''(0) = -24$, that is, the first of the derivatives which does not vanish when $x=0$ is of odd order.

As x increases from $-\infty$ to 1, $f'(x)$ is negative, and therefore $f(x)$ is a decreasing function; as x increases from 1 to $+\infty$, $f'(x)$ is positive, and therefore $f(x)$ is an increasing function. Hence $f(1)$ is not only a minimum value of $f(x)$, but it is also the least value $f(x)$ can take for any value of x; $f(x)$ is not negative for any value of x. The student should graph the function.

Ex. 2. Given the total surface, $2\pi a^2$, of a right circular cylinder, find the cylinder of maximum volume.

Denote the radius of the base by x, and the height by y; then

$$\text{volume} = \pi x^2 y; \quad \text{surface} = 2\pi xy + 2\pi x^2 = 2\pi a^2.$$

From the second equation $xy = a^2 - x^2$; therefore denoting the volume by $f(x)$, we get

$$f(x) = \pi x \cdot xy = \pi(a^2 x - x^3).$$

Therefore

$$f'(x) = \pi(a^2 - 3x^2); \quad f''(x) = -6\pi x; \quad f'(x) = 0 \text{ if } x = \pm a/\sqrt{3};$$

the negative root may be discarded as irrelevant. Now $f''(a/\sqrt{3})$ is negative, and therefore $f(a/\sqrt{3})$ is a maximum; the maximum volume is $2\pi a^3/3\sqrt{3}$.

The height is given by $y = (a^2 - x^2)/x$, and when $x = a/\sqrt{3}$, $y = 2a/\sqrt{3}$, so that the height of the cylinder of maximum volume is equal to the diameter of its base.

The student should observe how the given condition enables us to express $\pi x^2 y$ as a function of the one argument x.

Ex. 3. If $r = a\cos^2\theta + b\sin^2\theta$, find the maximum and minimum values of r, where a, b are positive constants.

Examples of this type are most simply solved without the use of derivatives. Thus,

$$r = \tfrac{1}{2}a(1 + \cos 2\theta) + \tfrac{1}{2}b(1 - \cos 2\theta) = \tfrac{1}{2}(a+b) + \tfrac{1}{2}(a-b)\cos 2\theta.$$

Now obviously r will be a maximum or a minimum according as $\tfrac{1}{2}(a-b)\cos 2\theta$ is so. If $a > b$, the greatest and least values of $\tfrac{1}{2}(a-b)\cos 2\theta$ are $\tfrac{1}{2}(a-b)$ and $-\tfrac{1}{2}(a-b)$, so that the greatest and least values of r are a and b. These values are reversed if $a < b$, since in that case the greatest and least values of $\tfrac{1}{2}(a-b)\cos 2\theta$ are $-\tfrac{1}{2}(a-b)$ and $\tfrac{1}{2}(a-b)$.

In a similar way we can find the maximum and minimum values of $x^2 + y^2$ when x and y are connected by the equation $ax^2 + 2hxy + by^2 = 1$. For put $x = r\cos\theta$, $y = r\sin\theta$, and then $x^2 + y^2$ becomes r^2, where

$$r^2(a\cos^2\theta + 2h\sin\theta\cos\theta + b\sin^2\theta) = 1.$$

Now r^2 will be a maximum or minimum according as $1/r^2$ is a minimum or maximum, and we may write, from the equation between r and θ,

$$1/r^2 = a\cos^2\theta + 2h\sin\theta\cos\theta + b\sin^2\theta$$
$$= \tfrac{1}{2}(a+b) + \tfrac{1}{2}(a-b)\cos 2\theta + h\sin 2\theta$$
$$= \tfrac{1}{2}(a+b) + R\cos(2\theta - \theta').$$

where $R\cos\theta' = \tfrac{1}{2}(a-b)$, $R\sin\theta' = h$, and $R = +\sqrt{\left\{\dfrac{(a-b)^2}{4} + h^2\right\}}$.

The maximum and minimum values of $1/r^2$ are given by

$$1/r^2 = \tfrac{1}{2}(a+b) \pm \tfrac{1}{2}\sqrt{\{(a-b)^2 + 4h^2\}}.$$

Geometrically, this example is the problem of finding the semi-axes of the conic whose equation is $ax^2 + 2hxy + by^2 = 1$. The values of θ that give the axes are determined by

$$\cos(2\theta - \theta') = \pm 1, \quad 2\theta = \theta' \text{ or } \pi + \theta', \quad \theta = \tfrac{1}{2}\theta' \text{ or } \tfrac{1}{2}\pi + \tfrac{1}{2}\theta',$$

so that the two axes are at right angles. The value of θ' is uniquely determined by the two equations $R\cos\theta' = \tfrac{1}{2}(a-b)$ and $R\sin\theta' = h$.

The solution of problems of this kind by use of derivatives is much more tedious.

Ex. 4. If $f(x) = e^{-ax}\sin(bx+c)$ where a, b are positive, find the turning values of $f(x)$.

$$f'(x) = -e^{-ax}\{a\sin(bx+c) - b\cos(bx+c)\}$$
$$= -Re^{-ax}\sin(bx+c-\theta),$$

where $R\cos\theta = a$, $R\sin\theta = b$, $R = +\sqrt{(a^2+b^2)}$,

$$f''(x) = R^2 e^{-ax}\sin(bx+c-2\theta).$$

Since e^{-ax} is not zero for any finite value of x, the roots of $f'(x) = 0$ are those of $\sin(bx+c-\theta) = 0$; therefore $f'(x)$ is zero when

$$bx + c - \theta = n\pi \quad (n = 0, \pm 1, \pm 2, \ldots).$$

Denoting by x_n the value of x corresponding to any n, we have

$$f''(x_n) = R^2 e^{-ax_n}\sin(bx_n + c - \theta - \theta) = R^2 e^{-ax_n}\sin(n\pi - \theta).$$

Now $\sin(n\pi - \theta) = -\cos n\pi \sin\theta$; and $\sin\theta$ and $R^2 e^{-ax_n}$ are positive, so that the sign of $f''(x_n)$ is the same as that of $-\cos n\pi$, that is, of $(-1)^{n+1}$.

Hence $f(x)$ is a maximum for $n = 0, 2, 4 \ldots$, but a minimum for $n = 1, 3, 5 \ldots$, limiting consideration to zero and positive values of n.

Now $$x_n = \frac{n\pi - c + \theta}{b}, \quad x_{n+1} - x_n = \frac{\pi}{b},$$

and $$f(x_n) = e^{-ax_n}\sin(bx_n + c) = e^{-ax_n}\sin(n\pi + \theta),$$

which may be put in the form

$$f(x_n) = (-1)^n e^{-\frac{n\pi a}{b}} \cdot e^{\frac{ac - a\theta}{b}} \cdot \sin\theta.$$

Thus the values of x for which $f(x)$ turns form an arithmetic progression with common difference π/b; the values of x for which $f(x)$ is a maximum (or a minimum) have the common difference $2\pi/b$. If e^{-ax_n} be called the amplitude of $f(x_n)$, the amplitudes of the maxima

WORKED EXAMPLES. 171

and minima values of $f(x)$ form a geometric progression with common ratio $e^{-\frac{2\pi a}{b}}$.

Since $D_x e^{-ax} = -ae^{-ax}$, the gradient of e^{-ax} is equal to that of $f(x)$ for those values of x, for which
$$-ae^{-ax} = -Re^{-ax}\sin(bx+c-\theta),$$
that is, for which
$$\sin(bx+c-\theta) = a/R = \sin\left(\frac{\pi}{2}+\theta\right);$$
that is, for which
$$bx+c-\theta = 2m\pi + \frac{\pi}{2} + \theta \text{ or } (2m+1)\pi - \frac{\pi}{2} - \theta \ (m=0, 1, 2 \ldots).$$

Now when $bx+c-\theta = (2m+1)\pi - \frac{\pi}{2} - \theta$, $\sin(bx+c) = 1$, and therefore for these values of x, $e^{-ax} = f(x)$.

Therefore when $bx+c = 2m\pi + \pi/2$, e^{-ax} and $f(x)$ have the same value and the same gradient, and therefore their graphs touch at the points whose abscissae are given by these values of x.

The discussion of $e^{-ax}\cos(bx+c')$ can be reduced to that of $e^{-ax}\sin(bx+c)$ by putting c' equal to $c-\frac{\pi}{2}$.

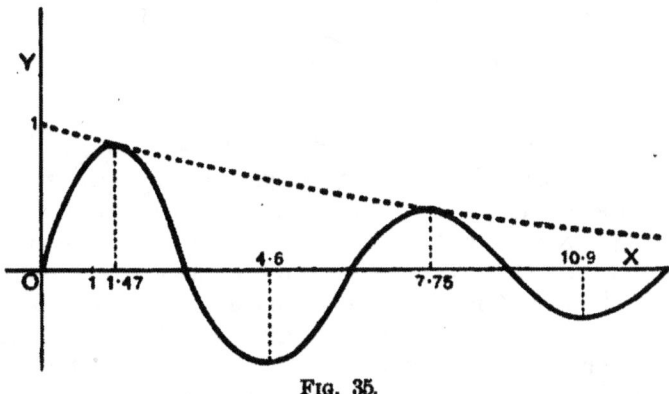

FIG. 35.

Fig. 35 shows the graph for $a = \cdot 1$, $b = 1$, $c = 0$. The dotted line is the graph of $e^{-\frac{1}{10}x}$.

§ 76. **Elementary Methods.** Certain types of problems can be solved very simply by elementary algebra or trigonometry.

The discussion of the quadratic function or the quotient of two quadratic functions will be found in any book on algebra; the turning values of y where
$$y = (ax^2 + bx + c)/(Ax^2 + Bx + C)$$

are found by writing the equation in the form
$$(Ay-a) x^2 + (By-b) x + Cy-c = 0,$$
and determining the values of y that make the discriminant, that is,
$$(By-b)^2 - 4(Ay-a)(Cy-c)$$
vanish. A little consideration distinguishes the maximum from the minimum if there are two values of y, and shows whether the solution is a maximum or a minimum when there is only one.

A more general case occurs when there are more variables than one and these are connected by a relation, all the quantities being positive. For two variables the 5th, 8th, and 9th propositions of Euclid's second book or their algebraic equivalents are fundamental.

(i) $xy = \left(\dfrac{x+y}{2}\right)^2 - \left(\dfrac{x-y}{2}\right)^2;$

(ii) $(x+y)^2 = 4xy + (x-y)^2;$

(iii) $x^2 + y^2 = \tfrac{1}{2}(x+y)^2 + \tfrac{1}{2}(x-y)^2.$

When the sum $(x+y)$ of two quantities is given, we see by (i) that their product is greatest and by (iii) that the sum of their squares is least when the two quantities are equal. When the product xy of two quantities is given we see by (ii) that their sum is least when they are equal.

These theorems may easily be extended. Thus let x, y, z, w ... be n positive quantities and let their sum (a) be given; then their product $xyzw$ ... will be greatest when they are all equal. For let $x_1, y_1, z_1, w_1, ...$ be a set of simultaneous values of these variables; then if any two of these, say x_1, y_1, are unequal it will be possible, without altering the sum of the n quantities, to get a greater product than $x_1 y_1 z_1 w_1 ...$ by replacing both x_1 and y_1 by $\tfrac{1}{2}(x_1+y_1)$ and leaving $z_1, w_1, ...$ unaltered, because the product of the two equal quantities $\tfrac{1}{2}(x_1+y_1)$, that is $\tfrac{1}{4}(x_1+y_1)^2$, is greater than $x_1 y_1$. So long, therefore, as any two are unequal the product has not reached its greatest value. When they are all equal each is equal to a/n, so that

$xyzw\ldots$ is less than $\left(\dfrac{a}{n}\right)^n$,

or $\quad xyzw\ldots$ is less than $\left(\dfrac{x+y+z+w\ldots}{n}\right)^n$,

unless $x=y=z=w\ldots=a/n$.

If we suppose p of the quantities equal to x, q of them equal to y, r of them equal to z, where $p+q+r=n$ then we may write the last inequality

$$x^p y^q z^r < \left(\dfrac{px+qy+rz}{p+q+r}\right)^{p+q+r}$$

except when $x=y=z$, and then the inequality becomes an equality. It is easy to see that this inequality is true even if p, q, r are positive fractions.

In the same way it may be seen that when the sum of the quantities is given the sum of their squares will be least when they are all equal, and when the product of the quantities is given their sum will be least when they are all equal.

These theorems may be again extended. For suppose $x, y, z, w\ldots$ connected by the linear equation,

$$ax+by+cz+dw+\ldots=k, \text{ a constant,}$$

the quantities being all positive. Then we may put

$$xyzw\ldots=\dfrac{(ax)(by)(cz)(dw)\ldots}{abcd\ldots}$$

and $xyzw\ldots$ will be greatest when the numerator of the fraction is greatest. But if we put x' for ax, y' for $by\ldots$ we reduce the case to that in which $x'+y'+z'+w'+\ldots$ is given. Hence the product is greatest when $x', y', z', w'\ldots$ are all equal, that is when

$$ax=by=\ldots=k/n.$$

By means of the above theorems a large number of the simpler problems of maxima and minima of functions of more than one variable may be solved. For a full discussion of the algebraic treatment, see Chrystal's *Algebra*, Vol. II. chap. xxiv.

Ex. 1. The equilateral triangle has maximum area for given perimeter.

In the usual notation for triangles,
$$\triangle = \sqrt{\{s(s-a)(s-b)(s-c)\}} = \sqrt{(sxyz)},$$
where $x = s-a$, $y = s-b$, $z = s-c$, $2s$ being constant.

Now $\qquad x + y + z = 3s - (a+b+c) = s.$

Hence xyz, and therefore $sxyz$, and therefore $\triangle$ is greatest when $x = y = z$, that is, when $a = b = c$.

Ex. 2. From the identity
$$\left(a^2x^2 + b^2y^2 + c^2z^2\right)\left(\frac{l^2}{a^2} + \frac{m^2}{b^2} + \frac{n^2}{c^2}\right)$$
$$= \left(lx + my + nz\right)^2 + \left(\frac{mcz}{b} - \frac{nby}{c}\right)^2 + \left(\frac{nax}{c} - \frac{lcz}{a}\right)^2 + \left(\frac{lby}{a} - \frac{max}{b}\right)^2,$$
deduce

(i) $a^2x^2 + b^2y^2 + c^2z^2 = $ minimum, if $lx + my + nz = $ const.,

(ii) $lv + my + nz = $ maximum, if $a^2x^2 + b^2y^2 + c^2z^2 = $ const.,

when $a^2x/l = b^2y/m = c^2z/n$.

These results are obvious. We might write A, B, C instead of a^2, b^2, c^2, but A, B, C must be positive. The student may prove a similar theorem for $\left(a^2x^2 + b^2y^2 + c^2z^2 + d^2w^2\right)\left(\frac{l^2}{a^2} + \frac{m^2}{b^2} + \frac{n^2}{c^2} + \frac{p^2}{d^2}\right)$, and extend to any number of variables.

§ 77. Variation near a Turning Value. When a function $f(x)$ and its first two derivatives are continuous near a we have
$$f(a+h) = f(a) + hf'(a) + \tfrac{1}{2}h^2 f''(a + \theta h),$$
$f''(a + \theta h)$ will be nearly equal to $f''(a)$ when h is small, and we may write as a very approximate equation
$$f(a+h) = f(a) + hf'(a) + \tfrac{1}{2}h^2 f''(a).$$

Hence when $f(a)$ is a turning value, so that $f'(a) = 0$, we have
$$f(a+h) = f(a) + \tfrac{1}{2}h^2 f''(a).$$

Thus when $f(a)$ is a turning value the change $f(a+h) - f(a)$ as x changes from a to $a+h$ is, approximately, proportional to h^2; if $f(a)$ is not a turning value the change is, approximately, proportional to h. Near a turning value therefore a function changes much more slowly than near a value for which it does not turn, since when h is small h^2 is much smaller than h.

If therefore in a physical application of the theory of maxima and minima it is not possible to make the arrangement that which corresponds to the exact solution, there will frequently be no great disadvantage in a slight departure from the theoretically best arrangement. Thus when a battery of mn cells is joined up so that m rows of n cells each, connected in series, are joined in parallel, the current γ is

$$\gamma = \frac{mnE}{mR+nr},$$

where E is the electromotive force of each cell, r the internal resistance of each cell and R the external resistance. Since mn is constant γ will be a maximum when $mR+nr$ is a minimum, that is, when $mR=nr$ or when $R=nr/m$, that is, when the total external resistance is equal to the internal resistance of the battery. It may not be possible to join up the battery so as exactly to satisfy this condition; but if the condition be very nearly satisfied the current will not fall far short of the maximum. In any case the nearer the arrangement can be brought to satisfy the condition the stronger will be the current.

Again in applying the theory of maxima and minima to physical problems great care is necessary in drawing conclusions; an arrangement that best satisfies one set of conditions may conflict with that which best satisfies another set of equally important conditions. Thus the above arrangement of cells gives the highest rate of working in the external part of the circuit but it is not the most economical. The student may with advantage read the remarks on pp. 85, 86 and chap. ix. of Gray's *Absolute Measurements in Electricity and Magnetism*. (London: Macmillan.) The theory of maxima and minima is of great value as a guide in all such investigations, but has to be applied with caution and not blindly.

EXERCISES XVI. a.

The cones and cylinders referred to in the examples are right circular cones and cylinders. For the mensuration of various solids see chap. iv.

Investigate the maxima and minima values of the functions in examples 1-12.

1. $2x^3 - 3x^2 - 12x + 2$. 2. $x^5 - 5x^4 + 5x^3 - 1$. 3. $x^4(x+1)^3$.
4. $x(a+x)^2(a-x)^3$. 5. $x/(1+x^2)$. 6. $(1+x)^2/(x-x^2)$.
7. $x/(ax^2 + 2bx + a)$. 8. $x(x^2+1)/(x^4 - x^2 + 1)$. 9. $(a+x)\sqrt{(a^2-x^2)}$.
10. $x/(1+x^2)^{\frac{5}{2}}$. 11. $a + b(x-c)^{\frac{2}{3}}$. 12. $a + b(x+c)^{\frac{5}{3}}$.

13. Find the maximum value of $x^m y^n$ if $x+y=k$, a constant, the quantities being all positive. Hence show that
$$a^m b^n < \left(\frac{ma + nb}{m+n}\right)^{m+n}$$
except when $a = b$.

14. From the inequality in example 13 deduce that $(1 + 1/z)^z$ constantly increases when z is positive and increases, but decreases when z is negative and increases numerically. Hence show that the limit of $(1 + 1/z)^z$ for $z = \pm \infty$ is a finite number greater than 2·5 but less than 3. $\left[\text{Put } a = 1 + \frac{1}{m}, b = 1; \text{ then } a = 1, b = 1 - \frac{1}{n}.\right]$

15. If $a/x + b/y = c$, find the least value of $ax + by$, the quantities being all positive. Find also the minimum value of xy.

16. For what value of x is
$$m_1(x - x_1)^2 + m_2(x - x_2)^2 + \ldots + m_n(x - x_n)^2$$
a minimum, $m_1, m_2, \ldots m_n$ being all positive.

In the following examples the methods of § 76 may be used; the quantities are understood to be all positive.

17. The equilateral triangle has minimum perimeter for given area.

18. The cube is the rectangular parallelepiped of maximum volume for given surface and of minimum surface for given volume.

19. Find the minimum value of $bcx + cay + abz$ if $xyz = abc$.

20. Find the maximum value of xyz when
$$x^2/a^2 + y^2/b^2 + z^2/c^2 = 1,$$
and the minimum value of $x^2/a^2 + y^2/b^2 + z^2/c^2$ when $xyz = d^3$.

21. If $xyz = a^2(x+y+z)$, then $yz + zx + xy$ is a minimum when $x = y = z = a\sqrt{3}$.

22. The electric time-constant of a cylindric coil of wire is approximately $u = mxyz/(ax + by + cz)$ where x is the mean radius, y the difference between the internal and external radii, z the axial

length, and m, a, b, c known constants; the volume of the coil is $nxyz$. Show that when the volume V is given, u will be a maximum when $ax = by = cz = \sqrt[3]{(abcV/n)}$.

23. If $u = (ax^2 + by^2)/\sqrt{(a^2x^2 + b^2y^2)}$ where $x^2 + y^2 = 1$ show that the minimum value of u is $2\sqrt{(ab)}/(a+b)$.

24. If P is a point within a triangle ABC such that $AP^2 + BP^2 + CP^2$ is a minimum, show that P is the centroid.

25. In any triangle the maximum value of $\cos A \cos B \cos C$ is $\frac{1}{8}$.

26. Find the greatest rectangle that can be inscribed in an ellipse whose semi-axes are a, b.

EXERCISES XVI. b.

1. $ABCD$ is a rectangle, and APQ meets BC in P and DC produced in Q. Find the position of APQ when the sum of the areas ABP, PCQ is a minimum.

2. Given one of the two parallel sides (a) and the two non-parallel sides (b) of an isosceles trapezium, find the length of the fourth side so that the area of the trapezium may be a maximum.

3. From a rectangular sheet of tin, the sides being a and b, equal squares are cut off at each corner, and a box with open top formed by turning up the sides. Find the side of the square so that the box may have maximum content.

4. An open tank is to be constructed with a square base and vertical sides to hold a given quantity of water; show that the expense of lining the tank with lead will be least if the depth be half the width.

If the tank be cylindrical show that the depth will be equal to the radius of the base. If the section of the cylinder is not circular but if its shape is given show that the curved surface will be twice the base.

5. Show that the altitude of the cone of maximum volume that can be inscribed in a sphere of radius R is $4R/3$.

Show that the curved surface of the cone is a maximum for the same value of R.

6. A cone is circumscribed about a sphere of radius R; show that when the volume of the cone is a minimum its altitude is $4R$ and its semi-vertical angle $\sin^{-1}\frac{1}{3}$.

7. Show that the altitude of the cylinder of maximum volume that can be inscribed in a sphere of radius R is $2R/\sqrt{3}$.

Show that when the curved surface is a maximum the altitude is $R\sqrt{2}$.

Show that when the whole surface is a maximum that surface is to the surface of the sphere in the ratio of $\sqrt{5}+1$ to 4.

8. A cylinder is inscribed in a cone; show that its volume is a maximum when its altitude is one-third that of the cone.

Show that the curved surface is a maximum when the altitude is half that of the cone. Show also that the total surface cannot have a maximum unless the semi-vertical angle of the cone is less than $\tan^{-1}\frac{1}{2}$.

9. Given the total surface of a cone show that when the volume of the cone is a maximum the semi-vertical angle will be $\sin^{-1}\frac{1}{3}$.

Given the volume of the cone show that the total surface will be a minimum for the same value of the semi-vertical angle.

10. PP' is a double ordinate of the ellipse whose equation is $x^2/a^2 + y^2/b^2 = 1$ and A is one end of the major axis; find when the triangle APP' has maximum area.

Find also when the cone formed by the revolution of the triangle about the major axis has maximum volume.

11. The strength of a rectangular beam varies as the product of the breadth and the square of the depth; find the breadth and the depth of the strongest rectangular beam that can be cut from a cylindrical log, the diameter of the cross-section being d inches.

12. The stiffness of a rectangular beam varies as the product of the breadth and the cube of the depth; find the breadth and the depth of the stiffest rectangular beam that can be cut from a cylindrical log, the diameter of the cross-section being d inches.

13. A person in a boat a miles from the nearest point A of the beach wishes to reach in the shortest time a place b miles from A along the beach; if he can row at u miles an hour and walk at v miles an hour ($u < v$) find how far from A he must land. Consider the cases in which the ratio of u to v is equal to or greater than that of b to $\sqrt{(a^2 + b^2)}$.

14. Assuming that the brightness of a small surface A varies inversely as the square of the distance r from the source of light and directly as the cosine of the angle between r and the normal to the surface at A, find at what height above the centre of a circle of radius a an electric light should be placed so that the brightness at the circumference should be greatest.

15. If the intensities of two sources A, B of light be a^3, b^3 respectively find the point on the line AB at which the brightness is least.

16. A, B are two points on opposite sides of a plane L and P a point in the plane; a particle travels from A to B by the path AP, PB its velocity along AP being constant (u) and its velocity along PB also constant (v) but the two velocities being different. Show that when the time of travelling from A to B is a minimum the plane through APB is normal to L and the sines of the angles that AP, PB make with the normal to L at P are in the ratio of u to v.

EXERCISES XVI. c.

1. Show that $x/(1+x\tan x)$ will be a maximum when $x=\cos x$; verify that x is very nearly ·739.

2. Show that $\sin x \sin 2x$ is a maximum or a minimum when $\sin x = \sqrt{(2/3)}$ according as the angle x is in the first or the second quadrant.

3. Show that $\sin x(1+\cos x)$ is a maximum when $x=\dfrac{\pi}{3}$.

4. Show that the minimum value of $\dfrac{a^2}{\sin^2 x}+\dfrac{b^2}{\cos^2 x}$ is $(a+b)^2$.

5. If $a\sec\theta + b\sec\phi = c$, show that $a\cos\theta + b\cos\phi$ is a minimum when $\theta = \phi$, a, b, c being positive and the angles θ, ϕ acute. (Compare ex. 15a.)

6. Given the length (a) of an arc of a circle, show that the segment of which a is the arc will be a maximum when a is the diameter of the circle.

7. A circular sector has a given perimeter; show that when the area is a maximum the arc is double the radius, and that the maximum area is equal to the square on the radius.

8. From a given circular sheet of metal it is required to cut out a sector so that the remainder can be formed into a conical vessel of maximum capacity; show that the angle removed must be $2(1-\tfrac{1}{3}\sqrt{6})\pi$ radians (66° 4′).

9. Draw a line through the vertex of a given triangle such that the sum of the projections upon it of the two sides which meet at that vertex may be a maximum.

10. The lower corner of a leaf, whose width is a, is folded over so as just to reach the inner edge of the page; find the width of the part folded over when the length of the crease is a minimum.

11. A ship sails from a given place A in a given direction AB at the same time that a boat sails from a given place C; supposing the speed of the ship to be u and that of the boat v (u, v constant), find in what direction the boat must sail so as to meet the ship. Discuss the condition that it shall be possible to meet. The course of the boat is understood to be rectilinear.

12. The distance between the centres of two spheres of radii a, b respectively is c; find at what point P on the line of centres AB the greatest amount of spherical surface is visible. *Note.*—The superficial area of a segment of height h is $2\pi a h$, a being the radius of the sphere (§ 85, ex. 2).

13. A straight line is drawn through a fixed point (a, b), meeting the axis OX at P and the axis OY at Q, the axes being rectangular and a, b positive; if the angle OPQ is equal to θ, find θ,

(i) when PQ, (ii) when $OP+OQ$, (iii) when $OP \cdot OQ$ is a minimum,

14. A tangent is drawn to an ellipse, whose axes are $2a$, $2b$, such that the part intercepted between the axes is a minimum; show that its length is $a+b$.

15. If $D=\phi-\phi'$ and $\sin\phi=\mu\sin\phi'$ where μ is greater than 1, and ϕ, ϕ' not greater than $\pi/2$, show that D increases as ϕ increases. Show also that the second and third derivatives of D with respect to ϕ are positive.

16. A ray of light travels in a plane perpendicular to the edge of a prism of angle i; if the angle of incidence is ϕ and the angle of emergence ϕ', show that the deviation $\phi+\phi'-i$ is a minimum when $\phi=\phi'$.

17. Find the maximum value of xe^{-x} and graph the function.

18. Find the minimum value of $x\log x$ and graph the function.

19. For what value of x is the ratio of $\log x$ to x greatest?

20. Find the maximum value of $x^2\log\dfrac{1}{x}$.

21. If a, b are positive and $a<b$, find the maximum value of $e^{-ax}-e^{-bx}$.

§ 78. Concavity and Convexity. Points of Inflexion.

A curve is said to be *concave upwards* at or near A when (Fig. 36, a, b) at *all* points near A it lies *above* the tangent at A; a curve is said to be *convex upwards* at or near A when (Fig. 36, c, d) at all points near A it lies *below* the tangent at A.

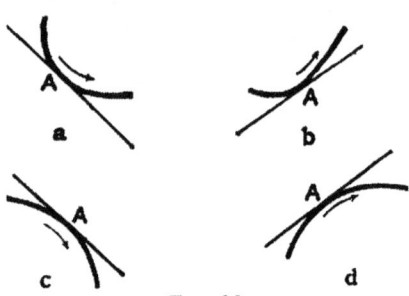

Fig. 36.

Let $y=f(x)$ be the equation of the curve and let h be a small positive number, a the abscissa of A. Then as x increases from $a-h$ to $a+h$, the gradient $f'(x)$ in the cases a, b steadily increases; as the graphic point moves to the right (the direction of the arrows) the tangent turns about its point of contact counter-clockwise, and therefore the angle it makes with the x-axis increases. But if $f'(x)$ is an increasing function its derivative $f''(x)$ must be positive; if $f''(a)$ is not zero then near a $f''(x)$ has the same sign as $f''(a)$. Hence the curve is concave upwards near A if $f''(a)$ is positive (not zero).

CONCAVITY AND CONVEXITY.

In the same way we see that the curve will be convex upwards near A if $f''(a)$ is negative (not zero).

Again, if A be a point of inflexion (§ 23) the gradient either increases as x increases from $a-h$ to a and then decreases as x increases from a to $a+h$ (Fig. 37, a) or else it decreases as x increases from $a-h$ to a and then increases as x increases from a to $a+h$ (Fig. 37, b). In both cases therefore $f'(x)$ turns for the value a of x. Hence

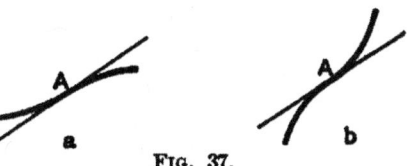

FIG. 37.

A will be a point of inflexion if and only if $f'(a)$ is a turning value of $f'(x)$; therefore, if $f'(x)$ and $f''(x)$ are continuous, $f''(a)$ must be zero in order that A may be a point of inflexion.

Conversely, if $f''(a)$ is zero A will, in general, be a point of inflexion; but, to make certain, the test that $f'(a)$ is a turning value of $f'(x)$ should be applied.

Ex. 1. $f(x) = 3x^4 - 4x^3 + 1$;

$f'(x) = 12(x^3 - x^2)$; $f''(x) = 12(3x^2 - 2x)$; $f'''(x) = 24(3x - 1)$;

$f''(x) = 0$, if $x = 0$ or $\frac{2}{3}$; $f'''(0) = -24$, $f'''(\frac{2}{3}) = 24$.

Therefore $f'(0)$, $f'(\frac{2}{3})$ are maximum and minimum values of $f'(x)$, and therefore the points $(0, 1)$, $(\frac{2}{3}, \frac{11}{27})$ are points of inflexion. The gradients at these points are 0 and $-16/9$ respectively.

Since $f''(x) = 36x(x - \frac{2}{3})$, we see that from $x = -\infty$ to $x = 0$, $f''(x)$ is positive, and therefore the graph is concave upwards for that range of x; from $x = 0$ to $x = 2/3$, $f''(x)$ is negative and the graph convex upwards; from $x = 2/3$ to $x = \infty$, $f''(x)$ is positive and the graph again concave upwards.

Ex. 2. $f(x) = 3x^4 - 8x^3 - 6x^2 + 1$;

$f''(x) = 12(3x^2 - 4x - 1) = 36\left(x - \frac{2-\sqrt{7}}{3}\right)\left(x - \frac{2+\sqrt{7}}{3}\right)$;

$f'''(x) = 24(3x - 2)$.

There are points of inflexion where $x = \frac{1}{3}(2 \pm \sqrt{7})$. From $x = -\infty$ to $x = \frac{1}{3}(2 - \sqrt{7})$, and again from $x = \frac{1}{3}(2 + \sqrt{7})$ to $x = +\infty$, the graph is concave upwards; from $x = \frac{1}{3}(2 - \sqrt{7})$ to $x = \frac{1}{3}(2 + \sqrt{7})$ it is convex upwards.

Ex. 3. $(y - 2)^3 = (x - 4)$.

$y = 2 + (x - 4)^{\frac{1}{3}}$; $y' = \frac{1}{3}(x - 4)^{-\frac{2}{3}}$; $y'' = -\frac{2}{9}(x - 4)^{-\frac{5}{3}}$.

When $x=4$, $y=2$, but y' and y'' are both infinite. When $x=4-h$, y'' is positive, but when $x=4+h$, y'' is negative. We may conclude therefore that the tangent at (4, 2) is perpendicular to the x-axis, that to the left of (4, 2) the curve is concave upwards, and that to the right of (4, 2) it is convex upwards. The point (4, 2) therefore must be considered a point of inflexion.

EXERCISES XVII.

1. Determine the points of inflexion of the graphs of the following functions, and state for what range of values of x they are concave upwards or convex upwards.

(i) x^3; (ii) x^4; (iii) x^5; (iv) x^{2n+1}; (v) x^{2n};

n being a positive integer.

2. Find the points of inflexion on the curve whose equation is $y=(x^2-1)^2$. Graph the curve.

3. Find the points of inflexion and graph the functions

(i) $\dfrac{1}{a^2+x^2}$; (ii) $\dfrac{x}{a^2+x^2}$; (iii) $\dfrac{x^2}{a^2+x^2}$; (iv) $\dfrac{x^3}{a^2+x^2}$.

4. Show that the curve whose equation is $y(x^2+a^2)=a^2(a-x)$ has three points of inflexion which lie on a straight line.

5. Find the points of inflexion on the curve whose equation is $a^2y^2=x^2(a^2-x^2)$, and trace the curve.

6. Show that the curve whose equation is $(a-x)y^2=x^3$ has no point of inflexion, and trace the curve.

7. Find the points of inflexion for values of x between 0 and 2π (0 included, 2π excluded) on the graphs of

(i) $\sin x$; (ii) $\cos x$; (iii) $\tan x$.

8. Show that the graphs of e^x and of $\log x$ have no points of inflexion.

9. Find the points of inflexion on the graphs of (i) xe^{-x}, (ii) e^{-x^2}. Trace the graph of e^{-x^2}.

10. Find the point of inflexion on the graph of $e^{-ax}-e^{-bx}$ where a, b are positive and a less than b.

11. Find the points of inflexion on the graph of $e^{-ax}\sin(bx+c)$.

12. When the equation of a curve is given in the form
$$x=f(t), \qquad y=\phi(t)$$
show that the points of inflexion will be determined by the equation
$$\dot{x}\ddot{y}-\dot{y}\ddot{x}=0.$$
Show that the curve whose equations are
$$x=a(t-\sin t), \qquad y=a(1-\cos t)$$
is everywhere convex upwards. (See § 68, ex. 2.)

13. Show that no conic section can have a point of inflexion.

CHAPTER X.

DERIVED AND INTEGRAL CURVES.
INTEGRAL FUNCTION.
DERIVATIVES OF AREA AND VOLUME OF A SURFACE OF REVOLUTION.
POLAR FORMULAE. INFINITESIMALS.

§ 79. Derived Curves. It is of some service in tracing the variation of a function $f(x)$ to draw the graph of the derived function $f'(x)$. The graph of $f'(x)$ may be called the derived graph or curve of $f(x)$, while in relation to the graph of $f'(x)$ that of $f(x)$ may be called the primitive curve or for a reason given in § 83, the integral curve of $f'(x)$.

It is usually most convenient to take a common axis of ordinates for the two curves, but to take the axis of abscissae of the derived curve at a convenient distance below that of the primitive curve. Assuming the unit segment for abscissae to be the same for both curves, but that for the ordinates to be the same or different as is most convenient, we may call those points and ordinates of the two curves which have the same abscissa "corresponding points and ordinates." Corresponding points on the two graphs may be denoted by unaccented and accented letters.

The student will easily prove that in general the following theorems hold :—

(i) To the turning points (T) of the primitive curve correspond points (T') at which the derived curve not only meets but crosses its axis of abscissae; and conversely.

(ii) To the points of inflexion (I) of the primitive curve correspond turning points (I') of the derived curve; and conversely.

The following geometrical construction may be given for the graphing of the derived curve when the functions $f(x)$, $f'(x)$ are not analytically expressed.

Let U (Fig. 38) be a point to the left of O_1 on the axis O_1X' of the derived curve, and draw Up parallel to the tangent PT meeting the common axis of ordinates at p. Draw PR and RT parallel and perpendicular respectively to OX; the triangles PRT, UO_1p will be similar. Hence

$$\frac{O_1p}{UO_1} = \frac{RT}{PR} = f'(x),$$

where $OM = O_1M' = x$, $MP = f(x)$; therefore

$$O_1p = f'(x) \cdot UO_1.$$

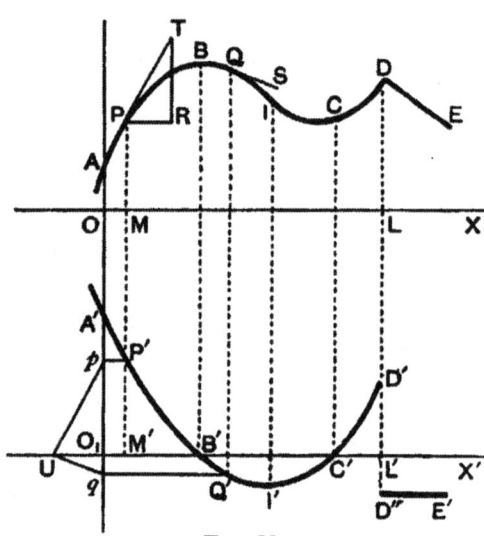

Fig. 38.

Draw pP' parallel to UO_1X' to meet $M'P$ at P'; then

$$M'P' = O_1p = f'(x) \cdot UO_1.$$

If we take the unit segment for the ordinates of the derived curve equal to UO_1 we shall have

$$M'P' = f'(x),$$

so that P' is the point corresponding to P.

Take any other point Q; draw Uq parallel to the tangent QS, and qQ' parallel to O_1X' to meet the ordinate through Q at Q'. Q' will correspond to Q, and in the same way any number of points may be found.

If the unit segment for the ordinates be not equal to UO_1 the ordinates will still be proportional to $f'(x)$.

To the turning points B, C correspond B', C'; to the point of inflexion I corresponds I' which is a turning point of the derived curve.

At D the derivative $f'(x)$ is discontinuous. As a point moves along the primitive curve from C to D the corresponding point moves from C' to D'; as the first point, however, changes from CD to DE the corresponding point passes abruptly from D' to D''. As x increases through the value OL or O_1L', $f'(x)$ suddenly changes from $L'D'$ to $L'D''$. LD is a maximum value of $f(x)$, but owing to the discontinuity of $f'(x)$ the derived curve does not (as at B' or C') meet the x-axis.

In a similar way the derived curve of $f'(x)$, that is the second derived curve of $f(x)$, may be formed, and so on.

§ 80. Derivative of an Area. Let CPD (Fig. 39) be the graph of a continuous function of x, $F(x)$;

$$OA = a, \ AC = F(a); \ OM = x, \ MP = F(x).$$

(i) Suppose the ordinates positive and AC to the left of MP. Let AC be fixed, MP variable, and let z be the measure of the area $AMPC$. We may consider the area as generated by a variable ordinate setting out from AC and moving to the right; z will be a function of x which is zero when $x = a$. Let us find the x-derivative of z, that is the x-rate of change of the area.

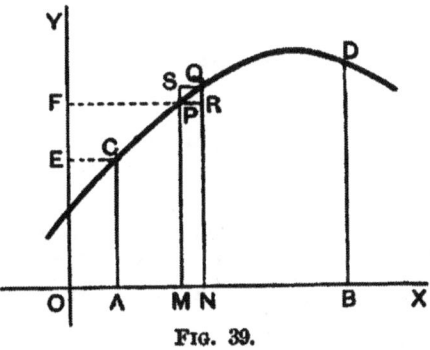

Fig. 39.

Let x take the increment δx or MN; z therefore will take an increment δz, the area $MNQP$. Complete the rectangles $MNRP$, $MNQS$; the area $MNQP$ will be greater than $MNRP$ but less than $MNQS$, therefore

$$MP \cdot \delta x < \delta z < NQ \cdot \delta x; \ MP < \frac{\delta z}{\delta x} < NQ.$$

In the figure MP is less than NQ; if MP is greater than NQ the signs of inequality will need to be reversed. As δx converges to zero MP remains fixed and NQ converges to MP. Hence

$$D_x z = MP = F(x) = \text{ordinate at } M,$$

or in the notation of differentials
$$dz = MP \cdot dx = F(x)\, dx.$$

(ii) Suppose the ordinates negative (Fig. 39a) and let z' be the *numerical* value of the area.

In this case the rectangle $MNRP$ is equal to $-MP \cdot \delta x$, since MP is negative, and we get by the same reasoning as before
$$D_x z' = -MP = -F(x).$$

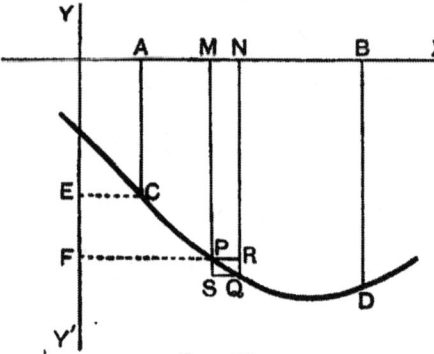

Fig. 39a.

It gives greater flexibility to the formulae to consider an area as a magnitude that, like a segment of a line, may be either positive or negative. If, therefore, the measure of the area be taken as negative, when the fixed ordinate is to the left of the variable one and the ordinates all negative, we may put z equal to $-z'$, the measure z being now negative; hence $D_x z = F(x)$ as in case (i).

(iii) Lastly, suppose the fixed ordinate to the right of MP, say at BD. The area $BMPD$ may be considered to be generated by a variable ordinate setting out from BD and moving to the left.

Let z' be the numerical value of the area $BMPD$; then z' is a decreasing function of x. By the same reasoning as before we see that the numerical value of $D_x z'$ is $F(x)$ for Fig. 39 but $-F(x)$ for Fig. 39a. Since z' is a decreasing function $D_x z'$ is algebraically negative, so that in sign and magnitude
$$D_x z' = -F(x) \text{ (Fig. 39)}; \quad D_x z' = F(x) \text{ (Fig. 39a)}.$$

If we take the measure z of the area $BMPD$ to be negative for Fig. 39, positive for Fig. 39a, we shall get for both cases $D_x z = F(x)$. The same formula therefore holds for all three cases (i), (ii), (iii).

Examination of the diagrams will show the truth of the following rule for determining the sign of the area.

DERIVATIVE OF AN AREA.

Let the boundary of the area be described in the order, x-axis, variable ordinate, curve, fixed ordinate; the sign will be positive or negative according as the area lies to the left or to the right during the description of the boundary.

Ex. 1. If the coordinate axes are inclined at an angle ω, show that $D_x z = F(x) \sin \omega$.

Ex. 2. If CE, PF (Figs. 39, 39a) are perpendicular to OY, and if w is the measure of the area $EFPC$, show that
$$D_y w = -FP = -OM; \quad dw = -xdy,$$
the sign of w being positive or negative according as the area is to the left or to the right when its boundary is described in the order $EFPCE$.

Consider the cases in which the abscissa is negative, and also the cases in which the fixed abscissa is on the opposite side of FP from that in the figures.

§ 81. Interpretation of Area.

The interpretation of the number z considered as the measure of an area will depend on the unit segments chosen for the abscissa and the ordinate. If the value 1 of x represents say 6 inches and the value 1 of y say 10 inches, then the value 1 of z will represent 60 square inches; if on the graph the value 1 of x is half an inch and the value 1 of y quarter of an inch, these representing 6 and 10 inches respectively, an area on the graph of one-eighth of a square inch will represent the area of 60 square inches.

The physical interpretation of the area will depend on the nature of the quantities represented by abscissa and ordinate.

Suppose that the ordinate represents the speed of a moving point and the corresponding abscissa the time at which the point has that speed; the graph is then the speed-curve of the motion. The speed is the time-rate of change of the distance, and the ordinate (which represents the speed) is the rate of change of the area with respect to the abscissa (which represents the time); hence the area $AMPC$ will represent the distance gone in the time represented by AM. If the value 1 of x represents 2 seconds and the value 1 of y a speed of 16 feet per second, then the value 1 of z will represent a distance of 32 feet.

If the ordinate represents a force that acts in a constant direction, and if the abscissa represents the distance through which the force has acted, the area $AMPC$ will represent the work done by the force acting through the distance represented by AM. If the force is not constant in direction the result holds provided the ordinate represents the component of the force along the tangent to the path of its point of application.

Ex. 1. If the ordinate represents acceleration and the abscissa time, what does the area represent ?

Ex. 2. If the ordinate represents the intensity of pressure of a gas, and the abscissa the volume, what does the area represent ?

§ 82. Integral Function. The fact that z in § 80 is a function of x which has $F(x)$, the ordinate of the curve CPD, as its derivative at once suggests the problem of finding a function which has a given continuous function as its derivative.

Now, if the derivative of $f(x)$ is $F(x)$ so is the derivative of $f(x)+C$ where C is any constant; further (§ 58, Th. VI.), every function which has $F(x)$ as its derivative must be of the form $f(x)+C$. The problem, therefore, as stated above, is indeterminate, since its solution involves a constant C which may have any value whatever; it becomes determinate, however, when stated in the form:—To find a function of x which shall have a given continuous function $F(x)$ as its derivative and which shall take a given value A when x has a given value a.

The solution is as follows:—Let CPD (§ 80) be the graph of $F(x)$, and let z be the measure of the area $AMPC$ where $OA=a$. z therefore is zero when $x=a$, and z has $F(x)$ as its derivative; the function $z+A$ gives the solution. We may, if we please, consider the constant A as the measure of an area.

It does not follow, however, that we can find an analytical expression for z in terms of known functions; thus, if $F(x)=\sqrt{(1+x^3)}$, we cannot find in the ordinary algebraic or transcendental functions one which has $F(x)$ as its derivative. The geometrical discussion shows, however, that in so far as we consider functions as being adequately represented by graphs, there always exists a function which is the solution of the problem, and it is possible by methods of approximation to get an analytical expression, for example, in the form of a series, that may be considered as a solution. Or, again, it may be possible by mechanical methods to get an approximate value of the area $AMPC$.

INTEGRAL FUNCTION.

Any function $f(x)$ which has $F(x)$ for its derivative is called an *Integral Function* or simply an *Integral* of $F(x)$. If $f(x)$ is one integral, $f(x)+C$ is called the *General Integral*, C being called the constant of integration.

To find $f(x)$ when $F(x)$ is given we fall back on the known results of differentiation. In the Integral Calculus the search for integral functions is systematically carried out, but from the nature of the case the process is largely tentative. The fundamental test that $f(x)$ should be an integral of $F(x)$ is that $D_x f(x)$ should be equal to $F(x)$.

Just as $\sin^{-1} x$ means a function whose sine is x so we may for the present use the symbol $D_x^{-1} F(x)$ or $D^{-1} F(x)$ as meaning *a function whose derivative is $F(x)$*, that is $D_x^{-1} F(x)$ is *an integral of $F(x)$*. We will suppose that $D^{-1} F(x)$ contains no constant of integration, so that the general integral is $D^{-1} F(x) + C$.*

We may now express the area $AMPC$ in the new notation. Since $D^{-1} F(x)$ is an integral of $F(x)$, the area z or $AMPC$ is a function of x of the form

$$z = D^{-1} F(x) + C.$$

Now when $x = a$, $z = 0$; denote by $[D^{-1} F(x)]_a$ the value of the integral when $x = a$; therefore,

$$0 = [D^{-1} F(x)]_a + C; \quad C = -[D^{-1} F(x)]_a,$$

and

$$z = D^{-1} F(x) - [D^{-1} F(x)]_a.$$

The area $ABCD$ is the value of z when $x = b$; therefore

$$\text{area } ABCD = [D^{-1} F(x)]_b - [D^{-1} F(x)]_a.$$

This symbol is usually contracted into

$$[D^{-1} F(x)]_a^b,$$

and this last symbol means "replace x by b, then replace x by a and subtract the second result from the former."

In the same way the function whose derivative is $F(x)$ and which is equal to A when x is equal to a is denoted by

$$D^{-1} F(x) - [D^{-1} F(x)]_a + A.$$

* For the ordinary notation for an integral see § 110.

Ex. 1. Find the area between the graph of x^2-3x+2, the x-axis, and the ordinates at $x=\frac{1}{2}$ and $x=1$.
$$F(x)=x^2-3x+2\,; \qquad D^{-1}F(x)=\tfrac{1}{3}x^3-\tfrac{3}{2}x^2+2x,$$
as may be tested by differentiation. Hence the required area is
$$[\tfrac{1}{3}x^3-\tfrac{3}{2}x^2+2x]_{\frac{1}{2}}^{1}=\tfrac{5}{6}-\tfrac{2}{3}=\tfrac{1}{6}.$$

Suppose the ordinates to be those at $x=\frac{1}{2}$ and $x=2$; then the area is
$$[\tfrac{1}{3}x^3-\tfrac{3}{2}x^2+2x]_{\frac{1}{2}}^{2}=\tfrac{2}{3}-\tfrac{2}{3}=0.$$

The reason for this apparently strange result is that from $x=\frac{1}{2}$ to $x=1$ the ordinates are positive while from $x=1$ to $x=2$ they are negative. From $x=1$ to $x=2$ the measure of the area is negative; numerically this area is equal to that for which the ordinates are positive.

Ex. 2. The area between the x-axis and the graph of $\sin x$ between the points $x=0$, $x=\pi$ at which it crosses the axis is
$$[D^{-1}\sin x]_0^\pi = [-\cos x]_0^\pi = -\cos\pi - (-\cos 0) = +1+1 = 2.$$

Ex. 3. A point moves on a straight line so that its velocity at time t is $V\cos nt$ ft./sec.; show that the space described from time $t=0$ till it first comes to rest is V/n ft.

Let x ft. be the distance described in time t seconds; then
$$D_t x = V\cos nt\,; \qquad x=\frac{V}{n}\sin nt + C.$$

When $t=0$, $x=0$ and therefore $C=0$. The point first comes to rest when t has increased from 0 to $\pi/2n$ because $\cos nt$ is first zero when $nt=\pi/2$. Hence we get for the distance required
$$\frac{V}{n}\sin\frac{\pi}{2}=\frac{V}{n}.$$

§ 83. Integral Curve. The graph of an integral function is called an integral curve. Since any two integral functions of $F(x)$ differ only by a constant C, the graph of the integral function $f(x)+C$ may be obtained from that of $f(x)$ by shifting the latter parallel to the y-axis through the distance C.

A geometrical construction may be given for graphing an integral curve based on that for the graphing of the derived curve (§ 79).

Divide O_1X' (Fig. 40) into equal short segments at the points $1_1, 2_1, 3_1, \ldots$ and draw the ordinates through these points. Let the ordinates at $2_1, 4_1, \ldots$ meet the graph of

INTEGRAL CURVE.

$F(x)$ at $2', 4', \ldots$ and let $2'', 4'', \ldots$ be the projections of these points on O_1Y, $1''$ being the point where the graph cuts O_1Y.

Let us take the integral curve that passes through O. Then the tangent at O is parallel to $U1''$. Let this tangent be drawn and let it cut the ordinate drawn from 1_1 at 1.

The tangent at the point on the integral curve corresponding to $2'$ is parallel to $U2''$. Draw 13 parallel to $U2''$ cutting $2_12'$ at 2, and the ordinate drawn from 3_1 at 3; 2 is the point corresponding to $2'$.

In the same way draw 35 parallel to $U4''$ cutting $4_14'$ at 4; 4 is the point corresponding to $4'$.

The construction may be repeated and we get a series

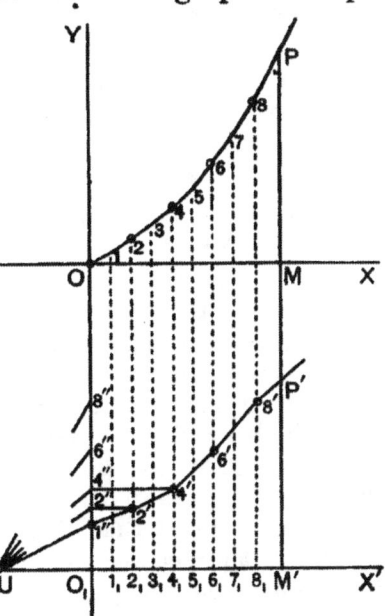

Fig. 40.

of lines $O1, 13, 35, \ldots$ which may be considered as, approximately, the tangents at $O, 2, 4, \ldots$ to the integral curve. That curve may now be drawn with a free hand through the points $O, 2, 4 \ldots$. The point O from which the construction begins is, of course, arbitrary, but when that is fixed the integral curve is determinate. The position of the other points $2, 4, \ldots$ is approximate; the nature of the approximation and the justification of the construction may be seen thus.

Let $f(x)$ be the integral function; the equations of the tangents at the points on the graph of $f(x)$ at which x is equal to a and b respectively, are

$$y = (x-a)f'(a) + f(a); \quad y = (x-b)f'(b) + f(b).$$

The abscissa of the point of intersection of these tangents is given by

$$\{f'(b) - f'(a)\} x = bf'(b) - af'(a) - f(b) + f(a).$$

Now, by the theorems of mean value, if $b=a+h$, we have
$$f(b)=f(a+h)=f(a)+hf'(a)+\tfrac{1}{2}h^2 f''(x_1),$$
$$f'(b)=f'(a+h)=f'(a)+hf''(x_2),$$
where x_1, x_2 are each greater than a but less than $a+h$. Substituting these values in the equation for x and reducing we get
$$x=a+h-\tfrac{1}{2}hf''(x_1)/f''(x_2).$$

Assuming the derivatives continuous, then if h is small $f''(x_1)$ and $f''(x_2)$ will differ very little from each other and from $f''(a)$. Therefore approximately $x=a+\tfrac{1}{2}h$; that is, the abscissa of the point of intersection of the tangents is very nearly that of a point half way between a and b.

Hence, in the figure the tangent at the point on the integral curve corresponding to $2'$ passes through 1; the point 2, which must lie on the ordinate $2_1 2'$, is therefore got as the intersection of the line through 1 parallel to $U2''$. Similarly for the other points.

It may be noticed that if $F(x)$ is of the first and, therefore, $f(x)$ of the second degree, the construction is exact since $f''(x)$ is constant.

§ 84. Graphical Integration. The area between $O_1 X'$, $O_1 Y$, the graph of $F(x)$ and the ordinate $M'P'$ (Fig. 40) is equal to $f(x)$ where $f(x)$ is that integral of $F(x)$ which is zero when $x=0$. But the ordinate MP of the integral curve is $f(x)$. Hence the area $O_1 M'P'1''$ is equal to the ordinate of the integral curve at the point corresponding to P'. We thus have a graphical method of finding the measure of an area and also of constructing an integral function even when the analytical form of the function $F(x)$ is not assigned.

The integral curve can be drawn with considerably greater accuracy than the derived curve. It is also possible to trace out the integral curve corresponding to a given curve by means of an instrument called an *Integraph*. For detailed description of the Integraph the reader is referred to the work of M. Abdank-Abakanowicz, *Les Intégraphes; la courbe intégrale et ses applications* (Paris: Gauthier-Villars), or to the German translation by

Bitterli, *Die Integraphen* (Leipzig: Teubner). The constructions given above are taken from this work; the notes of Bitterli contain several investigations on the properties of the integral curve and also numerous references to original memoirs. An article by Prof. W. F. Durand in the *Sibley Journal of Engineering*, January, 1897, will also be found serviceable.

§ 85. Surfaces of Revolution. Let V be the volume of the surface traced out by the revolution of the arc CP (Fig. 41) about OX; $OM = x$, $MP = F(x) = y$. To find $D_x V$.

When x increases by MN or δx, V increases by δV, the volume traced out by $MNQP$. Clearly, when δx is small, δV is greater than the cylinder of height MN and base the circle of radius MP, but less than the cylinder of height MN and base the circle of radius NQ; therefore

$$\pi MP^2 . MN < \delta V < \pi NQ^2 . MN; \quad \pi MP^2 < \delta V/\delta x < \pi NQ^2.$$

Hence taking the limit for $\delta x = 0$

$$D_x V = \pi MP^2 = \pi y^2; \quad dV = \pi y^2 dx.$$

Let S be the area of the surface traced out by the arc CP, and let CP be s. To find $D_x S$.

On the tangent at P take a length PT equal to the arc PQ, and let L be the foot of the ordinate to T. When x increases by MN or δx, CP increases by the arc PQ or δs; we may assume that the area traced out by the arc PQ is, when δx is small, greater than that traced out by the chord PQ but less than that traced out by the tangent PT. If the arc PQ lies below the chord PQ the inequalities will be reversed.

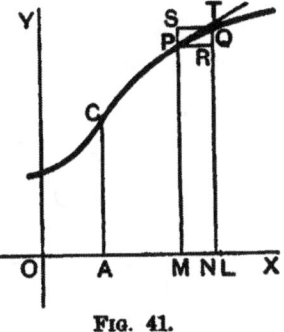

Fig. 41.

The curved surface of the conical frustum having MP, NQ as the radii of its circular ends and the chord PQ for slant side is $\pi(MP+NQ)PQ$; the surface traced out by PT is similarly $\pi(MP+LT)PT$. Hence

$$\pi(MP+NQ)\frac{PQ}{\delta x} < \frac{\delta S}{\delta x} < \pi(MP+LT)\frac{PT}{\delta x}.$$

The limit for $\delta x = 0$ of $PQ/\delta x$ and of $PT/\delta x$ is $D_x s$ (§ 62) and the limit of $MP + NQ$ and of $MP + LT$ is $2MP$; hence

$$D_x S = 2\pi MP \, D_x s = 2\pi y \, D_x s,$$

or
$$dS = 2\pi y \frac{ds}{dx} dx = 2\pi y \, ds.$$

The volume V is that integral of πy^2 which is zero when $x = OA$, and the surface S that integral of $2\pi y \, ds/dx$, which is zero when $x = OA$.

By § 62
$$\frac{ds}{dx} = \sqrt{\left\{1 + \left(\frac{dy}{dx}\right)^2\right\}}.$$

Ex. 1. If the curve revolves about OY show that

$$dV = \pi x^2 dy; \quad dS = 2\pi x \frac{ds}{dy} dy = 2\pi x \, ds.$$

Ex. 2. Show that the volume of a spherical cap of height h is $\pi h^2(R - \tfrac{1}{3}h)$ and that the area of the surface of the cap is $2\pi Rh$, R being the radius of the sphere.

The equation of CPQ is $y = \sqrt{(R^2 - x^2)}$; hence

$$D_x V = \pi(R^2 - x^2); \quad V = \pi(xR^2 - \tfrac{1}{3}x^3) + C$$

$V = 0$ when $x = OA = R - h$, and therefore

$$C = -\pi(\tfrac{2}{3}R^3 - Rh^2 + \tfrac{1}{3}h^3); \quad V = \pi(xR^2 - \tfrac{1}{3}x^3) - \pi(\tfrac{2}{3}R^3 - Rh^2 + \tfrac{1}{3}h^3).$$

The volume required is the value of V for $x = R$ and is therefore $\pi h^2(R - \tfrac{1}{3}h)$.

Again

$$\frac{ds}{dx} = \sqrt{\left\{1 + \left(\frac{dy}{dx}\right)^2\right\}} = \sqrt{\left\{1 + \frac{x^2}{R^2 - x^2}\right\}} = \frac{R}{\sqrt{(R^2 - x^2)}},$$

therefore

$$\frac{dS}{dx} = 2\pi\sqrt{R^2 - x^2} \cdot \frac{R}{\sqrt{(R^2 - x^2)}} = 2\pi R.$$

$$S = 2\pi Rx + C; \quad 0 = 2\pi R(R - h) + C.$$

So that $S = 2\pi R(x + h - R)$,
and when $x = R$, $S = 2\pi Rh$.

Ex. 3. If the area of a section of a surface by a plane perpendicular to OX is a known function of x, $F(x)$, and if V is the volume contained between a fixed plane perpendicular to OX and the plane which gives the section of area $F(x)$, show that

$$D_x V = F(x).$$

§ 86. Infinitesimals. The student will doubtless have noticed that in finding derivatives a good deal of work would have been saved had it been possible to reject at the outset those parts of an expression that had zero for limit. Thus in § 80 δz consists of the rectangle $MNRP$ and the curvilinear triangle PRQ, which is less than the rectangle $PRQS$. $\delta z/\delta x$ is therefore the sum of MP and of a line which is less than RQ. Since RQ converges to zero with δx we may, so far as the limit is concerned, throw aside RQ from the outset; we should thus at once obtain MP as the limit of $\delta z/\delta x$.

Now that the student has had so much practice in finding derivatives and limits generally, he will be ready to grasp the method which enables us to reject, at any stage, a quantity which we can see will not occur in the limit; the method is that of *Infinitesimals*.

DEFINITION. A variable quantity whose limit is zero is called an infinitesimal.

A constant, however small, is not an infinitesimal in the sense now defined; an infinitesimal is a *variable* quantity.

Let $a, \beta, \gamma \ldots$, be infinitesimals, and let $\beta, \gamma \ldots$, be such that when a converges to zero $\beta, \gamma \ldots$ also converge to zero; $\beta, \gamma \ldots$ are dependent on a and we can compare them with a and with one another. When a is taken as the standard of comparison a is usually called the principal infinitesimal.

β is said to be an infinitesimal of the same order as a when

$$\mathop{L}_{a=0} \frac{\beta}{a} = k,$$

where k is a finite number not zero. When k is zero β is said to be an infinitesimal of a higher order than a; when k is infinite β is said to be an infinitesimal of a lower order than a.

When the limit of β/a is infinite β is sometimes called an *infinite* with respect to a.

In practice one infinitesimal is chosen as principal infinitesimal and the other infinitesimals are said to be of a certain *order*, first, second, etc., the principal infinitesimal being either explicitly stated or sufficiently indicated by the context.

β is defined to be an infinitesimal of order n with respect to a, n being positive but not necessarily integral, when

$$\underset{a=0}{\mathrm{L}}\frac{\beta}{a^n}=k,$$

where k is a finite number, not zero.

By the definition of a limit we may write

$$\beta/a^n = k+\omega \quad \text{or} \quad \beta = ka^n + \omega a^n,$$

where ω is a variable that converges to zero with a, that is ω is an infinitesimal. The difference $\beta - ka^n$ or ωa^n is an infinitesimal of a higher order than a^n because the limit of $\omega a^n/a^n$, that is of ω, is zero.

ka^n is called the *principal* part of β; manifestly the ratio of an infinitesimal to its principal part has unity as its limit.

If
$$\underset{a=0}{\mathrm{L}}\,\beta a^n = k,$$

where k is finite, not zero, β is sometimes said to be infinite of order n with respect to a, n being positive.

If β, γ are infinitesimals of order m, n respectively, the product $\beta\gamma$ is an infinitesimal of order $m+n$, and the quotient β/γ is an infinitesimal of order $m-n$ if $m > n$, but an infinite of order $n-m$ if $m < n$. For

$$\beta = (k+\omega)a^m; \quad \gamma = (k'+\omega')a^n,$$

$$\underset{a=0}{\mathrm{L}}\left(\frac{\beta\gamma}{a^{m+n}}\right) = \underset{a=0}{\mathrm{L}}(k+\omega)(k'+\omega') = kk',$$

and in the same way the quotient theorem may be proved.

Ex. 1. $\sin a$, $1 - \cos a$, $\sin a\,(1-\cos a)$ are of the 1st, 2nd, 3rd order respectively with respect to a. For

$$\underset{a=0}{\mathrm{L}}\frac{\sin a}{a}=1;\quad \underset{a=0}{\mathrm{L}}\frac{1-\cos a}{a^2}=\tfrac{1}{2};\quad \underset{a=0}{\mathrm{L}}\frac{\sin a\,(1-\cos a)}{a^3}=\tfrac{1}{2},$$

and their principal parts are a, $\tfrac{1}{2}a^2$, $\tfrac{1}{2}a^3$ respectively.

Ex. 2. If $\beta = \sqrt{(9a - 2a^2 + 3a^3)}$, β is of order $\tfrac{1}{2}$ and its principal part is $3\sqrt{a}$. For

$$\underset{a=0}{\mathrm{L}}\frac{\beta}{a^{\frac{1}{2}}} = \underset{a=0}{\mathrm{L}}\sqrt{(9-2a+3a^2)}=3.$$

Ex. 3. $\tan a - \sin a$ is of the 3rd order and its principal part is $\tfrac{1}{2}a^3$. This follows at once from ex. 1.

§ 87. Fundamental Theorems.

The value of the explicit discussion of infinitesimals depends on the principle that so far as the *limit* of an expression is concerned we need only in general attend to the principal part; the other terms of the infinitesimal being of a higher order than the principal part will have to that part a ratio whose limit is zero, and may therefore be discarded from the outset.

If an expression contain a finite constant term A and infinitesimals $a, \beta, \gamma \ldots$, then so far as the *limit* is concerned we may, *in general*, at once replace $A + a + \beta + \gamma + \ldots$ by A. The essential thing is to find out the order of the expression; in comparison with infinitesimals the principal part alone need be retained, while in comparison with finite quantities no infinitesimal need be retained. The order of an infinitesimal $\beta + \gamma + \ldots$ is, of course, that of its principal part.

Care must, however, be exercised in applying this principle. Thus $1 - \cos a + \sin a$ contains the constant term 1; but $1 - \cos a$ is of the second order, $\sin a$ of the first. Hence the whole expression is an infinitesimal of the first order, its principal part being a.

The following are the two fundamental theorems.

THEOREM I. *The limit of the quotient of two infinitesimals is not altered by replacing each infinitesimal by another having the same principal part.*

Let β, γ be two infinitesimals. In order that their quotient should have a *finite* limit, not zero, each must be of the same order. We may therefore write, the order being n,
$$\beta = k a^n + \omega a^n; \quad \gamma = k' a^n + \omega' a^n.$$

Let β_1, γ_1 be two other infinitesimals having the same principal parts as β, γ respectively; then
$$\beta_1 = k a^n + \omega_1 a^n; \quad \gamma_1 = k' a^n + \omega_1' a^n,$$
where ω_1, ω_1' are infinitesimals different from ω, ω'. Now,
$$\underset{a=0}{\mathrm{L}} \frac{\beta_1}{\gamma_1} = \underset{a=0}{\mathrm{L}} \frac{k + \omega_1}{k' + \omega_1'} = \frac{k}{k'} = \underset{a=0}{\mathrm{L}} \frac{\beta}{\gamma}.$$

The reasoning would clearly hold if β were of higher order than γ, for the limit both of β/γ and of β_1/γ_1 would be zero. If β were of lower order the theorem would hold in

198 AN ELEMENTARY TREATISE ON THE CALCULUS.

the sense that the limit both of β/γ and of β_1/γ_1 would be infinite.

Ex. $\quad\underset{x=0}{L}\dfrac{\sin ax}{\tan bx}=\underset{x=0}{L}\dfrac{ax}{bx}=\dfrac{a}{b}.$

From its great use in the differential calculus this theorem is often called the fundamental theorem of the differential calculus.

THEOREM II. *The limit for n infinite of the sum of n infinitesimals is not altered by replacing each infinitesimal by another having the same principal part, provided all the infinitesimals are of the same sign.*

The theorem is not necessarily true if the infinitesimals are not all of the same sign.

Let $u_n = \beta_1 + \beta_2 + \ldots + \beta_n$; $v_n = \gamma_1 + \gamma_2 + \ldots + \gamma_n$, where β_1 has the same principal part as γ_1, β_2 as γ_2.... The principal infinitesimal, previously denoted by a, is here $1/n$, and therefore the limit of each of the quotients β_1/γ_1, β_2/γ_2... for $a=0$ or $n=\infty$ is unity. Of course the principal parts of $\beta_1, \beta_2, \beta_3$... are not necessarily the same.

It is a known theorem of algebra that when the quantities $\beta_1, \gamma_1 \ldots, \beta_2, \gamma_2 \ldots$, are all of the same sign the fraction u_n/v_n lies in value between the greatest and the least of the fractions $\beta_1/\gamma_1, \beta_2/\gamma_2 \ldots$. Hence, for every value of n the fraction u_n/v_n lies between two fractions each of which has the same limit, unity. Therefore,

$$\underset{n=\infty}{L}\dfrac{u_n}{v_n}=1,$$

and therefore if v_n converges to a limit, u_n will converge to the same limit, that is

$$\underset{n=\infty}{L}u_n = \underset{n=\infty}{L}v_n.$$

Ex. Let $\beta_p = n/(n+p)^2$, $\gamma_p = n/(n+p)(n+p+1)$; then the limit of β_p/γ_p for $n=\infty$ is unity for every integral value of p. But

$$\gamma_p = n/(n+p) - n/(n+p+1);$$

$$\therefore\; v_n = \left(\dfrac{n}{n+1} - \dfrac{n}{n+2}\right) + \left(\dfrac{n}{n+2} - \dfrac{n}{n+3}\right) + \ldots + \left(\dfrac{n}{n+n} - \dfrac{n}{n+n+1}\right)$$

$$= \dfrac{n}{n+1} - \dfrac{n}{2n+1};$$

$$\therefore\; L\, u_n = L\left\{\dfrac{n}{(n+1)^2} + \dfrac{n}{(n+2)^2} + \ldots + \dfrac{n}{(n+n)^2}\right\} = L\, v_n = 1 - \tfrac{1}{2} = \tfrac{1}{2}.$$

THEOREMS. WORKED EXAMPLES.

From its use in integration this theorem is often called the fundamental theorem of the integral calculus.

Ex. 1. When dx is the principal infinitesimal, then (§ 60) the principal part of $\delta f(x)$ is $df(x) = f'(x)dx$, and that of $\delta f'(x)$ is $df'(x) = f''(x)dx$. If $f'(x) = \tan\phi$ then the principal part of $\delta \tan\phi$ is $d\tan\phi = f''(x)dx$.

Ex. 2. Let PQ be the graph of $f(x)$; PT, TQ the tangents at P, Q. $OM = a$, $MN = PR = h$, $\angle RPS = \phi$, $\angle STQ = \delta\phi$, $\angle TPQ = a$, $\angle TQP = \beta$.

Let h or PR be the principal infinitesimal.
RS, PS, PQ are of the first order.

Let $f''(x)$ be finite, not zero, from $x = a$ to $x = a + h$; then by Th. III., § 72,

$$SQ = \tfrac{1}{2}h^2 f''(x_1); \quad \underset{h=0}{L}\frac{SQ}{h^2} = \tfrac{1}{2}f''(a).$$

Hence SQ is of the second order.

$\delta\phi$ is of the first order and its principal part is $h\cos^2\phi f''(a)$; for $d\tan\phi$ is equal to $\sec^2\phi\,d\phi$ and also (ex. 1) to $hf''(a)$, so that

$$d\phi = h\cos^2\phi f''(a).$$

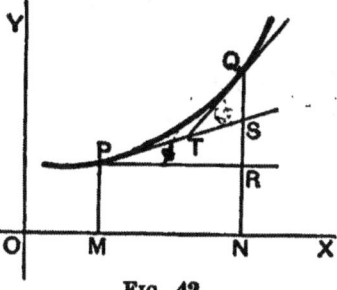

Fig. 42.

Again, a and β are of the first order. For $\sin PSR = \cos\phi$, and

$$\frac{\sin a}{h} = \frac{\sin a}{\sin PSR} \cdot \frac{\sin PSR}{h} = \frac{SQ}{PQ} \cdot \frac{\cos\phi}{h}; \quad L\frac{\sin a}{h} = \tfrac{1}{2}\cos^3\phi f''(a).$$

$\sin a$, and therefore a, is thus of the first order; the principal part of a is $\tfrac{1}{2}h\cos^3\phi f''(a)$, that is, *half the principal part of* $\delta\phi$. Since $\beta = \delta\phi - a$ its principal part is equal to that of a, that is, to half that of $\delta\phi$.

Again, $\qquad L\dfrac{PT}{PQ} = L\dfrac{\sin\beta}{\sin\delta\phi} = \tfrac{1}{2} = L\dfrac{TQ}{PQ}; \quad L\dfrac{PT}{TQ} = 1,$

so that PT, TQ are of first order. Also

$$PT + TQ - PQ = PT(1 - \cos a) + TQ(1 - \cos\beta);$$

so that the difference between $PT + TQ$ and PQ is of the third order, since PT and TQ are of the first and $(1-\cos a)$ and $(1-\cos\beta)$ of the second. Hence the difference between $PT + TQ$ and the arc PQ is at least of the third order since the arc PQ is greater than PQ.

The fact that the limit of PT/PQ is $1/2$ is sometimes expressed in the words "PT is *ultimately* equal to $\tfrac{1}{2}PQ$" or "PT is *in the limit* equal to $\tfrac{1}{2}PQ$." Similarly it is said that "the triangle PTQ is *ultimately* isosceles." This phraseology, though occasionally convenient, is apt to lead beginners astray.

If $f''(a) = 0$, SQ is of a higher order than the second, and $\delta\phi$, a, β are also of higher order than the first, and $PT + TQ - PQ$ of higher than the third.

Ex. 3. In Fig. 24, § 39, if MA be principal infinitesimal, prove
  (i) AT, arc AN of first order.
  (ii) MN, NT, MT of second order.
     Draw MC perpendicular to AT; then prove
  (iii) MC of second order.
  (iv) CT of third order.

Ex. 4. Show that (Fig. 42) if arc $PQ = \delta s$
$$\frac{d\phi}{ds} = \underset{\delta\phi=0}{L} \frac{\delta\phi}{\delta s} = \cos^3\phi \cdot f''(a) = f''(a)/\{1+(f'(a))^2\}^{\frac{3}{2}}.$$

§ 88. Polar Formulae. Let APQ (Fig. 43) be a curve whose polar equation is $r = f(\theta)$; let $\angle XOP = \theta$, $\angle POQ = \delta\theta$; $OP = r$, $OQ = r + \delta r$. Draw QR perpendicular to OP.

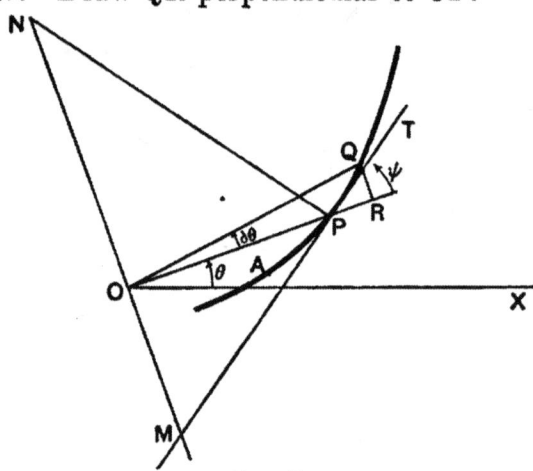

Fig. 43.

We will consider the arc PQ positive when the angle POQ is described by a positive rotation of the radius vector OP; the tangent PT is to be drawn towards the positive direction of PQ and by the angle, ψ say, between the tangent PT and the radius OP is meant the angle RPT between the outward drawn radius OP and the tangent PT.

(i) To find $\tan\psi$,
$$\tan RPQ = \frac{RQ}{PR} = \frac{(r+\delta r)\sin\delta\theta}{(r+\delta r)\cos\delta\theta - r}$$
$$= \frac{r\sin\delta\theta + \delta r\sin\delta\theta}{\delta r\cos\delta\theta - r(1-\cos\delta\theta)}.$$

POLAR FORMULAE.

If $\delta\theta$ be principal infinitesimal, δr is of first order since $dr/d\theta$ is in general finite; therefore $\delta r \sin \delta\theta$ and $(1-\cos \delta\theta)$ are of second order. We may, therefore, omit the quantities of the second order and put $\delta\theta$ for $\sin \delta\theta$ and 1 for $\cos \delta\theta$. Hence

$$\tan \psi = \mathrm{L} \tan RPQ = \mathrm{L}\, r\frac{\delta\theta}{\delta r} = r\frac{d\theta}{dr}.$$

(ii) *To find the derivative of the arc.*

Let $AP=s$, arc $PQ=\delta s$; then retaining only infinitesimals of the lowest order and remembering that PQ and arc PQ are of the same order we get

$$\left(\frac{ds}{d\theta}\right)^2 = \mathrm{L}\left\{\frac{(\delta r)^2+(r\delta\theta)^2}{(\delta\theta)^2}\right\} = r^2 + \left(\frac{dr}{d\theta}\right)^2,$$

or $\qquad ds = \sqrt{\{dr^2+r^2 d\theta^2\}}$,

and $\qquad \sin \psi = r d\theta/ds, \cos \psi = dr/ds$.

(iii) *To find the derivative of the area.*

Let sector $AOP=z$, sector $POQ=\delta z$; then δz is intermediate to the circular sectors of angle $\delta\theta$ and radii OP, OQ respectively. Hence $\delta z/\delta\theta$ lies between $\tfrac{1}{2}r^2$ and $\tfrac{1}{2}(r+\delta r)^2$, and therefore

$$\frac{dz}{d\theta} = \tfrac{1}{2}r^2; \quad dz = \tfrac{1}{2}r^2 d\theta.$$

(iv) *Polar subtangent and Polar subnormal.*

If PM, PN are the tangent and the normal at P and through O a line MON is drawn perpendicular to OP, meeting PM at M and PN at N, OM is called the polar subtangent and ON the polar subnormal. PM and PN are sometimes called the polar tangent and the polar normal respectively.

The lengths of these lines can be easily expressed when required in terms of r and ψ.

EXERCISES XVIII.

1. The equation $r=a\theta$ represents the curve called the *Spiral of Archimedes*. Prove $\tan \psi = \theta$ and show that the subnormal is constant. Sketch the curve.

2. The equation $r = a/\theta$ represents the *Reciprocal Spiral*. Show that the subtangent is constant.

Prove that the perpendicular from the point (r, θ) on the initial line OX is equal to $a \sin \theta/\theta$, and then show that the curve has an asymptote parallel to OX and at a distance a from OX.

3. The *Lituus* is the curve given by $r^2\theta = a^2$. Show, as in ex. 2, that OX is an asymptote and sketch the curve.

4. Show that ψ is constant for the curve given by $r = ae^{\theta \cot a}$. From this property the curve is called the *Equiangular Spiral*. Sketch the curve.

5. The curve given by $r = a(1 - \cos \theta)$ is called the *Cardioid*. Show that $\psi = \theta/2$ and sketch the curve.

6. If $r = 2a/(1 - \cos \theta)$, show that $\psi = \pi - \theta/2$. What is the curve?

7. If $r = a\theta$, $\dfrac{ds}{dr} = \dfrac{\sqrt{(r^2 + a^2)}}{a}$.

If $r = a/\theta$, $\dfrac{ds}{dr} = \dfrac{\sqrt{(r^2 + a^2)}}{r}$.

If $r = ae^{\theta \cot a}$, $\dfrac{ds}{d\theta} = r \operatorname{cosec} a$.

If $r^2 = a^2 \cos 2\theta$, $\dfrac{ds}{d\theta} = \dfrac{a^2}{r}$.

8. If, in the figure of § 88, PC is drawn perpendicular to OP and QC perpendicular to OQ, prove that the limit of PC as $\delta\theta$ converges to zero is $dr/d\theta$. If δz is the area of the sector CPQ, show that $dz/d\theta$ is equal to $\tfrac{1}{2}(dr/d\theta)^2$.

9. Find the area bounded by the curve and the radii whose vectorial angles are θ_1, θ_2 for the curves of examples 1-5.

10. The curve given by $r^2 = a^2 \cos 2\theta$ is called a Lemniscate; show that it consists of two loops of equal area and find the area of one loop.

11. APQ (Fig. 43) is the path of a moving point P. If u, v and a, β are the components of the velocity and of the acceleration of P along and perpendicular to the radius vector OP, show that

$$u = \dot{r}, \quad v = r\dot{\theta};$$

$$a = \ddot{r} - r\dot{\theta}^2, \quad \beta = r\ddot{\theta} + 2\dot{r}\dot{\theta} = \frac{1}{r}\frac{d}{dt}(r^2\dot{\theta}).$$

To prove these, note that (the limits being taken for $\delta t = 0$)

$$u = \mathrm{L}\frac{(r + \delta r)\cos \delta\theta - r}{\delta t}, \quad v = \mathrm{L}\frac{(r + \delta r)\sin \delta\theta}{\delta t};$$

and if u_1, v_1 are the values of u, v at Q,

$$a = \mathrm{L}\frac{u_1 \cos \delta\theta - v_1 \sin \delta\theta - u}{\delta t}, \quad \beta = \mathrm{L}\frac{u_1 \sin \delta\theta + v_1 \cos \delta\theta - v}{\delta t}.$$

12. If in example 11 the acceleration is always towards O, show that the radius vector sweeps out equal areas in equal times.

For $\beta=0$, and therefore $r^2\dot\theta = 2\dot{i}$ (§ 88) = constant.

13. If in Fig. 24 the tangent at N meets the tangents AT, BT at P, Q, show that the triangles PQT, ABT are each of the third order when MA is of the first, and that the limit of their ratio is $1/4$.

14. If in Fig. 42 the ordinate at T is LT, show that the limit of ML/MN is $1/2$. Show also that the principal part of the triangle PTQ is $\tfrac{1}{8}h^3 f''(a)$.

15. A circle is drawn touching PT at P, and passing through Q (Fig. 42); if ρ is the limit of the radius when Q converges to P, show that
$$\rho = \tfrac{1}{2}\mathrm{L}(PQ/\sin a) = ds/d\phi = \{1+(f'(a))^2\}^{\tfrac{3}{2}}/f''(a).$$

If SQ is produced to meet the circle at Q', show that the limit of SQ' is $2\sec^2\phi/f''(a)$.

16. A circle is described about the triangle PTQ (Fig. 42); if ρ_1 is the limit of the radius when Q converges to P, show that $\rho_1 = \tfrac{1}{2}\rho$ (ex. 15).

17. W is any point on the arc PQ (Fig. 42), and a circle is described about the triangle PWQ; show that when W and Q converge to P the radius of the circle converges to ρ (ex. 15). Show that the result is true if W and Q are on opposite sides of P, and W and Q both converge to P.

CHAPTER XI.

PARTIAL DIFFERENTIATION.

§ 89. Partial Differentiation. In the following chapter we will discuss very briefly functions of two or more independent variables; a thorough treatment of such functions is difficult, and we will restrict the discussion to the simpler properties of continuous functions.

DEFINITION. A function $f(x, y)$ of two independent variables x, y is defined to be continuous for the values a, b of x, y if the limit for $h = 0$ and $k = 0$ of
$$f(a+h, b+k) - f(a, b)$$
is zero, in whatever way h and k tend to zero.

A similar definition holds for a function of more than two variables.

Let u be a function of x and y, say $u = ax^2 + 2bxy + cy^2$. Since x and y are independent x may vary and y remain constant; the x-derivative of u when x varies and y does not vary is called *the partial x-derivative of u, or, the partial differential coefficient of u with respect to x*. In the same way the partial y-derivative of u is the derivative of u with respect to y on the supposition that x does not vary.

When u is a function of x alone its x-derivative is denoted by $D_x u$ or du/dx; the same notation is often used for the *partial* x-derivative of u, and the reader must infer from the context whether the derivative is partial or not. It has become customary, however, to represent partial derivatives by the notation
$$\frac{\partial u}{\partial x}, \frac{\partial u}{\partial y} \text{ or sometimes } \left(\frac{du}{dx}\right), \left(\frac{du}{dy}\right),$$

PARTIAL DIFFERENTIATION.

the form ∂ instead of d, or the bracket, indicating that the derivative is partial.

Notations analogous to $f'(x)$, $f_x'(x)$ are also in use. Thus

$$f_x'(x, y), \; f_y'(x, y), \; f_x(x, y), \; f_y(x, y), \; u_x, \; u_y,$$

denote partial derivatives of the functions $f(x, y)$ and u.

There is no notation, however, that is in itself quite free from ambiguity; the reader must usually infer from the context whether a derivative is partial or not.

The formal definition of $\partial u/\partial x$, $\partial u/\partial y$ where $u = f(x, y)$ is, therefore,

$$\frac{\partial u}{\partial x} = \underset{\delta x = 0}{\mathrm{L}} \frac{f(x + \delta x, y) - f(x, y)}{\delta x};$$

$$\frac{\partial u}{\partial y} = \underset{\delta y = 0}{\mathrm{L}} \frac{f(x, y + \delta y) - f(x, y)}{\delta y}.$$

Ex. 1. If $u = ax^2 + 2bxy + cy^2$.
$\partial u/\partial x = 2ax + 2by$; $\partial u/\partial y = 2bx + 2cy$.

Ex. 2. If $u = \sin(ax + by + c)$.
$\partial u/\partial x = a \cos(ax + by + c)$; $\partial u/\partial y = b \cos(ax + by + c)$.

§ 89a. Coordinate Geometry of Three Dimensions. A knowledge of coordinate geometry of three dimensions will greatly assist the reader in obtaining a clear conception of partial derivatives; we will therefore give in this article a few fundamental theorems regarding the representation of points, lines, and surfaces by means of three coordinates. In many cases the extension from two to three coordinates is extremely simple.

(i) *Coordinate Planes and Axes. Coordinates of a Point.* Through a point O let three planes YOZ, ZOX, XOY be drawn, the angle between each pair of planes being 90°, and suppose the planes to be produced indefinitely, their intersections being the lines $X'OX$, $Y'OY$, $Z'OZ$; these lines will be mutually at right angles. We will suppose $Y'OY$ and $Z'OZ$ to lie in the plane of the paper and the portion OX to be drawn upwards towards the reader (Fig. 44).

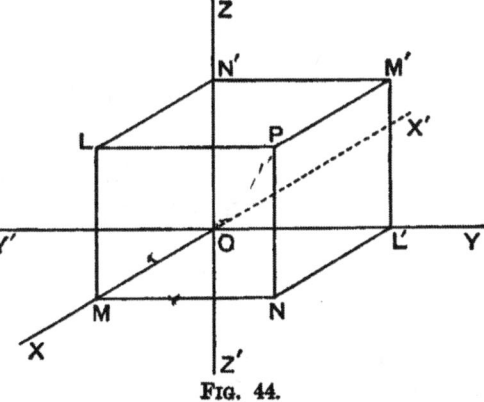

Fig. 44.

From any point P draw PN perpendicular to the plane XOY and NM perpendicular to the line $X'OX$; complete the parallelepiped. The position of P will be determined by the *segments* OM or $M'P$, OL' or LP, ON' or NP.

The three planes YOZ, ZOX, XOY are called the *coordinate planes*, the three lines $X'OX$, $Y'OY$, $Z'OZ$ the *coordinate axes* and the three segments OM, OL', ON' the *coordinates* of P; O is the origin of coordinates.

The positive directions of the axes and therefore of the segments or coordinates are from O to X, from O to Y, from O to Z respectively. P may be denoted *the point* (x, y, z) where

$$x = OM = L'N = M'P;\quad y = OL' = MN = LP;\quad z = ON' = L'M' = NP.$$

The coordinate planes divide space into eight portions (*octants*) and there will be eight arrangements of the signs $+$, $-$ corresponding to the octant in which the point is situated. Thus when the signs are $(+, +, +)$ P lies in the space bounded by YOZ, ZOX, XOY; when they are $(-, -, +)$ P lies in that bounded by $Y'OZ$, ZOX', $X'OY'$, and so on.

(ii) *Distance between two Points.* The geometry of Fig. 44 shows that
$$OP^2 = OM^2 + MN^2 + NP^2;\quad OP = \sqrt{(x^2 + y^2 + z^2)}\quad\ldots\ldots\ldots\ldots(1)$$

If P_1 is the point (x_1, y_1, z_1) and P_2 the point (x_2, y_2, z_2) draw through P_1, P_2 planes parallel to the coordinate planes (Fig. 45) forming the parallelepiped $P_1X_2Y_2Z_2P_2$; then

$$P_1X_2 = x_2 - x_1,\quad P_1Y_2 = y_2 - y_1,\quad P_1Z_2 = z_2 - z_1,$$
and
$$P_1P_2 = \sqrt{\{(x_2 - x_1)^2 + (y_2 - y_1)^2 + (z_2 - z_1)^2\}}\ldots\ldots\ldots\ldots\ldots(1')$$

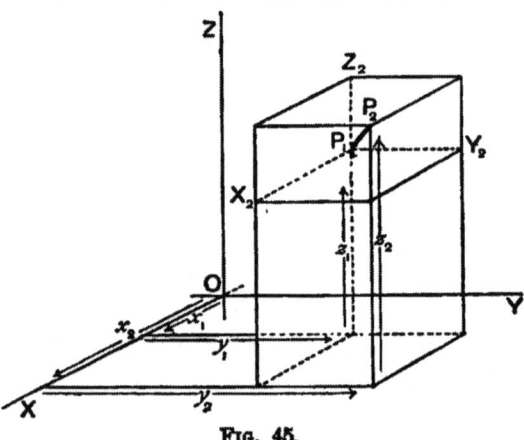

Fig. 45.

If we suppose the point P in Fig. 44 to vary its position, but always to remain at the same distance, a say, from O, it will lie upon a sphere; the coordinates of P will by (1) always satisfy the equation
$$x^2 + y^2 + z^2 = a^2,$$

which is therefore called the *equation of the sphere*. Similarly we see from (1′) that the equation of the sphere with centre $P_1(x_1, y_1, z_1)$ and radius a is
$$(x-x_1)^2+(y-y_1)^2+(z-z_1)^2=a^2. \quad\quad\quad\quad\quad\quad (2)$$

(iii) *Direction Cosines of a Line.* Let OP (Fig. 44) be any line through O; on the line take one direction, say, the direction from O to P as positive. The position of the line will be definitely fixed when the angles that the *positive direction* of the line makes with the positive directions of the coordinate axes are known. These angles, namely XOP, YOP, ZOP, are called the *direction angles* of the line, and the cosines of these angles are called the *direction cosines* of the line. Each of these angles may be taken as lying between 0° and 180° inclusive.

Thus the direction angles of OX are (0°, 90°, 90°), of OX' (180°, 90°, 90°), and the direction cosines are (1, 0, 0), (−1, 0, 0) respectively.

If a, β, γ are the direction angles of OP, then
$$\cos a = OM/OP, \quad \cos \beta = OL'/OP, \quad \cos \gamma = ON'/OP,$$
and
$$\cos^2 a + \cos^2 \beta + \cos^2 \gamma = \frac{OM^2 + OL'^2 + ON'^2}{OP^2} = 1.$$

If we write l, m, n in place of $\cos a$, $\cos \beta$, $\cos \gamma$, we see that the direction cosines (l, m, n) of a line are connected by the identical relation
$$l^2 + m^2 + n^2 = 1. \quad\quad\quad\quad\quad\quad (3)$$

When the line does not pass through the origin, draw a line through O parallel to the direction on the line that is taken as positive; the direction cosines of the line so drawn are those of the given line.

If the distance between P_1 and P_2 is r, r being considered positive, the direction cosines of the segment P_1P_2 are
$$(x_2-x_1)/r, \quad (y_2-y_1)/r, \quad (z_2-z_1)/r, \dots$$
and those of the segment P_2P_1 are
$$(x_1-x_2)/r, \quad (y_1-y_2)/r, \quad (z_1-z_2)/r. \dots \quad\quad (3')$$

(iv) *Cosine of the Angle between two Lines.* Let (l_1, m_1, n_1), (l_2, m_2, n_2) be the direction cosines of the lines, and draw OP, OQ (Fig. 46) parallel to the positive direction of the lines. Let OQ be the projection of OP on OQ, and let PN be perpendicular to the plane XOY and NM perpendicular to OX. Then
$$OM = l_1 OP, \quad MN = m_1 OP,$$
$$NP = n_1 OP, \quad OQ = OP \cos \theta,$$
where θ is the angle between the lines OP, OQ.

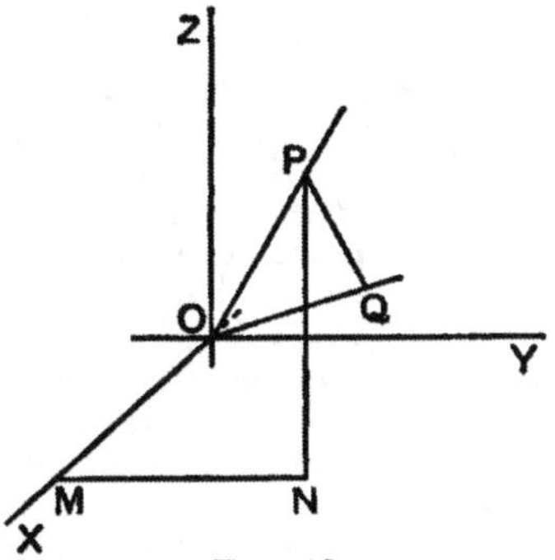

By the fundamental principles of projection, the projection of OP on OQ is equal to the sum of the projections of OM, MN, NP on OQ

FIG. 46.

But the projection of OM on OQ is $l_2 OM$, of MN is $m_2 MN$, and of NP is $n_2 NP$, since l_2 is the cosine of the angle between OM and OQ, etc.
Hence
$$OP \cos \theta = l_2 OM + m_2 MN + n_2 NP$$
$$= l_2 l_1 OP + m_2 m_1 OP + n_2 n_1 OP,$$
and therefore $\cos \theta = l_1 l_2 + m_1 m_2 + n_1 n_2.$ ....(4)

Since $\sin^2 \theta = 1 - \cos^2 \theta$ and $l_1^2 + m_1^2 + n_1^2 = 1$, $l_2^2 + m_2^2 + n_2^2 = 1$,
we have $\sin^2 \theta = (l_1^2 + m_1^2 + n_1^2)(l_2^2 + m_2^2 + n_2^2) - (l_1 l_2 + m_1 m_2 + n_1 n_2)^2$
$$= (m_1 n_2 - m_2 n_1)^2 + (n_1 l_2 - n_2 l_1)^2 + (l_1 m_2 - l_2 m_1)^2 \ldots\ldots\ldots (5)$$

The condition that two lines should be at right angles is, from (4),
$$l_1 l_2 + m_1 m_2 + n_1 n_2 = 0. \ldots\ldots\ldots\ldots\ldots\ldots (4')$$

(v) *Equations of a Straight Line.* Let the point P_1 (x_1, y_1, z_1) be a fixed point on the line and let P_2 (Fig. 45) be *any other* point (x, y, z) on the line. Let $P_1 P_2 = r$ and let (l, m, n) be the direction cosines of $P_1 P_2$; then
$$P_1 X_2 = x - x_1 = lr\ ;\quad y - y_1 = mr\ ;\quad z - z_1 = nr,$$
and therefore
$$\frac{x - x_1}{l} = \frac{y - y_1}{m} = \frac{z - z_1}{n} \ldots\ldots\ldots\ldots\ldots\ldots(6)$$

Equations (6) express the relations that hold between the co-ordinates of *any point* on the line and those of the fixed point, and are therefore called *the equations* of the line. Had a point P_3 been taken on the opposite side of P_1 from that on which P_2 lies the direction cosines of $P_1 P_3$ would have been $(-l, -m, -n)$ but the resulting equations would have been the same. If r be the absolute distance between the variable point (x, y, z) and the fixed point (x_1, y_1, z_1) we may write
$$(x - x_1)/l = \ldots = \pm r \ldots\ldots\ldots\ldots\ldots\ldots(6')$$
the + or − sign being taken according as the variable point lies to the positive or to the negative side of the fixed point.

(vi) *Equation of a Plane.* The equation $x = a$ is clearly true for every point on a plane parallel to the plane YOZ and distant a from that plane; in other words $x = a$ is *the equation of a plane* parallel to the plane YOZ. Similarly $y = b$, $z = c$ are the equations of planes parallel to the other two coordinate planes. The equations of the coordinate planes themselves are $x = 0$, $y = 0$, $z = 0$ respectively.

The equation $y = ax + b$ when considered *with reference solely to the coordinate plane XOY* represents a straight line, AB say. If through AB a plane is drawn parallel to $Z'OZ$ the coordinates of *every point* in that plane will still satisfy the equation $y = ax + b$. When considered with reference to space therefore the equation represents a plane *parallel to the axis of the omitted coordinate.* Similarly $z = ax + b$, $z = ay + b$ represent planes parallel to OY, OX respectively.

Let a plane meet the coordinate axes at A, B, C (Fig. 47); let OL be the perpendicular from O on the plane, (l, m, n) the direction cosines

EQUATIONS OF LINES AND PLANES

of OL. Take $P(x, y, z)$ any point on the plane and draw the coordinates OM, MN, NP.

The projection of OP on OL is OL itself which may be denoted by p; also the projection of OP is equal to the sum of the projections of OM, MN, NP on OL which are lOM, mMN nNP respectively. Hence

$$lx + my + nz = p \quad \ldots\ldots\ldots\ldots(7)$$

so that the equation of a plane is of the first degree in the coordinates.

If $D = \sqrt{(a^2 + b^2 + c^2)}$ the equation

$$ax + by + cz = d \quad \ldots\ldots\ldots\ldots(7')$$

may be written in the form

$$lx + my + nz = p$$

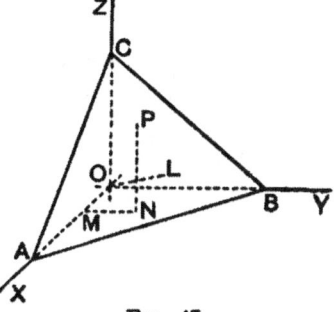

Fig. 47.

by putting l, m, n, p for $a/D, b/D, c/D$, d/D respectively, the sign of the root being chosen the same as that of d so that p or d/D may be positive. The quantities a/D, b/D, c/D are direction cosines since the sum of their squares is unity which is the condition required by (3) for direction cosines. These quantities are the direction cosines of the *normal* to the plane.

The direction cosines of the normal to the plane $x = 0$ are $(1, 0, 0)$; of the normal to the plane $y = ax + b$, that is, $-ax + y = b$ are $(-a/\sqrt{(a^2+1)}, 1/\sqrt{(a^2+1)}, 0)$, and so on.

(vii) *Equation of a Surface. Equations of a Curve.* In general an equation of the form $z = f(x, y)$ or $F(x, y, z) = 0$ represents a surface. Thus by (ii) the equation $x^2 + y^2 + z^2 - a^2 = 0$ represents a sphere of radius a.

Again, when the coordinates of a point satisfy *two* equations $F(x, y, z) = 0$, $\phi(x, y, z) = 0$, the point must lie *on each of the surfaces* represented by these equations, that is, the two equations, considered as simultaneous, are the equations of the *curve of intersection of the surfaces*. Thus the two equations

$$3x + 2y + z = 1, \quad 2x + 3y - z = 2$$

represent planes; the two, taken as simultaneous equations, represent their curve of intersection, that is, are the equations of a certain straight line. Or, again, equations (6) may be written

$$y - y_1 = \frac{m}{l}(x - x_1), \quad y - y_1 = \frac{m}{n}(z - z_1),$$

which are the equations of two planes; the intersection of the planes is the straight line given by (6).

The two equations $\quad x = 1, \quad x^2 + y^2 + z^2 = 9$

represent a circle which is the curve in which a plane intersects a sphere.

(viii) *Polar Coordinates.* In Fig. 44, let $OP=r$, $\angle ZOP=\theta$, $\angle XON=\phi$; then r, θ, ϕ are called the *polar* coordinates of P. The relations between the rectangular coordinates (x, y, z) and the polar coordinates (r, θ, ϕ) of the same point P are easily seen to be

$$x = r\sin\theta\cos\phi, \quad y = r\sin\theta\sin\phi, \quad z = r\cos\theta.$$

(ix) *Cylindrical coordinates.* In Fig. 44, let $ON=\rho$, $\angle XON=\phi$, $NP=z$; then ρ, ϕ, z are called the *cylindrical* coordinates of P. Evidently $\quad \rho = r\sin\theta\,; \quad x = \rho\cos\phi, \; y = \rho\sin\phi.$

ρ, ϕ are the *plane* polar coordinates of N, the projection of P on the plane XOY; r, θ, ϕ are sometimes called *spherical polar* coordinates.

Ex. 1. Find the equation of the plane through the three points $(1, 0, 0)$, $(0, 2, 0)$, $(0, 0, 3)$.

Let the equation be $ax+by+cz=d$; the coordinates of each point must satisfy the equation. Hence, to find a, b, c, d, we have

$$a=d\,; \quad 2b=d\,; \quad 3c=d,$$

that is, $\quad a/d = 1, \quad b/d = \tfrac{1}{2}, \quad c/d = \tfrac{1}{3}\,;$

and the required equation is

$$x + y/2 + z/3 = 1.$$

It will be noticed that only the *ratios* of a, b, c, d are required; the equation of the plane thus contains only *three* independent constants just as that of the straight line in Plane Geometry contains only *two*.

Ex. 2. The equation of the plane through $(a, 0, 0)$, $(0, b, 0)$, $(0, 0, c)$, is $\quad x/a + y/b + z/c = 1\,;$

a, b, c are *the intercepts* made by the plane on the coordinate axes.

Ex. 3. The equation of the plane through the three points $(2, 0, 3)$, $(-1, 5, 2)$, $(3, -4, -2)$ is

$$29x + 16y - 7z = 37.$$

Ex. 4. The equations of the line through (x_1, y_1, z_1), (x_2, y_2, z_2) are

$$\frac{x - x_1}{x_1 - x_2} = \frac{y - y_1}{y_1 - y_2} = \frac{z - z_1}{z_1 - z_2}.$$

By § 89a (v), the equations of the line through (x_1, y_1, z_1) in the direction (l, m, n) are

$$(x - x_1)/l = (y - y_1)/m = (z - z_1)/n.$$

Since (x_2, y_2, z_2) lies on the line, the ratios $l : m : n$ are determined by

$$(x_2 - x_1)/l = (y_2 - y_1)/m = (z_2 - z_1)/n,$$

from which the required equations follow.

Ex. 5. The direction cosines of the line through the points $(3, -4, -2)$, $(-1, 5, 2)$ are $\left(-\dfrac{4}{\sqrt{113}},\ \dfrac{9}{\sqrt{113}},\ \dfrac{4}{\sqrt{113}}\right)$, the positive direction of the line being from the first to the second of the points.

Ex. 6. The cosine of the (acute) angle between the planes
$$3x+y-2z=1, \quad 2x-3y+z=1$$
is 1/14.

Ex. 7. If $\delta\theta$ is the small angle between the lines whose direction cosines are l, m, n and $l+\delta l, m+\delta m, n+\delta n$, show that, approximately,
$$(\delta\theta)^2 = (\delta l)^2 + (\delta m)^2 + (\delta n)^2.$$

Both sets of cosines satisfy (iii) (3), and therefore
$$(l+\delta l)^2 + (m+\delta m)^2 + (n+\delta n)^2 = 1 = l^2 + m^2 + n^2,$$
and
$$2(l\delta l + m\delta m + n\delta n) = -\{(\delta l)^2 + (\delta m)^2 + (\delta n)^2\}.$$

Again $\quad 2\sin^2(\tfrac{1}{2}\delta\theta) = 1 - \cos\delta\theta = -(l\delta l + m\delta m + n\delta n),$
and the result follows at once.

§ 90. Total Derivatives. Complete Differentials.

Let $u = f(x, y)$ and let x and y be functions of a third variable t. To prove

$$\frac{du}{dt} = \frac{\partial u}{\partial x}\frac{dx}{dt} + \frac{\partial u}{\partial y}\frac{dy}{dt} \quad\quad\quad\quad\quad\quad (\text{A})$$

When t takes the increment δt let x, y, u take the increments $\delta x, \delta y, \delta u$ respectively; then
$$\delta u = f(x+\delta x, y+\delta y) - f(x, y),$$
and this equation may be written
$$\delta u = [f(x+\delta x, y+\delta y) - f(x, y+\delta y)]$$
$$+ [f(x, y+\delta y) - f(x, y)] \quad\quad\quad\quad (1)$$

By the mean value theorem § 72
$$f(x+\delta x, y+\delta y) - f(x, y+\delta y) = f_x(x+\theta_1\delta x, y+\delta y)\,\delta x \ldots (2)$$
$$f(x, y+\delta y) - f(x, y) = f_y(x, y+\theta_2\delta y)\,\delta y \ldots\ldots (3)$$

where θ_1, θ_2 are proper fractions. The coefficient of δx in (2) is the x-derivative of $f(x, y+\delta y)$ taken on the supposition that $y+\delta y$ does not vary and with x replaced by $x + \theta_1\delta x$; the coefficient of δy in (3) is the y-derivative of $f(x, y)$ taken on the supposition that x does not vary and with y replaced by $y + \theta_2\delta y$. Hence

$$\frac{\delta u}{\delta t} = f_x(x+\theta_1\delta x, y+\delta y)\frac{\delta x}{\delta t} + f_y(x, y+\theta_2\delta y)\frac{\delta y}{\delta t}.$$

Now, $\underset{\delta t=0}{\mathrm{L}} \dfrac{\delta u}{\delta t} = \dfrac{du}{dt}$, $\underset{\delta t=0}{\mathrm{L}} \dfrac{\delta x}{\delta t} = \dfrac{dx}{dt}$, $\underset{\delta t=0}{\mathrm{L}} \dfrac{\delta y}{\delta t} = \dfrac{dy}{dt}$,

$$\underset{\delta t=0}{\mathrm{L}} f_x(x + \theta_1 \delta x, y + \delta y) = f_x(x, y);$$

$$\underset{\delta t=0}{\mathrm{L}} f_y(x, y + \theta_2 \delta y) = f_y(x, y);$$

since δx, δy converge to zero with δt, and the functions are all supposed to be continuous. Writing $\partial u/\partial x$, $\partial u/\partial y$ in place of $f_x(x, y)$, $f_y(x, y)$ we get equation (A).

In the same way if $u = f(x, y, z)$ and x, y, z are all functions of a variable t we get

$$\dfrac{du}{dt} = \dfrac{\partial u}{\partial x}\dfrac{dx}{dt} + \dfrac{\partial u}{\partial y}\dfrac{dy}{dt} + \dfrac{\partial u}{\partial z}\dfrac{dz}{dt} \quad\ldots\ldots\ldots\ldots\ldots(B)$$

and so on for any number of variables.

In (A) we may suppose t to be the variable x; y is then a function of x, and u is really a function of the one variable x. Equation (A) becomes in that case

$$\dfrac{du}{dx} = \dfrac{\partial u}{\partial x} + \dfrac{\partial u}{\partial y}\cdot\dfrac{dy}{dx} \quad\ldots\ldots\ldots\ldots\ldots\ldots\ldots(A')$$

and in the same way, from (B)

$$\dfrac{du}{dx} = \dfrac{\partial u}{\partial x} + \dfrac{\partial u}{\partial y}\dfrac{dy}{dx} + \dfrac{\partial u}{\partial z}\dfrac{dz}{dx} \quad\ldots\ldots\ldots\ldots\ldots(B')$$

In these equations $\partial u/\partial x$ and du/dx have quite different meanings. The derivative $\partial u/\partial x$ is formed on the supposition that *an explicitly named* variable x alone varies; on the other hand du/dx is the limit of $\delta u/\delta x$ where δu is the change in u, due (i) to the change δx in the explicitly named variable x, and (ii) to the changes δy, δz, which are themselves due to the change δx.

du/dx, du/dt are called *total derivatives* with respect to x and t respectively.

Ex. If $u = x^2 + y^2$, then

$$\partial u/\partial x = 2x; \quad \partial u/\partial y = 2y.$$

But if y is a function of x, say $y = ax + b$, then

$$u = x^2 + (ax+b)^2; \quad \dfrac{du}{dx} = 2x + 2a(ax+b);$$

TOTAL DERIVATIVES.

or we may use (A'); then
$$\frac{du}{dx} = \frac{\partial u}{\partial x} + \frac{\partial u}{\partial y}\frac{dy}{dx} = 2x + 2y \cdot a,$$
since $dy/dx = a$, and we get the same result as before.

If x, y are independent, and if δu be the change in u, due to the independent changes δx, δy, equation (1) may be written
$$\delta u = f_x(x + \theta_1 \delta x, y + \delta y)\, \delta x + f_y(x, y + \theta_2 \delta y)\, \delta y$$
$$= [f_x(x, y) + \omega_1]\, \delta x + [f_y(x, y) + \omega_2]\, \delta y$$
where ω_1, ω_2 converge to zero with δx and δy. Hence if we take δx, δy as independent principal infinitesimals and write dx, dy in place of δx, δy the products $\omega_1 dx$, $\omega_2 dy$ will be of order higher than the first and the principal part du of δu will be given by
$$du = \frac{\partial u}{\partial x}\, dx + \frac{\partial u}{\partial y}\, dy \quad \ldots\ldots\ldots\ldots\ldots\ldots (\text{C})$$

Similarly for three (or more) independent variables
$$du = \frac{\partial u}{\partial x}\, dx + \frac{\partial u}{\partial y}\, dy + \frac{\partial u}{\partial z}\, dz \quad \ldots\ldots\ldots\ldots (\text{D})$$

du is called *a total differential* or *a complete differential*

and
$$\frac{\partial u}{\partial x}\, dx, \; \frac{\partial u}{\partial y}\, dy, \; \frac{\partial u}{\partial z}\, dz$$

are called *partial differentials*. These partial differentials are sometimes written $d_x u$, $d_y u$, $d_z u$.

If x, y, z are not independent but functions of t then, since $dx = (dx/dt)\,dt \ldots$, we should get equations of the same form as (C), (D) by multiplying (A), (B) by dt.

These equations (A)...(D) have important applications in geometry and mechanics. For plane geometry the equation (A') is very useful; the reader should study the following examples carefully.

Ex. 1. Let $u = ax^2 + by^2 - 1$; then, x and y being independent,
$$\partial u/\partial x = 2ax, \quad \partial u/\partial y = 2by.$$

Consider now the *equation* $u = 0$. The variables x, y are no longer independent; the point (x, y) must lie on the conic $u = 0$ and y may be considered a function of x, namely an ordinate of the conic. Since

u is now always zero for the admissible variations of x and y, the *total* x-derivative of u (not the *partial*) must be always zero. Hence by (A')

$$\frac{\partial u}{\partial x}+\frac{\partial u}{\partial y}\frac{dy}{dx}=0 \text{ and } \frac{dy}{dx}=-\frac{\partial u}{\partial x}\Big/\frac{\partial u}{\partial y}=-\frac{ax}{by}.$$

This equation gives the gradient at the point (x, y) on the conic.

Ex. 2. Let u be any function $f(x, y)$ of x and y; *the equation* $u=0$, that is $f(x, y)=0$ defines y as a function of x, namely y is the ordinate of the curve $f(x, y)=0$. As in ex. 1, the *total* x-derivative of u is zero and the gradient at the point (x, y) is given by

$$\frac{dy}{dx}=-\frac{\partial u}{\partial x}\Big/\frac{\partial u}{\partial y}=-\frac{\partial f}{\partial x}\Big/\frac{\partial f}{\partial y}$$

where f is written for brevity instead of $f(x, y)$.

Ex. 3. If $f(x, y)=x^3+y^3-3axy$, the gradient of the curve whose equation is $f(x, y)=0$ is

$$\frac{dy}{dx}=-\frac{3x^2-3ay}{3y^2-3ax}=\frac{ay-x^2}{y^2-ax}.$$

Ex. 4. If $pv=k\theta$ (k constant) find dp in terms of $dv, d\theta$.

$$\frac{\partial p}{\partial v}=-\frac{k\theta}{v^2}=-\frac{p}{v}\,;\quad \frac{\partial p}{\partial \theta}=\frac{k}{v}=\frac{p}{\theta}$$

$$dp=\frac{\partial p}{\partial v}dv+\frac{\partial p}{\partial \theta}d\theta=-\frac{p}{v}dv+\frac{p}{\theta}d\theta.$$

Ex. 5. If $u=\tan^{-1}(y/x)$ prove that
$$du=(xdy-ydx)/(x^2+y^2).$$

Ex. 6. If $x=r\cos\theta$, $y=r\sin\theta$, r and θ independent, show that
$$dx=\cos\theta\,dr-r\sin\theta\,d\theta, \quad dy=\sin\theta\,dr+r\cos\theta\,d\theta$$
$$xdy-ydx=r^2d\theta.$$

Ex. 7. Let $u=f(x, y)-z$, then

$$\frac{\partial u}{\partial x}=\frac{\partial f}{\partial x},\quad \frac{\partial u}{\partial y}=\frac{\partial f}{\partial y},\quad \frac{\partial u}{\partial z}=-1.$$

The *equation* $u=0$ defines a surface, and now z may be considered a function of two independent variables x, y, namely $z=f(x, y)$,

$$\frac{\partial z}{\partial x}=\frac{\partial f}{\partial x}=\frac{\partial u}{\partial x},\quad \frac{\partial z}{\partial y}=\frac{\partial f}{\partial y}=\frac{\partial u}{\partial y}.$$

§ 91. Geometrical Illustrations. Let P be the point (x, y, z) on the surface given by $z=f(x, y)$, and let APB, DPF be sections made by planes through P parallel to the planes YOZ, ZOX respectively (Fig. 48).

GEOMETRICAL ILLUSTRATIONS. 215

For points on the curve DPF, y is constant. Hence $\partial z/\partial x$ or $\partial f/\partial x$ is the gradient at P of the curve DPF. Similarly $\partial z/\partial y$ is the gradient at P of the curve APB.

If the equation of the surface is $u = 0$, where u is the function $F(x, y, z)$, the equation $u = 0$ defines z as a function of two independent variables x and y. Along the curve DPF, y is constant. Hence along that curve the total x-derivative of u or $F(x, y, z)$ must be zero, u being for that curve a function of x and z which is always zero. Therefore, as in § 90, ex. 1, 2,

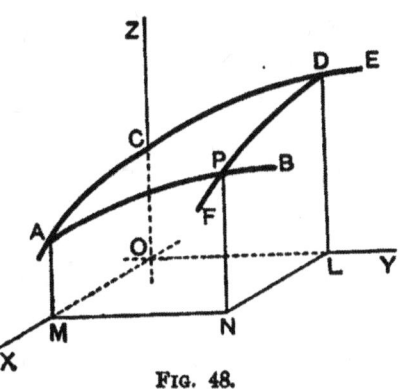

Fig. 48.

$$\frac{\partial u}{\partial x} + \frac{\partial u}{\partial z}\frac{\partial z}{\partial x} = 0 \text{ or } \frac{\partial z}{\partial x} = -\frac{\partial F}{\partial x}\bigg/\frac{\partial F}{\partial z} \quad \ldots\ldots\ldots\ldots(1)$$

and $\partial z/\partial x$ is the gradient at P of the curve DPF.

Similarly, the gradient at P of the curve APB is

$$\frac{\partial z}{\partial y} = -\frac{\partial F}{\partial y}\bigg/\frac{\partial F}{\partial z} \quad \ldots\ldots\ldots\ldots\ldots\ldots(1')$$

These expressions reduce to those first given if we put $u = f(x, y) - z$. (Compare § 90, ex. 7.)

Tangent Plane. In Fig. 49 let APP_2, BPP_1 be sections of the surface by planes parallel to YOZ, ZOX respectively. Let P be the point (x, y, z), $MM_1 = \delta x$, $MM_2 = \delta y$, M_3 the point $(x+\delta x, y+\delta y, 0)$ and P_3 the point on the surface $(x+\delta x, y+\delta y, z+\delta z)$. Let PT_1, PT_2 be the tangents at P to BPP_1, APP_2, T_1 lying on M_1P_1 produced and T_2 on M_2P_2 produced; $Pm_1m_3m_2$ is a rectangle parallel and congruent to $MM_1M_3M_2$; P_1P_3, P_2P_3 are the curves in which the planes M_1m_3, M_2m_3 cut the surface, and T_1T_3, T_2T_3 the straight lines in which the same planes cut the plane through PT_1T_2.

Since the gradient at P of the curve BPP_1 is $\partial z/\partial x$ and of the curve APP_2 is $\partial z/\partial y$ we have

$$m_1T_1 = \frac{\partial z}{\partial x}\delta x, \quad m_2T_2 = \frac{\partial z}{\partial y}\delta y.$$

Also the geometry of the figure shows that

$$m_1T_1 + m_2T_2 = m_3T_3,$$

so that

$$\frac{\partial z}{\partial x}\delta x + \frac{\partial z}{\partial y}\delta y = m_3T_3 \dots\dots\dots\dots\dots\dots(2)$$

But

$$\frac{\partial z}{\partial x}\delta x + \frac{\partial z}{\partial y}\delta y,$$

is the principal part of δz (§ 90, c); therefore, when Pm_1, Pm_2 represent δx, δy the line m_3T_3 represents the principal part of δz.

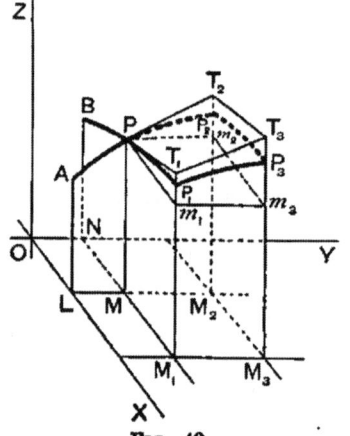

Fig. 49.

Again, if the plane PMM_3P_3 cut the surface in the curve PP_3 the gradient at P of PP_3 is the limit of $\delta z/MM_3$ or $\delta z/Pm_3$. But by the principles of infinitesimals the limit of $\delta z/Pm_3$ is the same as the limit of m_3T_3/Pm_3, since m_3T_3 is the principal part of δz. Hence the gradient at P of PP_3 is m_3T_3/Pm_3 and therefore PT_3 is the tangent at P to the arc PP_3.

The plane PT_1T_2 is completely determined by the two lines PT_1, PT_2, that is, by the point (x, y, z) and the derivatives $\partial z/\partial x$, $\partial z/\partial y$. By proper choice of the independent increments δx, δy we could get any point Q on the surface near P and the tangent to the arc PQ would lie in the plane PT_1T_2. This plane is therefore called the *tangent plane* to the surface at P, and the line through P perpendicular to the tangent plane is called the *normal* to the surface at P.

To find the equation of the tangent plane suppose T_3 to

TANGENT PLANE AND NORMAL. 217

be *any* point on it and let its coordinates be (X, Y, Z) those of P being (x, y, z); then

$$m_3 T_3 = Z - z,\ m_1 T_1 = \frac{\partial z}{\partial x} \delta x = \frac{\partial z}{\partial x}(X - x),\ m_2 T_2 = \frac{\partial z}{\partial y}(Y - y),$$

and therefore by (2)

$$Z - z = \frac{\partial z}{\partial x}(X - x) + \frac{\partial z}{\partial y}(Y - y)\dots\dots\dots\dots(3)$$

which is the equation of the tangent plane at the point (x, y, z) on the surface, X, Y, Z being the current coordinates of any point on the plane.

When the equation of the surface is $F(x, y, z) = 0$ we get by substituting the values of $\partial z/\partial x$, $\partial z/\partial y$ from (1) and (1′)

$$(X - x)\frac{\partial F}{\partial x} + (Y - y)\frac{\partial F}{\partial y} + (Z - z)\frac{\partial F}{\partial z} = 0 \dots\dots\dots(3')$$

The direction cosines of the normal are (89a (vi)) proportional to the coefficients of the current coordinates X, Y, Z and therefore the equations of the normal are

$$(X - x)\Big/\frac{\partial z}{\partial x} = (Y - y)\Big/\frac{\partial z}{\partial y} = -(Z - z)\dots\dots\dots\dots(4)$$

or $\qquad (X - x)\Big/\frac{\partial F}{\partial x} = (Y - y)\Big/\frac{\partial F}{\partial y} = (Z - z)\Big/\frac{\partial F}{\partial z}\dots\dots\dots(4')$

Ex. 1. The equation $F(x, y, z) = x^2 + y^2 + z^2 - a^2 = 0$ represents a sphere of radius a.

$$\frac{\partial F}{\partial x} = 2x,\quad \frac{\partial F}{\partial y} = 2y,\quad \frac{\partial F}{\partial z} = 2z.$$

Hence the tangent plane at (x, y, z) is
$$(X - x)2x + (Y - y)2y + (Z - z)2z = 0,$$
or $\qquad xX + yY + zZ = (x^2 + y^2 + z^2) = a^2,$
since (x, y, z) is on the sphere. If we take x, y, z as current coordinates and (x_1, y_1, z_1) as the point of contact, the equation is
$$x_1 x + y_1 y + z_1 z = a^2.$$
The equations of the normal are
$$(X - x)/2x = (Y - y)/2y = (Z - z)/2z,\ \text{ or }\ X/x = Y/y = Z/z.$$
With (x, y, z) as current coordinates the equations are
$$x/x_1 = y/y_1 = z/z_1.$$
The normal clearly passes through the origin which is the centre of the sphere.

Ex. 2. The equation $ax^2 + by^2 + cz^2 - 1 = 0$ represents a surface called a *central conicoid* (a plane section is in general a central conic). Find the equations of the tangent plane and the normal at (x_1, y_1, z_1).

Ex. 3. The equation $by^2 + cz^2 - 2x = 0$ represents a *non-central* conicoid. Find the equations of the tangent plane and the normal at (x_1, y_1, z_1).

Ex. 4. The equation $ax^2 + by^2 + cz^2 = 0$, where a, b, c are not all of the same sign, represents a cone with its vertex at the origin. Find the equations of the tangent plane and the normal at (x_1, y_1, z_1).

If $F(x, y, z) = ax^2 + by^2 + cz^2$, the derivatives $\partial F/\partial x$, $\partial F/\partial y$, $\partial F/\partial z$ are all zero when $x = y = z = 0$. Every tangent plane to the cone goes through the origin, and there is no definite normal at the origin; the equations of the tangent plane and normal are illusory if formed for the origin. At special points on a surface it may happen that the three partial derivatives are all zero; in that case there is no definite tangent plane or normal at the point. Such points are usually called *conical points*, the vertex of a cone being the simplest case.

§ 92. Rate of Variation in a given Direction. It is often necessary to find the rate at which a function of the coordinates of a point varies in a given direction. Thus at a point in a cooling solid the rate of diminution of temperature will usually be different along different lines issuing from the point.

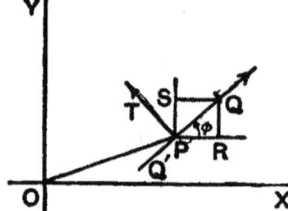

Fig. 50.

(i) Let u be a function $f(x, y)$ of two variables, and let u_P, u_Q denote the values of u at $P(x, y)$ and at $Q(x + \delta x, y + \delta y)$ respectively, where $PR = \delta x$, $RQ = PS = \delta y$ (Fig. 50). Then
$$u_P = f(x, y),$$
$$u_Q = f(x + \delta x, y + \delta y).$$

The average rate of increase of u in the direction PQ is $(u_Q - u_P)/PQ$ which may be written

$$\frac{u_Q - u_P}{PQ} = \frac{u_R - u_P}{PR} \cdot \frac{PR}{PQ} + \frac{u_Q - u_R}{RQ} \cdot \frac{RQ}{PQ}.$$

As in § 89a (iii) let the direction PQ be distinguished from that of PQ', and let PQ make with OX the angle ϕ (see note at end of this article); then

$$PR/PQ = \cos \phi, \quad RQ/PQ = \sin \phi.$$

VARIATION IN A GIVEN DIRECTION.

Exactly as in § 90 it may be seen that the limits for $PQ=0$ of $(u_R-u_P)/PR$ and $(u_Q-u_R)/RQ$ are $\partial u/\partial x$ and $\partial u/\partial y$ respectively. If the element PQ be denoted by δs (where s may represent the length of a line, straight or curved, measured from some point up to P) then the average rate of increase of u is $(u'_Q - u_P)/\delta s$ or $\delta u_P/\delta s$ and the rate of increase of u in the direction PQ is then

$$\frac{\partial u}{\partial s} = \frac{\partial u}{\partial x}\cos\phi + \frac{\partial u}{\partial y}\sin\phi \quad\ldots\ldots\ldots\ldots\ldots(1)$$

If the rate of increase of u in the direction PT perpendicular to PQ is denoted by $\partial u/\partial s'$, PT making the angle $\phi + \pi/2$ with OX

$$\frac{\partial u}{\partial s'} = -\frac{\partial u}{\partial x}\sin\phi + \frac{\partial u}{\partial y}\cos\phi \quad\ldots\ldots\ldots\ldots\ldots(2)$$

(ii) If u be a function $f(x, y, z)$ of three variables the rate of increase $\partial u/\partial s$ in the direction PQ may be proved in exactly the same way to be

$$\frac{\partial u}{\partial s} = l\frac{\partial u}{\partial x} + m\frac{\partial u}{\partial y} + n\frac{\partial u}{\partial z} \quad\ldots\ldots\ldots\ldots\ldots(3)$$

where (l, m, n) are the direction cosines of PQ.

If (1) and (2) be solved for $\partial u/\partial x$, $\partial u/\partial y$ we get

$$\frac{\partial u}{\partial x} = \frac{\partial u}{\partial s}\cos\phi - \frac{\partial u}{\partial s'}\sin\phi \quad\ldots\ldots\ldots\ldots\ldots(1')$$

$$\frac{\partial u}{\partial y} = \frac{\partial u}{\partial s}\sin\phi + \frac{\partial u}{\partial s'}\cos\phi \quad\ldots\ldots\ldots\ldots\ldots(2')$$

Equations (1), (2), (3) may be obtained at once from the equations of § 90 by taking t equal to s or s'. We have used the notation $\partial u/\partial s$ instead of du/ds since we wish to find the rate of variation of u in two (or three) independent directions. In this and similar cases the meaning of the symbols must be constantly attended to.

For examples on the use of these formulae, see the set at the end of the chapter (examples 9-13).

Note on Angles.—In earlier chapters it has been sufficient to consider the positive or negative *acute* angle that a line lying in the plane XOY makes with OX. In discussions like that of case (i), however, where only *half-lines* issuing from a point are dealt with, that restriction must be given up, and the angle may, like the angle θ of the polar coordinates, vary from 0 to 2π or from $-\pi$ to π. Thus PQ makes the angle $(\phi+\pi)$ or $(\phi-\pi)$ with OX. With this new convention $\cos\theta$ may have a negative value. For lines in space the determination specified in § 89a (iii) is always sufficient.

§ 93. Derivatives of Higher Orders. The derivatives of $u=f(x, y)$ will usually be functions of x and y, and will therefore have derivatives. Hence we have 2nd, 3rd, ... partial derivatives. The notation for these is similar to that for functions of one variable:

$$\frac{\partial^2 u}{\partial x^2}, \frac{\partial^3 u}{\partial y^3}, \ldots; u_{xx}, u_{yyy} \ldots; f_{xx}(x, y), f_{yyy}(x, y) \ldots.$$

The brackets and the letters within them are usually omitted and the last pair are written f_{xx}, f_{yyy}.

Again, the y-derivative of $\partial u/\partial x$ is

$$\frac{\partial}{\partial y}\frac{\partial u}{\partial x} \text{ or } \frac{\partial^2 u}{\partial y \partial x}$$

while the x-derivative of $\partial u/\partial y$ is

$$\frac{\partial}{\partial x}\frac{\partial u}{\partial y} \text{ or } \frac{\partial^2 u}{\partial x \partial y}.$$

When all the functions in question are continuous these two derivatives are equal (see below). For example, let $u=ax^m y^n$; then

$$\frac{\partial u}{\partial x}=max^{m-1}y^n; \quad \frac{\partial^2 u}{\partial y \partial x}=nmax^{m-1}y^{n-1};$$

$$\frac{\partial u}{\partial y}=nax^m y^{n-1}; \quad \frac{\partial^2 u}{\partial x \partial y}=mnax^{m-1}y^{n-1};$$

so that
$$\frac{\partial^2 u}{\partial y \partial x}=\frac{\partial^2 u}{\partial x \partial y}$$

when $u=ax^m y^n$. In other words the *order* of differentiating is indifferent; the operations of differentiating as to x and as to y are *commutative*.

Ex. Verify that these two derivatives are equal when

(i) $u=x\sin y+y\sin x$; (ii) $u=x\log y$; (iii) $u=\tan^{-1}\frac{y}{x}$.

HIGHER PARTIAL DERIVATIVES.

The symbol
$$\frac{\partial^5 u}{\partial x^2 \partial y^3}$$
means that u is to be differentiated first three times as to y then twice as to x; while the symbol
$$\frac{\partial^5 u}{\partial y^3 \partial x^2}$$
means that u is to be differentiated first twice as to x then thrice as to y.

Similar meanings and notations hold for the higher derivatives of a function of any number of variables.

A sound proof of the commutative property is somewhat difficult. Consider the expression
$$\frac{f(x+h, y+k) - f(x, y+k) - f(x+h, y) + f(x, y)}{hk} \quad \ldots\ldots\ldots\ldots(1)$$

By the definition of a derivative
$$\mathop{L}_{h=0} \frac{f(x+h, y+k) - f(x, y+k)}{h} = f_x(x, y+k),$$
$$\mathop{L}_{h=0} \frac{f(x+h, y) - f(x, y)}{h} = f_x(x, y).$$

Hence the limit of (1) for $h=0$ is
$$\{f_x(x, y+k) - f_x(x, y)\}/k \quad \ldots\ldots\ldots\ldots\ldots\ldots\ldots(2)$$

Again the limit of (2) for $k=0$ is the y-derivative of $f_x(x, y)$, that is f_{yx}.

By interchanging the second and third terms in the numerator of (1) and finding first the limit for $k=0$ and then the limit for $h=0$ we should get f_{xy}. Thus f_{yx} and f_{xy} are both derived as limits from the same expression. But the assumption that the limits will be the same in whatever order we make h and k tend to zero is equivalent to assuming the theorem to be proved. A simple example will show that the order of taking the limits is not necessarily indifferent.

Take the function $(h+2k)/(h+k)$

$$\mathop{L}_{h=0} \frac{h+2k}{h+k} = \frac{2k}{k}; \quad \mathop{L}_{k=0}\left\{\mathop{L}_{h=0}\frac{h+2k}{h+k}\right\} = \mathop{L}_{k=0}\frac{2k}{k} = 2,$$

$$\mathop{L}_{k=0} \frac{h+2k}{h+k} = \frac{h}{h}; \quad \mathop{L}_{h=0}\left\{\mathop{L}_{k=0}\frac{h+2k}{h+k}\right\} = \mathop{L}_{h=0}\frac{h}{h} = 1.$$

Of course neither in this expression nor in (1) must h or k become zero; zero is the *limit* not a *value* of h and k.

Assuming all the functions in question to be continuous we may proceed as follows. Let, for brevity,
$$F(x) = f(x, y+k) - f(x, y),$$

then the numerator of (1) is $F(x+h)-F(x)$. By the Mean Value Theorem
$$F(x+h)-F(x)=hF_x(x+\theta_1 h), \qquad 0<\theta_1<1,$$
or, returning to the function $f(x, y)$,
$$F(x+h)-F(x)=h\{f_x(x+\theta_1 h, y+k)-f_x(x+\theta_1 h, y)\},$$
so that (1) becomes
$$\{f_x(x+\theta_1 h, y+k)-f_x(x+\theta_1 h, y)\}/k \dots\dots\dots\dots\dots(2')$$
Now apply the Mean Value Theorem to the function of y in (2');
$$f_x(x+\theta_1 h, y+k)-f_x(x+\theta_1 h, y)=kf_{yx}(x+\theta_1 h, y+\theta_2 k), \quad 0<\theta_2<1,$$
and (1) becomes $\qquad f_{yx}(x+\theta_1 h, y+\theta_2 k).\dots\dots\dots\dots\dots(3)$

Again, taking $\phi(y)=f(x+h, y)-f(x, y)$ instead of $F(x)$, the numerator of (1) is $\phi(y+k)-\phi(y)$. Apply the Mean Value Theorem and proceed as before. We thus find that (1) is equal to
$$f_{xy}(x+\theta_3 h, y+\theta_4 k)\dots\dots\dots\dots\dots\dots(4)$$

The two expressions (3), (4) are therefore equal. Since the functions are continuous the limits are therefore equal in whatever way h and k tend to zero, that is $f_{yx}=f_{xy}$.

The commutative property may be easily extended by induction to higher derivatives, the functions being supposed all continuous. Thus, since
$$\frac{\partial^2 u}{\partial x \partial y}=\frac{\partial^2 u}{\partial y \partial x},$$
$$\frac{\partial^3 u}{\partial x^2 \partial y}=\frac{\partial}{\partial x}\left(\frac{\partial^2 u}{\partial x \partial y}\right)=\frac{\partial^3 u}{\partial x \partial y \partial x}=\frac{\partial^2}{\partial x \partial y}\left(\frac{\partial u}{\partial x}\right)=\frac{\partial^2}{\partial y \partial x}\left(\frac{\partial u}{\partial x}\right)=\frac{\partial^3 u}{\partial y \partial x^2}.$$

In general,
$$\frac{\partial^{p+q+r} u}{\partial x^p \partial y^q \partial z^r}=\frac{\partial^{p+q+r} u}{\partial x^p \partial z^r \partial y^q}=\frac{\partial^{p+q+r} u}{\partial z^r \partial x^p \partial y^q}=\dots,$$
as may be readily shown by induction.

Ex. 1. In Fig. 48, § 91, let V be the volume bounded by the surface $APDC$, the coordinate planes and the planes MP, LP.

Prove $\qquad$ (i) $\dfrac{\partial V}{\partial x}=$ area $MNPA$; $\dfrac{\partial^2 V}{\partial y \partial x}=NP$.

$\qquad\qquad$ (ii) $\dfrac{\partial V}{\partial y}=$ area $LNPD$; $\dfrac{\partial^2 V}{\partial x \partial y}=NP$.

If V be taken as the function $f(x, y)$, we get a geometrical proof of the commutative property.

COMMUTATIVE PROPERTY OF DERIVATIVES. 223

Ex. 2. If $u = \log r$ where $r^2 = (x-a)^2 + (y-b)^2$ and $(x-a), (y-b)$ are not simultaneously zero, show that

$$\frac{\partial^2 u}{\partial x^2} + \frac{\partial^2 u}{\partial y^2} = 0,$$

$$\frac{\partial (r^2)}{\partial x} = 2(x-a); \text{ also } \frac{\partial (r^2)}{\partial x} = 2r \frac{\partial r}{\partial x}; \therefore \frac{\partial r}{\partial x} = \frac{x-a}{r};$$

$$\frac{\partial u}{\partial x} = \frac{d \log r}{dr} \cdot \frac{\partial r}{\partial x} = \frac{1}{r} \cdot \frac{x-a}{r} = \frac{x-a}{r^2};$$

$$\frac{\partial^2 u}{\partial x^2} = \frac{1}{r^2} + (x-a) \frac{\partial \cdot r^{-2}}{\partial x} = \frac{1}{r^2} - \frac{2(x-a)^2}{r^4}.$$

Similarly,
$$\frac{\partial^2 u}{\partial y^2} = \frac{1}{r^2} - \frac{2(y-b)^2}{r^4},$$

and therefore by addition, since $r^2 = (x-a)^2 + (y-b)^2$, the result follows.

Ex. 3. If $u = 1/r$ where $r^2 = (x-a)^2 + (y-b)^2 + (z-c)^2$ and $(x-a), (y-b), (z-c)$ are not simultaneously zero, prove that

$$\frac{\partial^2 u}{\partial x^2} + \frac{\partial^2 u}{\partial y^2} + \frac{\partial^2 u}{\partial z^2} = 0.$$

A charge m of electricity concentrated at (a, b, c) has at (x, y, z) the potential m/r. The potential V therefore satisfies the equation last written, usually called *Laplace's Equation*.

If charges $m_1, m_2, \ldots$ are concentrated at $(a_1, b_1, c_1), (a_2, b_2, c_2), \ldots$ the potential V at (x, y, z) of these charges is $\Sigma (m/r)$ where

$$r_1^2 = (x-a_1)^2 + (y-b_1)^2 + (z-c_1)^2,$$

so that the potential at any point (x, y, z) not coincident with any of the masses also satisfies the same equation.

Ex. 4. If $u = f(x, y)$ and x, y are functions of t find d^2u/dt^2,

We have
$$\frac{du}{dt} = \frac{\partial u}{\partial x} \frac{dx}{dt} + \frac{\partial u}{\partial y} \frac{dy}{dt}, \quad \ldots \ldots \ldots \ldots \ldots \ldots (i)$$

$$\frac{d^2 u}{dt^2} = \frac{d}{dt}\left(\frac{du}{dt}\right) = \frac{\partial u}{\partial x} \frac{d^2 x}{dt^2} + \frac{dx}{dt} \frac{d}{dt}\left(\frac{\partial u}{\partial x}\right) + \frac{\partial u}{\partial y} \frac{d^2 y}{dt^2} + \frac{dy}{dt} \frac{d}{dt}\left(\frac{\partial u}{\partial y}\right).$$

Since $\partial u/\partial x$ is a function of x and y, its t-derivative is found in the same way as du/dt in (i); that is, write $\partial u/\partial x$ for u in (i),

$$\frac{d}{dt}\left(\frac{\partial u}{\partial x}\right) = \frac{\partial^2 u}{\partial x^2} \frac{dx}{dt} + \frac{\partial^2 u}{\partial y \partial x} \frac{dy}{dt}.$$

Similarly,
$$\frac{d}{dt}\left(\frac{\partial u}{\partial y}\right) = \frac{\partial^2 u}{\partial x \partial y} \frac{dx}{dt} + \frac{\partial^2 u}{\partial y^2} \frac{dy}{dt}.$$

Substituting these values and noting that $\dfrac{\partial^2 u}{\partial y \partial x} = \dfrac{\partial^2 u}{\partial x \partial y}$, we find

$$\frac{d^2 u}{dt^2} = \frac{\partial u}{\partial x} \frac{d^2 x}{dt^2} + \frac{\partial u}{\partial y} \frac{d^2 y}{dt^2} + \frac{\partial^2 u}{\partial x^2}\left(\frac{dx}{dt}\right)^2$$
$$+ 2\frac{\partial^2 u}{\partial x \partial y} \frac{dx}{dt} \frac{dy}{dt} + \frac{\partial^2 u}{\partial y^2}\left(\frac{dy}{dt}\right)^2.$$

Ex. 5. If $f(x, y) = 0$, show that d^2y/dx^2 is given by the equation

$$f_y \frac{d^2y}{dx^2} + f_{yy}\left(\frac{dy}{dx}\right)^2 + 2f_{xy}\frac{dy}{dx} + f_{xx} = 0.$$

This may be obtained directly; or in Ex. 4 put $t = x$ and note that $u = f(x, y) = 0$ for every value of x and y, and therefore du/dt and d^2u/dt^2 are both zero, while $dx/dt = 1$, $d^2x/dt^2 = 0$.

Deduce in this way the results of examples 26, 27, 28 of *Exercises XIV*.

Ex. 6. If $u = f(y + ax)$, prove

(i) $\dfrac{\partial u}{\partial x} = a\dfrac{\partial u}{\partial y}$; (ii) $\dfrac{\partial^2 u}{\partial x^2} = a^2\dfrac{\partial^2 u}{\partial y^2}$.

Ex. 7. If $u = f(x + at) + \phi(x - at)$, prove

$$\frac{\partial^2 u}{\partial t^2} = a^2 \frac{\partial^2 u}{\partial x^2}.$$

Verify for $u = A\cos(x + at) + B\sin(x - at)$.

§ 94. Complete Differentials. If u is a function of the two independent variables x and y the complete differential of u is (§ 90)

$$du = \frac{\partial u}{\partial x}dx + \frac{\partial u}{\partial y}dy \quad \ldots\ldots\ldots\ldots\ldots\ldots\ldots(1)$$

Now the question arises; given two functions $\phi(x, y)$, $\psi(x, y)$ of two independent variables x, y, is there always another function u which has

$$\phi(x, y)\,dx + \psi(x, y)\,dy \quad \ldots\ldots\ldots\ldots\ldots\ldots(2)$$

as its differential?

If x and y are not independent, say if y is a function $f(x)$ of x, we may replace y by $f(x)$ and dy by $f'(x)\,dx$. The expression (2) will thus become of the form $F(x)\,dx$ and in this case (§ 82) there is a function which has $F(x)$ as its x-derivative or $F(x)\,dx$ as its differential.

But if x and y are independent the case is altered. For suppose the expression (2) to be the complete differential of a function u; then the expressions (1) and (2) must be equal for all values of dx and dy. Since dx and dy are independent we may put $dy = 0$, $dx \neq 0$ and we get

$$\phi(x, y) = \partial u/\partial x,$$

and in the same way

$$\psi(x, y) = \partial u/\partial y$$

COMPLETE DIFFERENTIALS.

for all values of x and y. Therefore

$$\frac{\partial \phi}{\partial y} = \frac{\partial^2 u}{\partial y \partial x} = \frac{\partial^2 u}{\partial x \partial y} = \frac{\partial \psi}{\partial x} \quad \ldots\ldots\ldots\ldots\ldots(3)$$

Hence the expression (2) cannot be a complete differential unless $\partial \phi / \partial y = \partial \psi / \partial x$.

Condition (3) is therefore a *necessary* condition; it is also a sufficient condition, but for the proof of sufficiency we refer to treatises on Differential Equations.

If P, Q, R are functions of three independent variables x, y, z the necessary and sufficient conditions that

$$P\,dx + Q\,dy + R\,dz$$

should be a complete differential, that is, that there should be a function u of x, y, z such that

$$du = P\,dx + Q\,dy + R\,dz$$

are that
$$\frac{\partial Q}{\partial x} = \frac{\partial P}{\partial y}, \quad \frac{\partial R}{\partial y} = \frac{\partial Q}{\partial z}, \quad \frac{\partial P}{\partial z} = \frac{\partial R}{\partial x}.$$

The student may show that these conditions are *necessary*.

Ex. 1. $(3x^2 - 4xy)dx + (3y^2 - 2x^2)dy$ is a complete differential for

$$\frac{\partial}{\partial y}\left(3x^2 - 4xy\right) = -4x = \frac{\partial}{\partial x}\left(3y^2 - 2x^2\right)$$

and $u = x^3 - 2x^2y + y^3$.

Ex. 2. If $P = yz(2x + y + z)$, $Q = zx(x + 2y + z)$, $R = xy(x + y + 2z)$, show that $du = P\,dx + Q\,dy + R\,dz$ where $u = x^2yz + y^2zx + z^2xy$.

§ 95. Application to Mechanics.

Let the plane curve APQ be the path of a particle which is acted on by a force F, making an angle ϵ with the tangent PT, F and ϵ being functions of the coordinates x, y of P. Let W be the work done from the position $A(a, b)$ up to the position P, and let the arc AP be denoted by s. To the first order of infinitesimals the work done over the distance ds is

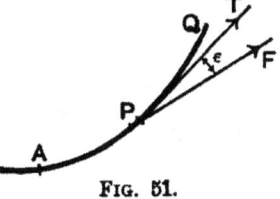

Fig. 51.

$$dW = F \cos \epsilon \, ds.$$

Let PT, PF make the angles ϕ, ψ with the x-axis; then $\cos \phi = dx/ds$, $\sin \phi = dy/ds$, and since $\cos \epsilon = \cos(\phi - \psi)$

$$F \cos \epsilon = F \cos \psi \frac{dx}{ds} + F \sin \psi \frac{dy}{ds} = X \frac{dx}{ds} + Y \frac{dy}{ds},$$

where $X = F \cos \psi$, $Y = F \sin \psi$, the components of F parallel to the axes. We thus get

$$dW = \left(X \frac{dx}{ds} + Y \frac{dy}{ds} \right) ds \ldots\ldots\ldots\ldots\ldots(1)$$

Suppose now that $Xdx + Ydy$ is the complete differential of a *single-valued* function $f(x, y)$. Therefore $X = \partial f/\partial x$ and $Y = \partial f/\partial y$, so that

$$dW = \left(\frac{\partial f}{\partial x} \frac{dx}{ds} + \frac{\partial f}{\partial y} \frac{dy}{ds} \right) ds = \frac{df}{ds} ds = df.$$

Hence, as the particle moves along the curve, the rate dW/ds at which W changes is equal to the rate df/ds at which the function $f(x, y)$ changes, and any change dW in W is equal to the corresponding change df in the function $f(x, y)$. As the particle moves from A to P the work done is therefore equal to the change in $f(x, y)$, so that

$$W = f(x, y) - f(a, b)\ldots\ldots\ldots\ldots\ldots(2)$$

If W' is the work from A to P when the particle moves along a *different* path of length s', we have as before

$$dW' = \frac{df}{ds'} ds' = df,$$

so that $\qquad W' = f(x, y) - f(a, b) = W.$

In this case, therefore, the work done by the force is independent of the path between A and P, and, when A is fixed and P variable, is a function simply of the coordinates of P. When P coincides with A, that is, when the path is a closed curve, the work done is zero (see Ex. 2 for an illustration in which $f(x, y)$ is *multiple-valued*).

Suppose on the other hand that $Xdx + Ydy$ is not a complete differential. In this case the coefficient of ds in (1) is not the total derivative of a function $f(x, y)$. To find the work from A to P we must express y in terms of x by

APPLICATION TO MECHANICS.

using the equation of the path. Equation (1) will then become

$$dW = \left(X + Y\frac{dy}{dx}\right)\frac{dx}{ds}ds = \left(X + Y\frac{dy}{dx}\right)dx \quad \ldots \ldots \ldots (1')$$

and the coefficient of dx in (1') is a function of x alone. For different paths the function $X + Y(dy/dx)$ will have different values, and therefore W will depend not merely on the coordinates of P but also on the path from A to P. (See Ex. 3).

If APQ is not a plane curve it is easy to prove by the same method as that of finding dx/ds, dy/ds for a plane curve (§ 62) that the direction cosines of the tangent PT are (§ 89a, iii. (3'))

$$dx/ds, \ dy/ds, \ dz/ds.$$

If l, m, n are the direction cosines of PF

$$\cos \epsilon = l\frac{dx}{ds} + m\frac{dy}{ds} + n\frac{dz}{ds}$$

and
$$dW = \left(X\frac{dx}{ds} + Y\frac{dy}{ds} + Z\frac{dz}{ds}\right)ds \quad \ldots \ldots \ldots \ldots (3)$$

where $X = lF$, $Y = mF$, $Z = nF$ are the components of F parallel to the axes.

Exactly as before we see that if $Xdx + Ydy + Zdz$ is the complete differential of a single-valued function $f(x, y, z)$

$$dW = df \quad \text{and} \quad W = f(x, y, z) - f(a, b, c)$$

where A is the point (a, b, c). In this case W is independent of the particular path from A to P.

If however $Xdx + Ydy + Zdz$ is not a complete differential it will be necessary to use the equations of the path and W will depend not merely on the coordinates of A and P but also on the particular path from A to P.

When $Xdx + Ydy + Zdz$ is a complete differential the force F is said to be *conservative*; the components are the derivatives of a force function u or a potential $-V$,

$$X = \frac{\partial u}{\partial x} \quad \text{or} \quad X = -\frac{\partial V}{\partial x}; \quad dW = du \quad \text{or} \quad dW = -dV.$$

Ex. 1. If $F = m/r^2$ where $r^2 = x^2 + y^2 + z^2 = OP^2$, and the direction of F is from O to P, then F is a conservative force.

For
$$X = \frac{x}{r}F = \frac{mx}{r^3}, \quad Y = \frac{my}{r^3}, \quad Z = \frac{mz}{r^3},$$

and
$$dW = Xdx + Ydy + Zdz = \frac{m}{r^3}\left(xdx + ydy + zdz\right)$$

or
$$dW = \frac{m}{r^2}dr = d\left(-\frac{m}{r}\right), \quad \text{since } xdx + ydy + zdz = rdr.$$

Hence $W = -m/r + \text{const}$, and if $V = m/r$,
$$X = -\partial V/\partial x, \quad Y = -\partial V/\partial y, \quad Z = -\partial V/\partial z.$$

The work from position P to position Q is
$$m/OP - m/OQ,$$
and is independent of the path between P and Q.

Ex. 2. Let $X = -y/r^2$, $Y = x/r^2$, where $r^2 = x^2 + y^2$. In this case, putting $y/x = \tan\theta$,
$$dW = (xdy - ydx)/r^2 = d.\tan^{-1}(y/x) = d\theta,$$
and therefore $W = \theta + \text{constant}.$

If the point P sets out from A and, after describing a closed curve within which the origin lies, returns to A, the angle θ and therefore W will increase by 2π. The work done is not zero, although dW is a complete differential $d\theta$; the function θ is multiple-valued.

If, however, the path is a closed curve within which the origin does not lie, the work done over that path will be zero.

Ex. 3. Let $X = -y$, $Y = x$. In this case $xdy - ydx$ is not a complete differential. Let A coincide with the origin O, and let the path be the parabola $y = cx^2$. Then, by (1'),
$$dW = (-cx^2 + x.2cx)dx = cx^2 dx; \quad W = \tfrac{1}{3}cx^3 = \tfrac{1}{3}xy.$$

If the path is $y = cx^3$, we find $W = \tfrac{1}{2}cx^4 = \tfrac{1}{2}xy$, the work being different for different paths.

§ 96. Applications to Thermodynamics.

The condition of a given mass of thermodynamic substance, say unit mass, is completely defined by three variables p, v, θ the intensity of pressure, the volume and the absolute temperature. p, v, θ are connected by an equation, the characteristic equation of the substance, $f(p, v, \theta) = 0$; for a perfect gas the equation is $pv = k\theta$, k being a constant. Of the three variables, therefore, only two are independent.

Since $f(p, v, \theta) = 0$ its total differential is zero; therefore

$$\frac{\partial f}{\partial p}dp + \frac{\partial f}{\partial v}dv + \frac{\partial f}{\partial \theta}d\theta = 0 \quad \ldots\ldots\ldots\ldots\ldots\ldots(1)$$

APPLICATION TO THERMODYNAMICS.

If p be constant, and v, θ vary, then $dp = 0$ and we have

$$\frac{\partial v}{\partial \theta} = -\frac{\partial f}{\partial \theta} \div \frac{\partial f}{\partial v}.$$

Forming in the same way $\partial \theta / \partial p$, $\partial p / \partial v$ and multiplying we get

$$\frac{\partial v}{\partial \theta} \cdot \frac{\partial \theta}{\partial p} \cdot \frac{\partial p}{\partial v} = -1 \quad \ldots\ldots\ldots\ldots\ldots\ldots\ldots(2)$$

Clearly

$$\frac{\partial v}{\partial \theta} \frac{\partial \theta}{\partial v} = +1 \quad \ldots\ldots\ldots\ldots\ldots\ldots\ldots(3)$$

It must be remembered that in all these expressions the derivative of one of the variables p, v, θ with respect to a second is formed on the supposition that the third variable is constant.

If a small quantity δQ of heat be communicated to the substance and change p, v, θ by δp, δv, $\delta \theta$ respectively, then δQ can be expressed in terms of any two of these increments. To the first order of infinitesimals we may write, with θ, v as the variables,

$$dQ = M d\theta + N dv \ldots\ldots\ldots\ldots\ldots\ldots(4)$$

It is to be most carefully noticed that $d\theta$, dv are *any* arbitrary small changes of temperature and volume. The three differentials $d\theta$, dv, dp are subject merely to the restriction expressed in equation (1), and any two of them may have values chosen at will.

The specific heat at constant volume (K_v) is the limit for $\delta \theta = 0$ of $\delta Q / \delta \theta$ on the supposition that the volume does not change when θ increases by $\delta \theta$, that is, on the supposition that $dv = 0$. But if $dv = 0$ equation (4) gives $dQ/d\theta = M$ so that $K_v = M$.

The specific heat at constant pressure (K_p) is the limit for $\delta \theta = 0$ of $\delta Q / \delta \theta$ on the supposition that p is constant, that is, that $dp = 0$. To find K_p equation (4) must be transformed so that θ and p shall be the independent variables. Since v is a function of θ and p we have

$$dv = \frac{\partial v}{\partial \theta} d\theta + \frac{\partial v}{\partial p} dp \ldots\ldots\ldots\ldots\ldots\ldots(5)$$

and (4) becomes
$$dQ = \left(M + N \frac{\partial v}{\partial \theta}\right) d\theta + N \frac{\partial v}{\partial p} dp.$$

230 AN ELEMENTARY TREATISE ON THE CALCULUS.

Therefore $\quad K_p = M + N\dfrac{\partial v}{\partial \theta} = K_v + N\dfrac{\partial v}{\partial \theta}$ ....(6)

The elasticity of the substance is $-v\,dp/dv$ (§ 70). Let E_θ denote the elasticity when the substance expands at constant temperature;

therefore, $\quad E_\theta = -v\dfrac{\partial p}{\partial v}.$

where $\partial p/\partial v$ is taken subject to the condition that θ is constant.

Let E_ϕ denote the elasticity when the substance expands adiabatically, that is, so that heat neither enters nor escapes. We must distinguish the v-derivative of p in the two cases. For the present denote the v-derivative of p for adiabatic expansion by $(\partial p/\partial v)_\phi$ and let $\partial p/\partial v$ retain its previous meaning. Therefore,

$$\dfrac{E_\phi}{E_\theta} = -v\left(\dfrac{\partial p}{\partial v}\right)_\phi \div \left\{-v\dfrac{\partial p}{\partial v}\right\} = \left(\dfrac{\partial p}{\partial v}\right)_\phi \div \dfrac{\partial p}{\partial v} \ldots\ldots\ldots(7)$$

To find $(\partial p/\partial v)_\phi$ we must transform (4) so that p and v shall be independent variables. Now

$$d\theta = \dfrac{\partial \theta}{\partial p}dp + \dfrac{\partial \theta}{\partial v}dv;$$

and therefore $\quad dQ = M\dfrac{\partial \theta}{\partial p}dp + \left(M\dfrac{\partial \theta}{\partial v} + N\right)dv.$............(8)

$(\partial p/\partial v)_\phi$ is the value deduced from (8) on the supposition that $dQ = 0$. Therefore

$$\left(\dfrac{\partial p}{\partial v}\right)_\phi = -\dfrac{M\dfrac{\partial \theta}{\partial v} + N}{M\dfrac{\partial \theta}{\partial p}} = -\dfrac{M + N\dfrac{\partial v}{\partial \theta}}{M\dfrac{\partial \theta}{\partial p}\dfrac{\partial v}{\partial \theta}} \text{ by (3).}$$

The numerator last written is K_p and $M = K_v$; therefore

$$\left(\dfrac{\partial p}{\partial v}\right)_\phi \div \left(\dfrac{\partial p}{\partial v}\right) = -K_p/K_v \dfrac{\partial \theta}{\partial p}\dfrac{\partial v}{\partial \theta}\dfrac{\partial p}{\partial v} = K_p/K_v \text{ by (2).}$$

Hence $\quad \dfrac{E_\phi}{E_\theta} = \dfrac{K_p}{K_v}.$ ....(9)

For a perfect gas K_p/K_v is a constant, γ; also for a perfect gas $pv = k\theta$, and therefore
$$\partial p/\partial v = -p/v.$$
Hence for adiabatic expansion, by (7) and (9),
$$\frac{dp}{dv} \div \left(-\frac{p}{v}\right) = \gamma, \text{ or } \frac{d}{dv}\log(pv^\gamma) = 0,$$
that is, $\qquad pv^\gamma = \text{constant}.$

The results (1)...(3), (5)...(9), are merely formal consequences of the definitions and the two equations
$$f(p, v, \theta) = 0 \text{ and } dQ = Md\theta + Ndv.$$

§ 97. Four Thermodynamic Relations. dQ in the previous article is not a complete differential; we cannot express Q in the form $F(\theta, v) - F(\theta_0, v_0)$ without assuming some further relation between θ and v. Physically, Q is not a function of θ and v; heat may be given to the substance and θ, v go through a range of values and return to their initial values, while the heat absorbed in the process is not equal to that given out. Compare § 95 when dW is a complete differential; when x, y return to their initial values a, b, $W = 0$, that is, the work done by the force F is equal to that done against it.

It is shown in treatises on thermodynamics that if we put $dQ = \theta d\phi$ where ϕ is the entropy we can replace (4) by
$$dE = \theta d\phi - pdv. \dots\dots\dots\dots\dots\dots(10)$$
E is the intrinsic energy and pdv the work done in the infinitesimal expansion dv. dE is a complete differential; that is, E is a function of the variables that define the state of the substance.

There are now four variables p, v, θ, ϕ, but of these only two are independent. If v, θ are chosen as independent the symbol $\partial p/\partial \theta$ is now not sufficiently clear; it means the θ-derivative of p when v is constant. But if ϕ, θ were the independent variables, $\partial p/\partial \theta$ would mean the θ-derivative when ϕ is constant. To avoid confusion we will, when there is doubt, enclose the derivative in a bracket and affix the independent variable which is supposed to be constant.

Thus $(\partial p/\partial\theta)_v$ means that v, θ are the independent variables, and that v is constant in forming $\partial p/\partial\theta$.

Since dE is a complete differential we have (§ 94 (3)) from equation (10)

$$\left(\frac{\partial\theta}{\partial v}\right)_\phi = \left(\frac{\partial(-p)}{\partial\phi}\right)_v, \text{ or } \left(\frac{\partial p}{\partial\phi}\right)_v = -\left(\frac{\partial\theta}{\partial v}\right)_\phi. \quad \ldots\ldots\ldots\ldots(4')$$

Let now v, θ be the independent variables, then since

$$d\phi = \frac{\partial\phi}{\partial v}dv + \frac{\partial\phi}{\partial\theta}d\theta,$$

(10) becomes $\quad dE = \theta\dfrac{\partial\phi}{\partial\theta}d\theta + \left(\theta\dfrac{\partial\phi}{\partial v} - p\right)dv,$

and therefore $\quad \dfrac{\partial}{\partial v}\left(\theta\dfrac{\partial\phi}{\partial\theta}\right) = \dfrac{\partial}{\partial\theta}\left(\theta\dfrac{\partial\phi}{\partial v} - p\right),$

or $\quad \theta\dfrac{\partial^2\phi}{\partial v\partial\theta} = \theta\dfrac{\partial^2\phi}{\partial\theta\partial v} + \dfrac{\partial\phi}{\partial v} - \dfrac{\partial p}{\partial\theta},$

that is, $\quad\left(\dfrac{\partial p}{\partial\theta}\right)_v = \left(\dfrac{\partial\phi}{\partial v}\right)_\theta, \quad\ldots\ldots\ldots\ldots\ldots\ldots\ldots\ldots\ldots(3')$

since the two derivatives of second order are equal.

In the same way, by taking p, ϕ as independent variables,

we get $\quad\left(\dfrac{\partial v}{\partial\phi}\right)_p = \left(\dfrac{\partial\theta}{\partial p}\right)_\phi \quad\ldots\ldots\ldots\ldots\ldots\ldots\ldots(2')$

and by taking p, θ as independent variables

$$\left(\frac{\partial v}{\partial\theta}\right)_p = -\left(\frac{\partial\phi}{\partial p}\right)_\theta \ldots\ldots\ldots\ldots\ldots\ldots\ldots(1')$$

Equations (1'), (2'), (3'), (4') are those numbered (1), (2), (3), (4) in Maxwell's *Heat*, p. 169.

In effecting the differentiations it must be borne in mind that for example when v, θ are the independent variables $\partial\theta/\partial v$ is zero. The careful working out of these four relations will give much information as to the meaning of partial derivatives; it is necessary at each step to attend to the meaning of the operations rather than to the notation.

Ex. 1. $d\phi = dQ/\theta = (Md\theta + Ndv)/\theta$.

For a perfect gas K_p, K_v are constant, and by § 96, $K_v = M$ and $K_p - K_v = N(\partial v/\partial \theta)_p$. But for a perfect gas $(\partial v/\partial \theta)_p = v/\theta$. Hence $N/\theta = (K_p - K_v)/v$ and

$$d\phi = K_v \frac{d\theta}{\theta} + (K_p - K_v)\frac{dv}{v} = d \cdot \log\left(\theta^{K_v} v^{K_p - K_v}\right)$$

as may be tested by differentiation. Therefore
$$\phi = \log(\theta^{K_v} v^{K_p - K_v}) + \text{const.}$$

For adiabatic expansion $dQ = 0$ and $d\phi = 0$; we therefore have
$$\theta^{K_v} v^{K_p - K_v} = \text{const.} \quad \text{or} \quad pv^\gamma = \text{const.},$$
as in § 96.

Ex. 2. The gain in energy dE due to a supply dQ of heat is given by
$$dE = dQ - p\,dv = (N - p)dv + Md\theta.$$

Show that if dE is a complete differential, dQ is not.

Since dE is a complete differential, we have
$$\frac{\partial(N-p)}{\partial \theta} = \frac{\partial M}{\partial v} \quad \text{or} \quad \frac{\partial N}{\partial \theta} = \frac{\partial M}{\partial v} + \frac{\partial p}{\partial \theta},$$

that is, $\partial N/\partial \theta$, $\partial M/\partial v$ are not equal and the result follows.

Ex. 3. Prove that
$$\frac{\partial p}{\partial \theta}\frac{\partial v}{\partial \phi} - \frac{\partial p}{\partial \phi}\frac{\partial v}{\partial \theta} = 1; \quad \frac{\partial \theta}{\partial p}\frac{\partial \phi}{\partial v} - \frac{\partial \theta}{\partial v}\frac{\partial \phi}{\partial p} = 1.$$

Ex. 4. Show that equation (10) may be written in the forms
$$dE = d(\theta\phi) - \phi d\theta - p\,dv\,; \quad dE = -d(pv) + \theta d\phi + v\,dp\,;$$
$$dE = d(\theta\phi) - d(pv) - \phi d\theta + v\,dp,$$
and then prove (1′), (2′), (3′).

Ex. 5. It is shown in works on Thermodynamics that $d\phi$ is a complete differential. Prove that
$$\frac{\partial}{\partial v}\left(\frac{M}{\theta}\right) = \frac{\partial}{\partial \theta}\left(\frac{N}{\theta}\right).$$

§ 98. Change of Variable. Differentials of Higher Orders. When the independent variable x of a function y is changed by a substitution, $x = \phi(t)$ say, to a new independent variable t the x-derivatives of y, $D_x y$, $D_x^2 y$..., must be expressed in terms of the t-derivatives of y, $\dot{y}$, $\ddot{y}$.... We have found (§ 68, ex. 2) that

$$D_x y = \dot{y}/\dot{x}, \quad D_x^2 y = (\dot{x}\ddot{y} - \dot{y}\ddot{x})/\dot{x}^3 \ldots\ldots\ldots\ldots(1)$$

and it is easy to find $D_x^3 y$, $D_x^4 y$... when these are required. Since $\phi(t)$ is supposed to be given, the values of $\dot{x}$, $\ddot{x}$..., can be calculated, and the substitution of $\phi(t)$ for x and the above values for Dy, $D^2 y$ changes any expression containing x, y, Dy, $D^2 y$... into one containing t, y, $\dot{y}$, $\ddot{y}$....

If we wish to make y the independent variable and x the dependent, then

$$D_x y = \frac{1}{D_y x}; \quad D_x^2 y = D_y\left(\frac{1}{D_y x}\right) \div D_y x = -\frac{D_y^2 x}{(D_y x)^3} \quad \ldots\ldots(2)$$

and so on.

Again if we change from rectangular to polar coordinates, an equation $f(x, y) = 0$ becomes an equation between r and θ and we may express $D_x y$, $D_x^2 y$..., in terms of $D_\theta r$, $D_\theta^2 r$, ... θ being the independent variable. For $x = r \cos\theta$, $y = r \sin\theta$ and we can differentiate the products $r \cos\theta$, $r \sin\theta$ with respect to θ, r being a function of θ,

$$\left. \begin{aligned} \frac{dx}{d\theta} &= \cos\theta \frac{dr}{d\theta} - r \sin\theta \\ \frac{d^2 x}{d\theta^2} &= \cos\theta \frac{d^2 r}{d\theta^2} - 2 \sin\theta \frac{dr}{d\theta} - r \cos\theta, \end{aligned} \right\} \quad (3)$$

with similar expressions for $dy/d\theta$, $d^2 y/d\theta^2$. In equations (1) we may suppose t replaced by θ, since of course t may represent any variable; $\dot{x}$ would be replaced by $dx/d\theta$, $\ddot{x}$ by $d^2 x/d\theta^2$ and so on. We should thus express Dy, $D^2 y$ in terms of r, θ, $dr/d\theta$, $d^2 r/d\theta^2$.

In geometry and mechanics differentials of order higher than the first are often required. When x is the independent variable, $dy = y' dx$ (§ 60). The second differential of y is denoted by $d^2 y$ and is defined by the equation

$$d^2 y = y'' dx^2 = (D_x^2 y) dx^2,$$

and in general the nth differential of y is denoted by $d^n y$ and is defined by the equation

$$d^n y = y^{(n)} dx^n = (D_x^n y) dx^n$$

where dx^n means $(dx)^n$.

If dx is an infinitesimal of the first order $d^n y$ is, in general, of the nth order.

DIFFERENTIALS OF HIGHER ORDERS.

In the second of equations (1) multiply the numerator and denominator of the fraction on the right by dt^3. Since t is the independent variable we have

$$dx = \dot{x}\,dt, \quad d^2x = \ddot{x}\,dt^2, \quad dy = \dot{y}\,dt, \quad d^2y = \ddot{y}\,dt^2,$$

and therefore $\quad D_x^2 y = (dx\,d^2y - dy\,d^2x)/dx^3 \quad \dotfill (4)$

$D_x^2 y$ is thus expressed as a quotient of differentials; the independent variable for the differentials is not x but t (or any other variable of which x and y are functions). If x is the independent variable, then *by definition*

$$d^2x = (D_x^2 x)dx^2 = 0 \times dx^2 = 0,$$

and similarly we see that d^3x, d^4x, ... are zero. In other words, *the differential of the independent variable is constant.*

From (4) we may easily derive (2). Take y as the independent variable; then $d^2y = 0$, $dx = (D_y x)dy$, $d^2x = (D_y^2 x)dy^2$ and (4) becomes

$$D_x^2 y = -dy(D_y^2 x)dy^2/(D_y x)^3 dy^3 = -D_y^2 x/(D_y x)^3.$$

For more than one independent variable the transformations are complicated. We will consider only one case that is of great importance in mathematical physics.

§ 99. Transformation of $\nabla^2 u$. Let u be a function of two independent variables x, y, and let x, y be changed to polar coordinates r, θ; we wish to express u_x, $u_{xx}\ldots$ in terms of u_r, $u_{rr}\ldots$. Of course a derivative u_r implies that x, y have been replaced in the function u by $r\cos\theta$, $r\sin\theta$.

$\partial u/\partial r$ is the rate of variation of u in the direction in which r increases, θ being constant. In § 92 put $\phi = \theta$, $s = r$ and we find

$$\frac{\partial u}{\partial r} = \cos\theta\frac{\partial u}{\partial x} + \sin\theta\frac{\partial u}{\partial y} \dotfill (1)$$

$\partial u/\partial s'$ in § 92 is the rate of variation of u in the direction $\phi + \pi/2$. Let $\phi = \theta$ so that PT is perpendicular to OP and $\delta s' = PT = r\tan\delta\theta$

$$\frac{\partial u}{\partial s'} = \mathop{\mathrm{L}}_{\delta s'=0} \frac{\delta u}{\delta s'} = \mathop{\mathrm{L}}_{\delta\theta=0} \frac{\delta u}{r\tan\delta\theta} = \frac{1}{r}\frac{\partial u}{\partial \theta}.$$

Hence $\quad \dfrac{1}{r}\dfrac{\partial u}{\partial \theta} = -\sin\theta\dfrac{\partial u}{\partial x} + \cos\theta\dfrac{\partial u}{\partial y} \dotfill (2)$

The element $\partial s'$ is replaced by $r\partial\theta$; $r\partial\theta$ is the element in the direction perpendicular to r just as ∂r is that in the direction of r.

Equations (1), (2) are so important that we give another proof of them. By § 90 (A), taking x and y as functions of r, θ being kept constant, we get by putting r for t

$$\frac{\partial u}{\partial r} = \frac{\partial u}{\partial x}\frac{\partial x}{\partial r} + \frac{\partial u}{\partial y}\frac{\partial y}{\partial r}. \quad\quad\quad (1')$$

Here $\partial u/\partial x$ means $(\partial u/\partial x)_y$ and $\partial x/\partial r$ means $(\partial x/\partial r)_\theta$ in the notation of § 97. Also

$$\left(\frac{\partial x}{\partial r}\right)_\theta = \frac{\partial(r\cos\theta)}{\partial r} = \cos\theta; \quad \left(\frac{\partial y}{\partial r}\right)_\theta = \sin\theta,$$

and the substitution of these values in (1') gives (1).

In the same way

$$\frac{\partial u}{\partial \theta} = \frac{\partial u}{\partial x}\left(\frac{\partial x}{\partial \theta}\right)_r + \frac{\partial u}{\partial y}\left(\frac{\partial y}{\partial \theta}\right)_r \quad\quad\quad (2')$$

and $\quad \left(\frac{\partial x}{\partial \theta}\right)_r = -r\sin\theta; \quad \left(\frac{\partial y}{\partial \theta}\right)_r = r\cos\theta,$

from which equation (2) follows.

Solving (1) and (2) for $\partial u/\partial x$, $\partial u/\partial y$ we get

$$\frac{\partial u}{\partial x} = \cos\theta \frac{\partial u}{\partial r} - \frac{\sin\theta}{r}\frac{\partial u}{\partial \theta} \quad\quad\quad (3)$$

$$\frac{\partial u}{\partial y} = \sin\theta \frac{\partial u}{\partial r} + \frac{\cos\theta}{r}\frac{\partial u}{\partial \theta} \quad\quad\quad (4)$$

The function $\quad \dfrac{\partial^2 u}{\partial x^2} + \dfrac{\partial^2 u}{\partial y^2} + \dfrac{\partial^2 u}{\partial z^2}$

is of very frequent occurrence in Physics and is usually denoted by $\nabla^2 u$. It is often necessary to transform $\nabla^2 u$ so that other variables shall be the independent variables.

First, let u be a function of the two variables x, y so that the third term is absent, and transform it so that r, θ, polar coordinates, shall be independent variables.

Denote $\partial u/\partial x$, $\partial u/\partial y$ by u_x, u_y; then we can find $\partial^2 u/\partial x^2$ in terms of r, θ by writing u_x in place of u in (3). We must calculate $\partial u_x/\partial r$, $\partial u_x/\partial \theta$. Now,

TRANSFORMATION OF $\nabla^2 U$.

$$\frac{\partial u_x}{\partial r} = \frac{\partial}{\partial r}\left\{\cos\theta\,\frac{\partial u}{\partial r} - \frac{\sin\theta}{r}\frac{\partial u}{\partial \theta}\right\}$$

$$= \cos\theta\,\frac{\partial^2 u}{\partial r^2} + \frac{\sin\theta}{r^2}\frac{\partial u}{\partial \theta} - \frac{\sin\theta}{r}\frac{\partial^2 u}{\partial r\partial\theta};$$

$$\frac{\partial u_x}{\partial \theta} = \cos\theta\,\frac{\partial^2 u}{\partial r\partial\theta} - \sin\theta\,\frac{\partial u}{\partial r} - \frac{\sin\theta}{r}\frac{\partial^2 u}{\partial \theta^2} - \frac{\cos\theta}{r}\frac{\partial u}{\partial \theta}.$$

Hence from (3)

$$\frac{\partial^2 u}{\partial x^2} = \frac{\partial u_x}{\partial x} = \cos\theta\,\frac{\partial u_x}{\partial r} - \frac{\sin\theta}{r}\frac{\partial u_x}{\partial \theta},$$

and when the above values of $\partial u_x/\partial r$, $\partial u_x/\partial \theta$ are substituted we get, after an easy reduction,

$$\frac{\partial^2 u}{\partial x^2} = \cos^2\theta\,\frac{\partial^2 u}{\partial r^2} - \frac{2\sin\theta\cos\theta}{r}\frac{\partial^2 u}{\partial r\partial\theta} + \frac{\sin^2\theta}{r^2}\frac{\partial^2 u}{\partial \theta^2}$$
$$+ \frac{\sin^2\theta}{r}\frac{\partial u}{\partial r} + \frac{2\sin\theta\cos\theta}{r^2}\frac{\partial u}{\partial \theta}\ldots\ldots(5)$$

In a similar way we find

$$\frac{\partial^2 u}{\partial y^2} = \sin^2\theta\,\frac{\partial^2 u}{\partial r^2} + \frac{2\sin\theta\cos\theta}{r}\frac{\partial^2 u}{\partial r\partial\theta} + \frac{\cos^2\theta}{r^2}\frac{\partial^2 u}{\partial \theta^2}$$
$$+ \frac{\cos^2\theta}{r}\frac{\partial u}{\partial r} - \frac{2\sin\theta\cos\theta}{r^2}\frac{\partial u}{\partial \theta}\ldots\ldots(6)$$

Adding (5) and (6) we get

$$\frac{\partial^2 u}{\partial x^2} + \frac{\partial^2 u}{\partial y^2} = \frac{\partial^2 u}{\partial r^2} + \frac{1}{r}\frac{\partial u}{\partial r} + \frac{1}{r^2}\frac{\partial^2 u}{\partial \theta^2}\ldots\ldots\ldots\ldots(7)$$

Next transform $\nabla^2 u$ from x, y, z to cylindrical coordinates

$$x = \rho\cos\phi,\ y = \rho\sin\phi,\ z = z.$$

Here z is not changed; we have merely to write ρ, ϕ for r, θ in (7), so that

$$\nabla^2 u = \frac{\partial^2 u}{\partial \rho^2} + \frac{1}{\rho}\frac{\partial u}{\partial \rho} + \frac{1}{\rho^2}\frac{\partial^2 u}{\partial \phi^2} + \frac{\partial^2 u}{\partial z^2}\ldots\ldots\ldots(8)$$

Lastly, transform to spherical polar coordinates

$$x = r\sin\theta\cos\phi,\ y = r\sin\theta\sin\phi,\ z = r\cos\theta.$$

The transformation may be effected in two steps. First transform to cylindrical coordinates ρ, ϕ, z where $\rho = r \sin \theta$; this change gives (8). Next transform from z, ρ to r, θ where $z = r \cos \theta$, $\rho = r \sin \theta$.

This change gives by writing z for x and ρ for y in (7),

$$\frac{\partial^2 u}{\partial z^2} + \frac{\partial^2 u}{\partial \rho^2} = \frac{\partial^2 u}{\partial r^2} + \frac{1}{r}\frac{\partial u}{\partial r} + \frac{1}{r^2}\frac{\partial^2 u}{\partial \theta^2} \quad \ldots \ldots \ldots (9)$$

Also by (4) replacing y by ρ,

$$\frac{\partial u}{\partial \rho} = \sin \theta \frac{\partial u}{\partial r} + \frac{\cos \theta}{r} \frac{\partial u}{\partial \theta} \quad \ldots \ldots \ldots (10)$$

Substitute from (9) and (10) in (8) and put $\rho = r \sin \theta$ and we get

$$\nabla^2 u = \frac{\partial^2 u}{\partial r^2} + \frac{2}{r}\frac{\partial u}{\partial r} + \frac{1}{r^2}\frac{\partial^2 u}{\partial \theta^2} + \frac{\cot \theta}{r^2}\frac{\partial u}{\partial \theta} + \frac{1}{r^2 \sin^2 \theta}\frac{\partial^2 u}{\partial \phi^2} \ldots (11)$$

It is sometimes useful to write the first two terms of (11) in the equivalent forms

$$\frac{1}{r^2}\frac{\partial}{\partial r}\left(r^2 \frac{\partial u}{\partial r}\right) \text{ or } \frac{1}{r}\frac{\partial^2(ru)}{\partial r^2},$$

and we may transform (11) to

$$\nabla^2 u = \frac{1}{r^3}\left\{r^2\frac{\partial^2(ru)}{\partial r^2} + \frac{\partial^2(ru)}{\partial \theta^2} + \cot \theta \frac{\partial(ru)}{\partial \theta} + \frac{1}{\sin^2 \theta}\frac{\partial^2(ru)}{\partial \phi^2}\right\} \ldots (12)$$

since $\dfrac{\partial^2(ru)}{\partial \theta^2} = r\dfrac{\partial^2 u}{\partial \theta^2}$, etc.

EXERCISES XIX.

1. If $x = r \cos \theta$, $y = r \sin \theta$, show that

(i) $\left(\dfrac{\partial x}{\partial r}\right)_\theta = \left(\dfrac{\partial r}{\partial x}\right)_y$; (ii) $\left(\dfrac{\partial x}{r\partial \theta}\right)_r = \left(\dfrac{r\partial \theta}{\partial x}\right)_y.$

The equation (i) is not in conflict with the theorem that when x is a function of the *single* variable r, the product of dx/dr and dr/dx is unity. The student should prove the equations by using a diagram, and he will see their meaning much more clearly.

EXERCISES XIX.

2. If $x = r\cos\theta$, $y = r\sin\theta$, prove

(i) $D_x^2 y = \{r^2 + 2(D_\theta r)^2 - rD_\theta^2 r\}/(\cos\theta\, D_\theta r - r\sin\theta)^3$;

(ii) $\{1 + (D_x y)^2\}^{\frac{3}{2}}/D_x^2 y = \{r^2 + (D_\theta r)^2\}^{\frac{3}{2}}/\{r^2 + 2(D_\theta r)^2 - rD_\theta^2 r\}$.

Deduce from (i) the condition for a point of inflexion on the curve given by the polar equation $r = f(\theta)$.

3. If $x = a(1 - \cos t)$, $y = a(nt + \sin t)$, express $D_x^2 y$ in terms of t.

4. If $x = r\cos\theta$, $y = r\sin\theta$, and x, y, r, θ functions of t, prove

(i) $\dot{x}\cos\theta + \dot{y}\sin\theta = \dot{r}$; (ii) $-\dot{x}\sin\theta + \dot{y}\cos\theta = r\dot{\theta}$;

(iii) $\ddot{x}\cos\theta + \ddot{y}\sin\theta = \ddot{r} - r\dot{\theta}^2$; (iv) $-\ddot{x}\sin\theta + \ddot{y}\cos\theta = r\ddot{\theta} + 2\dot{r}\dot{\theta}$.

If P is the point (x, y), equations (i) and (ii) give the velocity of P along and perpendicular to the radius vector while equations (iii) and (iv) give the acceleration of P in the same directions. It is easy to see that

$$r\ddot{\theta} + 2\dot{r}\dot{\theta} = \frac{1}{r}\frac{d}{dt}(r^2\dot{\theta}).$$

5. If s is the arc of a curve measured from a fixed point on it up to the point $P(x, y, z)$, prove, using accents to denote s-derivatives and dots to denote t-derivatives,

(i) $x'x'' + y'y'' + z'z'' = 0$; (ii) $\dot{x} = x'\dot{s}$; (iii) $\ddot{x} = x'\ddot{s} + x''\dot{s}^2$;

(iv) $x'\dot{x} + y'\dot{y} + z'\dot{z} = \dot{s}$; (v) $\ddot{x}^2 + \ddot{y}^2 + \ddot{z}^2 = \ddot{s}^2 + \dot{s}^4/\rho^2$;

where $1/\rho^2 = x''^2 + y''^2 + z''^2$.

Equation (i) is obtained by differentiating as to s the identity

$$x'^2 + y'^2 + z'^2 = 1,$$

this relation holding since x', y', z' are direction cosines. These results are important in Mechanics. Thus (iv) gives the tangential velocity, (v) the *total* acceleration.

6. If the axes are turned through an angle a, the old coordinates (x, y) of any point are connected with the new coordinates (ξ, η) of that point by the equations (§ 27)

$$x = \xi\cos a - \eta\sin a, \qquad y = \xi\sin a + \eta\cos a,$$

prove that

$$\frac{\partial^2 u}{\partial x^2} + \frac{\partial^2 u}{\partial y^2} = \frac{\partial^2 u}{\partial \xi^2} + \frac{\partial^2 u}{\partial \eta^2} \quad \text{(i)}.$$

For, $\dfrac{\partial u}{\partial \xi} = \dfrac{\partial u}{\partial x}\dfrac{\partial x}{\partial \xi} + \dfrac{\partial u}{\partial y}\dfrac{\partial y}{\partial \xi} = \dfrac{\partial u}{\partial x}\cos a + \dfrac{\partial u}{\partial y}\sin a$;

$\dfrac{\partial u}{\partial \eta} = \dfrac{\partial u}{\partial x}\dfrac{\partial x}{\partial \eta} + \dfrac{\partial u}{\partial y}\dfrac{\partial y}{\partial \eta} = -\dfrac{\partial u}{\partial x}\sin a + \dfrac{\partial u}{\partial y}\cos a$.

Solving for $\partial u/\partial x$, $\partial u/\partial y$, we find

$$\frac{\partial u}{\partial x} = \frac{\partial u}{\partial \xi}\cos\alpha - \frac{\partial u}{\partial \eta}\sin\alpha\,; \qquad \frac{\partial u}{\partial y} = \frac{\partial u}{\partial \xi}\sin\alpha + \frac{\partial u}{\partial \eta}\cos\alpha\,;$$

$$\frac{\partial^2 u}{\partial x^2} = \frac{\partial u_x}{\partial x} = \frac{\partial u_x}{\partial \xi}\cos\alpha - \frac{\partial u_x}{\partial \eta}\sin\alpha, \text{ etc.}$$

A similar equation to (i) holds for three variables x, y, z.

7. Prove $\dfrac{\partial}{\partial x}\left(\nabla^2 u\right) = \nabla^2\left(\dfrac{\partial u}{\partial x}\right).$

8. If in § 99 (12) θ be changed to μ where $\mu = \cos\theta$, show that $\nabla^2 u$ becomes

$$\frac{1}{r^2}\left[r\frac{\partial^2(ru)}{\partial r^2} + \frac{\partial}{\partial \mu}\left\{(1-\mu^2)\frac{\partial u}{\partial \mu}\right\} + \frac{1}{1-\mu^2}\frac{\partial^2 u}{\partial \phi^2}\right].$$

9. P, P' are the points (x, y, z), (x', y', z') and $PP' = r$, a *positive* number; $PQ = ds$, $P'Q' = ds'$; the direction cosines of PQ, $P'Q'$ are (l, m, n), (l', m', n'), and the angles QPP', $Q'P'P$ are θ, θ', while ϵ is the angle between the directions of PQ and $P'Q'$. Prove

(i) $\partial r/\partial s = -\cos\theta\,;$ \qquad (ii) $\partial r/\partial s' = -\cos\theta'\,;$

(iii) $r\dfrac{\partial^2 r}{\partial s\,\partial s'} + \dfrac{\partial r}{\partial s}\dfrac{\partial r}{\partial s'} = -\cos\epsilon\,;$ \qquad (iv) $\dfrac{\partial(r^{-1})}{\partial s} = \dfrac{\cos\theta}{r^2}\,;$

(v) $\dfrac{4}{\sqrt{r}}\dfrac{\partial^2(\sqrt{r})}{\partial s\,\partial s'} = -\dfrac{2\cos\epsilon + 3\cos\theta\cos\theta'}{r^2}.$

In § 92 (3) put $u = r$; then since $r^2 = (x'-x)^2 + (y'-y)^2 + (z'-z)^2$,

$$\partial r/\partial x = -(x'-x)/r, \quad \partial r/\partial x' = (x'-x)/r, \text{ etc.}$$

and the x-direction cosine of PP' is $(x'-x)/r$; of $P'P$, $(x-x')/r$.

Then $\dfrac{\partial r}{\partial s} = l\dfrac{\partial r}{\partial x} + m\dfrac{\partial r}{\partial y} + n\dfrac{\partial r}{\partial z} = -\left\{l\dfrac{x'-x}{r} + \ldots + \ldots\right\} = -\cos\theta.$

In finding $\partial^2 r/\partial s\,\partial s'$ by differentiating $\partial r/\partial s$ as to s', it is to be noted that l, m, n and x, y, z are independent of s'; so are l', m', n' and x', y', z' of s.

$$\frac{\partial}{\partial s'}\cdot\frac{l(x'-x)}{r} = \frac{l}{r}\frac{\partial x'}{\partial s'} - \frac{l(x'-x)}{r^2}\frac{\partial r}{\partial s'} = \frac{ll'}{r} - l\cdot\frac{x'-x}{r}\cdot\frac{1}{r}\frac{\partial r}{\partial s'},$$

since $l' = \partial x'/\partial s'$. Finding the derivatives of $m(y'-y)/r$, $n(z'-z)/r$ and adding

$$\frac{\partial^2 r}{\partial s\,\partial s'} = -\left(\frac{ll'+mm'+nn'}{r}\right) + \left(l\frac{x'-x}{r} + m\frac{y'-y}{r} + n\frac{z'-z}{r}\right)\frac{1}{r}\frac{\partial r}{\partial s'}$$

$$= -\frac{\cos\epsilon}{r} - \frac{\partial r}{\partial s}\cdot\frac{1}{r}\frac{\partial r}{\partial s'},$$

which gives (iii). Also

$\partial(r^{-1})/\partial s = -r^{-2}\partial r/\partial s = \cos\theta/r^2$, which is (iv).

EXERCISES XIX. 241

10. Let QP in ex. 9 be produced backward to Q_1, making $PQ_1 = QP$. Let $u = 1/PP' = 1/r$, and let u_Q, u_{Q_1} denote $1/QP'$, $1/Q_1P'$. Show (as in § 92) that

$$\frac{\partial u}{\partial s} = \operatorname*{L}_{Q_1Q=0} \frac{u_Q - u_{Q_1}}{Q_1Q}\ ;\quad \frac{\partial u}{\partial s} = \frac{\cos\theta}{r^2}.$$

11. With the notation of ex. 9, let P be the centre of an elementary magnet of moment M, whose axis is in the direction PQ; show that the potential V at P' of the magnet is

$$V = M\frac{\partial(1/r)}{\partial s} = \frac{M\cos\theta}{r^2}.$$

At Q, Q_1 (see ex. 10) let quantities m, $-m$ of magnetism be placed; the potential at P' of these quantities is

$$mu_Q - mu_{Q_1} = mQ_1Q(u_Q - u_{Q_1})/Q_1Q.$$

Let Q_1Q tend to zero while the product mQ_1Q remains constant and equal to M; then V is the limit of the fraction just written, which by ex. 10 is

$$M\frac{\partial u}{\partial s} = M\frac{\cos\theta}{r^2}.$$

12. The components of the magnetic force at P' (ex. 11) are $-\partial V/\partial x'$, $-\partial V/\partial y'$, $-\partial V/\partial z'$; show that

$$-\frac{\partial V}{\partial x'} = \frac{3M(x'-x)\cos\theta}{r^4} - \frac{Ml}{r^3},$$

with similar expressions for the other two components.

13. If an elementary magnet of moment M' is placed with its centre at P' and its axis along $P'Q'$, show that the mutual potential energy W of the magnets is

$$W = M'\frac{\partial V}{\partial s'} = MM'\frac{\partial^2(r^{-1})}{\partial s\,\partial s'} = \frac{MM'(\cos\epsilon + 3\cos\theta\cos\theta')}{r^3}.$$

Apply the method of ex. 11, taking V in place of u or $1/r$.

CHAPTER XII.

APPLICATIONS TO THE THEORY OF EQUATIONS.

§ 100. Rational Integral Functions. If $f(x)$ is a rational integral function of x of degree n, it is proved in treatises on the theory of equations that in general there are n values of x which make $f(x)$ zero; these values are called the *roots* of the *equation* $f(x)=0$, or the *zeroes* of the *function* $f(x)$. These values are not, however, necessarily real numbers, nor are they necessarily all different. Thus, if $f(x)=(x-1)^2(x-2)(x^2+1)$, $f(x)$ is of the 5th degree; two of the roots of $f(x)=0$ are equal to 1, one root is 2, and there are two imaginary roots $\pm\sqrt{(-1)}$.

a is called an r-ple root of $f(x)=0$, or an r-ple zero of $f(x)$ if $f(x)$ contain $(x-a)^r$, but no higher power of $(x-a)$. In this case $f(x)$ is of the form $(x-a)^r\phi(x)$, and $\phi(a)$ is not zero; if $\phi(a)$ were zero, then by the Remainder Theorem proved in Algebra $\phi(x)$ would contain $x-a$, and therefore $f(x)$ would contain a higher power than $(x-a)^r$.

When $f(x)=(x-a)^r\phi(x)$ it is obvious that the 1st, 2nd ... $(r-1)$th derivatives of $f(x)$ will contain $x-a$ as a factor, and will therefore vanish when $x=a$. We leave it as an exercise to the student to show that the necessary and sufficient conditions that a should be an r-ple zero of $f(x)$ are that $f(x)$ and its first $(r-1)$ derivatives should vanish when $x=a$, but that the rth derivative should not vanish when $x=a$. Also that the multiple roots of $f(x)=0$ are roots of $f'(x)=0$, and may therefore be obtained as the zeroes of the G.C.M. of $f(x)$ and $f'(x)$.

Manifestly the graph of $(x-a)^r\phi(x)$ will or will not cross the x-axis according as r is an odd or an even integer; if

THEORY OF EQUATIONS. 243

$r>1$ the x-axis will be a tangent at the point $(a, 0)$, since in that case $f'(a)$ will be zero.

Ex. 1. Show that 2 is a triple root of the equation
$$3x^4 - 16x^3 + 24x^2 - 16 = 0.$$
$f(2), f'(2), f''(2)$ are zero, but $f'''(2)$ is not zero; $f(2)$ is $(x-2)^3(3x+2)$ so that 2 is a triple root, and $-2/3$ is the remaining root.

Ex. 2. Find what relation must hold between q and r that the equation $x^3 + qx + r = 0$ should have a double root.

If the root be a, then $f(a) = 0, f'(a) = 0, f''(a) \neq 0$; therefore
$$a^3 + qa + r = 0 \text{ (i)}; \quad 3a^2 + q = 0 \text{ (ii)}; \quad 6a \neq 0 \text{ (iii)}.$$
From (ii) $a^2 = -q/3$, and therefore by (i) $2qa/3 + r = 0$. Hence
$$a^2 = -q/3, \text{ and } a^2 = 9r^2/4q^2,$$
so that $27r^2 + 4q^3 = 0$ is the required relation.

§ 101. Any Continuous Function. We will now suppose $f(x)$ to be any continuous function; it has always to be remembered that theorems proved on the assumption of the continuity of the function may cease to be true if the function be discontinuous.

If $f(a), f(b)$ are of opposite signs, then (§ 45, Th. II.) there is at least one root of $f(x) = 0$ in the interval (a, b); when it is said that a root *lies in the interval* (a, b), what is meant is that the root is greater than one of the numbers a, b and less than the other.

If $f'(x)$ is continuous and does not vanish for any value of x in the interval (a, b), then $f(x)$ is either an increasing or else a decreasing function in the interval (a, b), and therefore when $f(a)$ and $f(b)$ are of opposite signs, $f(x)$ vanishes once only; that is, there is only one root in the interval.

If $f(x)$ and $f'(x)$ are continuous, then between every two consecutive roots of $f(x) = 0$ there is at least one root of $f'(x) = 0$; and conversely, between two consecutive roots of $f'(x) = 0$ there cannot be more than one root of $f(x) = 0$ and there may be none.

The first part of this proposition is Rolle's Theorem (§ 72). To prove the converse, let α and β be the two roots of $f'(x) = 0$, and suppose if possible that there are two roots of $f(x) = 0$, say a and b, in the interval (α, β); we may assume $\alpha < a < b < \beta$. Since $f(a) = 0, f(b) = 0, f'(x)$ must vanish in the interval (a, b), contrary to the hypothesis that a, β are consecutive roots of $f'(x)=0$. It has already been pointed out that $f'(x)$ may vanish more than once between two consecutive roots of $f(x)=0$, and therefore it may happen that there is no root of $f(x)=0$ between two consecutive roots of $f'(x)=0$ (§ 72).

§ 102. Newton's Method of Approximating to the Roots of an Equation. Throughout the chapter we consider real roots alone, and we suppose that $f(x)$ and its first two derivatives are continuous within the range considered. When $f(x)$ is a rational integral function we will suppose that the multiple roots, if any, of $f(x)=0$ have been determined by the method of the G.C.M., and that the corresponding factors have been removed; hence $f(x)$ and $f'(x)$ will not vanish for the same value of x. (Of course it may quite well happen that the zeroes of the G.C.M. have to be determined by one of the methods about to be given for approximating to the roots of an equation.)

The following method of approximating to the roots is known as Newton's Method.

Suppose it has been found that $f(a)$ is numerically small; we can generally get a closer approximation than a as follows: Let α be the root to which a is an approximation, so that $f(\alpha)=0$. By the Mean Value Theorem,

$$f(\alpha) = f(a) + (\alpha - a)f'(a) + \tfrac{1}{2}(\alpha - a)^2 f''(x_1) \ldots\ldots\ldots(1)$$

where x_1 lies in the interval (a, α). If we neglect $(\alpha-a)^2$ in comparison with $(\alpha-a)$, equation (1) becomes, since $f(\alpha)=0$,

$$f(a) + (a_1 - a)f'(a) = 0, \text{ giving } a_1 = a - f(a)/f'(a)$$

where a_1 is the approximate value of α.

We may now use a_1 as we have just used a, and get another approximation a_2 where

$$a_2 = a_1 - f(a_1)/f'(a_1),$$

and so on. This process, however, does not show that a_1 is really closer to α than a is, and gives no criterion of the closeness of the approximation. We therefore investigate the conditions for the closeness.

NEWTON'S METHOD OF APPROXIMATION.

§ 103. Tests for Degree of Approximation. Let us suppose (i) that $f(a)$, $f(b)$ are of opposite signs, (ii) that $f'(x)$ does not vanish in the interval (a, b), (iii) that $f''(x)$ does not vanish in the interval.

Conditions (i), (ii) show that there is one and only one root, α say, in the interval (a, b); condition (iii) shows that the graph of $f(x)$ is either convex upwards or else concave upwards in the interval, that is, it has no point of inflexion.

Let a be that end of the interval at which $f(x)$ has the same sign as $f''(x)$; this choice of the end of the interval is essential. a may be either greater or less than b.

The figures (a), (b) show the graph when $f''(x)$ is negative, (c), (d) when $f''(x)$ is positive. The abscissae of A, B are a, b.

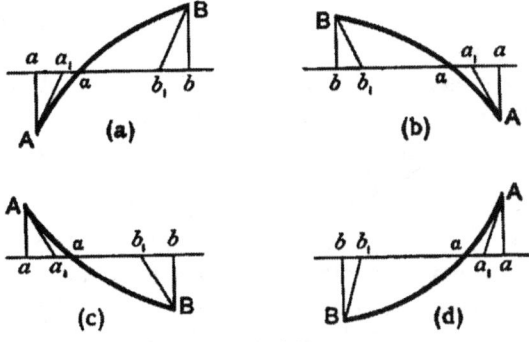

FIG. 52.

The graphs show that the tangent at A will cross the x-axis at a point a_1 between a and α; a_1 will therefore be a better approximation than a. Now the equation of the tangent at A is

$$y = f(a) + (x-a)f'(a),$$

and when $y = 0$, $x = a_1$. Hence

$$a_1 = a - f(a)/f'(a) \dots\dots\dots\dots\dots\dots(1)$$

Let the line through B parallel to the tangent at A cut the x-axis at the point b_1; the equation of the line is

$$y = f(b) + (x-b)f'(a).$$

Hence $\qquad b_1 = b - f(b)/f'(a) \dots\dots\dots\dots\dots\dots\dots(1')$

and b_1 lies between b and a, so that b_1 is a better approximation than b, though not necessarily better than a.

Now $b_1 - a_1 = - \{f(b) - f(a) - (b-a)f'(a)\}/f'(a)$,
which by the Mean Value Theorem may be written
$$b_1 - a_1 = -\tfrac{1}{2}(b-a)^2 f''(x_1)/f'(a)$$
where x_1 lies in the interval (a, b).

Let d be the numerical value of $(b-a)$, d_1 that of $(b_1 - a_1)$, and let G denote the greatest value of $f''(x)$, g the smallest value of $f'(x)$ in the interval (a, b); then
$$d_1 \leq d^2 G/2g, \text{ or } d_1 \leq d^2 k, \ k = G/2g.$$

Since $a - a_1$ is numerically less than $b_1 - a_1$, we have $a - a_1$ numerically less than d_1 or $d^2 k$, so that the error in taking a_1 instead of the root a is less than $d^2 k$. Similarly the error in b_1 is less than $d^2 k$.

We may repeat the process with a_1, b_1 instead of a, b; we should find, using a similar notation,
$$a_2 = a_1 - f(a_1)/f'(a_1); \ b_2 = b_1 - f(b_1)/f'(a_1)$$
$$d_2 \leq d_1^2 k, \text{ that is, } d_2 \leq d^4 k^3,$$
and the error in taking a_2 or b_2 is less than d_2 or $d^4 k^3$.

The process may be repeated. As soon as a, b are such that dk is less than 1, the approximation to a becomes very rapid. There is, as a rule, no need to calculate, b_1, b_2...

The student will see by examining figures that if a is not chosen as stated, the value of a_1 or b_1 may be further from a than a or b.

§ 104. Examples.

Ex. 1. If $f(x) = 3x^3 - 4x + 5$, find the roots of $f(x) = 0$.
$$f'(x) = 9(x + \tfrac{2}{3})(x - \tfrac{2}{3}); \ f''(x) = 18x; \ f'''(x) = 18.$$

$f(-\tfrac{2}{3}) = 6\tfrac{7}{9}$ is a maximum value of $f(x)$; $f(\tfrac{2}{3}) = 3\tfrac{2}{9}$ is a minimum. The point $(0, 5)$ is a point of inflexion.

It is easy to see that the graph of $f(x)$ crosses the x-axis once only, so that there is only one real root.

$f(-2) = -11$, $f(-1) = +6$, so that the root lies between -2 and -1; as $f(-2)$ and $f(-1)$ are large, we seek a closer approximation before choosing a, b. Now $f(-1\cdot 6) = -\cdot 888$, $f(-1\cdot 5) = +\cdot 875$. Since $f''(x)$ is negative when x is negative, we take $a = -1\cdot 6$, $b = -1\cdot 5$.

WORKED EXAMPLES.

G = numerically greatest value of $f''(x)$ in interval $(-1\cdot6, -1\cdot5) = 28\cdot8$.
g = numerically smallest value of $f'(x)$ in interval $= 16\cdot25$.
$k = G/2g = 14\cdot4/16\cdot25 < 1$; $d = \cdot1$; $d^2 k < \cdot01$.

$$a_1 = a - f(a)/f'(a) = -1\cdot6 + \cdot04 = -1\cdot56,$$

and a_1 differs from a by less than $\cdot01$.

$$a_2 = a_1 - f(a_1)/f'(a_1) = -1\cdot56 + \cdot0083 = -1\cdot5517,$$

and a_2 differs from a by less than $d^4 k^3$ or $\cdot0001$.

The values $\cdot04$ and $\cdot0083$ are of course approximations. Care must be taken that we do not go beyond the root. Thus

$$-f(a)/f'(a) = \cdot046...,$$

but if we take $\cdot05$ as the value, thus making $a_1 = -1\cdot55$, we find $f(-1\cdot55)$ to be *positive*. The reasoning, however, depends on having $f(a_1)$ of the same sign as $f''(x)$, that is in this case negative.

A closer approximation is

$$a_3 = -1\cdot551\;608\;12,$$

and the error is less than a unit of the last decimal place.

Ex. 2. Solve the equation $x + \sin x - \dfrac{\pi}{3} = 0$.

If A is a point on the circumference of a circle, and if AB, AC are two chords which trisect the area of the circle, then the angle between AB and the diameter through A is $\tfrac{1}{2}x$ radians.

$$f(x) = x + \sin x - \frac{\pi}{3}\,;\quad f'(x) = 1 + \cos x\,;\quad f''(x) = -\sin x.$$

It is easily found that x lies between $30°$ and $31°$, or in radians $\cdot5236$, and $\cdot5411$.

$$f(\cdot5236) = -\cdot0236\,;\quad f'(\cdot5236) = 1\cdot8660\,;$$
$$f(\cdot5411) = +\cdot0089\,;\quad f'(\cdot5411) = 1\cdot8572\,;$$
$$d = \cdot0175 < \cdot02\,;\quad k = G/2g < \cdot2,\; d^2 k < \cdot00008.$$

Since $f''(x)$ is negative, we take $a = \cdot5236$,

$$a_1 = a - f(a)/f'(a) = \cdot5236 + \cdot0126 = \cdot5362,$$

and the error is less than a unit of the fourth place.

The next approximation gives

$$a_2 = a_1 - f(a_1)/f'(a_1) = \cdot5362 + \cdot0000674 = \cdot5362674,$$

and the error is less than a unit of the last figure. In degrees the angle is $30°\;43'\;33''\cdot0$.

§ 105. Successive Approximations.

Suppose the equation to be of the form $x = \phi(x)$; let a be a root and a an approximation to a, $a = a + h$ say. Now

$$a = \phi(a) = \phi(a+h) = \phi(a) + h\phi'(a + \theta h);$$

and therefore $\qquad a - \phi(a) = h\phi'(a + \theta h)$.

248   AN ELEMENTARY TREATISE ON THE CALCULUS.

Using the terms "greater" and "less" to mean *numerically* "greater" and "less," we see that if, for every value of x that is nearer to a than a is, $\phi'(x)$ is less than a proper fraction m, the difference between a and $\phi(a)$ is less than mh; that is, the difference between a and $\phi(a)$ is less than that between a and a. Hence $\phi(a)$ is a closer approximation than a.

Denote $\phi(a)$ by a_1 and let $a = a_1 + h_1$ where h_1 is equal to $h\phi'(a + \theta h)$ and therefore less than mh. We find in the same way
$$a - \phi(a_1) = h_1 \, \phi'(a_1 + \theta_1 h_1) < h_1 m < h m^2.$$
So that $\phi(a_1) = a_2$ is a closer approximation than a_1. The upper limits of the errors hm, hm^2 usually decrease pretty rapidly as m is, in the cases to which the method applies, often a small fraction. We may proceed, of course, with a_2 and so on.

It is essential for the success of the method that $\phi'(x)$ be, near the root, a proper fraction. It may be proved that Newton's method is a particular case of that of Successive Approximations, and unless m be pretty small the latter method has no advantage over Newton's.

Ex. Solve the equation $10^x = 3456 \sqrt{x}$.
Take logarithms to the base 10, and we get
$$x = \tfrac{1}{2} \log x + 3 \cdot 538\,5737 = \phi(x).$$
If we draw the graph of $\tfrac{1}{2} \log x$ and of $x - 3\cdot538\,5737$ we see that they intersect for a value of x near 4 and also for a very small value of x. Take first $a = 4$, now
$$\phi'(x) = \frac{M}{2x} = \frac{\cdot 4343}{2x}, \quad M = \log_{10} e,$$
so that when x is nearly 4, $\phi'(x)$ is a proper fraction.

Take 4-figure logarithms for the first approximations,
$$a_1 = \phi(4) = 3\cdot5386 + \cdot 3010 = 3\cdot8396;$$
$$a_2 = \phi(a_1) = 3\cdot5386 + \cdot 2921 = 3\cdot8307;$$
$$a_3 = \phi(a_2) = 3\cdot5386 + \cdot 2916 = 3\cdot8302.$$
When $x = a_3$, $x - \phi(x) = \cdot 0005$, so that a_3 is a fairly close approximation. Take now 7-figure logarithms, and we find
$$a_4 = \phi(a_3) = 3\cdot538\,5737 + \cdot 291\,6107 = 3\cdot830\,1844;$$
$$a_5 = \phi(a_4) = 3\cdot830\,1835;$$
$$a_6 = \phi(a_5) = 3\cdot830\,1835;$$
to 7 decimals a_5 is correct.

SUCCESSIVE APPROXIMATIONS. EXPANSIONS. 249

For the other root the method is inapplicable, since near 0 $\phi'(x)$ is greater than 1. But since x is very small we get a good approximation by taking the value of x, which satisfies $\phi(x)=0$. Therefore
$$\log x = -7{\cdot}077\ 1474 = \bar{8}{\cdot}922\ 8526;$$
and $\qquad x = {\cdot}000\ 0000\ 8372\ 45.$

§ 106. Expansion of a Root in a Series. Reversion of Series.

Let the equation $x = \phi(x)$ be
$$x = Ay + Bx^2 + Cxy + Dy^2 + Ex^3 + Fx^2y + Gxy^2 + Hy^3 + \ldots,$$
or $\qquad x = Ay + u_2 + u_3 + \ldots,$

where $u_2, u_3 \ldots$ are of the 2nd, 3rd... degree in x and y.

If y is a small quantity one root will be approximately Ay, for this value of x makes u_2 of the second order in y, u_3 of the third.... Call this approximation a. Clearly for small values of x we may suppose $\phi'(x)$ a proper fraction.

The next approximation is $a_1 = \phi(a) = \phi(Ay)$. To the 2nd order in y we may neglect $u_3, u_4 \ldots$ and take
$$a_1 = \phi(Ay) = Ay + B(Ay)^2 + C(Ay)y + Dy^2$$
$$= Ay + B_1 y^2 \text{ say.}$$

The next approximation is $a_2 = \phi(a_1)$, and in forming $\phi(a_1)$ we need only retain terms of the third order in y. Hence in u_3 we need only substitute the first approximation a or Ay, since if we put $Ay + B_1 y^2$ all terms except those which come from Ay alone would be of a higher order than the third. In u_2 we substitute a_1 or $Ay + B_1 y^2$ but reject the term $B(B_1 y^2)^2$ which is of the fourth order. We thus get $\qquad a_2 = Ay + B_1 y^2 + C_1 y^3,$

and we proceed in a similar way to find $\phi(a_2)$.

The practical rule then may be stated as follows:

For the first approximation neglect $u_2, u_3 \ldots$; we get $Ay = a$.

For the second approximation neglect $u_3, u_4 \ldots$ and substitute a in u_2; we get $Ay + B_1 y^2 = a_1$.

For the third approximation neglect $u_4, u_5 \ldots$, substitute a in u_3, a_1 in u_2, and reject terms above the third order; we get $Ay + B_1 y^2 + C_1 y^3 = a_2$.

For the fourth approximation neglect $u_5, u_6 \ldots$, substitute a in u_4, a_1 in u_3, a_2 in u_2, and reject terms above the fourth order; we get $Ay + B_1 y^2 + C_1 y^3 + D_1 y^4 = a_3$ and so on.

Ex. 1. $\quad x = 2y + x^2 - xy + x^3 - x^4.$

1st App. $x = 2y$;

2nd App. $x = 2y + (2y)^2 - (2y)y = 2y + 2y^2$;

3rd App. $x = 2y + (2y + 2y^2)^2 - (2y + 2y^2)y + (2y)^3 = 2y + 2y^2 + 14y^3$;

4th App. $x = 2y + (2y + 2y^2 + 14y^3)^2 - (2y + 2y^2 + 14y^3)y + (2y + 2y^2)^3 - (2y)^4$
$= 2y + 2y^2 + 14y^3 + 54y^4.$

Ex. 2. $\phi(x)$ may be an infinite series, the usual conditions as to convergency being supposed satisfied. Thus if we put $e^x = 1 + y$, then
$$x + \tfrac{1}{2}x^2 + \tfrac{1}{6}x^3 + \tfrac{1}{24}x^4 + \ldots = y \; ;$$
or $\quad x = y - \tfrac{1}{2}x^2 - \tfrac{1}{6}x^3 - \tfrac{1}{24}x^4 \ldots \; ;$

and the student will readily find that to the fourth order
$$x = y - \tfrac{1}{2}y^2 + \tfrac{1}{3}y^3 - \tfrac{1}{4}y^4 \; ;$$
that is $\quad \log(1 + y) = y - \tfrac{1}{2}y^2 + \tfrac{1}{3}y^3 - \tfrac{1}{4}y^4.$

This is an example of Reversion of Series; the full discussion, however, of the subject of this article lies beyond our limits. The student is referred to Chrystal's *Algebra*, vol. ii., chap. 30, for an adequate treatment.

Ex. 3. Expand y in powers of x for large values of x when
$$y^3 + x^3 = 3axy.$$

When x and y are both large the product xy may be neglected in comparison with x^3 and y^3, hence a first approximation gives $y^3 + x^3 = 0$, that is $y = -x$. To get a second approximation write
$$y = -x + 3axy/(x^2 - xy + y^2),$$
and on the right side put $-x$ for y. We thus get

2nd App. $\quad y = -x + 3ax(-x)/(x^2 + x^2 + x^2) = -x - a.$

To get a third approximation put $-(x + a)$ for y and expand in powers of $1/x$, which by hypothesis is small since x is large. Then
$$y = -x - \frac{3a(x^2 + ax)}{3x^2 + 3ax + a^2} = -x - a\left(1 + \frac{a}{x}\right)\left(1 + \frac{a}{x} + \frac{a^2}{3x^2}\right)^{-1}$$
$$= -x - a\left(1 + \frac{a}{x}\right)\left\{1 - \left(\frac{a}{x} + \frac{a^2}{3x^2}\right) + \left(\frac{a}{x} + \frac{a^2}{3x^2}\right)^2\right\}$$
$$= -x - a\left(1 - \frac{a^2}{3x^2}\right) = -x - a + \frac{a^3}{3x^2}.$$

The line $y = -x - a$ is an asymptote of the curve; the term $a^3/3x^2$ shows that at both ends of the asymptote the curve is above the asymptote. (See Ex. 13, p. 62.)

The method of this article is of great service in finding the shape of a curve near any point on it. If the point is not the origin, we may shift the origin to the point, and then the equation will be of the form given at the beginning of the article. We may, of course, when we wish to expand

y in powers of x, write the equation in the form $y = \psi(y)$. For the application of the method to the finding of asymptotes and generally to the investigation of the shape of the curve at a great distance from the origin, example 3 may serve as an illustration. The student is referred to the admirable treatise on *Curve Tracing* by Frost (London: Macmillan) for a systematic exposition of the method in its applications to geometry; that book is, in the words of Professor Chrystal, "a work which should be in the hands of every one who aims at becoming a mathematician, either practical or scientific."

§ 107. The Equation $x = \tan x$. Equations of the form $mx = \tan x$ occur in the Theory of the Conduction of Heat and in the Theory of Vibrating Plates. For simplicity we take $m = 1$, but the discussion goes on similar lines when m is different from 1.

Obviously zero is a root, and the negative roots are equal in numerical value to the positive roots, so that we consider only the positive roots.

By drawing the graphs of $\tan x$ and of x we see that they intersect once and once only in the intervals $(\pi, 3\pi/2)$, $(2\pi, 5\pi/2)$ and in general $(n\pi, n\pi + \pi/2)$ where n is any positive integer. There is therefore one, and only one, root of the equation in each interval; there is no root between 0 and $\pi/2$.

Let $x - \tan x = f(x)$ and calculate by Newton's method the root in the interval $(\pi, 3\pi/2)$.

$$f'(x) = -\tan^2 x; \quad f''(x) = -2 \tan x \sec^2 x.$$

An inspection of the tables shows that the angle lies between $180° + 77°$ and $180° + 78°$. Expressing these angles in radians, we have to three decimals

$$x = 4\cdot485; \quad f(x) = \cdot154;$$
$$x = 4\cdot503; \quad f(x) = -\cdot202; \quad f'(x) = -22\cdot1.$$

Since $f''(x)$ is negative we take $a = 4\cdot503$, $b = 4\cdot485$, so that $d = \cdot018 < \cdot02$ and it is easily found that k is less than 5.

$$a_1 = a - f(a)/f'(a) = 4\cdot503 - \cdot009 = 4\cdot494,$$

and the error is less than $d^2 k$ or less than $\cdot002$.

In seeking a closer approximation care must be taken not to go beyond the root; if we do so $f(a_2)$ will be positive. Owing to the rapidity with which the tangent changes there is danger of doing so when using 4-figure tables; besides the next approximation will have an error less than d^4k^3 or 2×10^{-5}, so that we may use the ordinary 7-figure tables.

a_1 is not beyond the root, for $f(a_1) = -\cdot011\ 9542$. Again,
$$a_2 = a_1 - f(a_1)/f'(a_1) = a_1 - \cdot000\ 5888 = 4\cdot4934112,$$
so that if we take the root as $4\cdot49341$ the error is less than 2 units in the last place. A closer approximation is
$$4\cdot493\ 4095.$$

To get the other roots let $x = n\pi + \pi/2 - \theta$, then θ is an acute angle and
$$\tan x = \tan\left(\frac{\pi}{2} - \theta\right) = 1/\tan\theta\ ;$$
and since $x = \tan x$ we have $\tan\theta = \dfrac{1}{x}$, $\theta = \tan^{-1}\left(\dfrac{1}{x}\right)$. Hence putting c for $n\pi + \pi/2$ we have
$$x = c - \tan^{-1}\left(\frac{1}{x}\right).$$

It is shown in a later chapter that
$$\tan^{-1}\left(\frac{1}{x}\right) = \frac{1}{x} - \frac{1}{3x^3} + \frac{1}{5x^5} - \frac{1}{7x^7} + \cdots$$
so that
$$x = c - \frac{1}{x} + \frac{1}{3x^3} - \frac{1}{5x^5} + \frac{1}{7x^7} - \cdots.$$

The equation may be solved by the method of last article, since x is, even for $n = 2$, greater than $7\cdot5$, and therefore $1/x$ fairly small.

1st. App. $x = c$;

2nd. App. $x = c - \dfrac{1}{c}$;

3rd. App. $x = c - \left(c - \dfrac{1}{c}\right)^{-1} + \dfrac{1}{3c^3} = c - \dfrac{1}{c} - \dfrac{2}{3c^3}$;

4th. App. $x = c - \left(c - \dfrac{1}{c} - \dfrac{2}{3c^3}\right)^{-1} + \dfrac{1}{3}\left(c - \dfrac{1}{c}\right)^{-3} - \dfrac{1}{5c^5}$;

$= c - \dfrac{1}{c}\left(1 + \dfrac{1}{c^2} + \dfrac{5}{3c^4}\right) + \dfrac{1}{3c^3}\left(1 + \dfrac{3}{c^2}\right) - \dfrac{1}{5c^5}$;

$= c - \dfrac{1}{c} - \dfrac{2}{3c^3} - \dfrac{13}{15c^5}$;

5th. App. $x = c - \dfrac{1}{c} - \dfrac{2}{3c^3} - \dfrac{13}{15c^5} - \dfrac{146}{105c^7}.$

For $n = 2, 3, 4, \ldots$, this last approximation is amply sufficient for all practical purposes. The student may show that x/π has the values $1\cdot4303$, $2\cdot4590$, $3\cdot4709$, $4\cdot4747$, $5\cdot4818$, $6\cdot4844$, for $n = 1, 2, 3, 4, 5, 6$. [Rayleigh's *Sound*, I., p. 334 (2nd Ed.).]

Many equations involving trigonometric and exponential functions were discussed by Euler, and the general solution of the equation $x = \tan x$ is due to him.

EXERCISES XX.

In the following examples it will usually be sufficient to calculate the root to 3 or 4 decimal places; in some cases the results are given to more figures.

1. Find the real root of $3x^3 + 5x - 40 = 0$.

2. A sphere of radius 1 is divided by a plane into two parts whose volumes are in the ratio of 1 to 2; the distance x of the plane from the centre of the sphere is a root of the equation $3x^3 - 9x + 2 = 0$. Find x.

3. Find the root of $x^3 - 4x^2 - 7x + 24 = 0$ that lies between 2 and 3.

4. If $(1 + x)^x = 27\cdot34$, find x.

5. If $10^x = 20x$, find x.

6. The chord AB of a circle, centre C, bisects the sector ACB; if the angle ACB is x radians, show that $x = 2\sin x$ and find x.

7. Solve the equation $x = \cos x$.

8. The equation $2x = \tan x$ has one root between 0 and $\pi/2$ and another between π and $3\pi/2$; find both roots.

9. Show how to solve the equation

$$l = a\left(e^{\frac{c}{2a}} - e^{-\frac{c}{2a}}\right)$$

for a when l and c are given, l being not much greater than c; for example, $c = 100$, $l = 105$. The value of a determines the catenary assumed by a string of length l hanging from two points in a horizontal line distant c from each other.

10. Find the least roots of

(i) $(e^x + e^{-x})\cos x - 2 = 0$; (ii) $(e^x + e^{-x})\cos x + 2 = 0$.

Obviously zero is a root of (i); find the next smallest root.

11. Solve $\quad\quad x - a \sin x = b$

where $\quad\quad a = \cdot 245316,\ b = 5\cdot 755067$.

12. Show that the approximations to the root a of $x = \phi(x)$ given by the method of § 105 are alternately greater and less than a if $\phi'(a)$ is negative.

13. If $f(x, y) = 0$ and $F(x, y) = 0$ have as an approximate pair of solutions $x = a$, $y = b$ show that in general the values $a + h$, $b + k$ will be closer approximations if h, k satisfy the equations

$$f(a, b) + h\frac{\partial f}{\partial x} + k\frac{\partial f}{\partial y} = 0,\ F(a, b) + h\frac{\partial F}{\partial x} + k\frac{\partial F}{\partial y} = 0$$

where in the derivatives x, y are replaced by a, b.

If $\quad f(x, y) = x^3 + 3xy^2 - y - 12,\ F(x, y) = 2x^2y + y^3 - 8$,

find closer approximations to the roots near $x = 2$, $y = 1$.

14. If $\quad\quad (y - x)(y - 2x) = x^3 + 2x^3y + x^2y^3$

show that when x is small there are two values of y given, as far as terms of the third order in x, by the equations

$$y = x - x^2 - x^3 \text{ and } y = 2x + x^2 + 3x^3.$$

Show that the curve given by the equation has two branches that pass through the origin and that the tangents at the origin are $y = x$ and $y = 2x$. Sketch the curve for small values of x.

[Write $y = x + \dfrac{x^3}{y - 2x} + \dfrac{2x^3y}{y - 2x} + \dfrac{x^2y^3}{y - 2x}$ and proceed as in § 106; then write $y = 2x + x^3/(y - x) +$, etc.]

15. If $a^2(y^2 - x^2) = x^4 + y^4$, show that for small values of x there are two values of y given by

$$y = x + x^3/a^2 \text{ and } y = -x - x^3/a^2.$$

Show also that near $(0, a)$ the shape of the curve is given by

$$x^2 + 2a(y - a) = 0.$$

Graph the curve.

16. If $(y - x)^2 = x^3 + x^2y + x^4$ show that for small values of x

$$y = x \pm \sqrt{2} \cdot x^{\frac{3}{2}}.$$

Graph the curve near the origin.

§ **108. Proportional Parts.** In the use of Logarithmic and similar Tables it is often necessary to find the value of the function for a value of the argument not given exactly in the Tables. It becomes necessary, therefore, to interpolate, and the ordinary rule is based on the assumption that the difference in the function is proportional to the difference in the argument. We will now examine the assumption.

Let h and z be small quantities having the same sign, but z being numerically less than h; then by the Mean Value Theorem, $f(x)$, $f'(x)$, $f''(x)$ being assumed continuous, the following equations are approximately correct.

$$f(a+h) - f(a) = hf'(a) + \tfrac{1}{2}h^2 f''(a) \quad \ldots \ldots \ldots \ldots (1)$$

$$f(a+z) - f(a) = zf'(a) + \tfrac{1}{2}z^2 f''(a) \quad \ldots \ldots \ldots \ldots (2)$$

Let $D = f(a+h) - f(a)$ and eliminate $f'(a)$; therefore

$$f(a+z) - f(a) = \frac{z}{h}D + \tfrac{1}{2}z(z-h)f''(a) \quad \ldots \ldots \ldots \ldots (3)$$

Equation (3) is approximate, but by following the lines of the proof of the Mean Value Theorem we can show it to be exact if in place of $f''(a)$ we write $f''(a + \theta h)$ where θ is a proper fraction.

For let
$$f(a+z) - f(a) = \frac{z}{h}D + \tfrac{1}{2}z(z-h)P \quad \ldots \ldots \ldots \ldots (\text{A})$$

and let $F(x) = f(x) - f(a) - \dfrac{x-a}{h}D - \tfrac{1}{2}(x-a)(x-a-h)P$.

Now $F(a) = 0$ identically; $F(a+z) = 0$ by (A); $F(a+h) = 0$ identically, remembering the value of D. Hence $F'(x)$ must vanish for a value of x between a and $a+z$, and again for a value of x between $a+z$ and $a+h$; therefore $F''(x)$ must vanish for a value of x between these two values, and therefore between a and $a+h$. But

$$F''(x) = f''(x) - P,$$

and therefore
$$P = f''(a + \theta h).$$

Hence, instead of (3) we get the exact equation,

$$f(a+z) - f(a) = \frac{z}{h}D + \tfrac{1}{2}z(z-h)f''(a+\theta h) \quad \ldots \ldots (4)$$

where
$$D = f(a+h) - f(a).$$

In the figure (Fig. 53),

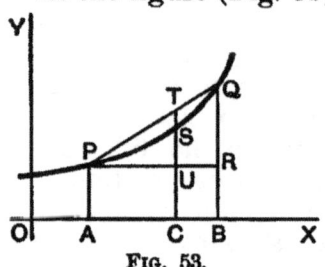

Fig. 53.

$OA = a$, $AC = z$, $AB = h$.
$RQ = f(a+h) - f(a) = D$.
$US = f(a+z) - f(a)$.
$UT = zD/h$.
$ST = \tfrac{1}{2}z(h-z)f''(a+\theta h)$.

The error committed in replacing the arc PSQ by the chord PTQ is measured by ST. z is by hypothesis less than h, and the numerically greatest value of $z(h-z)$ is $\tfrac{1}{4}h^2$. Hence, if G be the numerically greatest value of $f''(x)$ in the interval $(a, a+h)$, the numerically greatest value of ST or of

$$\tfrac{1}{2}z(h-z)f''(a+\theta h)$$

will be $\tfrac{1}{8}h^2 G$.

Suppose now that $f(x)$ is tabulated for a series of equidistant values of x, the difference between successive values being h. Let $a+z$ be a value of x between a and $a+h$, and therefore not given in the Table. The ordinary rule is to calculate $f(a+z)$ from (4), neglecting the second term on the right; that is,

$$f(a+z) = f(a) + \frac{z}{h}D.$$

For a given value of a, the amount by which $f(a)$ is increased to find $f(a+z)$, namely zD/h, is therefore proportional to z; the error in following the rule is therefore not greater than $h^2 G/8$.

Exceptions to the application of the rule occur in the following cases:

I. *G may be such that $h^2 G/8$ can not be neglected in comparison with zD/h; in this case the difference D is said to be irregular.*

II. *D may be so small that it vanishes to the number of figures in the Table; in this case the difference is said to be insensible. The difference will be insensible when $f'(a)$ is very small, since*

$$D = f(a+h) - f(a) = hf'(a) + \tfrac{1}{2}h^2 f''(a+\theta h).$$

Example. $f(x) = \log_{10} \sin x$.

Let $M = \log_{10} e = \cdot 434\,2945$;

then $f'(x) = M \cot x$; $f''(x) = -M \operatorname{cosec}^2 x$.

If x is small, $f''(x)$ is large, and the differences are irregular; since $\cot x$ is not large the differences are not insensible.

If x is nearly 90°, $\cot x$ is small, and the differences are insensible; though $f'''(x)$ is not large the ratio $f'''(x)/f'(x) = -2/\sin 2x$ is numerically large and therefore $h^2 G/8$ can not be neglected in comparison with zD/h. Near 90° therefore the differences are both insensible and irregular.

For tables that proceed at differences of 1', h is 1' or in radians
$$h = \cdot 000\,2909,$$
and $\quad \tfrac{1}{8} M h^2 = \cdot 000\,0000046$.

To find when $\tfrac{1}{8} M h^2 \operatorname{cosec}^2 x$ would affect the seventh figure we may put
$$\tfrac{1}{8} M h^2 \operatorname{cosec}^2 x = 5 \times 10^{-8},$$
and we find from this equation that x is about 18°. Hence, apart altogether from errors due to neglected figures in carrying out the numerical work which may easily amount to more than a unit in the seventh place, the error due to neglecting the term $h^2 G/8$ will amount to half a unit in the seventh place for angles less than 18°.

If h is equal to 10" the student may show that the seventh figure will not be affected by the neglect of $h^2 G/8$ till the angle is about 3°.

The student may with advantage consult Hobson's *Trigonometry*, Chap. 9. The advanced student will find a thorough discussion of all the principles involved in numerical approximations and the use of tables in Lüroth's *Vorlesungen über numerisches Rechnen* (Leipzig: Teubner, 1900).

Ex. 1. Show that for $\log \cos x$ the differences are insensible and irregular when x is small, and irregular when x is near 90°.

Ex. 2. Show that for $\log \tan x$ the differences are irregular when x is small and when x is near 90°. Show also that the maximum error is least when x is near 45°.

Ex. 3. In a 7-figure table of the logarithms of numbers, show that the term $h^2 G/8$ is most important when the number is 10000, and that the greatest error arising from the neglect of that term is about $5\cdot 5 \times 10^{-10}$, and is therefore negligible for these tables.

§ 109. Small Corrections. In practice all measurements are subject to errors, and it is therefore of importance to determine the influence on the result of a calculation when the argument or arguments of the calculated function are given by measurements whose errors are approximately known.

Let a quantity x be determined by measurement and let y be a function $f(x)$ of x. Suppose that the value x given by the measurement differs from the true value by δx, then the true value of y is $f(x+\delta x)$ and the error δy is

$$\delta y = f(x+\delta x) - f(x) = f'(x+\theta \delta x)\delta x$$
or $\qquad \delta y = f'(x)\delta x$ approximately.

The relative error $\delta y/y$ is, approximately,

$$\frac{\delta y}{y} = \frac{f'(x)}{f(x)}\delta x.$$

As a rule it is the relative error that is important; of the two factors δx and $f'(x)/f(x)$ the first depends solely on the accuracy of the measurements while the second is conditioned by the general arrangements of the investigation.

If there are two or more variables, x, y, z say, then the error δu in the function $u = f(x, y, z)$ is

$$\delta u = \frac{\partial f}{\partial x}\delta x + \frac{\partial f}{\partial y}\delta y + \frac{\partial f}{\partial z}\delta z$$

as far as quantities of the first order in δx, δy, δz. Since the value of δu is of the first degree in δx, δy, δz the joint effect of the individual errors δx, etc., is obtained by addition of the effects due to each separately. This principle of "the superposition of small errors" is of great importance in practice.

Ex. 1. The side a and the angles B, C of a triangle ABC are measured; if these be liable to the errors δa, δB, δC, to find the error in the calculated value of the area S.

SMALL CORRECTIONS. 259

Denote by $(\delta S)_a$ the error in S due to the error δa taken by itself, and use a similar notation for the other errors. In finding the derivative of S it is most convenient to differentiate logarithmically.

$$S = \tfrac{1}{2}a^2 \sin B \sin C / \sin(B+C),$$

$$(\delta S)_a / S = 2\delta a / a,$$

$$(\delta S)_B / S = \{\cot B - \cot(B+C)\}\delta B,$$

$$(\delta S)_C / S = \{\cot C - \cot(B+C)\}\delta C.$$

The total error δS is got by adding these separate errors.

As an example, let $a = 250$ (feet), $B = 27° \ 12'$, $C = 45° \ 18'$, $\delta a = \cdot 25$, $\delta B = 10'$, $\delta C = 20'$. The percentage errors in a, B, C are

$$100\frac{\delta a}{a} = \cdot 1 \ ; \quad 100\frac{\delta B}{B} = \cdot 6 \ ; \quad 100\frac{\delta C}{C} = \cdot 7.$$

It is sufficient therefore to use 5-figure logarithms. We find

$$\frac{\delta S}{S} = \cdot 002 + \cdot 00474 + \cdot 00392 = \cdot 01066.$$

$$S = 10646 \ ; \quad \delta S = 113\cdot49 \ ; \quad \frac{100\delta S}{S} = 1\cdot1.$$

The calculation of S from the values $a + \delta a$, $B + \delta B$, $C + \delta C$ gives, if S' be the new value of S,

$$S' = 10760 \ ; \quad S' - S = 114.$$

Since $b = a \sin B / \sin(B+C)$, we have for the error in b

$$\delta b / b = \delta a / a + (\cot B - \cot(B+C))\delta B - \cot(B+C)\delta C,$$

so that $\quad \delta b / b = \cdot 00390, \ 100 \delta b / b = \cdot 4, \ \delta b = \cdot 5$ nearly,

and in the same way

$$\delta c / c = \cdot 0040, \ 100 \delta c / c = \cdot 4, \ \delta c = \cdot 75.$$

Ex. 2. The sides a, b, c of a triangle ABC are measured; to find the error δA in A due to errors δa, δb, δc in a, b, c.

We may take the value of $\cos A$ given by

$$\cos A = (b^2 + c^2 - a^2)/2bc$$

and differentiate; but the result may be obtained more quickly, thus:

$$a = b \cos C + c \cos B \ ;$$

therefore $\quad \delta a = \cos C \, \delta b + \cos B \, \delta c - (b \sin C \, \delta C + c \sin B \, \delta B)$

$$= \cos C \, \delta b + \cos B \, \delta c - b \sin C (\delta C + \delta B)$$

$$= \cos C \, \delta b + \cos B \, \delta c + b \sin C \, \delta A,$$

since $b \sin C = c \sin B$ and $A + B + C = 180°$, so that $\delta A + \delta B + \delta C$ is zero.

Hence $\quad \delta A = (\delta a - \cos C \, \delta b - \cos B \, \delta c)/b \sin C,$

and the trigonometrical functions may easily be expressed in terms of the sides if required.

EXERCISES XXI.

1. The area S of a triangle ABC is determined by a, b, C; show that the relative error in the area is given by
$$\frac{\delta S}{S} = \frac{\delta a}{a} + \frac{\delta b}{b} + \cot C\, \delta C.$$
Show that the error in the side c is given by
$$\delta c = \cos B\, \delta a + \cos A\, \delta b + a \sin B\, \delta C.$$

2. At a distance of 120 feet from the foot of a tower the elevation of its top is 40° 16′; if the distance and the elevation are measured to within 1 inch and 1 minute, find the greatest error in the calculated height.

3. If the density (ρ) of a body be inferred from its weights W, w in air and in water respectively, show that the relative error in ρ due to errors $\delta W, \delta w$ in W, w is
$$\frac{\delta \rho}{\rho} = \frac{-w}{W-w}\frac{\delta W}{W} + \frac{\delta w}{W-w}.$$

4. The side a and the opposite angle A of a triangle ABC remain constant; show that when the other sides and angles are slightly varied,
$$\frac{\delta b}{\cos B} + \frac{\delta c}{\cos C} = 0.$$

5. If a triangle ABC be slightly varied but so as to remain inscribed in the same circle, show that
$$\frac{\delta a}{\cos A} + \frac{\delta b}{\cos B} + \frac{\delta c}{\cos C} = 0.$$

6. In a tangent galvanometer the tangent of the deflection of the needle is proportional to the current; show that the relative error in the value of the current due to an error in the reading of the deflection is least when the deflection is 45°.

7. If ordinates which differ by less than one-hundredth of the unit line are considered to be equal, show that the parabola $y = x + 2x^2$ will coincide with the graph of
$$x + 2x^2 + 3x^3 + 4x^4 + 5x^5$$
for values of x between $-\cdot 14$ and $+\cdot 14$.

8. Show that the curve $x^3 + y^3 = 3axy$ has two branches which pass through the origin and that the equations of these branches near the origin are
$$x^2 = 3ay, \quad y^2 = 3ax.$$
Show that closer approximations are given by
$$y = x^2/3a + x^5/81a^4, \quad x = y^2/3a + y^5/81a^4.$$

EXERCISES XXI.

9. Show that, near the points stated, the curve $x^3 + y^3 = 2ax^2$ is given by the respective equations, a being positive.

Near (o, o) $y^3 = 2ax^2$; near $(2a, 0)$ $y^3 = -4a^2(x-2a)$;

At infinity $y = -x + 2a/3 + 4a^2/9x$.

Show that y is a maximum when $x = 4a/3$ and graph the curve.

10. Show that for the curve $xy^2 - x^3 - 6x^2 + xy + y^2 = 0$ the following approximations hold :—

Near (o, o) $y = 2x - \tfrac{3}{2}x^2$ and $y = -3x + \tfrac{9}{2}x^2$.

At infinity $x + 1 = 1/y$, $y = x + 2 - 5/x$, $y = -x - 3 + 6/x$.

Show that the asymptote $x + 1 = 0$ crosses the curve at $(-1, -5)$, the asymptote $y = x + 2$ crosses at $(-\tfrac{2}{5}, \tfrac{8}{5})$ and the asymptote $y = -x - 3$ crosses at $(-\tfrac{2}{3}, -\tfrac{7}{3})$. Graph the curve.

11. Show that the curves

$$\text{(i) } (y - x^2)^2 = x^3, \quad \text{(ii) } (y - x^2)^2 = x^5$$

have each a cusp at the origin but that *both* branches of (ii) lie above the x-axis near the origin. Graph the curves.

In case (ii) the cusp is called a cusp *of the second kind* or a *ramphoid* cusp while the ordinary cusp is called for distinction a cusp *of the first kind* or a *ceratoid* cusp.

CHAPTER XIII.

INTEGRATION.

§ 110. Integration. In § 82 the general problem of the Integral Calculus has been stated, namely:—Given a continuous function $F(x)$, to find another function which (i) has $F(x)$ as its derivative and (ii) takes a given value A when x takes a given value a.

When condition (i) alone is given there is an indefinite number of solutions. These solutions, however, differ only by a constant; any one of them is called an *indefinite integral* of $F(x)$ and the constant is called the *constant of integration*. This constant is sometimes called an *arbitrary* constant since it may have any value whatever. If $f(x)$ is an indefinite integral, $f(x)+C$ is called the *general* integral, C being an arbitrary constant.

Instead of the notation of inverse functions $D_x^{-1} F(x)$ it is customary to denote the indefinite integral of $F(x)$ by the symbol

$$\int F(x)\, dx; \quad \dots\dots\dots\dots\dots\dots\dots\dots(1)$$

read, "the integral of $F(x)$ with respect to x," or "integral of $F(x)\, dx$." The differential dx indicates *the variable of integration*, namely x, and the joint symbol $\int \dots dx$ means "integral of ... with respect to x." $F(x)$ is called the *integrand*.

What was in § 82 denoted by $[D^{-1} F(x)]_a^b$ is now denoted by

$$\int_a^b F(x)\, dx; \quad \dots\dots\dots\dots\dots\dots\dots\dots(2)$$

read, "the integral from a to b of $F(x)\, dx$." The function

INTEGRALS. NOTATION. TERMINOLOGY.

denoted by the symbol is called a *definite* integral, and a, b are called the *limits* of the integral, a being the *lower* limit and b the *upper*. (The word "limit" in this use of it means merely "value of the variable of integration at one end of its range," "end-value"; this use of the word must not be confused with the technical sense employed in other connections.) The interval $(b-a)$ is called the range of integration.

Geometrically, the symbol (2) denotes the area, in *sign* and in *magnitude*, swept out by an ordinate of the graph of $F(x)$ as x varies from the lower limit a to the upper limit b. If $f(x)$ is an indefinite integral of $F(x)$ then as in § 82

$$\int_a^b F(x)\, dx = [D^{-1} F(x)]_a^b = f(b) - f(a)\dots\dots\dots(3)$$

We may, if we please, use the general integral $f(x) + C$ instead of $f(x)$; the result will be the same since C, being a constant, will disappear in the subtraction.

It follows at once from the geometrical meaning or from (3) that $\int_b^a F(x)\, dx = -\int_a^b F(x)\, dx = f(a) - f(b)\dots\dots\dots(4)$ that is, the limits a, b may be interchanged if at the same time the sign of the integral is changed.

Again, the form $f(b) - f(a)$, or the geometrical meaning, shows that the *definite* integral is a function of its *limits*, not of the variable of integration. Thus $\int_a^b F(u)\, du$ has precisely the same value as $\int_a^b F(x)\, dx$.

From the point of view of a rate, $F(x)$ when it is the derivative of $f(x)$ measures the rate at which $f(x)$ increases with respect to x; the amount, positive or negative, by which $f(x)$ increases as x varies from a to b is $f(b) - f(a)$. Hence the definite integral (3) measures the amount by which a function $f(x)$ increases for a given change $(b-a)$ of its argument when the rate of change, $F(x)$, of the function is known.

The function which has $F(x)$ as its derivative, and which is equal to A when x is equal to a, is (§ 82)

$$D^{-1} F(x) - [D^{-1} F(x)]_a + A,$$

and is, in the present notation, represented by

$$\int_a^x F(x)\,dx + A \text{ or by } \int_a^x F(u)\,du + A. \quad\ldots\ldots\ldots\ldots(5)$$

Here the upper limit x denotes the particular value of the argument for which the function is calculated. In the geometrical representation of § 82 the upper limit x is the abscissa OM of the point P. From the point of view of rates the symbol (5) denotes the function which is equal to A when its argument is equal to a and which increases at the rate $F(x)$.

The subject of definite integrals will be more fully considered in Chapter XIV.; enough, however, has been given in this article and in Chapter X. to enable the student to solve the simple examples on areas, etc., which are given in the exercises of this chapter.

111. Standard Forms. Integration from the point of view from which it is now being considered is simply the inverse of differentiation and the first requisite for the calculation of an integral, definite or indefinite, is a table of known integrals; the table will be formed from an examination of the known results of differentiation. Various methods will then be given for reducing, if possible, an integrand not found in the table to a form that may be integrated by means of the standard forms. In all cases of indefinite integrals the test to be applied is that *the derivative of the integral must be equal to the integrand.*

In symbols

$$f(x) = \int F(x)\,dx \text{ if } \frac{df(x)}{dx} = F(x).$$

so that the equation that defines an integral is

$$\frac{d}{dx}\left[\int F(x)\,dx\right] = F(x).$$

Considered as symbols of operation d/dx and $\int \ldots dx$ are *inverse* to each other.

In the language of differentials $F(x)\,dx$ is the differential

STANDARD FORMS OF INTEGRALS.

of $f(x)$ when $f(x)$ is the integral of $F(x)$; $f(x)$ is often called the integral of the differential $F(x)dx$. Since

$$F(x)dx = df(x) = d\left[\int F(x)dx\right]$$

the operators d and $\int$ are inverse to each other.

The following table contains what may be called the fundamental standard forms; other important forms will be given later. Most of the forms are given twice; the argument occurs so often in the combination $ax+b$ that the student should from the outset make himself familiar with the corresponding integral. The results should of course be tested by differentiation.

1. If $n \neq -1$.

$$\int x^n dx = \frac{x^{n+1}}{n+1}; \qquad \int (ax+b)^n dx = \frac{(ax+b)^{n+1}}{(n+1)a}.$$

2. If $n = -1$.

$$\int \frac{1}{x} dx = \log x; \qquad \int \frac{1}{ax+b} dx = \frac{1}{a} \log(ax+b).$$

3. $\int e^x dx = e^x; \qquad \int e^{ax} dx = \frac{1}{a} e^{ax}.$

4. $\int \sin x \, dx = -\cos x; \qquad \int \sin(ax+b) \, dx = -\frac{1}{a} \cos(ax+b).$

5. $\int \cos x \, dx = \sin x; \qquad \int \cos(ax+b) \, dx = \frac{1}{a} \sin(ax+b).$

6. $\int \sec^2 x \, dx = \tan x; \qquad \int \sec^2(ax+b) \, dx = \frac{1}{a} \tan(ax+b).$

7. $\int \operatorname{cosec}^2 x \, dx = -\cot x;$

$$\int \operatorname{cosec}^2(ax+b) \, dx = -\frac{1}{a} \cot(ax+b).$$

8. $\int \frac{dx}{\sqrt{(1-x^2)}} = \sin^{-1} x; \quad \int \frac{dx}{\sqrt{(a^2-x^2)}} = \sin^{-1}\left(\frac{x}{a}\right);$

$\qquad \text{or} = -\cos^{-1} x; \qquad \text{or} = -\cos^{-1}\left(\frac{x}{a}\right).$

9. $\int \dfrac{dx}{\sqrt{(x^2 \pm 1)}} = \log(x + \sqrt{x^2 \pm 1})$;

$\int \dfrac{dx}{\sqrt{(x^2 \pm a^2)}} = \log(x + \sqrt{x^2 \pm a^2})$.

10. $\int \dfrac{dx}{1+x^2} = \tan^{-1} x$; $\int \dfrac{dx}{a^2+x^2} = \dfrac{1}{a} \tan^{-1}\left(\dfrac{x}{a}\right)$;

or $= -\cot^{-1} x$;   or $= -\dfrac{1}{a}\cot^{-1}\left(\dfrac{x}{a}\right)$.

11. $\int \dfrac{dx}{x^2 - a^2} = \dfrac{1}{2a} \log\left(\dfrac{x-a}{x+a}\right)$ if $x^2 > a^2$ (§ 112, Ex. 2),

$= \dfrac{1}{2a} \log\left(\dfrac{a-x}{a+x}\right)$ if $x^2 < a^2$.

Since $\sin^{-1}x - (-\cos^{-1}x)$ is equal to $\pi/2$, both $\sin^{-1}x$ and $-\cos^{-1}x$ are integrals of $1/\sqrt{(1-x^2)}$; a similar observation holds for the integral of $1/(1+x^2)$. An indefinite integral may often be expressed in different forms, any two of which must however differ only by a constant. Particular care is required in dealing with the inverse trigonometric functions since these are many-valued; the restriction on the range of the angle (§§ 28, 64) must always be attended to.

If x is negative, the integral of $1/x$ is not $\log x$ but $\log(-x)$; if x is less than a, the integral of $1/(x-a)$ is $\log(a-x)$. Form 11 is inserted for the sake of comparison with 10; for a similar reason forms 8 and 9 are brought together.

Again, if x is negative, it may be verified that the integral of

$1/\sqrt{(x^2+k)}$ is $-\log\{-x + \sqrt{(x^2+k)}\}$.

Instead of the logarithms in form 9 inverse hyperbolic functions may be used (§ 66).

9a. $\int \dfrac{dx}{\sqrt{(x^2+a^2)}} = \sinh^{-1}\left(\dfrac{x}{a}\right)$; $\int \dfrac{dx}{\sqrt{(x^2-a^2)}} = \cosh^{-1}\left(\dfrac{x}{a}\right)$;

and it should be remembered that $\cosh^{-1}x$ is two-valued. The forms $\tanh^{-1}x$, $\coth^{-1}x$, are of less importance.

Ex. 1. Integrate with respect to x

$\sqrt{x}$; $\dfrac{1}{\sqrt{x}}$; $\sqrt{(3x-4)}$; $\dfrac{1}{\sqrt{(3x-4)}}$; $\dfrac{1}{\sqrt{(3-x^3)}}$.

Ex. 2. Evaluate

$\int_0^{\pi} \sin x\, dx$; $\int_0^{\pi} \cos x\, dx$; $\int_0^{\frac{\pi}{4}} \dfrac{dx}{\cos^2 x}$; $\int_{a^2}^{b^3} \dfrac{dx}{x}$; $\int_{-3}^{-1} \dfrac{dx}{x}$.

§ 112. Algebraic and Trigonometric Transformations.

By applying the definition of an integral and Theorems II., III. of § 58 the following theorems are easily proved:

(i) $\int c\,F(x)\,dx = c\int F(x)\,dx$ if c is a constant.

(ii) $\int (u - v + \ldots + z)\,dx = \int u\,dx - \int v\,dx + \ldots + \int z\,dx,$

where $u, v, \ldots z$ are functions of x or constants.

Thus the derivative of the integral on the left of (ii) is by definition
$$u - v + \ldots + z;$$
by Theorem III., § 58, the derivative of the sum on the right of (ii) is the sum of the derivatives of the terms, and that sum is by the definition of an integral $u - v + \ldots + z$. Hence, apart from constants of integration, which are not considered, equation (ii) is seen to be true.

Ex.
$$\int (3x^4 - 5x^2 + 1)\,dx = \int 3x^4\,dx - \int 5x^2\,dx + \int 1\,dx \quad \text{by (ii)}$$
$$= 3\int x^4\,dx - 5\int x^2\,dx + \int dx \quad \text{by (i)}$$
$$= \tfrac{3}{5}x^5 - \tfrac{5}{3}x^3 + x.$$

Integration is essentially a tentative process, and it often happens that among the known functions there is none of which a given function is the derivative (see § 82). Two general methods of integration will be given (§§ 113, 118) which are of great use in the search for integrals; but usually some simple algebraic or trigonometric transformation of the integrand will be of great assistance in reducing it to a sum of terms each of which is a standard form. Some of the results are so important as to be included among the standard forms, but the student should rather try to seize the spirit of the transformations than burden his memory with a mass of isolated results. (See the remarks in § 123.)

We now take one or two examples of such transformations.

Ex. 1. Integrate $(2x^3 - 7x^2 + 1)/(2x - 1)$.

By division,
$$(2x^3 - 7x^2 + 1)/(2x - 1) = x^2 - 3x - \tfrac{3}{2} - \tfrac{1}{2}/(2x - 1).$$

Hence the integral is
$$\tfrac{1}{3}x^3 - \tfrac{3}{2}x^2 - \tfrac{3}{2}x - \tfrac{1}{4}\log(2x - 1).$$

Any fraction in which the numerator is a rational integral function of x and the denominator a linear function may be integrated in the same way.

Ex. 2. Integrate $1/(x^2 - a^2)$.

Resolve the fraction into partial fractions:
$$\frac{1}{x^2 - a^2} = \frac{1}{2a}\left(\frac{1}{x - a} - \frac{1}{x + a}\right)$$
$$\int \frac{dx}{x^2 - a^2} = \frac{1}{2a}\Big(\log(x - a) - \log(x + a)\Big)$$
$$= \frac{1}{2a}\log\frac{x - a}{x + a}.$$

This is the proper form if $x^2 > a^2$, because then, and only then, is $(x - a)/(x + a)$ positive; if $x^2 < a^2$ the integral is
$$\frac{1}{2a}\log\frac{a - x}{a + x},$$
because in that case the integral of $1/(x - a)$ is $\log(a - x)$.

The transformation is a particular case of the method of partial fractions, and the student should refer to some text-book of algebra for an account of the method; see also § 120.

Since
$$\frac{3x - 5}{(x - 1)(x - 2)} = \frac{2}{x - 1} + \frac{1}{x - 2},$$
we find
$$\int \frac{3x - 5}{(x - 1)(x - 2)}dx = 2\log(x - 1) + \log(x - 2).$$

Ex. 3. The forms $\dfrac{1}{a + bx^2}$, $\dfrac{1}{\sqrt{(a + bx^2)}}$.

If a, b are both positive, we have
$$\int \frac{dx}{a + bx^2} = \frac{1}{b}\int \frac{dx}{\frac{a}{b} + x^2} = \frac{1}{\sqrt{(ab)}}\tan^{-1}\left(\frac{x\sqrt{b}}{\sqrt{a}}\right).$$

If a is negative, b positive, we reduce the integrand to the form of ex. 2; thus
$$\int \frac{dx}{3x^2 - 5} = \frac{1}{3}\int \frac{dx}{x^2 - \frac{5}{3}} = \frac{1}{2\sqrt{15}}\log\left(\frac{x - \sqrt{5}/\sqrt{3}}{x + \sqrt{5}/\sqrt{3}}\right).$$

In a similar way $1/\sqrt{(a + bx^2)}$ may be treated; thus
$$\int \frac{dx}{\sqrt{(3 - 2x^2)}} = \frac{1}{\sqrt{2}} \cdot \int \frac{dx}{\sqrt{(\frac{3}{2} - x^2)}} = \frac{1}{\sqrt{2}}\sin^{-1}\left(\frac{x\sqrt{2}}{\sqrt{3}}\right).$$

With a little practice the student should be able to do many of the steps mentally; the full process for the first case is

$$\frac{1}{b}\int\frac{dx}{\frac{a}{b}+x^2}=\frac{1}{b}\cdot\frac{1}{\frac{\sqrt{a}}{\sqrt{b}}}\tan^{-1}\left(\frac{x}{\frac{\sqrt{a}}{\sqrt{b}}}\right)=\frac{1}{\sqrt{(ab)}}\tan^{-1}\frac{x\sqrt{b}}{\sqrt{a}}.$$

Ex. 4. $\sin^n x$, $\cos^n x$, $\sin mx \cos nx$.

When n is a small positive integer, $\sin^n x$, $\cos^n x$ may, without any difficulty, be expressed in terms of sines or cosines of multiples of x; for other values of n it is best to take the method of successive reduction (§ 119) or the method of ex. 4, § 114,

$$\sin^2 x = \tfrac{1}{2}(1-\cos 2x);\quad \sin^3 x = \tfrac{3}{4}\sin x - \tfrac{1}{4}\sin 3x;$$

$$\int \sin^2 x\, dx = \tfrac{1}{2}x - \tfrac{1}{4}\sin 2x;\quad \int \sin^3 x\, dx = -\tfrac{3}{4}\cos x + \tfrac{1}{12}\cos 3x;$$

$$\int_0^{\frac{\pi}{2}} \sin^2 x\, dx = \frac{\pi}{4};\quad \int_0^{\frac{\pi}{2}} \sin^3 x\, dx = 0 - [-\tfrac{3}{4}+\tfrac{1}{12}] = \tfrac{2}{3}.$$

In the same way powers of $\cos x$ may be treated.

Again, a product of a sine and a cosine, or of two sines or of two cosines, may be expressed as a sum or a difference of sines or cosines and then integrated. Thus

$$\sin mx \cos nx = \tfrac{1}{2}\{\sin (m+n)x + \sin(m-n)x\};$$

hence, if $m \neq n$,

$$\int \sin mx \cos nx\, dx = -\frac{\cos(m+n)x}{2(m+n)} - \frac{\cos(m-n)x}{2(m-n)}$$

but if $m=n$, then the integral is

$$-\frac{1}{4m}\cos 2mx.$$

EXERCISES XXII.

Integrate, with respect to x, examples 1-15.

1. $\dfrac{3x^4-4x^3+2x^2-3}{x-3}$; 2. $\dfrac{2x+1}{2x-1}$; 3. $\dfrac{2x-3}{x^2-3x+2}$;

4. $\dfrac{3x-7}{(x-1)(x-2)(x-3)}$; 5. $\dfrac{1}{7-3x^2}$; 6. $\dfrac{1}{7+3x^2}$;

7. $\dfrac{1}{\sqrt{(3-4x^2)}}$; 8. $\dfrac{1}{\sqrt{(3+4x^2)}}$; 9. $\cos^2 x$; 10. $\cos^3 x$;

11. $\cos^2(ax+b)$; 12. $\sin^4 x$; 13. $\sin 3x \sin 4x$;

14. $\sin(3x+2)\cos(4x+3)$; 15. $\cos x \cos 2x \cos 3x$.

Find the value of the integrals in examples 16-21.

16. $\int_0^{\frac{\pi}{2}} \cos^2 x\, dx$; 17. $\int_0^{\frac{\pi}{2}} \sin^2 2x\, dx$; 18. $\int_0^2 \dfrac{dx}{4+x^2}$;

19. $\int_0^1 \dfrac{dx}{4-x^2}$; 20. $\int_3^4 \dfrac{dx}{4-x^2}$; 21. $\int_0^{\frac{\sqrt{3}}{2}} \dfrac{dx}{\sqrt{(3-x^2)}}$.

22. If m, n are unequal positive integers, prove that

$$\int_0^\pi \cos mx \cos nx\, dx = 0 = \int_0^\pi \sin mx \sin nx\, dx,$$

and find the value of each integral when m, n are equal positive integers.

23. Show by considering the graphs of the integrands that the following equations are true :

(i) $\int_0^{\frac{\pi}{2}} \cos^n x\, dx = \int_0^{\frac{\pi}{2}} \sin^n x\, dx$ where n is positive,

(ii) $\int_0^\pi \sin^n x\, dx = 2\int_0^{\frac{\pi}{2}} \sin^n x\, dx$ where n is positive,

(iii) $\int_0^\pi \cos^n x\, dx = 2\int_0^{\frac{\pi}{2}} \cos^n x\, dx$ if n is an even integer,

but $\qquad = 0 \qquad$ if n is an odd integer.

24. The area bounded by the parabola $y^2 = 4ax$, and the double ordinate through the point (b, c) on it is $\tfrac{4}{3}bc$.

25. If a, b are positive and $a < b$, the area between the hyperbola $xy = c^2$, the x-axis and the ordinates at a, b is $c^2 \log(b/a)$.

If instead of a hyperbola the curve is that given by $y = x^n/c^{n-1}$, then the area is
$$(b^{n+1} - a^{n+1})/(n+1)c^{n-1}.$$

26. The area between the x-axis and one arch of the harmonic curve $y = b\sin(x/a)$ is $2ab$.

27. An ellipse revolves about its major axis; show that the volume of the spheroid generated by a complete revolution is $\tfrac{4}{3}\pi ab^2$.

If the axis of revolution be the minor axis, the volume is $\tfrac{4}{3}\pi a^2 b$.

28. The area of the section of certain surfaces made by a plane through the point whose abscissa is x perpendicular to the x-axis is $A + Bx + Cx^2$ where A, B, C are constants. Show that the volume intercepted between two planes perpendicular to the x-axis is

$$A(b-a) + \tfrac{1}{2}B(b^2 - a^2) + \tfrac{1}{3}C(b^3 - a^3)$$

where a, b are the abscissae of the points where the planes cut the x-axis, $(a < b)$.

Apply the result to find (i) the volume of a cone; (ii) the volume of a segment of a sphere; (iii) the volume of the ellipsoid whose equation is
$$x^2/a^2 + y^2/b^2 + z^2/c^2 = 1.$$

29. If in ex. 28 S_1, S_2, M are the areas of the sections through a, b and the point midway between a and b, and if $b - a = 2h$, show that the volume is
$$\tfrac{1}{3}h(S_1 + S_2 + 4M).$$

§ 113. Change of Variable. The rule for differentiating a function of a function (§ 59) leads to one of the two general methods of integration referred to in the preceding article, namely, that of changing the variable of integration.

Take first the simple example

$$y = \int \frac{dx}{x^2 + 2x + 2}, \quad \frac{dy}{dx} = \frac{1}{x^2 + 2x + 2}.$$

Let y be made a function of u by the substitution $x = u - 1$; by § 59,

$$\frac{dy}{du} = \frac{dy}{dx}\frac{dx}{du} = \frac{1}{x^2 + 2x + 2} = \frac{1}{u^2 + 1}.$$

Hence, considered as a function of u, the integral is

$$y = \int \frac{du}{u^2 + 1} = \tan^{-1} u, \text{ that is, } y = \tan^{-1}(x+1).$$

The change of variable has enabled us to reduce the integrand to a known form, and thus to integrate it.

Take now the general case in which the integrand is $F(x)$. Let y be made a function of u by the substitution $x = \phi(u)$; then

$$\frac{dy}{du} = \frac{dy}{dx}\frac{dx}{du} = F(x)\frac{dx}{du}. \quad \ldots\ldots\ldots\ldots\ldots(1)$$

In (1) let dx/du be found from the equation $x = \phi(u)$ and then express the new integrand $F(x)\, dx/du$ in terms of u by means of the same equation. Equation (1) will now be free from x and we shall have

$$y = \int F(x) \frac{dx}{du} du. \quad \ldots\ldots\ldots\ldots\ldots(2)$$

It may happen that the new integrand is, as in the above example, a standard form; if not, it may perhaps be more easily reduced to one than the old integrand $F(x)$.

Expressing y as an integral with respect to x, and equating it to the value given by (2), we have

$$y = \int F(x)\, dx = \int F(x)\frac{dx}{du}\, du \ldots\ldots\ldots\ldots\ldots(3)$$

The simple rule then for changing the variable is: *Replace dx by $(dx/du)\, du$ and by means of the equation*

272   AN ELEMENTARY TREATISE ON THE CALCULUS.

between x and u express the new integrand $F(x)\,dx/du$ in terms of u; the integral will then be a function of the new variable u.

When the integration has been effected the integral should be expressed in terms of the old variable.

If when $x=a$, $u=a$, and when $x=b$, $u=\beta$, the relation being such that as x varies continuously from a to b, u also varies continuously from a to β, then

$$\int_a^b F(x)\,dx = \int_a^\beta F(x)\frac{dx}{du}\,du. \qquad\ldots\ldots\ldots\ldots(4)$$

In this case, of course, there is no need for returning to the old variable.

In applying the transformations (3) and (4) it is essential that to each value of x there should correspond one and only one value of u, and to each value of u one and only one value of x, within the ranges $b-a$, $\beta-a$ of integration. When the equation between x and u gives u as a multiple-valued function of x, or x as a multiple-valued function of u, care must be taken to choose the proper value. (See § 117, Ex. 3, § 123.)

§ 114.   Examples of Change of Variable.

Ex. 1. $F(x)$ of the form $\psi(ax+b)$.

Let $u=ax+b$; $du=a\,dx$, $dx=\dfrac{1}{a}du$

$$\int \psi(ax+b)\,dx = \frac{1}{a}\int \psi(u)\,du.$$

This type constantly occurs. Thus if $u=x-1/4$,

$$\int \frac{dx}{2x^2-x+1} = \tfrac{1}{2}\int \frac{dx}{(x-\tfrac{1}{4})^2+\tfrac{7}{16}} = \tfrac{1}{2}\int \frac{du}{u^2+\tfrac{7}{16}};$$

so that the integral is

$$\tfrac{1}{2}\cdot\frac{4}{\sqrt{7}}\tan^{-1}\!\left(\frac{4u}{\sqrt{7}}\right) = \frac{2}{\sqrt{7}}\tan^{-1}\!\left(\frac{4x-1}{\sqrt{7}}\right).$$

$$\int \frac{dx}{\sqrt{(2x^2-x+1)}} = \frac{1}{\sqrt{2}}\int \frac{dx}{\sqrt{\{(x-\tfrac{1}{4})^2+\tfrac{7}{16}\}}} = \frac{1}{\sqrt{2}}\log\{x-\tfrac{1}{4}+\sqrt{x^2-\tfrac{1}{2}x+\tfrac{1}{2}}\}.$$

A constant factor, like 2, can be taken outside the integral sign when necessary; similarly a constant factor may be introduced, as in ex. 3.

WORKED EXAMPLES.

Ex. 2. $F(x)$ of the form $\psi(x^n)x^{n-1}$.

Let $u = x^n$; $du = nx^{n-1}dx$, $x^{n-1}dx = \frac{1}{n}du$

$$\int \psi(x^n)x^{n-1}dx = \frac{1}{n}\int \psi(u)du.$$

Thus, if $u = x^2$,

$$\int \sqrt{(ax^2+b)}\,x\,dx = \frac{1}{2}\int \sqrt{(au+b)}\,du = \frac{1}{3a}(au+b)^{\frac{3}{2}},$$

and the given integral is $(ax^2+b)^{3/2}/3a$.

The integral may also be found by putting $u = ax^2 + b$, or by putting $u^2 = ax^2 + b$. The last substitution gives

$$x\,dx = \frac{1}{a}u\,du\,;\quad \int \sqrt{(ax^2+b)}\,x\,dx = \frac{1}{a}\int u^2 du = \frac{u^3}{3a},$$

leading to the same value as before.

Ex. 3. $F(x)$ of the form $[\psi(x)]^n \psi'(x)$.

Let $u = \psi(x)$; $du = \psi'(x)dx$; $F(x)dx = u^n du$,

and the integral is a power or a logarithm according as n is different from or equal to -1. We have

(3a) $\quad \int [\psi(x)]^n \psi'(x)dx = \frac{1}{n+1}[\psi(x)]^{n+1}, \quad n \neq -1.$

(3b) $\quad \int \frac{\psi'(x)}{\psi(x)}dx = \log [\psi(x)].$

From (3b) we see that when the integrand is a fraction whose numerator is the derivative of the denominator, the integral is the logarithm of the denominator.

The introduction of a factor is sometimes needed to make the integrand of the form 3. Thus

(i) $\int \frac{(x-1)dx}{\sqrt{(2x^2-4x+1)}} = \frac{1}{4}\int (2x^2-4x+1)^{-\frac{1}{2}}(4x-4)dx$

$\qquad\qquad\qquad\qquad = \frac{1}{2}\sqrt{(2x^2-4x+1)}.$

(ii) $\int \frac{(x+a)dx}{(x+a)^2+\beta^2} = \frac{1}{2}\log \{(x+a)^2+\beta^2\}.$

(iii) $\int \tan x\,dx = -\int \frac{-\sin x}{\cos x}dx = -\log \cos x.$

(iv) $\int \tan^3 x\,dx = \int \tan x\,(\sec^2 x - 1)dx = \int \tan x \sec^2 x\,dx - \int \tan x\,dx,$

and therefore $\quad = \frac{1}{2}\tan^2 x + \log \cos x.$

Ex. 4. $F(x) = \sin^m x \cos^n x$.

(i) When either m or n is an odd positive integer the integration can be effected by substituting u for $\cos x$ when m is odd, but u for $\sin x$ when n is odd. For example, take $F(x) = \sin^{\frac{1}{2}} x \cos^5 x$.

Let $u = \sin x$; $du = \cos x \, dx$; $\cos^4 x = (1-u^2)^2$

$$\int \sin^{\frac{1}{2}} x \cos^5 x \, dx = \int (u^{\frac{1}{2}} - 2u^{\frac{5}{2}} + u^{\frac{9}{2}}) du$$

$$= \tfrac{2}{3} u^{\frac{3}{2}} - \tfrac{4}{7} u^{\frac{7}{2}} + \tfrac{2}{11} u^{\frac{11}{2}}$$

$$= \sin^{\frac{3}{2}} x (\tfrac{2}{3} - \tfrac{4}{7} \sin^2 x + \tfrac{2}{11} \sin^4 x).$$

$$\int_0^{\frac{\pi}{2}} \sin^{\frac{1}{2}} x \cos^5 x \, dx = \int_0^1 (u^{\frac{1}{2}} - 2u^{\frac{5}{2}} + u^{\frac{9}{2}}) du = \tfrac{2}{3} - \tfrac{4}{7} + \tfrac{2}{11} = \tfrac{64}{231}.$$

Again, if $u = \cos x$, $du = -\sin x \, dx$

$$\int \sin^5 x \, dx = -\int (1-u^2)^2 du = -(u - \tfrac{2}{3} u^3 + \tfrac{1}{5} u^5),$$

and
$$\int \sin^5 x \, dx = -\cos x + \tfrac{2}{3} \cos^3 x - \tfrac{1}{5} \cos^5 x.$$

(ii) When $m+n$ is an even negative integer, let $u = \tan x$ (or $\cot x$); the new integrand can be expanded by the Binomial Theorem. Thus,

$$\int \frac{dx}{\sin^5 x \cos^3 x} = \int \frac{(1+u^2)^4}{u^5} \frac{du}{1+u^2} = \int \left(\frac{1}{u^5} + \frac{3}{u^3} + \frac{3}{u} + u \right) du,$$

and the integral is readily found in terms of x.

Ex. 5. If $F(x)$ is a rational function of x and of $\sqrt{(ax+b)}$, the substitution $ax+b = u^2$ will make the new integrand a rational function of u. Thus, if $x+1 = u^2$,

$$\int x^2 \sqrt{(x+1)} dx = 2\int (u^2-1)^2 u^2 du = 2(\tfrac{1}{7} u^7 - \tfrac{2}{5} u^5 + \tfrac{1}{3} u^3),$$

and after a little reduction we get for the integral

$$2\sqrt{(x+1)}(15x^3 + 3x^2 - 4x + 8)/105.$$

The forms just given include many of the most important cases in elementary work, and the student should at once try the earlier examples in Exercises XXIII. Only through practice will he gain facility in making the transformations.

§ 115. Quadratic Functions. If $R = ax^2 + bx + c$ and if $f(x)$ is rational and integral, the fraction $f(x)/R$ can be expressed as the sum of an integral function and a proper fraction $(Ax+B)/R$. We will now consider the forms $(Ax+B)/R$ and $(Ax+B)/\sqrt{R}$.

QUADRATIC FUNCTIONS. 275

For beginners the simplest method is to write R in the form
$$R = a\left(x + \frac{b}{2a}\right)^2 + \frac{4ac - b^2}{4a};$$
when a is positive we may take it as equal to $+1$, and when negative as equal to -1: there is no loss of generality in so doing since a constant factor may always be taken outside the integral sign.

If $4ac - b^2$ is positive the factors of R are imaginary; R is then of the form
$$R = (x + a)^2 + \beta^2 \quad \text{...................(i)}$$

If $4ac - b^2$ is negative the factors of R are real, and

for $a = +1$, $\qquad R = (x+a)^2 - \beta^2$ ....(ii)

for $a = -1$, $\qquad R = \beta^2 - (x+a)^2$ ....(iii)

I. $(Ax+B)/R$.

(i) If the factors of R are real resolve the fraction into partial fractions as in § 112, Ex. 2.

(ii) If the factors of R are imaginary then $R = (x+a)^2 + \beta^2$ and we can transform the fraction so that the substitutions of Ex. 3 and Ex. 1 of § 114 can be used. Choose λ and μ so that
$$Ax + B = \lambda(2x + 2a) + \mu; \quad \lambda = \tfrac{1}{2}A, \ \mu = B - aA.$$

Hence $\quad \dfrac{Ax+B}{R} = \lambda \dfrac{2x + 2a}{(x+a)^2 + \beta^2} + \mu \dfrac{1}{(x+a)^2 + \beta^2};$

and $\quad \displaystyle\int \frac{(Ax+B)\,dx}{R} = \lambda \log\{(x+a)^2 + \beta^2\} + \frac{\mu}{\beta} \tan^{-1}\left(\frac{x+a}{\beta}\right),$

the first integral being a case of § 114, Ex. 3, the second of § 114, Ex. 1.

II. $(Ax+B)/\sqrt{R}$.

(i) Let R be either $(x+a)^2 + \beta^2$ or $(x+a)^2 - \beta^2$. Make the same transformation of $Ax+B$; then

$$\int \frac{(Ax+B)\,dx}{\sqrt{R}} = \lambda \int \frac{(2x+2a)\,dx}{\sqrt{\{(x+a)^2 \pm \beta^2\}}} + \mu \int \frac{dx}{\sqrt{\{(x+a)^2 \pm \beta^2\}}}$$
$$= 2\lambda\sqrt{R} + \mu \log\{(x+a) + \sqrt{(x+a)^2 \pm \beta^2}\}.$$

(ii) Let $R = \beta^2 - (x+a)^2$; then

$$\int \frac{(Ax+B)\,dx}{\sqrt{R}} = -\lambda \int \frac{-(2x+2a)\,dx}{\sqrt{\{\beta^2-(x+a)^2\}}} + \mu \int \frac{dx}{\sqrt{\{\beta^2-(x+a)^2\}}}$$

$$= -2\lambda\sqrt{R} + \mu \sin^{-1}\left(\frac{x+a}{\beta}\right),$$

when $A=0$, $\lambda=0$ and the integrand is of the type § 114, Ex. 1.

In working numerical examples it is best to find first the derivative of R; it is then easy to write $Ax+B$ in the required form.

Ex. 1. $\qquad (3x+1)/(2x^2+x+3).$

$D_x(2x^2+x+3) = 4x+1$; $\qquad 3x+1 = \tfrac{3}{4}(4x+1) + \tfrac{1}{4}$;

$2x^2+x+3 = 2\{(x+\tfrac{1}{4})^2 + \tfrac{23}{16}\}.$

Integral $= \dfrac{3}{4}\int \dfrac{(4x+1)\,dx}{2x^2+x+3} + \dfrac{1}{8}\int \dfrac{dx}{(x+\tfrac{1}{4})^2 + \tfrac{23}{16}}$

$= \tfrac{3}{4}\log(2x^2+x+3) + \dfrac{1}{2\sqrt{23}}\tan^{-1}\left(\dfrac{4x+1}{\sqrt{23}}\right).$

Ex. 2. $\qquad (3x+1)/\sqrt{(-2x^2+x+3)}.$

$3x+1 = -\tfrac{3}{4}(-4x+1) + \tfrac{7}{4}$;

$\sqrt{(-2x^2+x+3)} = \sqrt{2}\cdot\sqrt{\{\tfrac{25}{16}-(x-\tfrac{1}{4})^2\}}.$

Integral $= -\dfrac{3}{4}\int \dfrac{(-4x+1)\,dx}{\sqrt{(-2x^2+x+3)}} + \dfrac{7}{4\sqrt{2}}\int \dfrac{dx}{\sqrt{\{\tfrac{25}{16}-(x-\tfrac{1}{4})^2\}}}$

$= -\tfrac{3}{2}\sqrt{(-2x^2+x+3)} + \dfrac{7}{4\sqrt{2}}\sin^{-1}\left(\dfrac{4x-1}{5}\right).$

The types

$$\frac{1}{x\sqrt{(ax^2+bx+c)}}, \qquad \frac{1}{(mx+n)\sqrt{(ax^2+bx+c)}},$$

can be reduced to the cases just discussed by the substitutions $x=1/u$, $mx+n=1/u$ respectively. These give by logarithmic differentiation

$$\frac{dx}{x} = -\frac{du}{u}; \qquad \frac{dx}{mx+n} = -\frac{1}{m}\cdot\frac{du}{u}.$$

The substitution of $1/u$ for x is effective in other cases; thus
$$\int \frac{dx}{(a^2+x^2)^{\frac{3}{2}}} = -\int \frac{u\,du}{(a^2u^2+1)^{\frac{3}{2}}} = \frac{1}{a^2(a^2u^2+1)^{\frac{1}{2}}},$$

which, expressed in terms of x, is $x/a^2(a^2+x^2)^{\frac{1}{2}}$.

TRIGONOMETRIC SUBSTITUTIONS.

The more general form $1/(ax^2+bx+c)^{\frac{3}{2}}$ can be treated in a similar way after expressing the quadratic in the form given at the beginning of this article.

§ 116. Trigonometric and Hyperbolic Substitutions. Another method of treating the quadratic function is to transform it by a trigonometric or hyperbolic substitution. The particular transformation is suggested by the form of the quadratic.

a^2-x^2, $\sqrt{(a^2-x^2)}$; $x=a\sin\theta$ or $x=a\cos\theta$;

x^2+a^2, $\sqrt{(x^2+a^2)}$; $x=a\tan\theta$ or $x=a\sinh\theta$;

x^2-a^2, $\sqrt{(x^2-a^2)}$; $x=a\sec\theta$ or $x=a\cosh\theta$;

$\sqrt{\{\beta^2-(x+a)^2\}}$; $x+a=\beta\sin\theta$; etc.

Ex. 1. If $x=a\sin\theta$; $dx=a\cos\theta\,d\theta$.

$$\int\sqrt{(a^2-x^2)}\,dx = a^2\int\cos^2\theta\,d\theta = \frac{a^2}{2}\Big(\theta+\sin\theta\cos\theta\Big),$$

and therefore

$$\int\sqrt{(a^2-x^2)}\,dx = \tfrac{1}{2}x\sqrt{(a^2-x^2)} + \frac{a^2}{2}\sin^{-1}\frac{x}{a}.$$

Ex. 2. If $x=a\sinh\theta$; $dx=a\cosh\theta\,d\theta$.

$$\int\sqrt{(x^2+a^2)}\,dx = a^2\int\cosh^2\theta\,d\theta = \frac{a^2}{2}\Big(\theta+\sinh\theta\cosh\theta\Big),$$

and therefore

$$\int\sqrt{(x^2+a^2)}\,dx = \tfrac{1}{2}x\sqrt{(x^2+a^2)} + \frac{a^2}{2}\sinh^{-1}\Big(\frac{x}{a}\Big).$$

By putting $x=a\cosh\theta$ we find

$$\int\sqrt{(x^2-a^2)}\,dx = \tfrac{1}{2}x\sqrt{(x^2-a^2)} - \frac{a^2}{2}\cosh^{-1}\Big(\frac{x}{a}\Big).$$

Ex. 3. If $x+2=\sqrt{3}\tan\theta$; $dx=\sqrt{3}\sec^2\theta\,d\theta$.

$$\int\frac{dx}{(x^2+4x+7)^2} = \int\frac{\sqrt{3}\sec^2\theta\,d\theta}{(3\sec^2\theta)^2} = \frac{\sqrt{3}}{9}\int\cos^2\theta\,d\theta,$$

and the integral is

$$\frac{1}{6}\frac{x+2}{x^2+4x+7} + \frac{\sqrt{3}}{18}\tan^{-1}\Big(\frac{x+2}{\sqrt{3}}\Big).$$

For definite integrals trigonometric substitutions are of great importance.

§ 117. **Some Trigonometric Integrands.** The integration of powers and products of sines and cosines can often be effected by the methods of § 112, Ex. 4, § 114, Ex. 4, and § 119, Ex. 2, 3. There is another method, however, that is frequently useful. When the integrand is a rational function of $\sin x$ and of $\cos x$ the substitution $u = \tan \frac{1}{2} x$ will reduce the integral to that of a rational algebraic function of u, for

$$\sin x = \frac{2u}{1+u^2}, \quad \cos x = \frac{1-u^2}{1+u^2}, \quad dx = \frac{2du}{1+u^2}.$$

Examples 1-3 may almost be reckoned among the standard forms; the substitution is for each $u = \tan \frac{1}{2} x$.

Ex. 1. $\quad \int \dfrac{dx}{\sin x} = \int \dfrac{du}{u} = \log u = \log \tan \tfrac{1}{2} x.$

Ex. 2. $\quad \int \dfrac{dx}{\cos x} = \int \dfrac{2du}{1-u^2} = \log \dfrac{1+u}{1-u} = \log \dfrac{1+\tan \tfrac{1}{2} x}{1-\tan \tfrac{1}{2} x}.$

The integral can be put in several forms as

$$\log \tan \left(\frac{x}{2} + \frac{\pi}{4} \right) \text{ or } \tfrac{1}{2} \log \frac{1+\sin x}{1-\sin x}.$$

The substitution $v = \dfrac{\pi}{2} - x$ or $v = x - \dfrac{\pi}{2}$ will reduce the integral of $1/\cos x$ to that of $1/\sin x$.

Ex. 3. $\quad \int \dfrac{dx}{a+b\cos x} = \int \dfrac{2du}{a(1+u^2)+b(1-u^2)} = 2\int \dfrac{du}{(a+b)+(a-b)u^2}.$

Let $a+b$ be positive; then there are three cases according as b is numerically less than or greater than or equal to a.

(i) $b^2 < a^2$ and therefore $b < a$, numerically

$$\int \frac{dx}{a+b\cos x} = \frac{2}{\sqrt{(a^2-b^2)}} \tan^{-1}\left(u\sqrt{\frac{a-b}{a+b}} \right), \quad u = \tan \tfrac{1}{2} x.$$

(ii) $b^2 > a^2$ and therefore $b - a$ positive,

$$\int \frac{dx}{a+b\cos x} = \frac{1}{\sqrt{b^2-a^2}} \log \frac{\sqrt{b+a}+u\sqrt{b-a}}{\sqrt{b+a}-u\sqrt{b-a}}$$

(iii) $b^2 = a^2$,

$$\int \frac{dx}{a+a\cos x} = \frac{1}{a} \tan \tfrac{1}{2} x; \quad \int \frac{dx}{a-a\cos x} = -\frac{1}{a} \cot \tfrac{1}{2} x.$$

Case (ii) is of less importance than (i). A more easily remembered form of the integral (i) is obtained by writing

TRIGONOMETRIC INTEGRANDS.

$$\theta = 2\tan^{-1}\left\{\tan\frac{x}{2} \cdot \sqrt{\left(\frac{a-b}{a+b}\right)}\right\},$$

whence $\qquad \cos\theta = (a\cos x + b)/(a + b\cos x),$

or $\qquad (a - b\cos\theta)(a + b\cos x) = a^2 - b^2.$

x is the true and θ the eccentric anomaly in an ellipse of eccentricity b/a (Godfray's *Astronomy*, § 186; Gray's *Physics*, § 520.)

$a + b\cos x$ goes through its complete range of values if x varies from 0 to π or again if x varies from $-\pi$ through negative values to 0. If x lies between 0 and π, θ is positive and lies between 0 and π; but if x lies between $-\pi$ and 0, θ is negative and lies between $-\pi$ and 0. Hence bearing in mind the restriction on the inverse cosine (§§ 28, 64).

$$\int\frac{dx}{a+b\cos x} = \frac{1}{\sqrt{(a^2-b^2)}}\cos^{-1}\left(\frac{a\cos x+b}{a+b\cos x}\right) \text{ if } 0 \leqq x \leqq \pi.$$

but $\qquad = \frac{-1}{\sqrt{(a^2-b^2)}}\cos^{-1}\left(\frac{a\cos x+b}{a+b\cos x}\right)$ if $-\pi \leqq x \leqq 0.$

There is no ambiguity when the integral is expressed in terms of the inverse tangent. See also Examples 11, 12, p. 135.

Ex. 4. $\qquad \int\frac{dx}{a+b\sin x}, \ a \text{ positive}, \ -\frac{\pi}{2} < x < \frac{\pi}{2}.$

The integral $= \int\frac{2du}{a+2bu+au^2}.$

If $b^2 < a^2$, the substitution $x = \pi/2 - v$ or $x = \pi/2 + v$ will reduce it to Ex. 3 (i); the student should make both of the latter substitutions. He will thus see that it is not sufficient to consider only the one value of θ as determined by $\cos\theta$. The substitution furnishes a good instance of the care needed in dealing with inverse functions. There is no ambiguity if the integral of Ex. 3 (i) in terms of the inverse tangent is used.

Ex. 5. $\qquad \int\frac{dx}{a+b\cos x+c\sin x}, \ a \text{ positive}.$

If $b^2 + c^2 = k^2$ we may write

$$a + b\cos x + c\sin x = a + k\cos(x - a)$$

and the integral reduces to Ex. 3. For $k^2 < a^2$ the integral is

$$\frac{\pm 1}{\sqrt{(a^2-b^2-c^2)}}\cos^{-1}\left(\frac{a\cos(x-a)+b}{a+b\cos(x-a)}\right),$$

the sign being $+$ or $-$ according as $x - a$ lies between 0 and π or between $-\pi$ and 0.

EXERCISES XXIII.

Integrate with respect to x examples 1-22.

1. $\dfrac{1}{2x^2+3x+4}$;
2. $\dfrac{1}{\sqrt{(ax-x^2)}}$;
3. $\dfrac{1}{\sqrt{(x^2-ax)}}$;

4. $\dfrac{1}{\sqrt{(x-a)(b-x)}}$;
5. $\dfrac{x}{a^2+x^2}$;
6. $\dfrac{x}{\sqrt{(a^2+x^2)}}$;

7. $\dfrac{x^2}{x^6-1}$;
8. $\dfrac{x}{x^4+x^2+1}$;
9. $\dfrac{x+1}{\sqrt{(x^2+2x-3)}}$;

10. $\cot x$;
11. $\dfrac{\cos x}{1+\sin x}$;
12. $\dfrac{1+\cos x}{x+\sin x}$;

13. $\tan^4 x$;
14. $\cot^5 x$;
15. $\dfrac{1}{a^2\cos^2 x+b^2\sin^2 x}$;

16. $\sin^7 x$;
17. $\sin^5 x \cos^4 x$;
18. $\dfrac{1}{\sin^2 x \cos^2 x}$;

19. $\dfrac{\sin x}{\cos^5 x}$;
20. $x\sqrt{(a-x)}$;
21. $x/\sqrt{(a-x)}$;

22. $\dfrac{1}{x+\sqrt{(x-1)}}$.

23. Find the value of the integrals

(i) $\displaystyle\int_0^{\pi/2}\sin^5 x\,dx$; (ii) $\displaystyle\int_0^{\pi/2}\sin^4 x\cos^5 x\,dx$; (iii) $\displaystyle\int_0^{\pi}\dfrac{dx}{a^2\cos^2 x+b^2\sin^2 x}$;

(iv) $\displaystyle\int_0^{\pi/2}\dfrac{\cos x\,dx}{4-\sin^2 x}$; (v) $\displaystyle\int_0^{\pi/4}\tan x\,dx$; (vi) $\displaystyle\int_0^1\dfrac{dx}{1+x+x^2}$;

(vii) $\displaystyle\int_0^1\dfrac{dx}{1-x+x^2}$; (viii) $\displaystyle\int_0^{\pi}\dfrac{\sin x\,dx}{1+\cos^2 x}$.

Integrate with respect to x examples 24-41.

24. $\dfrac{x+1}{x^2+x+1}$;
25. $\dfrac{x^2-1}{x^2+1}$;
26. $\dfrac{x^4+7}{x^2+2x+3}$;

27. $\dfrac{x^3+x}{x^4-4}$;
28. $\dfrac{(x+1)^2}{(x-1)^2}$;
29. $\sqrt{\left(\dfrac{1+x}{1-x}\right)}$;

30. $\sqrt{\left(\dfrac{x+1}{x-1}\right)}$;
31. $\sqrt{\left(\dfrac{a+x}{x}\right)}$;
32. $\sqrt{\left(\dfrac{a-x}{x}\right)}$;

33. $\dfrac{1}{x\sqrt{(8x^2+2x-1)}}$;
34. $\dfrac{1}{(x+1)\sqrt{(x^2-1)}}$;
35. $\dfrac{1}{(x-1)\sqrt{(x^2-1)}}$;

36. $\dfrac{1}{(x+1)\sqrt{(1-x^2)}}$;
37. $\dfrac{1}{(x-1)\sqrt{(1-x^2)}}$;
38. $\dfrac{1}{x(x^n+1)}$;

39. $\dfrac{1}{(x^2+2x+3)^{3/2}}$;
40. $\dfrac{1}{1+\tan x}$;
41. $\dfrac{\cos x+\sin x}{\sin x+2\cos x}$.

EXERCISES XXIII.

42. Find the value of the integrals

(i) $\int_0^\pi \dfrac{dx}{5+3\cos x}$; (ii) $\int_0^{2\pi} \dfrac{dx}{5+3\cos x}$; (iii) $\int_{-\frac{\pi}{2}}^{\frac{\pi}{2}} \dfrac{dx}{5+3\sin x}$;

(iv) $\int_0^\pi \dfrac{dx}{1-2r\cos x+r^2}(r^2+1)$; (v) $\int_0^{\frac{\pi}{2}} \dfrac{dx}{1+\cos a\cos x}$ $(0<a<\pi)$;

(vi) $\int_0^{\frac{\pi}{2}} \dfrac{dx}{1-k^2\sin^2 x}(k^2<1)$; (vii) $\int_0^{\frac{\pi}{2}} \dfrac{dx}{3+5\cos x}$.

43. Find the value of $\int_{-1}^{+1} \dfrac{\sin a\, dx}{1-2x\cos a+x^2}$,

 (i) when $0<a<\pi$; (ii) when $\pi<a<2\pi$.

44. If a is positive, and b numerically less than a, prove, by the substitution $\cos\theta=(a\cos x+b)/(a+b\cos x)$, that

$$\int_0^\pi \dfrac{dx}{(a+b\cos x)^n} = \dfrac{1}{(a^2-b^2)^{n-\frac{1}{2}}}\int_0^\pi (a-b\cos\theta)^{n-1}d\theta.$$

45. Trace the curve given by $ay^2=x^2(a-x)$, $a>0$, and find the area of the loop.

46. Trace the curve given by $a^2y^2=x^2(a^2-x^2)$, and find the area of both loops.

47. Trace the curve whose polar equation is $r=a+b\cos\theta$, $a>b>0$, and find the area enclosed by it.

48. By transferring to polar coordinates, find the area of the ellipse whose equation is $ax^2+2hxy+by^2=1$.

The area is

$$2\int_{-\frac{\pi}{2}}^{\frac{\pi}{2}} \tfrac{1}{2}r^2 d\theta = \int_{-\frac{\pi}{2}}^{\frac{\pi}{2}} \dfrac{d\theta}{a\cos^2\theta+2h\sin\theta\cos\theta+b\sin^2\theta} = \dfrac{\pi}{\sqrt{(ab-h^2)}}.$$

§ 118. Integration by Parts. The second of the general methods of Integration is that called "Integration by Parts"; it corresponds to the theorem for the differentiation of a product.

For the moment denote integration by a suffix and differentiation by an accent; thus

$$u_1 = \int u\, dx;\quad u' = \dfrac{du}{dx}.$$

By the rule for differentiating a product we have

$$\dfrac{d(u_1 v)}{dx} = \dfrac{du_1}{dx}v + u_1\dfrac{dv}{dx},$$

that is, $\quad \dfrac{d(u_1 v)}{dx} = uv + u_1 v',\quad$ since $\dfrac{du_1}{dx} = u.$

Hence
$$u_1 v = \int (uv + u_1 v') dx \quad \text{.........................(1)}$$
$$= \int uv\, dx + \int u_1 v'\, dx;$$
and therefore
$$\int uv\, dx = u_1 v - \int u_1 v'\, dx \quad \text{.........................(2)}$$

Equation (2) gives the theorem in question. It may happen that the integral of $u_1 v'$ can be more easily determined than that of uv.

For a definite integral, lower limit a upper limit b, we get instead of (1)
$$[u_1 v]_a^b = \int_a^b (uv + u_1 v')\, dx \quad \text{.....................(3)}$$
and instead of (2)
$$\int_a^b uv\, dx = [u_1 v]_a^b - \int_a^b u_1 v'\, dx \quad \text{..................(4)}$$

where the symbol $[u_1 v]_a^b$ means as usual that x is to be first replaced by b, then by a, and the second result subtracted from the first.

The examples will show the great power of the theorem.

Ex. 1. Find $\int x \cos x\, dx$.

Here both x and $\cos x$ can be immediately integrated; but we take $v = x$ since then $v' = 1$.
$$\int x \cos x\, dx = x \cdot \sin x - \int 1 \cdot \sin x\, dx = x \sin x + \cos x.$$

Ex. 2. Find $\int x^2 \cos x\, dx$.

Again we put $v = x^2$ since $v' = 2x$, and the new integrand will therefore be simpler than the old.
$$\int x^2 \cos x\, dx = x^2 \cdot \sin x - \int 2x \cdot \sin x\, dx.$$

The theorem may be again applied
$$\int 2x \cdot \sin x\, dx = 2x(-\cos x) - \int 2(-\cos x)\, dx = -2x \cos x + 2 \sin x.$$

Hence $\int x^2 \cos x\, dx = x^2 \sin x + 2x \cos x - 2 \sin x.$

INTEGRATION BY PARTS.

Ex. 3. Find $\int e^{ax}\cos(bx+c)dx$ and $\int e^{ax}\sin(bx+c)\,dx$.

In finding one of these integrals we also find the other.

Let $\quad P=\int e^{ax}\cos(bx+c)dx, \quad Q=\int e^{ax}\sin(bx+c)\,dx$.

In this case it does not matter which factor is taken for v.

$$P=\frac{e^{ax}}{a}\cdot \cos(bx+c)-\int \frac{e^{ax}}{a}\cdot[-b\sin(bx+c)]\,dx = \frac{e^{ax}\cos(bx+c)}{a}+\frac{bQ}{a}.$$

Hence $\qquad aP-bQ=e^{ax}\cos(bx+c).\ \dotfill\text{(i)}$

In the same way by operating on Q we find
$$bP+aQ=e^{ax}\sin(bx+c).\ \dotfill\text{(ii)}$$

Solving (i) and (ii) for P and Q we find

$$P=\int e^{ax}\cos(bx+c)\,dx = \frac{e^{ax}[a\cos(bx+c)+b\sin(bx+c)]}{a^2+b^2}.$$

$$Q=\int e^{ax}\sin(bx+c)\,dx = \frac{e^{ax}[a\sin(bx+c)-b\cos(bx+c)]}{a^2+b^2}.$$

These two integrals are of great importance in mathematical physics.

Ex. 4. Find $\int \sqrt{(a^2-x^2)}dx$ and $\int \sqrt{(x^2\pm a^2)}dx$.

Here the integrand has but one factor; but we may take unity as a factor and put $u=1$. Hence

$$\int \sqrt{(a^2-x^2)}dx = x\sqrt{(a^2-x^2)}-\int x\cdot \frac{-x}{\sqrt{(a^2-x^2)}}dx$$

$$= x\sqrt{(a^2-x^2)}-\int \frac{-x^2}{\sqrt{(a^2-x^2)}}dx.\ \dotfill\text{(1)}$$

We now write
$$\frac{-x^2}{\sqrt{(a^2-x^2)}} = \frac{(a^2-x^2)-a^2}{\sqrt{(a^2-x^2)}} = \sqrt{(a^2-x^2)} - \frac{a^2}{\sqrt{(a^2-x^2)}}.$$

The first term on the right is the given integrand while the integral of the second term is $-a^2\sin^{-1}(x/a)$.

Substitute in (1), transfer the integral to the left side, and divide by 2; we thus get

$$\int \sqrt{(a^2-x^2)}dx = \tfrac{1}{2}x\sqrt{(a^2-x^2)} + \tfrac{1}{2}a^2\sin^{-1}\!\left(\frac{x}{a}\right),$$

the same result as in § 116, ex. 1.

In the same way it may be shown that

$$\int \sqrt{(x^2\pm a^2)}dx = \tfrac{1}{2}x\sqrt{(x^2\pm a^2)} \pm \tfrac{1}{2}a^2\log(x+\sqrt{x^2\pm a^2}).$$

Compare § 116, ex. 2.

The algebraic transformation used above is often useful; a similar transformation occurs in integrating circular functions (§ 119, 2, 3).

The quadratic $\sqrt{(ax^2+bx+c)}$ can be integrated by expressing it as in § 115, and putting $x+a=u$.

Ex. 5. Find $\int \log x \, dx$.

$$\int \log x \, dx = x \log x - \int x \frac{1}{x} dx = x \log x - x.$$

§ 119. Successive Reduction.

Ex. 1. Let $u_n = \int x^n e^x dx$; then, integrating by parts,

$$u_n = \int x^n e^x dx = x^n e^x - \int n x^{n-1} e^x dx = x^n e^x - n \int x^{n-1} e^x dx,$$

that is, $u_n = x^n e^x - n u_{n-1}.$

Writing $n-1$ in place of n, we find

$$u_{n-1} = x^{n-1} e^x - (n-1) u_{n-2},$$

that is, $u_n = x^n e^x - n x^{n-1} e^x + n(n-1) u_{n-2}.$

Proceeding in this way, we see that if n is a positive integer, u_n may be made to depend on u_0, that is, $\int e^x dx$ or e^x. If n is not a positive integer but is still positive, u_n may be made to depend on an integral in which the integrand contains x with a positive proper fraction as index. The integral cannot in that case be expressed in finite terms by means of known functions, but it is reduced to the most convenient form for studying.

The above method of making an integral depend on another of the same form is called that of *Successive Reduction*.

The integrals of $x^n \sin x$, $x^n \cos x$ may be treated in the same way.

Ex. 2. $\qquad u_n = \int \sin^n x \, dx.$

$$u_n = \int \sin^n x \, dx = \int \sin^{n-1} x \cdot \sin x \, dx$$

$$= \sin^{n-1} x (-\cos x) - \int (n-1) \sin^{n-2} x (-\cos^2 x) \, dx$$

$$= -\sin^{n-1} x \cos x + (n-1) \int \sin^{n-2} x \cos^2 x \cdot dx.$$

Now $\cos^2 x = 1 - \sin^2 x$; $\sin^{n-2} x \cos^2 x = \sin^{n-2} x - \sin^n x.$

Hence $u_n = -\sin^{n-1} x \cos x + (n-1) u_{n-2} - (n-1) u_n,$

and therefore $u_n = -\dfrac{\sin^{n-1} x \cos x}{n} + \dfrac{n-1}{n} u_{n-2}.$ ....(i)

The index n has thus been reduced by 2. Writing $n-2$ in place of n, we get

$$u_{n-2} = -\frac{\sin^{n-3} x \cos x}{n-2} + \frac{n-3}{n-2} u_{n-4},$$

and therefore

$$u_n = -\frac{\sin^{n-1} x \cos x}{n} - \frac{n-1}{n} \frac{\sin^{n-3} x \cos x}{n-2} + \frac{(n-1)(n-3)}{n(n-2)} u_{n-4}.$$

If n is a positive integer, we can repeat the reduction until the index is 1 if n be odd, or 0 if n be even; $u_1 = -\cos x$ and $u_0 = x$, since

SUCCESSIVE REDUCTION.

$\sin^0 x = 1$. If n is positive but not integral, u_n may be reduced to depend on an integral in which the index is either a positive or a negative proper fraction. For negative values of n see ex. 4.

The most useful case of the formula (i) is that in which n is a positive integer and the integral is taken between the limits 0 and $\pi/2$. In this case (i) becomes

$$\int_0^{\frac{\pi}{2}} \sin^n x \, dx = \Big[u_n\Big]_0^{\frac{\pi}{2}} = \Big[-\frac{\sin^{n-1} x \cos x}{n}\Big]_0^{\frac{\pi}{2}} + \frac{n-1}{n}\Big[u_{n-2}\Big]_0^{\frac{\pi}{2}}$$

$$= \frac{n-1}{n} \int_0^{\frac{\pi}{2}} \sin^{n-2} x \, dx,$$

since the integrated term vanishes at both limits.

When n is odd, the last term of u_n is

$$\frac{(n-1)(n-3)\ldots 4 \cdot 2}{n(n-2)\ldots 5 \cdot 3}\Big(-\cos x\Big);$$

but when n is even,

$$\frac{(n-1)(n-3)\ldots 3 \cdot 1}{n(n-2)\ldots 4 \cdot 2} x.$$

Hence

$$\int_0^{\frac{\pi}{2}} \sin^n x \, dx = \frac{(n-1)(n-3)\ldots 4 \cdot 2}{n(n-2)\ldots 5 \cdot 3} \cdot 1 \quad (n \text{ odd integer});$$

$$\int_0^{\frac{\pi}{2}} \sin^n x \, dx = \frac{(n-1)(n-3)\ldots 3 \cdot 1}{n(n-2)\ldots 4 \cdot 2} \frac{\pi}{2} \quad (n \text{ even integer}).$$

If $v_n = \int \cos^n x \, dx$, then

$$v_n = \frac{\cos^{n-1} x \sin x}{n} + \frac{n-1}{n} v_{n-2},$$

and it is easy to prove from the formula or, better, directly from the meaning of the definite integral that

$$\int_0^{\frac{\pi}{2}} \cos^n x \, dx = \int_0^{\frac{\pi}{2}} \sin^n x \, dx.$$

A simple inspection of the graphs of $\sin^n x$ and $\cos^n x$ will show that

$$\int_0^{\pi} \sin^n x \, dx = 2 \int_0^{\frac{\pi}{2}} \sin^n x \, dx.$$

$$\int_0^{\pi} \cos^n x \, dx = 2 \int_0^{\frac{\pi}{2}} \cos^n x \, dx \quad (n \text{ even integer}),$$

but $\qquad = 0 \quad (n \text{ odd integer}).$

In a similar way such results as

$$\int_0^{2\pi} \sin^3 x \, dx = 0; \quad \int_0^{2\pi} \cos^6 x \, dx = 4 \int_0^{\frac{\pi}{2}} \cos^6 x \, dx$$

are readily proved. See also the rule given in ex. 3.

286   AN ELEMENTARY TREATISE ON THE CALCULUS.

Ex. 3. $\quad f(m, n) = \int \sin^m x \cos^n x \, dx.$

In $f(m, n)$ the first letter is the index of $\sin x$, the second that of $\cos x$. For brevity denote $\sin x$ by s, $\cos x$ by c. Then

$$f(m, n) = \int s^m c^n \, dx = \int s^m c \cdot c^{n-1} \, dx.$$

Since c is the derivative of s, the integral of $s^m c$ is $s^{m+1}/(m+1)$. Thus,

$$f(m, n) = \frac{s^{m+1}}{m+1} c^{n-1} - \int \frac{s^{m+1}}{m+1} (n-1) c^{n-2} (-s) \, dx$$

$$= \frac{s^{m+1} c^{n-1}}{m+1} + \frac{n-1}{m+1} \int s^{m+2} c^{n-2} \, dx. \quad \ldots\ldots\ldots\ldots\ldots\ldots\ldots\text{(i)}$$

But $\quad s^{m+2} c^{n-2} = s^m (1 - c^2) c^{n-2} = s^m c^{n-2} - s^m c^n.$

The first term is the integrand of $f(m, n-2)$, and the second that of $f(m, n)$. Substitute in (i), transfer $f(m, n)$ to the left side, and then multiply by $(m+1)/(m+n)$. Therefore

$$f(m, n) = \frac{s^{m+1} c^{n-1}}{m+n} + \frac{n-1}{m+n} f(m, n-2). \quad \ldots\ldots\ldots\ldots\ldots\text{(A)}$$

The integral thus depends on another of the same form with m unchanged but the other index reduced by 2.

Had we begun by writing $s^{m-1} \cdot sc^n$ and integrating the cosine, we should have got

$$f(m, n) = -\frac{s^{m-1} c^{n+1}}{m+n} + \frac{m-1}{m+n} f(m-2, n), \quad \ldots\ldots\ldots\ldots\text{(B)}$$

and now m is reduced by 2, n unchanged.

We will continue the reduction for the case in which m, n are *positive integers*, so as to obtain *the definite integral from 0 to $\pi/2$*.

If n is odd, (A) makes $f(m, n)$ depend on $f(m, 1)$; (B) then makes $f(m, 1)$ depend on $f(1, 1)$ or on $f(0, 1)$ according as m is odd or even.

If n is even, (A) makes $f(m, n)$ depend on $f(m, 0)$; but $f(m, 0)$ is the integral of ex. 2, with m in place of n. Thus, by ex. 2 (i), $f(m, 0)$ depends on $f(1, 0)$ or on $f(0, 0)$ according as m is odd or even.

Thus, $f(m, n)$ may be reduced to depend on one of the four

$$f(1, 1) = \int sc \, dx = \tfrac{1}{2} \sin^2 x \; ; \quad f(0, 1) = \int c \, dx = \sin x \; ;$$

$$f(1, 0) = \int s \, dx = -\cos x \; ; \quad f(0, 0) = \int 1 \, dx = x.$$

When the integral is taken between 0 and $\pi/2$ the values of these are $1/2$, 1, 1, $\pi/2$ respectively.

The student may now show that the following rule is correct:

$$\int_0^{\frac{\pi}{2}} \sin^m x \cos^n x \, dx = \frac{(m-1)(m-3) \ldots \times (n-1)(n-3) \ldots}{(m+n)(m+n-2) \ldots} \times a,$$

THE INTEGRAL $\int_0^{\frac{\pi}{2}} \sin^m x \cos^n x\, dx$.

where $a=1$ except when m and n are *both* even integers, in which case $a = \pi/2$; each of the three series of factors is to be continued so long as the factors are *positive*.

It will be noticed that the factors of each series decrease by 2. The rule includes the integral of ex. 2, putting m (or n) zero and omitting *negative* factors.

$$\int_0^{\frac{\pi}{2}} \sin^2 x \cos^4 x\, dx = \frac{1 \times 3 \cdot 1}{6 \cdot 4 \cdot 2} \times \frac{\pi}{2} = \frac{\pi}{32};$$

$$\int_0^{\frac{\pi}{2}} \sin^6 x \cos^4 x\, dx = \frac{5 \cdot 3 \cdot 1 \times 3 \cdot 1}{10 \cdot 8 \cdot 6 \cdot 4 \cdot 2} \times \frac{\pi}{2} = \frac{3\pi}{512};$$

$$\int_0^{\frac{\pi}{2}} \sin^7 x \cos^5 x\, dx = \frac{6 \cdot 4 \cdot 2 \times 4 \cdot 2}{12 \cdot 10 \cdot 8 \cdot 6 \cdot 4 \cdot 2} \times 1 = \frac{1}{120};$$

$$\int_0^{\frac{\pi}{2}} \sin^8 x\, dx = \frac{7 \cdot 5 \cdot 3 \cdot 1}{8 \cdot 6 \cdot 4 \cdot 2} \times \frac{\pi}{2} = \frac{35\pi}{256}.$$

The great importance of the results of ex. 2 and 3 arises from the fact that many integrals are, by a proper substitution, easily reduced to these forms. For example, if we put $x = a \sin \theta$, so that when $x=0$, $\theta = 0$, and when $x=a$, $\theta = \pi/2$, we get

$$\int_0^a x^2 (a^2 - x^2)^{\frac{3}{2}}\, dx = a^6 \int_0^{\frac{\pi}{2}} \sin^2 \theta \cos^4 \theta\, d\theta = \frac{\pi a^6}{32}.$$

If we put $x = a \sin^2 \theta$, then

$$\int_0^a x^2 (a-x)^{\frac{5}{2}}\, dx = 2a^{\frac{13}{2}} \int_0^{\frac{\pi}{2}} \sin^6 \theta \cos^6 \theta\, d\theta = \frac{16 a^{\frac{13}{2}}}{315}.$$

Ex. 4. If n is negative the index of u_{n-2} is numerically greater than that of n. In ex. 2 (i) let $n = -m$, where m is positive; then

$$\int \frac{dx}{s^m} = \frac{c}{m s^{m+1}} + \frac{m+1}{m} \int \frac{dx}{s^{m+2}};$$

therefore $\quad \int \frac{dx}{s^{m+2}} = -\frac{c}{(m+1) s^{m+1}} + \frac{m}{m+1} \int \frac{dx}{s^m}.$

Now put $m+2 = n$, where n is positive, and we get

$$\int \frac{dx}{\sin^n x} = -\frac{\cos x}{(n-1) \sin^{n-1} x} + \frac{n-2}{n-1} \int \frac{dx}{\sin^{n-2} x}.$$

In many cases the integration will be simplified by writing

$$\frac{1}{\sin^n x} = \frac{\sin^2 x + \cos^2 x}{\sin^n x} = \frac{1}{\sin^{n-2} x} + \frac{\cos x}{\sin^n x} \cdot \cos x.$$

But these integrals are of small importance for elementary work. The key to the transformations is that after one integration by parts the new and the old indices differ by 2; when an index is negative it is simpler to begin by integrating the integrand with the reduced index.

Ex. 5.
$$u_n = \int \tan^n x \, dx,$$
$$u_n = \int \tan^{n-2} x (\sec^2 x - 1) \, dx = \int \tan^{n-2} x \cdot \sec^2 x \, dx - u_{n-2},$$
so that
$$u_n = \frac{1}{n-1} \tan^{n-1} x - u_{n-2}.$$

Other examples of reduction formulae will be found in the Exercises, but in many cases a trigonometric substitution will reduce the integral to one of the forms just discussed.

EXERCISES XXIV.

Integrate with respect to x examples 1-24.

1. xe^{-x};
2. $x^3 e^{-x}$;
3. $x \sin x$;
4. $x \cos x$;
5. $x \sin x \cos x$;
6. $x^2 \sin x$;
7. $x^n \log x \, (n \neq -1)$;
8. $\frac{1}{x} \log x$;
9. $e^{-x} \sin^2 x$;
10. $\frac{xe^x}{(1+x)^2}$;
11. xe^{-x^2};
12. $\sin^{-1} x$;
13. $\tan^{-1} x$;
14. $x \sin^{-1} x$;
15. $x \tan^{-1} x$;
16. $\sqrt{(3+2x-x^2)}$;
17. $\sqrt{(3+2x+x^2)}$;
18. $\sqrt{(2ax-x^2)}$;
19. $\sqrt{(2ax+x^2)}$;
20. $\frac{x^2}{\sqrt{(1-x^2)}}$;
21. $\frac{x+\sin x}{1+\cos x}$;
22. $e^{-3x} \cos 4x$;
23. $\cosh x \cos x$;
24. $\sinh x \sin x$.

25. Find the value of the integrals

(i) $\int_0^{\frac{\pi}{2}} \cos^8 x \, dx$;   (ii) $\int_0^{\pi} \sin^6 x \, dx$;

(iii) $\int_0^{\pi} \sin^6 x \cos^4 x \, dx$;   (iv) $\int_0^{\pi} \sin^3 x \cos^4 x \, dx$;

(v) $\int_0^{4\pi} \sin^8 x \cos^4 x \, dx$;   (vi) $\int_0^{\frac{\pi}{4}} \tan^6 x \, dx$.

26. Find by a trigonometric substitution the value of

(i) $\int_0^a x^2 \sqrt{(a^2-x^2)} \, dx$;   (ii) $\int_0^{2a} x \sqrt{(2ax-x^2)} \, dx$;

(iii) $\int_0^{2a} x^2 \sqrt{(2ax-x^2)} \, dx$.

27. Integrate $\int_0^a \frac{x \sqrt{(a^2-x^2)}}{\sqrt{(a^2+x^2)}} \, dx$ by the substitution $x^2 = a^2 \cos 2\theta$.

EXERCISES XXIV.

28. If $f(m, n) = \int x^m (1-x)^n \, dx$ show that
$$f(m, n) = \frac{x^{m+1}(1-x)^n}{m+n+1} + \frac{n}{m+n+1} f(m, n-1).$$
Hence, or by the substitution $x = \sin^2\theta$, find the value of
$$\int_0^1 x^m (1-x)^n \, dx,$$
m, n being positive integers.

29. If $u_n = \int dx/(a^2 + x^2)^n$, prove that
$$u_n = \frac{x}{(2n-2) a^2 (a^2 + x^2)^{n-1}} + \frac{2n-3}{(2n-2) a^2} u_{n-1}.$$

30. If $u_n = \int x^n \sqrt{(a^2 - x^2)} \, dx$, prove that
$$u_n = -\frac{x^{n-1}(a^2 - x^2)^{\frac{3}{2}}}{n+2} + \frac{n-1}{n+2} a^2 u_{n-2}.$$

31. If $u_n = \int x^n \sqrt{(2ax - x^2)} \, dx$, show that
$$u_n = -\frac{x^{n-1}(2ax - x^2)^{\frac{3}{2}}}{n+2} + \frac{2n+1}{n+2} a u_{n-1}.$$
Write $u_n = \int x^{n-1}\{a - (a-x)\} R^{\frac{1}{2}} \, dx = a u_{n-1} - \frac{1}{2}\int x^{n-1} \frac{dR}{dx} R^{\frac{1}{2}} dx$
where $R = 2ax - x^2$, and then integrate by parts.

32. If $u_n = \int x^n \, dx / \sqrt{(2ax - x^2)}$, show that
$$u_n = -\frac{x^{n-1}\sqrt{(2ax - x^2)}}{n} + \frac{2n-1}{n} a u_{n-1}.$$

33. If m, n are positive integers find the value of
$$\int_0^1 (1 - x^{\frac{1}{n}})^m \, dx.$$

34. Find the value of
(i) $\int_0^a x^4 \sqrt{(a^2 - x^2)} \, dx$; (ii) $\int_0^a x^4 \sqrt{(2ax - x^2)} \, dx$.

35. OM is the abscissa and MP the ordinate at the point $P(\xi, \eta)$ on the hyperbola $x^2/a^2 - y^2/b^2 = 1$, ξ, η being both positive. If A is the vertex nearest P show that the area AMP is equal to
$$\tfrac{1}{2}\xi\eta - \tfrac{1}{2}ab \log\left(\frac{\xi}{a} + \frac{\eta}{b}\right),$$
and that the area of the sector OAP is
$$\tfrac{1}{2}ab \log\left(\frac{\xi}{a} + \frac{\eta}{b}\right).$$

36. Trace the curve given by $y^2 = (x-1)(x-3)^2$ and find the area of the loop.

37. Trace the curve given by $a^2y^2 = x^3(2a-x)$, a being positive; find the whole area enclosed by it.

38. Find the length of an arc of the catenary $y = \frac{1}{2}a(e^{\frac{x}{a}} + e^{-\frac{x}{a}})$ measured from the point C where $x=0$. Show that the area between the two axes, the curve and the ordinate at a point P is a times the arc CP.

39. Find the length of an arc of the cardioid $r = a(1-\cos\theta)$, the arc being measured from the origin.

40. Find the length of an arc of the spiral $r = a\theta$, taking $s=0$, when $r=0$.

41. Find the length of an arc of the spiral $r = ae^{\theta \cot a}$, taking $s=0$, when $\theta = 0$.

§ 120. Partial Fractions. The method of resolving a rational fraction into partial fractions is now found in most text-books of Algebra. We will therefore refer the student to Chrystal's *Algebra*, Vol. I., Chap. viii., for a full discussion of the theory, and will merely work out a few examples. The fraction will be supposed to be a proper fraction, that is to have the degree of its numerator in the variable x less than that of its denominator, and to be at its lowest terms.

Let the fraction be $F(x)/f(x)$ where $F(x)$ and $f(x)$ are rational integral functions of x. $f(x)$ can be resolved into a product of *real prime factors*, each of which is a linear or else a quadratic function of x, but a factor, linear or quadratic, may be repeated several times.

$F(x)/f(x)$ can be resolved in one and in only one way into a sum of proper partial fractions; these partial fractions are of the following types:

(i) To every non-repeated linear factor $x - a$ of $f(x)$ corresponds a partial fraction of the form $A/(x-a)$.

(ii) To every r-fold linear factor $(x-\beta)^r$ of $f(x)$ correspond r partial fractions of the form

$$\frac{B_r}{(x-\beta)^r} + \frac{B_{r-1}}{(x-\beta)^{r-1}} + \ldots + \frac{B_2}{(x-\beta)^2} + \frac{B_1}{x-\beta}.$$

(iii) To every non-repeated quadratic factor $x^2 + \gamma x + \delta$ of $f(x)$ corresponds a partial fraction of the form $(Cx+D)/(x^2+\gamma x+\delta)$.

(iv) To every r-fold quadratic factor $(x^2+\gamma x+\delta)^r$ of $f(x)$ correspond r partial fractions of the form

$$\frac{C_r x + D_r}{(x^2+\gamma x+\delta)^r} + \frac{C_{r-1}x + D_{r-1}}{(x^2+\gamma x+\delta)^{r-1}} + \ldots + \frac{C_1 x + D_1}{x^2+\gamma x+\delta}.$$

The method of determining the coefficients $A, B, \ldots$ will be learned from the examples.

PARTIAL FRACTIONS.

Ex. 1. $x^3/(x^2-1)(x-2)$.

No factor of the denominator is repeated; therefore

$$\frac{x^3}{(x+1)(x-1)(x-2)} = \frac{A}{x+1} + \frac{B}{x-1} + \frac{C}{x-2}.$$

Clear of fractions; therefore

$$x^3 = A(x-1)(x-2) + B(x+1)(x-2) + C(x+1)(x-1).$$

This equation being an identity, we may give to x any value we please. Put $x+1=0$, that is, $x=-1$, and the terms in B and C vanish, and we get
$$1 = A(-1-1)(-1-2) \quad \text{or} \quad A = 1/6.$$

Similarly, by putting $x=1$ we get $B=-1/2$, and by putting $x=2$ we get $C=4/3$ and

$$\frac{x^3}{(x^2-1)(x-2)} = \frac{1}{6} \cdot \frac{1}{x+1} - \frac{1}{2} \cdot \frac{1}{x-1} + \frac{4}{3} \cdot \frac{1}{x-2}.$$

Or, to find A, multiply both sides by its denominator $x+1$ and then put $x+1=0$;

$$A = \left[\frac{x^3}{(x-1)(x-2)} \right]_{x=-1}.$$

In the same way, if $x-a$ is a non-repeated factor of $f(x)$ and $A/(x-a)$ the corresponding partial fraction

$$A = \left[\frac{(x-a)F(x)}{f(x)} \right]_{x=a}.$$

If $f(x) = (x-a)\phi(x)$, then

$$f'(x) = \phi(x) + (x-a)\phi'(x) \quad \text{and} \quad f'(a) = \phi(a),$$

so that
$$A = \left[\frac{(x-a)F(x)}{(x-a)\phi(x)} \right]_{x=a} = \frac{F(a)}{\phi(a)} = \frac{F(a)}{f'(a)}.$$

Ex. 2. $(x^3+x+2)/(x-1)^2(x^2-x+1)$.

The repeated factor $(x-1)^2$ gives two fractions, and the factor x^2-x+1, since it has no real linear factors, gives a fraction of the type (iii); hence

$$\frac{x^3+x+2}{(x-1)^2(x^2-x+1)} = \frac{A}{(x-1)^2} + \frac{B}{x-1} + \frac{Cx+D}{x^2-x+1}.$$

Clearing of fractions, we get

$$x^3+x+2 = A(x^2-x+1) + B(x-1)(x^2-x+1) + (Cx+D)(x-1)^2.$$

Putting $x=1$, we get $A=4$. Now bring the term in A to the left side and reduce after putting 4 for A. The right side will contain $(x-1)$ as a factor, and therefore, since the equation is an identity, the left side must also contain $(x-1)$ as a factor. If it does not, there is an error in the work. We get

$$-(3x-2)(x-1) = B(x-1)(x^2-x+1) + (Cx+D)(x-1)^2.$$

Divide by $(x-1)$, and then by putting $x=1$ we find $B=-1$. Now take the term in B to the left, and again divide by $(x-1)$. Then
$$x-3 = Cx+D$$
so that, since the equation is an identity, $C=1$, $D=-3$, and therefore
$$\frac{x^2+x+2}{(x-1)^2(x^2-x+1)} = \frac{4}{(x-1)^2} - \frac{1}{x-1} + \frac{x-3}{x^2-x+1}.$$

Ex. 3. $\qquad (x^3-2)/(x^2+x+2)^2(x^2+x+1)$.

By (iv) and (iii), since there are no real linear factors of the denominator,
$$\frac{x^3-2}{(x^2+x+2)^2(x^2+x+1)} = \frac{Ax+B}{(x^2+x+2)^2} + \frac{Cx+D}{x^2+x+2} + \frac{Ex+F}{x^2+x+1}.$$

Clearing of fractions,
$$x^3-2 = (Ax+B)(x^2+x+1) + (Cx+D)(x^2+x+2)(x^2+x+1) + (Ex+F)(x^2+x+2)^2.$$

Put $x^2+x+2=0$ and reduce x^2 and x^3 to linear functions by means of this equation. It gives
$$x^2+x+1 = -1; \quad x^2 = -x-2,\ x^3 = -x^2-2x = -x+2,$$
and therefore $\qquad -x = -Ax - B,$

so that $A=1$, $B=0$. Take the term in A and B to the left and divide by x^2+x+2 which must be a factor. Hence
$$-1 = (Cx+D)(x^2+x+1) + (Ex+F)(x^2+x+2).$$

Put $x^2+x+2=0$ and proceed as before. We get $C=0$, $D=1$. Hence, after dividing by x^2+x+2,
$$-1 = Ex+F; \quad E=0,\ F=-1,$$
and the fraction is equal to
$$\frac{x}{(x^2+x+2)^2} + \frac{1}{x^2+x+2} - \frac{1}{x^2+x+1}.$$

These examples show sufficiently the method of determining the coefficients; other methods will suggest themselves to the student, and he will find full details in the chapter of Chrystal's *Algebra* referred to above.

§ 121. **Integration of Rational Functions.** If $F(x)/f(x)$ is not a proper fraction it may by division be expressed as the sum of a rational integral function and of a rational proper fraction.

The integral of a rational integral function is a rational integral function.

The integral of $A/(x-a)$ is $A \log(x-a)$.

The integral of $B/(x-\beta)^r$ where r is different from unity is $-B/(r-1)(x-\beta)^{r-1}$.

The integral of $(Cx+D)/(x^2+\gamma x+\delta)$ has been discussed in § 115 and is of the form

$$\lambda \log(x^2+\gamma x+\delta) + \mu \tan^{-1} \frac{2x+\gamma}{\sqrt{(4\delta-\gamma^2)}}.$$

We have, therefore, only to consider $(Cx+D)/(x^2+\gamma x+\delta)^r$. Writing the quadratic in the form $R=(x+a)^2+\beta^2$ the integral is

$$\tfrac{1}{2} C \int \frac{2(x+a)\,dx}{R^r} + (D-aC) \int \frac{dx}{R^r}$$

$$= \frac{-C}{2(r-1)R^{r-1}} + (D-aC) \int \frac{dx}{R^r}.$$

In practice it is usually simplest to integrate $1/R^r$ by the substitution $x+a=\beta \tan\theta$; but it is of some theoretical interest to get a formula of reduction. If we differentiate $(x+a)/R^{r-1}$ we find

$$\frac{d}{dx}\left(\frac{x+a}{R^{r-1}}\right) = \frac{1}{R^{r-1}} - \frac{2(r-1)(x+a)^2}{R^r},$$

$$= \frac{-(2r-3)}{R^{r-1}} + \frac{2(r-1)\beta^2}{R^r},$$

by putting $(x+a)^2 = R-\beta^2$. Integrating and rearranging we get
$$\int \frac{dx}{R^r} = \frac{x+a}{2(r-1)\beta^2 R^{r-1}} + \frac{2r-3}{2(r-1)\beta^2} \int \frac{dx}{R^{r-1}}.$$

Hence the integral of $(Cx+D)/R^r$ can be made to depend on that of $1/R$, which is an inverse trigonometric function.

Thus the integral of any Rational Function of x can be expressed in terms of rational functions, logarithms and inverse circular functions.

There is always a considerable amount of labour in integrating by the method of partial fractions. The student should, before resolving into partial fractions, examine whether the integral may be simplified by a substitution.

Thus, $$\int \frac{x^3 dx}{x^4-x^2+1} = \tfrac{1}{2}\int \frac{u\,du}{u^2-u+1}, \quad u=x^2,$$

and the fraction in u is easier to handle than that in x.

§ 122. Irrational Functions. We consider one or two cases in which the integrand is an irrational function.

(i) When the integrand contains only fractional powers of x let n be the L.C.D. of the fractions; then the substitution $x = u^n$ will make the new integrand rational in u.

Thus, if $x = u^6$

$$\int \frac{x^{\frac{7}{6}} dx}{1 + x^{\frac{1}{3}}} = 6 \int \frac{u^8 du}{1 + u^2}$$
$$= 6(\tfrac{1}{7}u^7 - \tfrac{1}{5}u^5 + \tfrac{1}{3}u^3 - u + \tan^{-1} u)$$
$$= 6\left(\tfrac{1}{7}x^{\frac{7}{6}} - \tfrac{1}{5}x^{\frac{5}{6}} + \tfrac{1}{3}x^{\frac{1}{2}} - x^{\frac{1}{6}} + \tan^{-1}(x^{\frac{1}{6}})\right).$$

(ii) When the integrand contains $\sqrt{(ax+b)}$ but no other irrationality the substitution $ax + b = u^2$ will make the new integrand rational in u.

(iii) When the integrand contains $\sqrt{(ax^2 + bx + c)}$ but no other irrationality the integral may be reduced to that of a rational function as follows:

First, let a be positive and write the root in the form

$$y = \sqrt{a}\sqrt{(x^2 + px + q)}, \quad p = b/a, \; q = c/a.$$

Let $\sqrt{(x^2 + px + q)} = u - x$ so that, squaring and solving for x,
$$x = \frac{u^2 - q}{2u + p}, \quad \frac{dx}{du} = \frac{2(u^2 + pu + q)}{(2u + p)^2}.$$

The new integrand will clearly be rational in u.

Second, let a be negative. In order that y may be real the linear factors of $ax^2 + bx + c$ must be real; if they were not real the quadratic would be negative for every real value of x and therefore y would be imaginary. We may therefore write, since $(-a)$ is positive,

$$y = \sqrt{(-a)}\sqrt{(x-a)(\beta-x)}.$$

For definiteness suppose $\beta > a$ (algebraically) and let $u = +\sqrt{\{(x-a)/(\beta-x)\}}$.

Then, $u^2 = (x-a)/(\beta-x)$;
$$x - a = \frac{(\beta - a) u^2}{1 + u^2}; \quad \beta - x = \frac{\beta - a}{1 + u^2}; \quad x = \frac{a + \beta u^2}{1 + u^2};$$
$$y = \sqrt{(-a)} \cdot (\beta - a) \cdot \frac{u}{1 + u^2}; \quad \frac{dx}{du} = \frac{2(\beta - a) u}{(1 + u^2)^2}.$$

The new integrand will clearly be rational in u.

In (ii), (iii) we may suppose all the roots to be *positive*. (See § 123, *end*.)

The above analysis shows that if y be either $\sqrt{(ax+b)}$ or $\sqrt{(ax^2+bx+c)}$, and if the integrand be a rational function $f(x, y)$ of x and of y, the integration of $f(x, y)$ can always be reduced to that of a rational function, and therefore (§ 121) requires for its integration only rational functions, logarithms, or inverse circular functions.

(iv) Let the integrand be $x^m(a+bx^n)^p$.

(*a*) If p is a positive integer expand $(a+bx^n)^p$.

(*b*) Try the substitution $u = a + bx^n$ which gives

$$x = \frac{1}{b^{\frac{1}{n}}}(u-a)^{\frac{1}{n}}; \quad \frac{dx}{du} = \frac{(u-a)^{\frac{1}{n}-1}}{nb^{\frac{1}{n}}},$$

and the integral becomes

$$\frac{1}{nb^{\frac{m+1}{n}}} \int u^p (u-a)^{\frac{m+1}{n}-1} du,$$

so that if $(m+1)/n$ is a positive integer the binomial may be expanded and the integral obtained in finite terms.

(*c*) If $(m+1)/n$ is not a positive integer let $x = 1/v$ and the integral becomes

$$-\int v^{-m-np-2}(b+av^n)^p dv.$$

Instead of m we have now $-(m+np+2)$ and therefore by (*b*) if $-(m+np+1)/n$ be a positive integer, that is, if $(m+1)/n + p$ be a negative integer the integral may be got in finite terms. The substitution is

$$u = b + av^n = b + ax^{-n}.$$

§ 123. **General Remarks.** From the discussion now given it will be seen that integration is a somewhat haphazard process. The only general results obtained are those of §§ 121, 122; in most cases the integration, when it is possible at all, has to be effected by reducing the given integrand by various methods to a few standard forms. Even for the cases discussed in § 122 it is frequently simpler to take a special method for a given case than to apply the general theorem.

Much of the difficulty beginners find in integration is due to a deficiency in power of algebraic and trigonometric manipulations. When the standard forms have been committed to memory the next step is to master the two principles of change of variable and of integration by parts; but the student who has not a thorough mastery of elementary algebraic and trigonometric transformations will often fail to see the reasons that suggest the particular devices adopted and will have to struggle with difficulties that are due, not to the nature of the calculus but to his own deficient algebraic training.

Integral dependent on the range of the variable. Another source of difficulty requires special notice, namely that the integral may have one form for one range of the variable and a different form for another range. Thus the integral of $1/x$ is $\log x$ or $\log(-x)$ according as x is positive or negative; in this case the integral may be written $\frac{1}{2}\log(x^2)$, a form which covers both cases. See § 117, Ex. 3, for another case.

Again, difficulty may arise from the ambiguity of the *square root*; in that ambiguity the explanation of the two forms for the integral of $1/(a+b\cos x)$ is to be found when the inverse cosine is derived from the inverse tangent. Thus, if it be agreed that the root is always to be taken with the positive sign, the transformation $P\sqrt{Q} = \sqrt{(P^2 Q)}$ would only be correct if P were positive; if P were negative we should have $P\sqrt{Q} = -\sqrt{(P^2 Q)}$.

EXERCISES XXV.

Integrate with respect to x examples 1-24.

1. $\dfrac{x^2 - 6x - 4}{(2x+1)(x+2)(3x+2)}$;
2. $\dfrac{30x^5}{(x^2-1)(x^2-4)}$;
3. $\dfrac{x^2}{(x-a)(x-b)(x-c)}$;
4. $\dfrac{x}{(x+1)^2(x-1)}$;
5. $\dfrac{1}{x^3(x-1)}$;
6. $\dfrac{1}{(x^2-1)^3}$;
7. $\dfrac{x}{(x^2-1)^3}$;
8. $\dfrac{x^2+1}{x^3+1}$;
9. $\dfrac{x^3}{x^6+1}$;
10. $\dfrac{x^2+x+1}{x^2-x+1}$;

EXERCISES XXV.

11. $\dfrac{x^3}{x^4+x^2-2}$; 12. $\dfrac{1}{(x^2+a^2)(x^2+b^2)}$; 13. $\dfrac{x}{(x^2+a^2)(x^2+b^2)}$;

14. $\dfrac{x^2}{(x^2+a^2)(x^2+b^2)}$; 15. $\dfrac{1}{(x-1)(x^2+1)^2}$; 16. $\dfrac{1}{x^4+1}$;

17. $\dfrac{x^3}{(x^2+2x+5)^2}$; 18. $\dfrac{1}{x^4(ax+b)}$; 19. $\dfrac{\cos x}{16+9\sin^2 x}$;

20. $\dfrac{1}{3\sin x+\sin^3 x}$; 21. $\dfrac{x+1}{(x^2+4x+6)^3}$; 22. $\dfrac{x^2-a^2}{x^4+a^2x^2+a^4}$;

23. $\dfrac{1}{(x+1)(4x^4+17x^2+4)}$; 24. $\dfrac{1}{x^2(1+x^2)^2}$;

25. Transform the integral
$$\int \dfrac{dx}{(x-a)^m(x-b)^n}$$
by the substitution $u=(x-a)/(x-b)$; find its value when $m=3$, $n=2$.

Integrate with respect to x examples 26–37.

26. $\dfrac{\sqrt{x}}{1+x}$; 27. $\dfrac{1}{\sqrt{x}+\sqrt[3]{x}}$; 28. $\dfrac{x^3}{\sqrt{(x-1)}}$;

29. $\dfrac{x}{(a+bx)^{3/2}}$; 30. $\dfrac{1}{(1+x^2)\sqrt{(1-x^2)}}$; 31. $\dfrac{1}{(1-x^2)\sqrt{(1+x^2)}}$;

32. $\dfrac{1}{x+\sqrt{(x^2-a)}}$; 33. $x^3\sqrt{(a+bx^2)}$; 34. $x^4(1+x^{\frac{2}{3}})^{\frac{1}{2}}$;

35. $\dfrac{1}{x^4\sqrt{(1+x^2)}}$; 36. $\dfrac{2a+x}{a+x}\sqrt{\left(\dfrac{a-x}{a+x}\right)}$; 37. $\sqrt{\tan x}$.

CHAPTER XIV.

DEFINITE INTEGRALS. GEOMETRICAL APPLICATIONS.

§ 124. Definite Integrals. In this and the two following articles we will state a few of the more important theorems respecting definite integrals.

THEOREM I. *A definite integral is a function of its limits, not of the variable of integration.*

This theorem is obvious from the geometrical meaning of the integral; so long as the symbol F denotes the same function the graph of $F(x)$ with x for abscissa is the same as that of $F(u)$ with u for abscissa, and therefore

$$\int_a^b F(x)\,dx = \int_a^b F(u)\,du.$$

Or, again, if $F(x) = D_x f(x)$, then $F(u) = D_u f(u)$ and each symbol represents $f(b) - f(a)$.

THEOREM II. $\quad \int_b^a F(x)\,dx = -\int_a^b F(x)\,dx.\quad$ See § 110.

THEOREM III. *If $a < b$ and if $F(x)$ is positive for every value of x within the range of integration, the integral*

$$\int_a^b F(x)\,dx$$

is positive, not zero; if $F(x)$ is negative, the integral is negative.

For the area represented by the integral is positive in the first case, negative in the second. Obviously the theorem will still be true if $F(x)$ is zero for some but not all of the

DEFINITE INTEGRALS. GENERAL THEOREMS.

values of x in the interval (a, b), and a similar observation is true in Theorems V., VI, VII.

Thus such an equation as
$$\int_0^2 \frac{dx}{(x-1)^2} = \left[\frac{-1}{x-1}\right]_0^2 = -2$$
is absurd. The contradiction arises from the fact that the positive integrand $1/(x-1)^2$ is discontinuous when $x=1$, the value 1 lying in the interval $(0, 2)$.

THEOREM IV. $\int_a^b F(x)\,dx = \int_a^c F(x)\,dx + \int_c^b F(x)\,dx.$

For the area represented by the integral on the left, sign as well as magnitude of the areas being taken into account, is equal to the sum of the areas represented by the integrals on the right. In the same way,
$$\int_a^b F(x)\,dx = \int_a^c F(x)\,dx + \int_c^g F(x)\,dx + \int_g^b F(x)\,dx,$$
and so on for any number of subdivisions of the interval (a, b). Of course one or more of the numbers $c, g, \ldots$, may be greater than the greater or less than the smaller of the two numbers a, b, provided $F(x)$ is continuous for all the values considered.

THEOREM V. *If $a < b$ and if G is the (algebraically) greatest and L the (algebraically) least value of $F(x)$ in the interval (a, b), then $\int_a^b F(x)\,dx < G(b-a)$ but $> L(b-a)$.*

For $G - F(x)$ and $F(x) - L$ are positive; hence by Th. III. the integrals
$$\int_a^b [G - F(x)]\,dx \text{ and } \int_a^b [F(x) - L]\,dx,$$
that is $\int_a^b G\,dx - \int_a^b F(x)\,dx$ and $\int_a^b F(x)\,dx - \int_a^b L\,dx,$

or $G(b-a) - \int_a^b F(x)\,dx$ and $\int_a^b F(x)\,dx - L(b-a),$

are both positive, so that the integral is less than $G(b-a)$ but greater than $L(b-a)$.

The integral will be equal to $H(b-a)$ where H is a number less than G but greater than L; but since $F(x)$ is continuous it must, for at least one value x_1 of x between a and b, be equal to H. The value x_1 is of the form $a+\theta(b-a)$ where $0<\theta<1$ (§ 73). Hence,

$$\int_a^b F(x)\,dx = F(x_1)(b-a) = F\{a+\theta(b-a)\}(b-a).$$

The theorem is evident from the figure; for the area $ABDEC$ is less than the rectangle $G.AB$, greater than the rectangle $L.AB$, equal to the rectangle $H.AB$ or $MP.AB$ where MP is an ordinate less than G but greater than L.

The value H or $F(x_1)$ is sometimes called the *Mean Value* or the *Average Value* of the function $F(x)$ over the range $(b-a)$. (See § 134.)

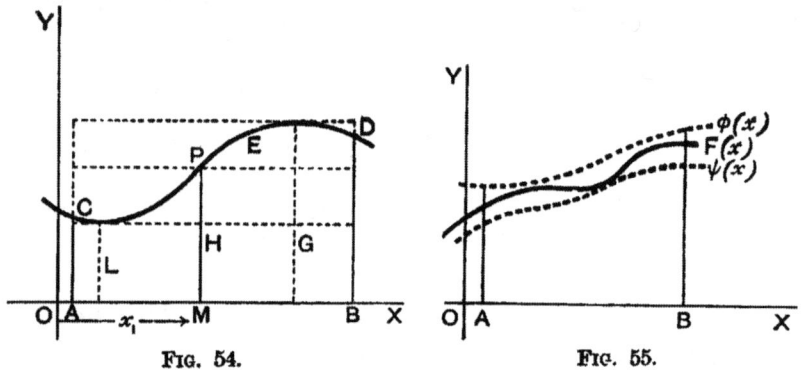

Fig. 54.   Fig. 55.

THEOREM VI. *If $a<b$ and if for every value of x in the interval (a, b), $F(x)$ is (algebraically) less than $\phi(x)$ but (algebraically) greater than $\psi(x)$, then*

$$\int_a^b F(x)\,dx < \int_a^b \phi(x)\,dx \text{ but } > \int_a^b \psi(x)\,dx.$$

Proved in the same way as Th. V. since $\phi(x)-F(x)$ and $F(x)-\psi(x)$ are positive. For geometrical proof see the figure (Fig. 55).

THEOREM VII. *If $a<b$ and if $F(x)$ is the product of two functions $\phi(x)$, $\psi(x)$, one of which, $\phi(x)$, is positive for every value of x in the interval (a, b), then*

$$\int_a^b \phi(x)\psi(x)\,dx < G\int_a^b \phi(x)\,dx \text{ but } > L\int_a^b \phi(x)\,dx,$$

INTEGRAL THEOREM OF MEAN VALUE.

where G, L are the (algebraically) greatest and least values of $\psi(x)$ in the interval (a, b).

Proved in the same way as Th. V., since $G-\psi(x)$, $\psi(x)-L$, and therefore $(G-\psi(x))\phi(x)$ and $(\psi(x)-L)\phi(x)$ are positive.

If $\phi(x)$ is *negative* for every value of x in the interval (a, b) we shall have

$$\int_a^b \phi(x)\psi(x)\,dx > G\int_a^b \phi(x)\,dx \text{ but } < L\int_a^b \phi(x)\,dx.$$

In both cases, the function $\psi(x)$ being continuous, we may as in Th. V. write

$$\int_a^b \phi(x)\psi(x)\,dx = \psi(x_1)\int_a^b \phi(x)\,dx \quad \ldots\ldots\ldots\ldots(A)$$

where $a < x_1 < b$.

The theorem expressed in equation (A) is called *The First (Integral) Theorem of Mean Value*. (See Exercises XXVI., 29-31.)

Ex. Show that if $n > 2$, the integral

$$\int_0^{\frac{1}{2}} \frac{dx}{\sqrt{(1-x^n)}}$$

is greater than ·5 but less than ·524.

For every value of x within the range of integration, the value 0 excepted,

$$x^2 > x^n > 0\,; \quad 1-x^2 < 1-x^n < 1\,; \quad 1/\sqrt{(1-x^2)} > 1/\sqrt{(1-x^n)} > 1,$$

so that the integral is less than

$$\int_0^{\frac{1}{2}} \frac{1}{\sqrt{(1-x^2)}}\,dx = \sin^{-1}\tfrac{1}{2} = \frac{\pi}{6} = \cdot 523 \ldots$$

but greater than $\quad\int_0^{\frac{1}{2}} 1\,dx = \cdot 5.$

§ 125. Related Integrals.

THEOREM I. $\quad \int_0^a F(x)\,dx = \int_0^a F(a-x)\,dx.$

Let $x = a - u$; then $dx = -du$, and when $x=0, u=a$, when $x=a, u=0$, so that

$$\int_0^a F(x)\,dx = -\int_a^0 F(a-u)\,du = \int_0^a F(a-u)\,du,$$

and in the integral last written we may put x for u (§ 124 Th. 1).

A useful case is
$$\int_0^{\frac{\pi}{2}} f(\sin x)\,dx = \int_0^{\frac{\pi}{2}} f(\sin[\tfrac{\pi}{2}-x])\,dx = \int_0^{\frac{\pi}{2}} f(\cos x)\,dx.$$

THEOREM II. $\quad \int_{-a}^{a} F(x)\,dx = \int_0^{a} \{F(-x)+F(x)\}\,dx.$

For $\quad \int_{-a}^{a} F(x)\,dx = \int_{-a}^{0} F(x)\,dx + \int_0^{a} F(x)\,dx.$

In the first integral let $x = -u$ and it becomes
$$-\int_a^0 F(-u)\,du = \int_0^a F(-u)\,du = \int_0^a F(-x)\,dx,$$
from which the result follows. Hence
$$\int_{-a}^{a} F(x)\,dx = 2\int_0^{a} F(x)\,dx, \text{ if } F(-x)=F(x),$$
$$= 0, \text{ if } F(-x)=-F(x).$$

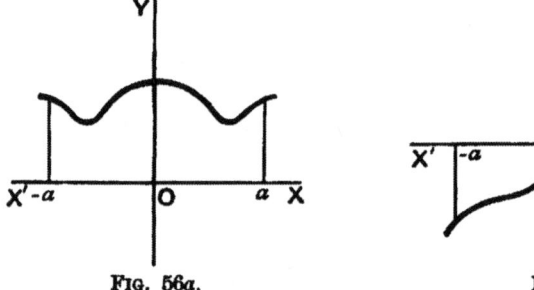
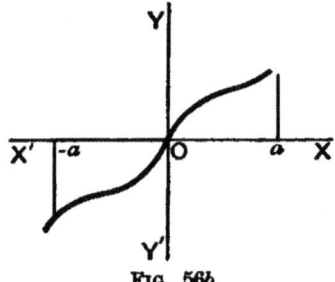

FIG. 56a.   FIG. 56b.

The last results are evident geometrically from the figures.

THEOREM III. $\quad \int_0^{a} F(x)\,dx = \int_0^{\frac{1}{2}a} \{F(x)+F(a-x)\}\,dx,$

so that $\quad \int_0^{a} F(x)\,dx = 2\int_0^{\frac{1}{2}a} F(x)\,dx, \text{ if } F(a-x)=F(x).$
$$= 0, \text{ if } F(a-x)=-F(x).$$

The proof is the same as for Th. II.; divide the interval into $(0,\tfrac{1}{2}a)$ and $(\tfrac{1}{2}a, a)$, and in the second integral put $x = a - u$.

RELATED INTEGRALS.

As a particular case

$$\int_0^\pi f(\sin x)\, dx = 2\int_0^{\frac{\pi}{2}} f(\sin x)\, dx.$$

THEOREM IV. *If $F(x)$ is periodic, with the period a, that is, if $F(x+na)$ is equal to $F(x)$ for every integral value of n*

$$\int_0^{pa} F(x)\, dx = p\int_0^a F(x)\, dx,$$

where p is any positive integer.

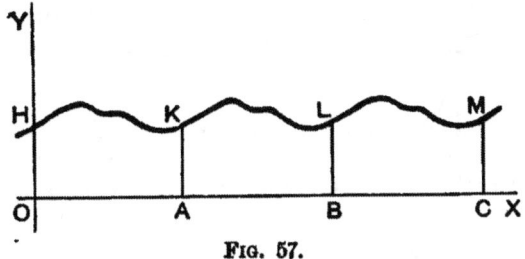

Fig. 57.

If $OA = a = AB = BC = \ldots$ then from the nature of the graph the areas $OAKH$, $ABLK$, $BCML$, ... are all equal, so that if $OC = p \cdot OA$ the area $OCMH$ is p times $OAKH$.

Or divide the range pa into p parts each equal to a, then

$$\int_0^{pa} F(x)\, dx = \int_0^a F(x)\, dx + \ldots + \int_{ka}^{(k+1)a} F(x)\, dx$$
$$+ \ldots + \int_{(p-1)a}^{pa} F(x)\, dx.$$

In the integral having ka, $(k+1)a$ for limits let $x = u + ka$; then $dx = du$, and when $x = ka$, $u = 0$, when $x = (k+1)a$ $u = a$, so that

$$\int_{ka}^{(k+1)a} F(x)\, dx = \int_0^a F(u+ka)\, du = \int_0^a F(u)\, du = \int_0^a F(x)\, dx,$$

since $F(u+ka) = F(u)$. Thus each of the p integrals has the same value and the result follows.

Similar reasoning shows that the theorem is also true when p is a *negative* integer.

As a particular case
$$\int_0^{2p\pi} f(\sin x)\,dx = p\int_0^{2\pi} f(\sin x)\,dx.$$

These theorems are of great service in the evaluation of integrals.

§ 126. Infinite Limits. Infinite Integrand. Up to this point the limits of the integral have been assumed to be finite, and the integrand has been supposed continuous and therefore finite for every value of the variable within the range of integration. It is, however, possible in certain cases to remove these restrictions by the use of limits.

A. Infinite Limits. An integral with one of its limits infinite is defined as follows:
$$\int_a^\infty F(x)\,dx = \underset{b=\infty}{L}\int_a^b F(x)\,dx;\quad \int_{-\infty}^b F(x)\,dx = \underset{a=-\infty}{L}\int_a^b F(x)\,dx,$$
provided the limits for $b=\infty$ and for $a=-\infty$ are definite quantities.

Ex. 1. $\int_1^\infty \dfrac{dx}{x^2} = \underset{b=\infty}{L}\int_1^b \dfrac{dx}{x^2} = \underset{b=\infty}{L}\left(1 - \dfrac{1}{b}\right) = 1.$

Ex. 2. $\int_1^\infty \dfrac{dx}{x} = \underset{b=\infty}{L}\int_1^b \dfrac{dx}{x} = \underset{b=\infty}{L}\log b.$

In this case the limit of $\log b$ is not a definite number, and the integral is therefore a meaningless symbol.

Ex. 3. $\int_0^\infty e^{-x}\cos x\,dx.$

By § 118, ex. 3, the indefinite integral is $\tfrac{1}{2}e^{-x}(-\cos x + \sin x)$, and we have to find the limit for $b=\infty$ of
$$\tfrac{1}{2} + \tfrac{1}{2}e^{-b}(-\cos b + \sin b).$$

Now $\cos b$, $\sin b$ are each never greater than 1, and the limit of e^{-b} is zero so that the integral is equal to $\tfrac{1}{2}$.

The limit for $x=\infty$ of $x^n e^{-ax}$, where a is positive, is often needed in dealing with these integrals. It is easy, by § 49, to see that
$$\underset{x=\infty}{L} x^n e^{-ax} = 0.$$

See also Exercises VII. ex. 9.

B. Infinite Integrand. If $F(x)$ is continuous for all values of x between a and b except for $x=a$ when it is infinite, then the integral of $F(x)$ between a and b is defined thus, a being less than b and ϵ being positive,

INFINITE LIMITS. INFINITE INTEGRAND. 305

$$\int_a^b F(x)\,dx = \underset{\epsilon=0}{L}\int_{a+\epsilon}^b F(x)\,dx,$$

provided the limit is a definite quantity.

If $F(x)$ is continuous except at b then, ϵ being positive,

$$\int_a^b F(x)\,dx = \underset{\epsilon=0}{L}\int_a^{b-\epsilon} F(x)\,dx.$$

Ex. 1. $\int_0^1 \dfrac{dx}{\sqrt{x}} = \underset{\epsilon=0}{L}\int_\epsilon^1 \dfrac{dx}{\sqrt{x}} = \underset{\epsilon=0}{L}(2 - 2\sqrt{\epsilon}) = 2.$

Ex. 2. $\int_0^a \dfrac{dx}{\sqrt{(a^2-x^2)}} = \underset{\epsilon=0}{L}\int_0^{a-\epsilon}\dfrac{dx}{\sqrt{(a^2-x^2)}} = \underset{\epsilon=0}{L}\sin^{-1}\left(1-\dfrac{\epsilon}{a}\right);$

the limit is obviously $\sin^{-1} 1$ or $\pi/2$.

Ex. 3. $\int_0^1 \dfrac{dx}{x^2} = \underset{\epsilon=0}{L}\int_\epsilon^1 \dfrac{dx}{x^2} = \underset{\epsilon=0}{L}\left(\dfrac{1}{\epsilon} - 1\right).$

In this case there is no definite limit and the integral therefore does not exist.

If $a < c < b$ and if $F(x)$ is continuous except when $x = c$, then the integral between a and b is defined thus, ϵ, ϵ' being positive,

$$\int_a^b F(x)\,dx = \underset{\epsilon=0}{L}\int_a^{c-\epsilon} F(x)\,dx + \underset{\epsilon'=0}{L}\int_{c+\epsilon'}^b F(x)\,dx,$$

provided each limit is separately a definite quantity.

Ex. 4. $\int_{-1}^1 \dfrac{dx}{\sqrt[3]{x^2}} = \underset{\epsilon=0}{L}(-3\sqrt[3]{\epsilon}+3) + \underset{\epsilon'=0}{L}(3 - 3\sqrt[3]{\epsilon'}).$

Here the first limit is 3, the second is also 3, and the integral is 6.

Ex. 5. $\int_0^2 \dfrac{dx}{(x-1)^2} = \underset{\epsilon=0}{L}\left(\dfrac{1}{\epsilon}-1\right) + \underset{\epsilon'=0}{L}\left(-1+\dfrac{1}{\epsilon'}\right).$

In this case there is no definite limit and the integral does not exist.

A change of variable will often remove the difficulty of an infinite integrand or an infinite limit; thus, in ex. 2, we might put $x = a\sin\theta$. The change of variable is specially useful for the forms given in § 116.

These exceptional cases of integrals may be illustrated by consideration of the graph of $F(x)$. Let $F(x) = 1/x^n$ where n is positive; then the x-axis is an asymptote and the area $ABDC$ is ($n \neq 1$)

$$\int_a^b \dfrac{dx}{x^n} = \dfrac{1}{n-1}\left(\dfrac{1}{a^{n-1}} - \dfrac{1}{b^{n-1}}\right).$$

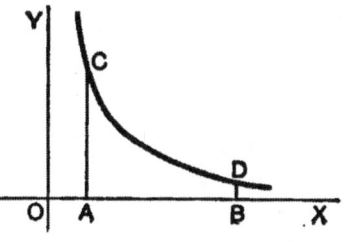

Fig. 58.

G.C. U

Hence, if $n>1$, the area $ABDC$ tends to the value $1/(n-1)a^{n-1}$ as b tends to ∞; while if $0<n<1$ the area tends to ∞ since $1/b^{n-1}$, that is, b^{1-n} tends to ∞. If $n=1$ the area $ABDC$ is equal to $\log(b/a)$ and therefore tends to ∞ with b.

On the other hand, consider $F(x)=1/(x-a)^n$ where n is positive. If $OA=a$, $AE=\epsilon$, $OB=b$, then the area $EBDF$ is equal to $(n \neq 1)$

$$\int_{a+\epsilon}^{b} \frac{dx}{(x-a)^n} = \frac{1}{1-n}\{(b-a)^{1-n} - \epsilon^{1-n}\}.$$

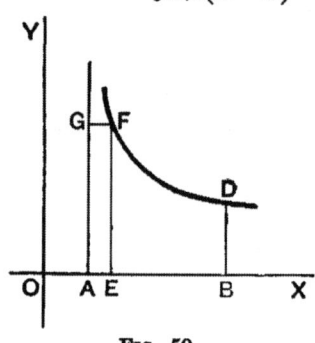

FIG. 59.

Hence, if $0<n<1$ the area tends to $(b-a)^{1-n}/(1-n)$ as ϵ tends to zero; while if $n>1$ the area tends to ∞ since ϵ^{1-n}, that is, $1/\epsilon^{n-1}$ tends to infinity as ϵ tends to zero. If $n=1$ the area is $\log\{(b-a)/\epsilon\}$ and therefore tends to ∞ as ϵ tends to zero.

It is easy to show by the use of Th. VII., § 124, that if near a, $F(x)$ is of the form $\phi(x)/(x-a)^n$, where $\phi(x)$ is continuous, the area $EBDF$ and the corresponding integral tend to a finite limit if n is a positive proper fraction, but that *when $\phi(a)$ is not zero* the limit is infinite if n is equal to or greater than 1.

It is beyond the scope of this book to enter further into these exceptional cases.

EXERCISES XXVI.

Evaluate the following integrals:

1. $\int_0^\infty e^{-ax} \cos bx \, dx \, (a>0)$;   2. $\int_0^\infty e^{-ax} \sin bx \, dx \, (a>0)$;

3. $\int_{-\infty}^\infty \frac{dx}{x^2+2x+2}$;   4. $\int_0^\infty \frac{dx}{(x^2+a^2)^3}$;

5. $\int_0^a \frac{x^{\frac{5}{2}} dx}{\sqrt{(a-x)}}$;   6. $\int_0^{2a} \frac{x \, dx}{\sqrt{(2ax-x^2)}}$;

7. $\int_a^b \frac{dx}{\sqrt{\{(x-a)(b-x)\}}}$.   [Let $x = a\cos^2\theta + b\sin^2\theta$].

EXERCISES XXVI.

8. $\int_a^b \sqrt{\{(x-a)(b-x)\}}\,dx$;

9. $\int_{-1}^{1} \frac{dx}{(a-bx)\sqrt{(1-x^2)}} \quad (a > b > 0)$.

10. $\int_0^{\frac{\pi}{2}} \frac{dx}{a^2\cos^2 x + b^2\sin^2 x}$;

11. $\int_0^{\frac{\pi}{2}} \frac{dx}{(a^2\cos^2 x + b^2\sin^2 x)^2}$;

12. $\int_0^{\frac{\pi}{2}} \frac{\cos^2 x \sin x\,dx}{1 + e^2\cos^2 x}$;

13. $\int_0^{\frac{\pi}{2}} \frac{\cos^2 x \sin x\,dx}{\sqrt{(1 + e^2\cos^2 x)}}$;

14. $\int_{-\frac{\pi}{4}}^{\frac{\pi}{4}} \tan x\,dx$;

15. $\int_0^1 \log x\,dx$;

16. $\int_0^1 x^2 \log x\,dx$.

17. Prove that if m and n are positive,
$$\int_0^1 x^m(1-x)^n dx = \int_0^1 x^n(1-x)^m dx.$$

18. Prove that if n is positive,
$$\int_0^\infty e^{-x} x^n dx = n \int_0^\infty e^{-x} x^{n-1} dx.$$

Find the value of the integral if n is a positive integer.

19. If
$$u = \int_0^\pi \frac{x \sin x\,dx}{1 + \cos^2 x},$$
prove that
$$u = \int_0^\pi \frac{\pi \sin x\,dx}{1 + \cos^2 x} - u,$$
and then find the value of u.

20. If
$$u = \int_0^\pi \frac{x\,dx}{1 + e \sin x} \quad \text{where } 0 < e < 1,$$
prove that
$$u = \int_0^\pi \frac{\pi\,dx}{1 + e \sin x} - u,$$
and then find the value of u.

21. Show that $\int_a^b F(x)\,dx = \int_a^b F(a + b - x)\,dx$.

22. Prove that if n is a positive integer
$$\int_0^{\frac{\pi}{2}} \sin^n x > \int_0^{\frac{\pi}{2}} \sin^{n+1} x\,dx.$$

Hence, show that $\pi/2$ lies between
$$\frac{2 \cdot 2 \cdot 4 \cdot 4 \cdot 6 \cdot 6 \ldots 2n \cdot 2n}{1 \cdot 3 \cdot 3 \cdot 5 \cdot 5 \cdot 7 \ldots (2n-1)(2n+1)}$$
and the fraction obtained by omitting the last factor in numerator and denominator. (This is often quoted as Wallis's value of π.)

23. Prove that, n being a positive integer,

$$\int_0^{n\pi} \frac{\sin x}{x}\,dx = u_0 - u_1 + u_2 - u_3 + \ldots + (-1)^k u_k + \ldots \pm u_m$$

where
$$u_k = \int_0^{\pi} \frac{\sin u\,du}{u + k\pi}.$$

Show that $u_0, u_1, u_2\ldots$ are positive, and that if k is equal to or greater than 1, u_k is less than $1/k$. Interpret these results by considering the graph of $\sin x/x$, and show that the integral has a finite limit for $n = \infty$. The limit is $\pi/2$ but the proof can not be given here.

24. Prove $\displaystyle\int_0^1 \frac{dx}{\sqrt{(4 - x^2 + x^3)}} > \frac{1}{2}$ but $< \frac{\pi}{6}$.

25. Prove $\displaystyle\int_0^1 \frac{dx}{\sqrt{(4 - 3x + x^3)}} < \int_0^1 \frac{dx}{\sqrt{(4-3x)}}$ but $> \int_0^1 \frac{3x+8}{16}dx$,

that is, $< \frac{2}{3}$, but $> 19/32$.

26. Prove $\displaystyle\int_1^2 \frac{dx}{\sqrt{(4 - 3x + x^3)}} > \cdot 573$, but $< \cdot 595$.

Put $x = 1 + u$; then replace $u^3 + 3u^2 + 2$ by $4u^2 + 2$ and by $3u^2 + 2$.

27. If a and ϕ are positive acute angles, prove

$$\int_0^{\phi} \frac{dx}{\sqrt{(1 - \sin^2 a \sin^2 x)}} > \phi \text{ but } < \frac{\phi}{\sqrt{(1 - \sin^2 a \sin^2 \phi)}}.$$

If $a = \phi = \pi/6$, show that the integral lies between $\cdot 523$ and $\cdot 541$. More accurate methods give $\cdot 52943$ as an approximate value of the integral.

28. Prove

(i) $\displaystyle\int_1^{\infty} e^{-x^2}dx < \int_1^{\infty} xe^{-x^2}dx$; (ii) $\displaystyle\int_0^{\infty} e^{-x^2}dx < 1 + \frac{1}{2e}$.

29. Give a geometrical interpretation of Th. VII., § 124, by considering the volume of the solid bounded by the coordinate planes, the planes through $x = a$ and $x = b$ perpendicular to the x-axis, and the cylinders $y = \phi(x)$ and $z = \psi(x)$.

30. If $\psi(x)$ is positive, and if $\phi(x)$ is a *positive decreasing* function in the interval (a, b), show by considering the volume of the solid of ex. 29 that

(i) $\displaystyle\int_a^b \phi(x)\psi(x)dx = \phi(a)\int_a^{\xi} \psi(x)dx$, where $a < \xi < b$;

but that if $\phi(x)$ is a *positive increasing* function,

(ii) $\displaystyle\int_a^b \phi(x)\psi(x)dx = \phi(b)\int_{\xi}^b \psi(x)dx$ where $a < \xi < b$.

31. If $\phi(x)$ increases (algebraically) as x increases from a to b, show that in ex. 30 (i) we may put $\phi(b) - \phi(x)$ in place of $\phi(x)$, while if

STANDARD AREAS AND VOLUMES. 309

$\phi(x)$ decreases (algebraically) we may in ex. 30 (ii) put $\phi(a)-\phi(x)$ in place of $\phi(x)$. Show that, when these substitutions are made, both (i) and (ii) become

$$\int_a^b \phi(x)\psi(x)dx = \phi(a)\int_a^\xi \psi(x)dx + \phi(b)\int_\xi^b \psi(x)dx.$$

In this case $\phi(x)$ may be either positive or negative. The theorem expressed by the equation is called *The Second (Integral) Theorem of Mean Value*; it is true even if $\psi(x)$ take both positive and negative values, though the illustration would require more careful elaboration to show this.

Illustrate by an area when $\psi(x)=1$.

§ 127. Some Standard Areas and Volumes. In this article we collect some of the more important results already obtained or easily proved.

1. *The right Circular Cylinder.* Let the radius of the base be a and the height h.

$$\text{volume} = \pi a^2 h\,;\quad \text{curved surface} = 2\pi ah.$$

2. *The right Circular Cone.* Let the radius of the base be a, the height h, and the slant side $l = \sqrt{(a^2+h^2)}$.

$$\text{volume} = \tfrac{1}{3}\pi a^2 h\,;\quad \text{curved surface} = \pi al.$$

For a frustum of height h, slant side l, and with radii of ends a, b,

$$\text{volume} = \tfrac{1}{3}\pi(a^2+ab+b^2)h\,;\quad \text{curved surface} = \pi(a+b)l.$$

Let A be the base, h the height, and X the section parallel to the base at distance x from the vertex of *any* cone; then

$$X : A = x^2 : h^2,$$

since parallel sections are similar figures. Let V be the volume of the portion having X for base and height x; then to the first order of infinitesimals $\delta V = X\delta x$, and $D_x V$ is equal to X. Hence the volume of the whole cone is

$$\int_0^h X dx = \frac{A}{h^2}\int_0^h x^2 dx = \tfrac{1}{3}Ah.$$

For a frustum of height h, the areas of its ends being A and B, the volume is

$$\tfrac{1}{3}\{A + \sqrt{(AB)} + B\}h.$$

3. *The Sphere.* Let the radius be R; then, by § 85, ex. 2, the volume of a spherical cap of height h is

$$\pi h^2(R - \tfrac{1}{3}h),$$

and the curved surface of the cap is $2\pi Rh$. By putting $h = 2R$ we get for the volume and the surface of the sphere $\tfrac{4}{3}\pi R^3$ and $4\pi R^2$ respectively.

It will be noticed that the surface of the cap is equal to the curved surface of a cylinder of the same height whose base is equal to a great circle of the sphere.

To find the volume of a spherical sector, add to the volume of the cap that of the cone whose vertex is at the centre of the sphere and whose height is $R-h$. The result is
$$\pi h^2(R-\tfrac{1}{3}h)+\tfrac{1}{3}\pi(2Rh-h^2)(R-h)=\tfrac{1}{3}2\pi R^2h=\tfrac{1}{3}SR,$$
where S is the surface of the cap. The result is more easily obtained by supposing the surface of the cap divided into a large number of small areas; the sector may then be considered as made up of a large number of cones having the same height R, and the volume of the sector will therefore be $\tfrac{1}{3}SR$.

4. *The Ellipse.* The area of an ellipse whose axes are $2a$, $2b$ is
$$4\int_0^a y\,dx = \frac{4b}{a}\int_0^a \sqrt{(a^2-x^2)}\,dx = \pi ab.$$

The volume of the spheroid generated by the revolution of the ellipse about its major axis $2a$ is
$$2\int_0^a \pi y^2 dx = 2\pi \frac{b^2}{a^2}\int_0^a (a^2-x^2)dx = \tfrac{4}{3}\pi ab^2.$$

This spheroid is called "prolate." When the axis of revolution is the minor axis $2b$, the spheroid is called "oblate." The volume of the oblate spheroid is
$$2\int_0^b \pi x^2 dy = 2\pi \frac{a^2}{b^2}\int_0^b (b^2-y^2)dy = \tfrac{4}{3}\pi a^2 b.$$

The surface of the prolate spheroid is
$$2\int_0^a 2\pi y \frac{ds}{dx}dx,$$
where
$$\left(\frac{ds}{dx}\right)^2 = 1 + \left(\frac{dy}{dx}\right)^2 = \frac{a^4-(a^2-b^2)x^2}{a^2(a^2-x^2)}.$$

Let ϵ be the eccentricity of the ellipse; then $a^2\epsilon^2 = a^2 - b^2$, and the integral may be written, since $b = a\sqrt{(1-\epsilon^2)}$,
$$4\pi\sqrt{(1-\epsilon^2)}\int_0^a \sqrt{(a^2-\epsilon^2 x^2)}\,dx,$$
and the value is easily found to be
$$2\pi a^2 \left\{1-\epsilon^2 + \sqrt{(1-\epsilon^2)}\frac{\sin^{-1}\epsilon}{\epsilon}\right\}.$$

The limit of this expression for $\epsilon = 0$ is $4\pi a^2$, which gives the surface of the sphere of radius a.

For the oblate spheroid the student will readily prove that the surface is
$$2\int_0^b 2\pi x \frac{ds}{dy}dy = \frac{4\pi}{1-\epsilon^2}\int_0^{a\sqrt{1-\epsilon^2}} \sqrt{\{a^2(1-\epsilon^2)^2 + \epsilon^2 y^2\}}\,dy,$$
$$= 2\pi a^2\left\{1 + \frac{1-\epsilon^2}{2\epsilon}\log\frac{1+\epsilon}{1-\epsilon}\right\}.$$

Since $\qquad \dfrac{1}{\epsilon}\log\dfrac{1+\epsilon}{1-\epsilon}=\log(1+\epsilon)^{\frac{1}{\epsilon}}+\log(1-\epsilon)^{-\frac{1}{\epsilon}},$

we find $\qquad \mathop{\mathrm{L}}\limits_{\epsilon=0}\dfrac{1}{\epsilon}\log\dfrac{1+\epsilon}{1-\epsilon}=\log e+\log e=2\,;\quad$ (§ 48, Cor.)

so that the limit for $\epsilon=0$ of this area is also $4\pi a^2$.

5. *The Ellipsoid* $x^2/a^2+y^2/b^2+z^2/c^2=1$.

The traces of this surface on the coordinate planes are ellipses; the section MPQ by a plane parallel to the plane YOZ is an ellipse. If $OM=x$ then

$$MP=\dfrac{b}{a}\sqrt{(a^2-x^2)}\,;\;\; MQ=\dfrac{c}{a}\sqrt{(a^2-x^2)},$$

and the area X of the quarter-ellipse MPQ is

$$X=\dfrac{\pi}{4}MP\cdot MQ=\dfrac{\pi bc}{4a^2}(a^2-x^2).$$

If V is the volume bounded by the coordinate planes, the surface $BCQP$ and the section MPQ, then to the first order of infinitesimals $\delta V=X\delta x$ and $D_x V=X$. Hence the volume of the octant $OABC$ is

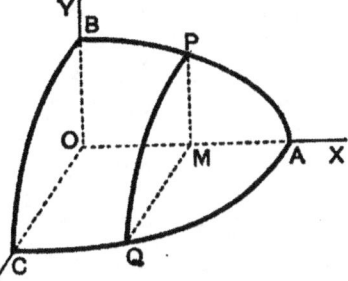

Fig. 60.

$$\int_0^a X dx=\dfrac{\pi bc}{4a^2}\int_0^a(a^2-x^2)\,dx=\dfrac{\pi abc}{6},$$

so that the volume of the ellipsoid is $4\pi abc/3$.

The method of finding the volume illustrated in examples 2 and 5 is obviously applicable whenever the area of a section perpendicular to the x-axis is a known function $F(x)$ of x; the volume is simply the integral of $F(x)$ between proper limits. (See ex. 3, § 85.) The modification needed when the axes are not rectangular is plain.

Curve Tracing.—Before proceeding to the next set of Exercises the student should read over carefully the hints given in the earlier chapters for tracing curves; these, with the additional help furnished by the first and second derivatives, should enable him to graph the more elementary curves. In general he should proceed in some such way as the following:

(i) Examine the equation for symmetry.
(ii) Find where the curve crosses the axes.
(iii) Find the *finite* values of x (or of y) that make y (or x) infinite; these values usually show the asymptotes that are parallel to the axes. Asymptotes inclined to the

312   AN ELEMENTARY TREATISE ON THE CALCULUS.

axes may in the simpler cases be found as in §24 or by the method of §106; but such cases lie outside elementary work.

(iv) Find the values of the one coordinate that make those of the other coordinate imaginary.

(v) Find the gradient (see §54); note the turning points.

(vi) Find the second derivative; it determines the convexity or concavity of the arc and the points of inflexion. It is often laborious, however, to find the second derivative, and general considerations will frequently show the course of the curve without its use.

For polar coordinates the procedure is similar. It is often convenient, however, to suppose that the radius vector may take *negative* values; thus the point $(-1, -1)$ in the third quadrant may be given in polar coordinates as $(\sqrt{2}, 5\pi/4)$ or as $(-\sqrt{2}, \pi/4)$. In the second form $(-\sqrt{2}, \pi/4)$, if $\angle XOP$ is $\pi/4$ and OP equal to $\sqrt{2}$, produce PO beyond O to P' so that $OP' = PO$ and P' is the point $(-\sqrt{2}, \pi/4)$. See Exer. XXVII., ex. 23.

The general course of the curve should always be found before attempting to find an area, or arc, etc. In evaluating the integrals substitutions will usually be necessary, and the student will find that sometimes a considerable amount of labour will be saved by choosing a good substitution.

Even though the curve is given in rectangular coordinates a change to polars will sometimes simplify the integrations.

EXERCISES XXVII.

1. The parabola $y^2 = 4ax$ revolves about the x-axis; find the volume and the surface of the segment cut off from the solid by a plane perpendicular to the x-axis through the point where $x = h$.

2. Find the volume cut off from the paraboloid
$$y^2/b + z^2/c = 2x,$$
by a plane perpendicular to the x-axis through the point where $x = h$.

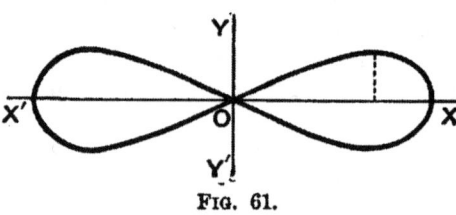

Fig. 61.

3. Find the area enclosed by the curve (Fig. 61)
$$a^4 y^2 + b^2 x^4 = a^2 b^2 x^2.$$
Symmetry about both axes; $x^2 \leqq a^2$; max. of $y = b/2$.

Find also the volume of the solid generated by the revolution of the curve about the x-axis.

EXERCISES XXVII.

4. Find the area enclosed by the curve $c^2y^2 = x^2(x-a)(b-x)$, where $b>a>0$.

If x is less than a or greater than b, y is imaginary *except* when $x=0$ and then $y=0$. The curve is therefore a closed curve symmetrical about the x-axis; the origin is called an *isolated* point because its coordinates satisfy the equation, while there is no other point nearer the origin than $(a, 0)$ which lies on the curve.

5. Find the area of the curve
$$(x^2+y^2)^2 = a^2x^2 + b^2y^2.$$
Change to polar coordinates. The origin is an isolated point.

6. Trace the curve
$$by^2 = x(x-a)(2a-x)$$
where a and b are positive..

y is imaginary (i) if $x>2a$; (ii) if $0<x<a$.

The curve consists of an infinite branch and an oval as in Fig. 62.

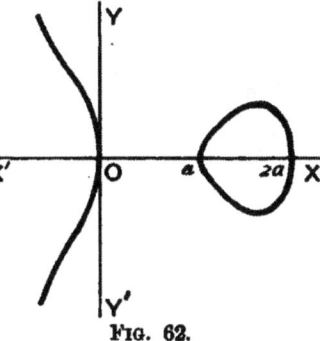

Fig. 62.

7. Find the area of the loop of the curve $16a^3y^2 = b^2x^2(a-2x)$ where a, b are positive.

8. Trace the curve $ky^2 = (x-a)(x-b)(x-c)$ where $c>b>a>0$, $k>0$. Consider the forms for which (i) $a=b$; (ii) $b=c$; (iii) $a=b=c$.

The general form consists of an oval and an infinite branch like Ex. 6, only the oval lies to the left of the infinite branch. When $a=b$ the oval shrinks up to an isolated point at $(a, 0)$; when $a=b=c$ the curve is the semi-cubical parabola, the point $(a, 0)$ being a cusp. The area of the oval in the general case, a, b, c unequal, cannot be expressed in terms of the elementary integrals.

9. Trace the curve $y^2(a-x) = x^2(a+x)$; find (i) the area of the loop, (ii) the area between the curve and the asymptote (Fig. 63).

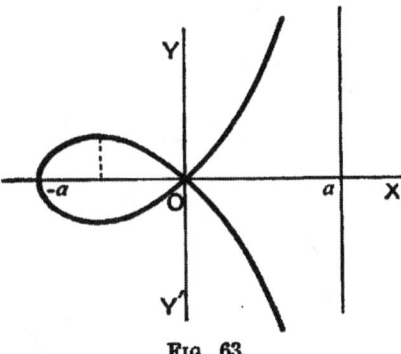

Fig. 63.

Here the gradient is zero when x is $(1 \pm \sqrt{5})a/2$, but the value $(1+\sqrt{5})a/2$ makes y imaginary.

10. The "cissoid" is the curve given by the equation $y^2(2a-x)=x^3$; find the whole area between the curve and its asymptote (Fig. 64).

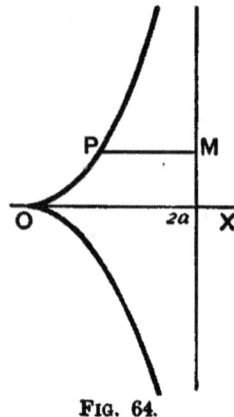

Fig. 64.

Find also the volume of the solid generated by the revolution of the cissoid about its asymptote.

If PM is perpendicular to the asymptote the volume is

$$2\int_0^\infty \pi PM^2\, dy = 2\pi\int_0^\infty (2a-x)^2\, dy.$$

To integrate let $x = 2a\sin^2\theta$, then

$$y = 2a\sin^3\theta/\cos\theta,$$

and the limits for θ are 0 and $\pi/2$.

11. Find the area between the curve

$$xy^2 = a^2(a-x)$$

and its asymptote; also the volume of the solid generated by the revolution of the curve about its asymptote.

12. Find the area of a loop of the curve $y^2(a^2+x^2)=x^2(a^2-x^2)$.

13. The figure bounded by a quadrant of a circle of radius a, and the tangents at its ends revolves about one of these tangents; find the volume of the solid.

14. An arc of a circle of radius a revolves about its chord; if the length of the arc is $2a\alpha$ show that the volume of the solid is

$$2\pi a^3(\sin\alpha - \tfrac{1}{3}\sin^3\alpha - \alpha\cos\alpha),$$

and that the surface of the solid is

$$4\pi a^2(\sin\alpha - \alpha\cos\alpha).$$

15. If s is an arc of the curve $a^{n-1}y = x^n$ show that

$$\frac{ds}{dx} = \frac{1}{a^{n-1}}\sqrt{(n^2x^{2n-2}+a^{2n-2})}.$$

Show that the arc can be expressed by means of the elementary functions when n is of either of the forms $(2k+1)/2k$ or $2k/(2k-1)$ where k is any integer, positive or negative.

16. Find the area between the graph of $4/(e^x+e^{-x})^2$ and the x-axis.

17. Find the whole area enclosed by the curve

$$(x/a)^{\frac{2}{3}}+(y/b)^{\frac{2}{3}}=1.$$

Put $x = a\sin^3\theta$, then $y = b\cos^3\theta$, and the area is

$$4\int_0^a y\, dx = 12ab\int_0^{\frac{\pi}{2}}\sin^2\theta\cos^4\theta\, d\theta = \tfrac{3}{8}\pi ab.$$

EXERCISES XXVII.

18. The cycloid is the curve given by the equations (§ 146)
$$x = a(\theta - \sin\theta); \quad y = a(1 - \cos\theta).$$
Find (i) the area between the x-axis and one arch of the curve; (ii) the length of the arch from $\theta = 0$ to $\theta = a$; (iii) the volume of the solid generated by the revolution of the arch about the x-axis; (iv) the volume of the solid generated by the revolution of the arch about the tangent at the highest point (or vertex) of the arch, namely, where $\theta = \pi$.

Here $\quad \int y\,dx = a^2 \int (1 - \cos\theta)^2 d\theta; \quad \dfrac{ds}{d\theta} = 2a\sin\dfrac{\theta}{2}.$

19. Find the volume of the tetrahedron formed by the coordinate planes and the plane
$$x/a + y/b + z/c = 1.$$

20. Find the volume of the cono-cuneus determined by the equation
$$z^2 + a^2 y^2 / x^2 = c^2,$$
which is contained between the planes $x = 0$ and $x = a$.

21. Find the perimeter of the curve
$$x^{\frac{2}{3}} + y^{\frac{2}{3}} = a^{\frac{2}{3}}.$$
If $x = a\sin^3\theta$, then $y = a\cos^3\theta$ and $ds/d\theta = 3a\sin\theta\cos\theta$; the perimeter is
$$4\int_0^{\frac{\pi}{2}} 3a\sin\theta\cos\theta\,d\theta = 6a.$$

22. The polar equation of a conic, the focus being the pole, is $r(1 + e\cos\theta) = l$. Find the area bounded by the initial line, the curve and the radius vector for which $\theta = a$, where $a < \pi$, (i) for the parabola, (ii) for the ellipse.

23. Show that the curve $r = a\sin 3\theta$ consists of three loops of equal area lying within a circle of radius a, and find the area of a loop.

As θ increases from 0 to $\pi/3$, the graphic point describes the loop $OABCO$; as θ increases from $\pi/3$ to $2\pi/3$, r is negative and the graphic point describes the loop $ODEFO$; as θ increases from $2\pi/3$ to π, r is again positive and the graphic point describes the loop $OGHKO$. A further increase of θ gives no new arc.

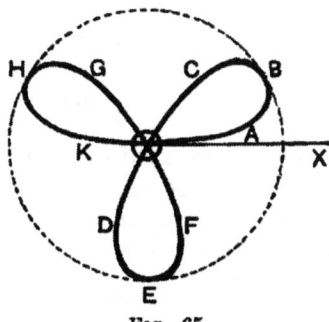

Fig. 65.

24. Find the area enclosed by all the loops of the curve $r = a\sin n\theta$ (i) when n is an odd integer, (ii) when n is an even integer.

25. Find the area of a loop of the curve $r^2 \cos\theta = a^2 \sin 3\theta$.

26. Find the area of the loop of the curve $r\cos\theta = a\cos 2\theta$.

§ 128. Closed Curves. Let CP_1DP_2 be a curve that can not be cut by a straight line in more than two points, and let each ordinate be positive; let AC, BD be the tangents parallel to the y-axis, $OA = a$, $OB = b$.

The area enclosed by the curve is

$$\int_a^b MP_1\,dx - \int_a^b MP_2\,dx \quad\quad\quad\quad\quad (1)$$

where P_1 and P_2 move along CP_1D and CP_2D respectively as x increases from a to b.

The integrals (1) may be written

$$\int_a^b MP_1\,dx + \int_b^a MP_2\,dx \quad\quad\quad\quad\quad (2)$$

Suppose now that the coordinates x and y of a point on the curve can be expressed as functions of a variable, t say,

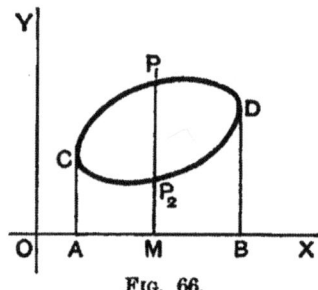

Fig. 66.

such that as t increases from t_1 to t_2 the point (x, y) travels completely round the curve. As t increases from t_1 to t' let the point (x, y) travel from C to D along the arc CP_1D; as t increases from t' to t_2 let the point (x, y) travel from D to C along the arc DP_2C. We might, for example, suppose t to be an arc of the curve measured from C; then $t_1 = 0$, $t' = $ arc CP_1D, $t_2 = $ whole perimeter. If we make t the variable of integration, (2) becomes

$$\int_{t_1}^{t'} MP_1 \frac{dx}{dt}\,dt + \int_{t'}^{t_2} MP_2 \frac{dx}{dt}\,dt \quad\quad\quad\quad (3)$$

The second integral in (3) is negative, since MP_2 is positive and dx/dt is negative as t increases from t' to t_2. When t represents an arc of the curve dx/dt is the cosine of the angle which the tangent at (x, y) makes with the x-axis, the angle being measured as in § 92. We may combine the two integrals of (3) into one and write as the expression for the area of the closed curve

$$\int_{t_1}^{t_2} y \frac{dx}{dt}\,dt \quad\quad\quad\quad\quad\quad (4)$$

AREA OF A CLOSED CURVE.

As an example, let CP_1DP_2 be the ellipse
$$(x-h)^2/a^2+(y-k)^2/\beta^2=1.$$
Put $x=h-a\cos t$, $y=k+\beta \sin t$; as t varies from 0 to 2π, the point (x, y) travels round the curve in the direction CP_1DP_2. The area is
$$\int_0^{2\pi}(k+\beta\sin t)a\sin t\,dt=a\beta\int_0^{2\pi}\sin^2 t\,dt=\pi a\beta.$$

The restriction that the curve is to be cut in not more than two points by a straight line is easily removed.

Thus, when the point (x, y) travels in the direction shown by the arrows, the area swept out by the ordinate of the point is

$$ACEM-NFEM+NFDB-ACGDB,$$

which is clearly the area enclosed by the curve. Along the arcs EF, DGC, dx/dt and the corresponding integrals are negative; the areas $NFEM$, $ACGDB$ are therefore to be subtracted.

We might have written (1) in the form

$$-\int_a^b MP_2\,dx-\int_b^a MP_1\,dx.$$

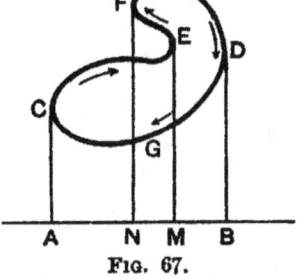

Fig. 67.

If as t increases from t_1' to t_2' the point (x, y) travels completely round the curve in the direction CP_2DP_1, the area will be

$$-\int_{t_1'}^{t_2'} y\frac{dx}{dt}dt. \quad\quad\quad\quad\quad\quad\quad (4')$$

The area, as given by (4) or (4') is a *positive* number; if, however, we agree to give the area a sign, the integral

$$\int y\frac{dx}{dt}dt \quad\quad\quad\quad\quad\quad\quad (5)$$

taken round the curve, that is, the range of t being such that the point (x, y) travels completely round the curve, will always give the *algebraical* measure of the area.

In exactly the same way as (4), (4') are established, it may be proved that the integral

$$\int x\frac{dy}{dt}dt \quad\quad\quad\quad\quad\quad\quad (6)$$

taken round the curve will give the algebraical measure of the area. If, as t increases from t_1 to t_2, the point travels in the direction CP_1DP_2, the integral (5) is positive and (6) is negative, and

$$\int y \frac{dx}{dt} dt = -\int x \frac{dy}{dt} dt \, ;$$

if the point travel in the direction CP_2DP_1 it is (5) that is negative and (6) that is positive.

The direction of motion of the point (x, y) is of course arbitrary; in mathematical physics it is customary to choose the number that measures the area to be positive when the area lies to the *left* of an observer who moves round the curve in the direction corresponding to increasing t. If we adopt this convention we find for the area A of a closed curve

$$A = \int x \frac{dy}{dt} dt = -\int y \frac{dx}{dt} dt = \tfrac{1}{2} \int \left(x \frac{dy}{dt} - y \frac{dx}{dt} \right) dt, \ldots \ldots (7)$$

the integral being taken round the curve in the direction in which t increases. The integrals in (7) are often abbreviated to

$$A = \int x\,dy = -\int y\,dx = \tfrac{1}{2} \int (x\,dy - y\,dx).$$

There is no difficulty now in removing the restriction that the coordinates are to be positive; the expressions (7) always give the algebraical measure of the area. Of course

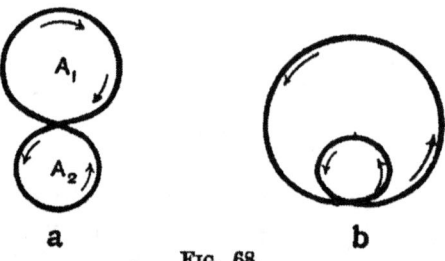

Fig. 68.

it is understood that the point (x, y) travels round the curve in a direction determined once for all; the sign of A given by (7) is positive for the direction CP_2DP_1; if the direction be CP_1DP_2 the sign will be negative.

AREA SWEPT OUT BY A MOVING LINE.

The theorem includes cases in which the curve cuts itself; thus, if the point travels round the figure of eight in the direction of the arrows, the integral (7) is equal to $A_2 - A_1$. For the other figure the integral gives the sum of the areas of the two loops; for the inner area is taken twice.

§ 129. Area swept out by a moving Line. Let AB be a straight line of length l, and let it be displaced to a close position $A'B'$, sweeping out an area $ABB'A'$; this area will be taken as positive or negative according as it lies to the left or to the right of an observer moving round the boundary in the direction $ABB'A'$.

Draw $A'C$, BC parallel to AB and to the chord AA' respectively; let AX' be parallel to a fixed line and let the angles $X'AB$, $CA'B'$ be a and δa. To the first order of infinitesimals the area $ABB'A'$, δz say, is equal to the sum of the parallelogram AC and the triangle $A'CB'$.

The motion of AB may be resolved into (i) a translation to $A'C$, (ii) a rotation about A' to the position $A'B'$. Let h be the altitude of the parallelogram, then to the first order of infinitesimals

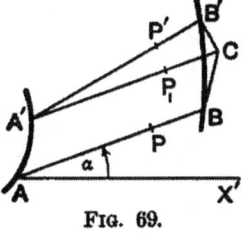

FIG. 69.

$$dz = lh + \tfrac{1}{2}l^2 da \dots\dots\dots\dots\dots(1)$$

Let P be a fixed point in AB; $AP = a = A'P_1 = A'P'$, and consider the displacement of P normal to AB. For the translation the normal displacement is (not PP_1 but) h; for the rotation it is $a\,da$. The total normal displacement, ds say, of P is therefore

$$ds = h + a\,da \dots\dots\dots\dots\dots(2)$$

From (2) $h = ds - a\,da$; therefore (1) becomes

$$dz = l\,ds + (\tfrac{1}{2}l^2 - al)\,da \dots\dots\dots\dots(3)$$

If we suppose the variables to be functions of t, as in § 128, we have

$$\frac{dz}{dt} = l\frac{ds}{dt} + (\tfrac{1}{2}l^2 - al)\frac{da}{dt} \dots\dots\dots\dots(4)$$

320   AN ELEMENTARY TREATISE ON THE CALCULUS.

Equation (4) is general, provided the variables are given the proper signs. ds and ds/dt will be taken positive when the motion of P is to the left of an observer looking along AB from A to B; positive rotation (a) is counter clockwise. The constant a will be positive when P lies in AB or in AB produced beyond B; negative when it lies in BA produced beyond A.

As t increases from t_1 to t_2 the area swept out by AB is

$$z = l\int_{t_1}^{t_2}\frac{ds}{dt}dt + (\tfrac{1}{2}l^2 - al)\int_{t_1}^{t_2}\frac{da}{dt}dt$$
$$= ls + (\tfrac{1}{2}l^2 - al)(a_2 - a_1)\ldots\ldots\ldots\ldots\ldots(5)$$

where s is the total normal displacement of P during the motion and a_1, a_2 are the initial and final values of a. s is not, in general, the same thing as the length of P's path.

Suppose now that B describes a closed curve C and let the area of the curve be also denoted by C.

(i) When B makes a complete circuit of C let A move to and fro along an arc EF, returning to its initial position when B returns to its initial position; in (5) $a_2 = a_1$ and z is simply equal to C, so that

$$C = ls\ldots\ldots\ldots\ldots(6)$$

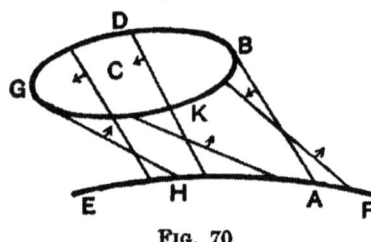

FIG. 70.

where s is the total normal displacement of P. For, clearly, the integral (5) gives the area $ABDGH$ diminished by the area $ABKGH$. In this case s is independent of a, that is of the position of P on AB.

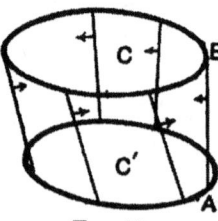

FIG. 71.

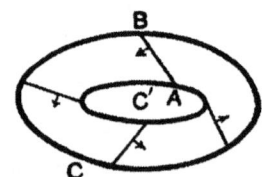

FIG. 72.

(ii) Suppose that while B makes a complete circuit of C A travels round a closed curve C'. If C' is outside C

THEORY OF THE PLANIMETER.

(Fig. 71) a_1 and a_2 will be equal; (5) will be ls but the area swept out by AB will be $C-C'$, so that

$$C-C'=ls \dots\dots\dots\dots\dots\dots\dots\dots(7)$$

If, however, C completely encloses C' (Fig. 72) then a_2-a_1 will be 2π and we shall have

$$C-C'=ls+2\pi(\tfrac{1}{2}l^2-al)\dots\dots\dots\dots\dots(8)$$

The signs of the numbers C, C' are supposed to be determined by the convention of § 128 (7).

§ 130. **Planimeters.** The investigations in the last two articles contain the theory of several instruments that have been devised for mechanically evaluating the area of a closed curve; the best known of these is Amsler's Polar-Planimeter.

Essentially the polar-planimeter consists of two bars OA, AB freely jointed at A, the bar OA rotating about a fixed point O. If B is made to describe a closed curve, A will move along the circumference of a circle. When A merely oscillates along the circumference, not making a complete revolution, the area enclosed by the curve which B describes is, by (6) of § 129, ls. In this case s is independent of the position of P on the bar AB.

To find s a wheel with axis parallel to AB is attached to AB; the wheel, as B describes its curve, partly slides and partly rolls. The sliding and the rolling motions are independent, and the sliding motion has no effect in the way of turning the wheel. The normal displacement of P is therefore equal to the circumference, $2\pi r$ say, of the wheel multiplied by n, the number of turns made by the wheel while B describes its curve; that is, $s=2\pi rn$. A counter is provided that registers n; n of course may be integral or fractional.

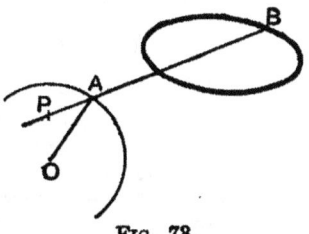

Fig. 73.

If we suppose the curve C so large that the circle of radius OA lies wholly inside it then, by (8) of § 129,

$$C-\pi OA^2=ls+2\pi(\tfrac{1}{2}l^2-al),$$

that is, $\qquad C=2\pi lrn+2\pi(\tfrac{1}{2}l^2-al)+\pi OA^2,$

since $s = 2\pi r n$. All the numbers except n are constants of the instrument.

For information on Planimeters the student is referred to Henrici's "Report on Planimeters," *Brit. Ass. Rep.* 1894. The method of proof followed in §§ 128, 129, is essentially that given by Appell in his *Eléments d'Analyse Mathématique*.

EXERCISES XXVIII.

1. Show that in polar coordinates the area of a closed curve is given by the integral
$$\tfrac{1}{2}\int r^2 \frac{d\theta}{dt} dt$$
taken round the curve. Prove the result (i) by use of the polar formula for area; (ii) by transformation of the last integral in (7), § 128, by putting $x = r\cos\theta$, $y = r\sin\theta$. (See Exer. XII., ex. 15.)

2. If the coordinates of the vertices of the triangle OAB are, when taken in the order O, A, B, $(0, 0)$, (x, y), $(x + \delta x, y + \delta y)$ respectively, prove geometrically that the area of the triangle is $\tfrac{1}{2}(x\delta y - y\delta x)$, in sign and in magnitude. Apply the result to establish the theorem of ex. 1.

3. Find the area common to the two parabolas $y^2 = 4ax$, $x^2 = 4ay$.

4. Find the area between the asymptote $y = a$, the y-axis and the branch of the curve $y^2(a^2 + x^2) = a^2 x^2$ that lies in the first quadrant.
The area is equal to
$$\underset{b=\infty}{\mathrm{L}}\int_0^b (a-y)dx = a^2 + \underset{b=\infty}{\mathrm{L}}\, a\{b - \sqrt{(a^2 + b^2)}\} = a^2.$$
Find the area by integrating with respect to y.

5. The "tore" or the "anchor-ring" is the solid formed by the revolution of a circle about a straight line in its plane. Let a be the radius of the circle, the y-axis the axis of revolution, and let the centre of the circle be on the x-axis at a distance c from the origin. The coordinates of any point on the circle may be taken as
$$x = c + a\cos t, \quad y = a\sin t.$$
If V is the volume and S the surface of the tore, then, when $c \geqq a$, prove

(i) $V = \pi \int_0^{2\pi} (c + a\cos t)^2 a\cos t\, dt = 2\pi^2 a^2 c = AL$;

(ii) $S = 2\pi \int_0^{2\pi} (c + a\cos t)a\, dt = 4\pi^2 ac = CL$,

where A is the area and C the perimeter of the circle, and L is the circumference $2\pi c$ of the circle described by the centre of the revolving circle.

EXERCISES XXVIII.

6. The curve $r = 3 + 2\cos\theta$ consists of a single oval; trace the curve and find its area.

7. The curve $r = 2 + 3\cos\theta$ consists of two ovals (Fig. 74); if $\cos a = -\tfrac{2}{3}$ ($0 < a < \pi$), show that the area of the large oval is

$$A = \tfrac{17}{2}a + 12\sin a + \tfrac{9}{2}\sin a \cos a,$$

and of the small oval is

$$\tfrac{17}{2}\pi - A.$$

Show also that the integral of $\tfrac{1}{2}r^2$ from $\theta = 0$ to $\theta = 2\pi$ gives the sum of these two areas.

Examples 6, 7 show the nature of the curve $r = a + b\cos\theta$ for $a > b$ and $a < b$ respectively.

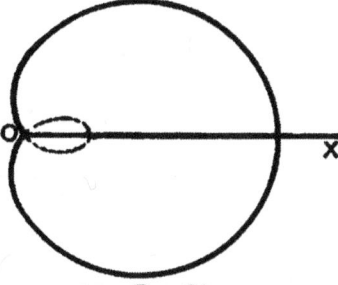

FIG. 74.

8. How may the curve given by the equation $f(mx, ny) = 0$, where m and n are constants, be deduced from that given by $f(x, y) = 0$? If the second curve is closed, show that the first is also closed and that the area of $f(mx, ny) = 0$ is equal to that of $f(x, y) = 0$ divided by mn.

Let $mx = x'$, $ny = y'$, and therefore $x'dy' = mn x dy$. Now apply (7), § 128; the integral of $x'dy'$ round the curve $f(x', y') = 0$, (which is *the same thing* as the integral of xdy round the curve $f(x, y) = 0$), will be equal to the integral of $mn x dy$ round the curve $f(mx, ny) = 0$, that is, to mn times the area enclosed by that curve (since mn is constant and the integral of xdy is the area).

9. Apply the method of ex. 8 to deduce from Exer. XXVII., ex. 5, the area of the curve $(m^2x^2 + n^2y^2)^2 = a^2x^2 + b^2y^2$.

10. When AB (§ 129) describes one complete revolution, show that P describes a curve which encloses an area C'' given by

(i) $C'' = (aC + bC')/(a + b) - \pi ab,$

where $PB = b$ and a, C, C' denote the same quantities as in § 129.

Show also that if the ends A, B move on a closed oval curve C

(ii) $C - C'' = \pi ab.$ (Holditch's Theorem.)

Use equation (8), § 129. Put $l = a + b$ and we get $C - C''$; then put $l = a$ and we get $C'' - C'$. The elimination of s gives (i). To find (ii), consider the areas swept out by AP and BP.

CHAPTER XV.

INTEGRAL AS LIMIT OF A SUM.
DOUBLE INTEGRALS.

§ 131. Integral as the Limit of a Sum. It is instructive and for some applications necessary to consider an integral as the limit of a sum. $F(x)$ is, as usual, understood to be continuous.

In the first place, suppose $a<b$ and $F(x)$ a positive increasing function; these restrictions will afterwards be removed. Between a and b insert $(n-1)$ values in ascending order of magnitude, $x_1, x_2, x_3, \ldots, x_{n-1}$, and form the differences $(x_1-a), (x_2-x_1), (x_3-x_2), \ldots, (b-x_{n-1})$; these n differences are all of the same sign, in this case positive, and their sum is $b-a$. The interval $b-a$ is thus divided into n sub-intervals.

Now multiply each sub-interval by the value of $F(x)$ at the beginning of that sub-interval and add the n products. We get the sum

$$F(a)(x_1-a)+F(x_1)(x_2-x_1)+F(x_2)(x_3-x_2)+\ldots \\ +F(x_{n-1})(b-x_{n-1})\ldots\ldots\ldots(1)$$

or, in the ordinary notation of differences $\delta a, \delta x_1 \ldots$

$$F(a)\delta a + F(x_1)\delta x_1 + F(x_2)\delta x_2 + \ldots + F(x_{n-1})\delta x_{n-1}\ldots\ldots(1')$$

The sum (1') may be more compactly written

$$\sum_{x=a}^{x=b} F(x)\delta x\ldots\ldots\ldots\ldots\ldots\ldots(2)$$

The symbol $\Sigma F(x)\delta x$ means "the sum of all the terms of the type $F(x)\delta x$" and is read "sigma $F(x)\delta x$." In interpreting the symbol, the manner in which the interval $b-a$ has been divided has to be gathered from the context; the

INTEGRAL AS LIMIT OF A SUM.

end of the interval from which the division begins is indicated by "$x=a$," the other end by "$x=b$," and each difference δx has the same sign as $b-a$, in this case positive.

We wish to find the limit of the sum (1) or (2) for n increasing indefinitely, each difference $\delta a, \delta x_1, \ldots$, at the same time diminishing indefinitely. To find the limit consider the graph of $F(x)$, (Fig. 75).

Let $OA = a$, $OA_1 = x_1 \ldots$, $OB = b$; then $AC = F(a)$, $A_1C_1 = F(x_1) \ldots$, $A_{n-1}C_{n-1} = F(x_{n-1})$, $BD = F(b)$. CE_1, $C_1E_2, \ldots, C_{n-1}E_n$ are parallel to the x-axis. The sum (1) is clearly the area enclosed by the rectangles AE_1, A_1E_2, $\ldots$, $A_{n-1}E_n$ and differs from the area $ABDC$ by the sum of the curvilinear triangles CE_1C_1, $C_1E_2C_2, \ldots, C_{n-1}E_nD$.

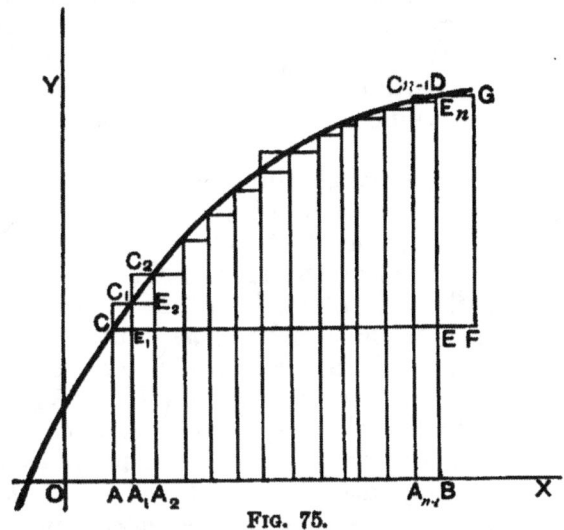

FIG. 75.

Draw CE parallel to AB to cut BD at E and produce CE to F so that EF may be equal to the greatest of the sub-intervals AA_1, $A_1A_2, \ldots$, and complete the rectangle $EFGD$. Let z denote the area $ABDC$; then the difference between z and the sum (1) is less than the sum of the n rectangles $CE_1 . E_1C_1$, $C_1E_2 . E_2C_2, \ldots, C_{n-1}E_n . E_nD$, and therefore less than the rectangle

$$EF(E_1C_1 + E_2C_2 + \ldots + E_nD)$$

or the rectangle $EF . ED$, that is, $EF\{F(b) - F(a)\}$.

If n increases indefinitely and if at the same time each sub-interval diminishes indefinitely, the limit of EF will be zero and therefore the limit of (1) will be z. Hence

$$\underset{n=\infty}{\mathrm{L}} \sum_{x=a}^{x=b} F(x)\delta x = z = \text{area } ABDC \ldots \ldots \ldots \ldots \ldots (3)$$

We may, of course, write

$$\sum_{x=a}^{x=b} F(x)\delta x = z, \text{ approximately.}$$

It is easy now to remove the restriction that $F(x)$ should be positive and increasing or that a should be less than b.

If $a<b$ and $F(x)$ positive and decreasing the only change is that z is less than the sum (1); if $F(x)$ is sometimes increasing and sometimes decreasing we can combine the results for the cases of increasing and of decreasing $F(x)$.

If $a>b$ and $F(x)$ positive, each of the differences (x_1-a), (x_2-x_1), ... is negative and the limit gives the area with negative sign.

Lastly, if $F(x)$ is negative the limit is still the area if the appropriate sign be chosen as in § 80.

In regard to the sub-intervals we may if we please suppose them all equal, each therefore being $(b-a)/n$; the only restriction on the sub-intervals is that *each* must have zero for limit when n tends to infinity as limit.

We have supposed $F(x)$ in the sum (1) to have its value at the beginning of each interval; but the *limit* will be the same if we take the value at the *end* or at *any intermediate point* of each interval, as may be proved by § 87, Th. II. For, restricting attention to the case $a<b$, $F(x)$ positive, since the others can be easily deduced from this, if a', x_1', x_2'... are values of x within or at the end of the intervals (x_1-a), (x_2-x_1), (x_3-x_2)... respectively, we may take

$$\beta_1 = F(a')(x_1-a), \quad \beta_2 = F(x_1')(x_2-x_1),\ldots$$
$$\gamma_1 = F(a)(x_1-a), \quad \gamma_2 = F(x_1)(x_2-x_1), \ldots$$

and the conditions of that theorem apply since, $F(x)$ being continuous, the limit for $n=\infty$ of β_1/γ_1, β_2/γ_2... is unity.

Having proved that the limit of (1) is the area z, we can now show, as in § 80, that the derivative of that limit with respect to b is BD, that is $F(b)$, and therefore we can apply all the theorems respecting integrals to the limit of the sum (1). The origin of the ordinary notation for integrals is also obvious, the $\int$ being a form of the initial letter of the word "sum"; it will be remembered, however, that *the integral is not a sum but the limit of a sum*. (See § 132, ex. 2).

§ 132. Examples.

Ex. 1. Evaluate $\int_0^b x^2 dx$.

Divide the interval b into n *equal* parts; in the notation of § 131,
$$OA=0, \quad OA_1=b/n, \quad OA_2=2b/n, \quad \ldots, \quad OA_{n-1}=(n-1)b/n.$$
The sum (1) becomes
$$0 \cdot \frac{b}{n} + \left(\frac{b}{n}\right)^2 \frac{b}{n} + \left(\frac{2b}{n}\right)^2 \frac{b}{n} + \ldots + \left\{\frac{(n-1)b}{n}\right\}^2 \frac{b}{n}$$
$$= \frac{b^3}{n^3}\{1^2 + 2^2 + \ldots + (n-1)^2\}$$
$$= \frac{b^3}{n^3} \cdot \frac{(n-1)n(2n-1)}{6}, \quad \text{or} \quad \frac{b^3}{3}\left\{1 - \frac{3}{2n} + \frac{1}{2n^2}\right\},$$
and the limit is clearly $b^3/3$.

Ex. 2. Show that if in § 131, (2), we put $F(x)=f'(x)$, the limit will be $f(b)-f(a)$.

By the definition of a derivative,
$$\frac{f(x+\delta x)-f(x)}{\delta x} = f'(x) + a; \quad f(x+\delta x) - f(x) = f'(x)\delta x + a\delta x, \ldots (\text{A})$$
where a vanishes with δx. Give successively to x and δx in (A) the values in § 131; a will not usually have the same value for all values of x, and we therefore use suffixes. Hence
$$f(x_1) - f(a) = f'(a)\delta a + a_1 \delta a;$$
$$f(x_2) - f(x_1) = f'(x_1)\delta x_1 + a_2 \delta x_1;$$
$$f(x_3) - f(x_2) = f'(x_2)\delta x_2 + a_3 \delta x_2;$$
$$\ldots\ldots\ldots\ldots\ldots\ldots\ldots\ldots\ldots\ldots\ldots\ldots$$
$$f(b) - f(x_{n-1}) = f'(x_{n-1})\delta x_{n-1} + a_n \delta x_{n-1}.$$

Add:
$$\therefore f(b) - f(a) = \sum_{x=a}^{x=b} f'(x)\delta x + R,$$
where $R = a_1 \delta a + a_2 \delta x_1 + \ldots + a_n \delta x_{n-1}.$

Let a' be the greatest, numerically, of the quantities $a_1, a_2, \ldots$; then, numerically,
$$R < a'(\delta a + \delta x_1 + \ldots + \delta x_{n-1}) \quad \text{or} \quad a'(b-a).$$
Since every a, and therefore a', has zero for limit, R will have zero for limit and the result follows.

Ex. 3. Find the limit for $n = \infty$ of
$$\frac{1}{n+1} + \frac{1}{n+2} + \frac{1}{n+3} + \ldots + \frac{1}{2n}.$$

We may write this sum
$$\frac{1}{1+\frac{1}{n}} \cdot \frac{1}{n} + \frac{1}{1+\frac{2}{n}} \cdot \frac{1}{n} + \frac{1}{1+\frac{3}{n}} \cdot \frac{1}{n} + \ldots + \frac{1}{1+\frac{n}{n}} \cdot \frac{1}{n}$$

or
$$\sum_{r=1}^{r=n} \frac{1}{\left(1+\frac{r}{n}\right)} \cdot \frac{1}{n}.$$

Consider the function $F(x) = 1/x$; in § 131, let each difference be $1/n$, let $a = 1$, $b = 2$, and the above sum will be the same as (1), § 131, if we suppose the values of $F(x)$ to be those at the *end* of each interval. Hence the required limit is

$$\int_1^2 \frac{dx}{x} = \Big[\log x\Big]_1^2 = \log 2 = \cdot 693.$$

§ 133. Approximations. The method of evaluating an integral by first finding the function of which the integrand is the derivative would fail if we could not find such a function. An important case in which that method can not be used is that in which the integrand is given only by its graph, as often happens in physical applications. Methods have therefore been devised for determining approximately the value of the integral when only a limited number of values of the integrand are known; it is assumed that the integrand may be treated as a continuous function, though if only a limited number of values of the integrand are known, the analytical expression for the function can not be given. The rules now to be stated can be applied even when the analytical form of the function is known, though in general more powerful methods are available in that case, in particular the method of expansion in series.

Let AB be divided into n *equal* parts, each part being equal to h, and suppose the $(n+1)$ ordinates at A, B and the points of division to be known; let these be $y_1, y_2, y_3, \ldots$. The calculation of the integral

$$\int_a^b F(x)dx \ldots\ldots\ldots\ldots\ldots\ldots\ldots (1)$$

is then equivalent to finding the area $ABDC$ (Fig. 76).

APPROXIMATIONS. 329

The most obvious method is to replace the graph by the inscribed polygon $CC_1C_2\ldots$. The area of the first trapezium is $\tfrac{1}{2}h(y_1+y_2)$, and this area may be assumed to differ but little from that of the corresponding strip of $ABDC$. Adding together all the trapeziums, we get, as an approximation to the area, and therefore to the integral (1)

$$A_1 = \tfrac{1}{2}h(y_1+y_2) + \tfrac{1}{2}h(y_2+y_3) + \ldots + \tfrac{1}{2}h(y_n+y_{n+1})$$
$$= \tfrac{1}{2}h\{y_1+y_{n+1}+2(y_2+y_3+\ldots+y_n)\}\ldots\ldots\ldots\ldots(2)$$

If the graph is, as in the figure, convex upwards throughout the value A_1 is in defect; if the graph is concave upwards, A_1 is in excess.

Through the ends of the even ordinates $y_2, y_4 \ldots$ let tangents be drawn and produced to meet the adjacent odd ordinates; if the number of ordinates is *odd*, $2n+1$ say, we shall get n trapeziums whose sum exceeds

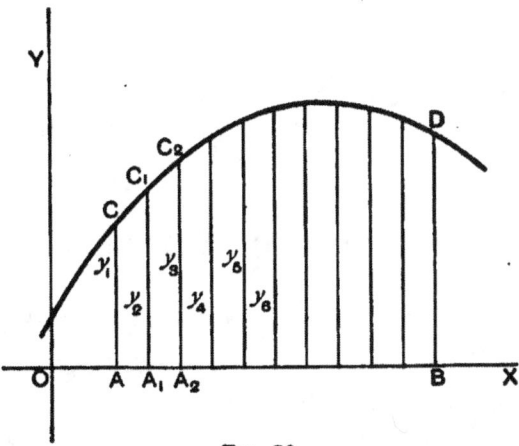

Fig. 76.

$ABDC$ in area when the graph is convex upwards throughout. The area of the first trapezium is $2hy_2$, of the second $2hy_4$, and so on. Hence we get another approximation

$$A_2 = 2h(y_2+y_4+\ldots+y_{2n})\ldots\ldots\ldots\ldots\ldots(3)$$

The value of the integral (1) always lies between A_1 and A_2 when there is no point of inflexion on the arc CD, and the difference $\pm(A_1-A_2)$ gives a measure of the error involved in either approximation. The formula (2) is usually referred to as the Trapezoidal Rule.

A formula that is in practice more accurate than (2) or (3) is got as follows: By § 72 we may write

$$F(x) = F(c) + (x-c)F'(c) + \tfrac{1}{2}(x-c)^2 F''(x_1).$$

If $x-c$ is small we may assume that $F'''(x_1)$ differs but little from $F'''(c)$; if $F(x)$ were of the second degree $F'''(x_1)$ would be simply $F'''(c)$. The equation

$$y = F(c) + (x-c)F'(c) + \tfrac{1}{2}(x-c)^2 F''(c) \dots \dots (4)$$

represents a parabola; we therefore replace a short length of the graph of $F(x)$ by this parabola.

Now consider the double strip AA_2C_2C; for convenience let $OA_1=c$, $OA=c-h$, $OA_2=c+h$; then using (4) as the value of $F(x)$ along the arc CC_1C_2 we find for the area of AA_2C_2C

$$\int_{c-h}^{c+h} F(x)\,dx = \int_{-h}^{h} F(x'+c)\,dx' = 2hF(c) + \tfrac{1}{3}h^3 F'''(c) \dots (5)$$

where, to integrate, we put $x = x' + c$. We can now express (5) in terms of h and y_1, y_2, y_3, assuming $F(x)$ to be given by (4). For $F(c) = y_2$ and

$$y_1 = F(c-h) = F(c) - hF'(c) + \tfrac{1}{2}h^2 F''(c),$$
$$y_3 = F(c+h) = F(c) + hF'(c) + \tfrac{1}{2}h^2 F''(c).$$

By addition

$$h^2 F''(c) = y_1 + y_3 - 2F(c) = y_1 + y_3 - 2y_2,$$

and (5) becomes $\quad \tfrac{1}{3}h(y_1 + 4y_2 + y_3) \dots \dots \dots (6)$

Suppose now $ABDC$ divided into an even number, $2n$, of strips by an odd number, $2n+1$, of equidistant ordinates. The formula (6) may be applied in succession to the n double strips; the sum of the n expressions is, the terms being rearranged,

$$A_2 = \tfrac{1}{3}h\{y_1 + y_{2n+1} + 2(y_3 + y_5 + \dots + y_{2n-1})$$
$$\qquad + 4(y_2 + y_4 + \dots + y_{2n})\} \dots \dots \dots (7)$$

Formula (7) is known as Simpson's Rule, which may be stated thus: *Let the area be divided into an even number of strips by equidistant ordinates; find* (i) *the sum of the extreme ordinates,* (ii) *twice the sum of the other odd ordinates,* (iii) *four times the sum of the even ordinates; add the three sums thus obtained and multiply this total sum by one-third of the common distance between the ordinates.*

Let $$u = y_1 + y_{2n+1}$$
$$v = y_2 + y_4 + \ldots + y_{2n}; \quad w = y_3 + y_5 + \ldots + y_{2n-1},$$
then in terms of h, u, v, w, we find
$$A_1 = \tfrac{1}{2}h(u+2v+2w); \quad A_2 = 2hv; \quad A_3 = \tfrac{1}{3}h(u+4v+2w),$$
and therefore
$$A_3 = \tfrac{2}{3}A_1 + \tfrac{1}{3}A_2 \ldots\ldots\ldots\ldots\ldots\ldots\ldots(8)$$

Suppose the graph convex upwards and the ordinates positive, so that $A_1 < \text{area } ABDC < A_2$; then
$$A_3 - A_1 = \tfrac{1}{3}(A_2 - A_1); \quad A_2 - A_3 = \tfrac{2}{3}(A_2 - A_1).$$
The error in the Simpson Rule is therefore less than
$$\tfrac{2}{3}(A_2 - A_1) \text{ or } \tfrac{1}{3}h(2v - 2w - u) \ldots\ldots\ldots\ldots(9)$$

Formula (8) shows that in Simpson's Rule greater weight is given to the inscribed than to the circumscribed polygon.

These methods of approximation apply of course to a definite integral, whether $F(x)$ be considered as the ordinate of a curve or not; for example, $F(x)$ might be a radius vector and x the vectorial angle in a curve given by its polar equation. The values of the function for equidifferent values of the argument then take the place of the ordinates $y_1, y_2, \ldots$. A very important practical case is that of the mensuration of solids; $y_1, y_2, \ldots$ are then the areas of equidistant sections. (See, for a good statement of Simpson's Rule for practical mensuration, Lodge's *Mensuration for Senior Students*: London, Longmans.)

Ex. Calculate $\int_1^2 \frac{dx}{x}$.

Let $2n+1 = 11$; $h = \cdot1$; $a = 1$; $b = 2$. An easy calculation gives
$$u = 1\cdot5; \quad v = 3\cdot459\,5394; \quad w = 2\cdot728\,1746.$$
$$A_1 = \cdot693\,771; \quad A_2 = \cdot691\,908; \quad A_3 = \cdot693\,150.$$

The exact value of the integral is $\log 2$, that is, $\cdot693\,147$. The value of $2(A_1 - A_2)/3$ is $\cdot001\,242$, while $A_3 - \log 2$ is $\cdot000\,003$. As a rule, the error in Simpson's formula is considerably less than that given by (9).

EXERCISES XXIX.

1. If in § 133 $F(x)$ is the area of a section of a surface made by a plane perpendicular to the x-axis, and if the ordinates $y_1, y_2, \ldots$ be replaced by the sections $S_1, S_2, \ldots$, the expressions (2), (3), (6), (7) give the volume intercepted between the surface and the corresponding planes. Thus (6) gives for the volume

$$V = \tfrac{1}{3}h(S_1 + 4S_2 + S_3), \quad \dots \dots \dots \dots \dots \dots \dots \dots \dots \dots \dots \dots \text{(i)}$$

where S_1, S_3 are the areas of the end sections, S_2 that of the mid-section, and $2h$ the distance between the end sections. The value (i) is exact when $F(x)$ is a quadratic function of x.

Apply the formula to obtain the results regarding volumes in § 127. Apply it also to the solid formed by the revolution of a parabola about its axis.

2. Show that the formula (i) holds for a *prismoid*.

A prismoid is a solid whose lower and upper bounding surfaces are polygons with the same number of sides and with corresponding sides parallel, and whose lateral bounding surfaces are trapeziums.

3. If d_1 is the head diameter, d_2 the bung diameter, and h the depth of a cask, show that when the curve of the cask is a parabola, the volume is

$$\frac{\pi}{12}h\left\{d_1^2 + 2d_2^2 - \frac{4}{10}(d_2 - d_1)^2\right\}.$$

When the upper and lower halves of the cask are equal frustums of a paraboloid of revolution, the greatest bases being joined in the middle of the cask, show that the volume is

$$\frac{\pi}{8}h(d_1^2 + d_2^2).$$

4. If $F(x) = A + B(x-c) + C(x-c)^2 + D(x-c)^3$, show that formula (6) of § 133 still holds.

5. If $F(x) = A + Bx + Cx^2 + Dx^3$ and if y_1, y_2, y_3, y_4 are the values of $F(x)$ when x has the values a, $a+h$, $a+2h$, $a+3h$ respectively, show that the area between the curve, the x-axis, and the ordinates y_1, y_4 is

$$\tfrac{3}{8}h(y_1 + 3y_2 + 3y_3 + y_4).$$

The formula is sometimes called *Simpson's Second Rule*. To prove it most simply, put $x = a + ht$; then $F(x)$ takes the form

$$\phi(t) = P + Qt + Rt^2 + St^3,$$

and y_1, y_2, y_3, y_4 are the values of $\phi(t)$ for t equal to 0, 1, 2, 3, and the area is

$$\int_a^{a+3h} F(x)\,dx = h\int_0^3 \phi(t)\,dt.$$

6. Show that

$$\int_0^{\frac{\pi}{2}} \log \sin x\,dx = -\int_0^{\frac{\pi}{2}} x \cot x\,dx,$$

and calculate the value of the integral by Simpson's rule. The exact value of the integral is $-\tfrac{1}{2}\pi \log 2$. For let the integral be u; then

$$u = \int_0^{\frac{\pi}{2}} \log \sin x\,dx = \int_0^{\frac{\pi}{2}} \log \cos x\,dx = \tfrac{1}{2}\int_0^{\frac{\pi}{2}} (\log \sin x + \log \cos x)\,dx,$$

MEAN VALUES. 333

so that
$$2u = \int_0^{\frac{\pi}{2}} \log(\tfrac{1}{2}\sin 2x)\,dx = \frac{\pi}{2}\log\tfrac{1}{2} + \int_0^{\frac{\pi}{2}} \log\sin 2x\,dx\,;$$

also $\int_0^{\frac{\pi}{2}}\log\sin 2x\,dx = \tfrac{1}{2}\int_0^{\pi}\log\sin z\,dz = \int_0^{\frac{\pi}{2}}\log\sin z\,dz = u,$

from which the result follows.

7. Show that $\int_0^{\frac{\pi}{2}} \log\tan x\,dx = 0.$ (No integration is necessary.)

8. Show that the limit when n is ∞ of
$$\sum_{r=0}^{r=n-1} \frac{1}{\sqrt{(n^2 - r^2)}}$$
is $\pi/2$.

9. Show that the limit when n is ∞ of
$$\sum_{r=0}^{r=n-1} \frac{n}{n^2 + r^2}$$
is $\pi/4$.

§ 134. Mean Values. The arithmetic mean of n quantities $y_1, y_2, \ldots, y_n$ is $(y_1 + y_2 + \ldots + y_n)/n$. Let $y_1, y_2, \ldots, y_n$ be the values of $F(x)$ for x equal to $a, a+h, \ldots, b-h$, the interval $b-a$ being divided into n parts each equal to h; the limit for $n = \infty$ of the arithmetic mean of $y_1, y_2, \ldots, y_n$ is called the *mean value of the function $F(x)$ over the range* $b-a$.

The mean value may be expressed as an integral; for
$$(y_1 + y_2 + \ldots + y_n)/n = (y_1 h + y_2 h + \ldots + y_n h)/(b-a). \ldots (1)$$
The numerator of the fraction on the right is
$$F(a)h + F(a+h)h + \ldots + F(b-h)h,$$
and the limit of it for $n = \infty$ (and therefore $h = 0$) is
$$\int_a^b F(x)\,dx\,;$$
and the Mean Value is
$$\frac{1}{b-a}\int_a^b F(x)\,dx. \ldots\ldots\ldots\ldots\ldots\ldots\ldots(2)$$

Ex. 1. The mean value of the ordinate of a semicircle of radius a is
$$\frac{1}{2a}\int_{-a}^{a}\sqrt{(a^2 - x^2)}\,dx = \frac{\pi}{4}a = \cdot 7854a.$$

In this case the *diameter* is divided into n equal parts. If, however, the *semi-circumference* is divided into n equal parts, so that the independent variable of the function is the arc $a\theta$ from one end of the

diameter to the point from which the ordinate is drawn, the mean value is, since the ordinate is $a \sin \theta$,
$$\frac{1}{\pi a}\int_0^\pi a \sin \theta \, a d\theta = \frac{2}{\pi}a = \cdot 6366a.$$

In speaking of mean values, therefore, it is essential that the independent variable should be clearly indicated.

Ex. 2. For the harmonic curve $y = a \sin x$, find (i) the mean ordinate, (ii) the square root of the mean of the square of the ordinate for the range from $x = 0$ to $x = \pi$.

(i) mean ord. $= \dfrac{1}{\pi}\displaystyle\int_0^\pi a \sin x \, dx = \dfrac{2}{\pi}a = \cdot 6366a.$

In case (ii) the function is y^2, and the mean value of y^2 is
$$\frac{1}{\pi}\int_0^\pi a^2 \sin^2 x \, dx = \tfrac{1}{2}a^2,$$
and the square root of this mean is $a/\sqrt{2}$ or $\cdot 7071a$.

In the theory of alternating currents the important mean is not (i), but (ii); the latter is sometimes called *the mean-square value of the ordinate*.

If the interval $b-a$ is divided into n sub-intervals $h_1, h_2, \ldots$, and if $y_1, y_2, \ldots$ are the values of $F(x)$ at *any* point of the intervals $h_1, h_2, \ldots$ respectively, the limit for n infinite (and *each* sub-interval $h_1, h_2, \ldots$ zero) of
$$(y_1 h_1 + y_2 h_2 + \ldots + y_n h_n)/(b-a)$$
is still given by (2). The integral (2) may be taken as the general definition of the mean value of $F(x)$.

§ 135. Double Integrals. Let $EFGH$ (Fig. 77) be a plane curve, and let $f(x, y)$ be a single-valued continuous function of x and y for all points within or on the curve. Let AH, BF, and CE, DG be the tangents parallel to the axes; we suppose that no straight line cuts the curve

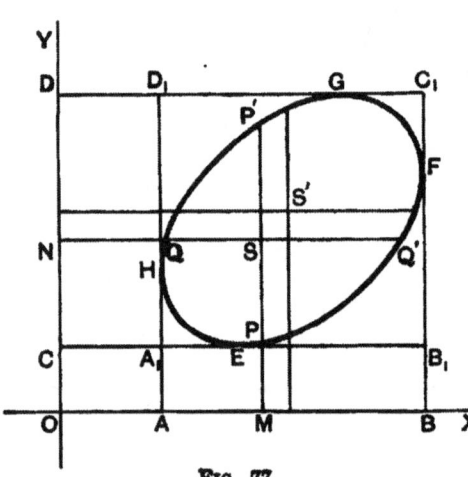

Fig. 77.

DOUBLE INTEGRALS.

in more than two points; any curve that does not satisfy this condition may be divided into partial areas, each of which satisfies it.

Let AB be divided into m and CD into n sub-intervals, and through the points of division let parallels be drawn to the axes. The area bounded by $EFGH$ will thus be divided into partial areas; these areas are rectangles, though near the boundary $EFGH$ the rectangles will contain points that lie outside the curve.

Let x_r, $x_r + \delta x_r$ be the abscissae of two consecutive points of division on AB and y_s, $y_s + \delta y_s$ the ordinates of two consecutive points of division on CD; and let S, S' be the points (x_r, y_s), $(x_r + \delta x_r, y_s + \delta y_s)$.

Multiply $\delta x_r \, \delta y_s$, the area of the rectangle SS', by $f(x_r, y_s)$, the value of $f(x, y)$ at S, and form the sum

$$\sum f(x_r, y_s) \, \delta x_r \delta y_s \quad \ldots\ldots\ldots\ldots\ldots\ldots(1)$$

for all points such as S within or on the boundary of $EFGH$.

Geometrically, $z = f(x, y)$ represents a surface; the typical term $f(x_r, y_s) \, \delta x_r \, \delta y_s$ of the sum (1) is the volume of a parallelepiped whose base is the rectangle $\delta x_r \, \delta y_s$, and height the z-coordinate $f(x_r, y_s)$ of the point in which the normal from S to the rectangle meets the surface; the sum (1) is therefore approximately equal to the volume of the solid bounded by the surface, the plane XOY and the cylinder formed by a straight line which moves round the boundary $EFGH$, remaining always perpendicular to the plane XOY. (Compare Figs. 48, 49.)

We wish to find the limit of (1) for m and n each increasing indefinitely, each element δx_r, δy_s, and therefore each area $\delta x_r \delta y_s$ at the same time diminishing indefinitely. Seeing that there are two sets of increments we may appropriately represent (1) as a double summation

$$\sum \sum f(x_r, y_s) \, \delta x_r \delta y_s \quad \ldots\ldots\ldots\ldots\ldots(2)$$

the one Σ referring to δy_s and the other to δx_r.

First, keep x_r and δx_r constant, that is, find the limit for $n = \infty$;

$$\underset{n=\infty}{\mathrm{L}} \sum f(x_r, y_s) \, \delta y_s = \int_{MP}^{MP'} f(x_r, y) \, dy \ldots\ldots\ldots\ldots(3)$$

by the definition of the integral of a function of one variable y. The integral (3) will contain x_r, MP, MP'; MP and MP' are functions of OM or x_r determined by the equation of the curve $EFGH$. Hence (3) is a function of x_r and may be denoted by $\phi(x_r)$.

Geometrically, $\phi(x_r)$ is the area of the curve of section of the solid defined above, made by a plane through PP' perpendicular to XOY; and $\phi(x_r)\,\delta x_r$ is, to the first order of infinitesimals, the volume of the slice of the solid of thickness δx_r.

Next find the limit for $m = \infty$. We get

$$\mathop{L}_{m=\infty} \sum \delta x_r\, \phi(x_r) = \int_{OA}^{OB} \phi(x)\, dx \ldots\ldots\ldots\ldots\ldots(4)$$

Hence, finally, the limit of (1) is expressed by (4) and that limit is the volume of the solid already mentioned.

Since $\phi(x)$ is itself an integral the expression (4) is a *double integral* and this double integral is denoted by the symbol

$$\int_{OA}^{OB} dx \int_{MP}^{MP'} f(x, y)\, dy \ldots\ldots\ldots\ldots\ldots(5)$$

The mode of establishing (4) shows that (5) which is merely the fuller symbol for (4) means, *integrate $f(x, y)$ as to y from $y = MP$ to $y = MP'$, treating x as a constant during this integration; then integrate the result as to x from $x = OA$ to $x = OB$*.

We might also find the limit of (1) by making first m, then n infinite; the result would be stated in the form

$$\int_{OC}^{OD} dy \int_{NQ}^{NQ'} f(x, y)\, dx \ldots\ldots\ldots\ldots\ldots(6)$$

In (6) the integration is first carried out as to x, treating y as a constant during this operation; then the result is integrated as to y. Clearly the double integrals (5) and (6) are equal since they represent the same volume.

When the area is the rectangle $A_1B_1C_1D_1$ the limits MP, MP' of y in (5) are constant and equal to OC, OD respectively, and the limits NQ, NQ' of x in (6) are also constant and equal to OA, OB respectively. Hence, writing a, b, a', b' for OA, OB, OC, OD,

$$\int_a^b dx \int_{a'}^{b'} f(x, y)\, dy = \int_{a'}^{b'} dy \int_a^b f(x, y)\, dx, \ldots\ldots\ldots\ldots(7)$$

NOTATIONS. POLAR ELEMENTS. 337

that is, *when the limits are all constants* the limits of y and the limits of x are the same in whatever order the integrations are effected. *When the limits are not all constants* the limits of y (or of x) in (5) are not the same as the limits of y (or of x) in the equal integral (6).

The geometrical representation of the meaning of the double integral is very helpful. Other illustrations might of course be given; for example, $f(x, y)$ might be taken as representing the (variable) density of a surface distribution of matter over the area $EFGH$, and then the integral would give the total mass.

§ 136. Notations for Double Integrals. Polar Elements.

The forms (5), (6) indicate clearly the order in which the integrations are to be carried out. Other notations are, however, in use which, though not so expressive, are often convenient. Thus the form

$$\iint f(x, y) dx\, dy \quad \ldots\ldots\ldots\ldots\ldots\ldots\ldots\ldots (8)$$

with the addition "the integration being extended over the area $EFGH$" (or a similar phrase) is used as an equivalent either of (5) or of (6).

Instead of (5) we also find

$$\int_{OA}^{OB} \int_{MP}^{MP'} f(x, y) dx\, dy$$

with the convention that the first integration is made with respect to the variable on the *right*, namely y, between the limits named on the symbol $\int$ that stands next the integrand, that is, MP, MP'. But there is not complete agreement as to this convention.

Again, we might suppose the area enclosed by $EFGH$ to be divided into partial areas other than rectangles. If δS be the type of such an area, and if (x, y) be the coordinates of any point within or on the boundary of δS, the sum

$$\sum f(x, y) \delta S \quad \ldots\ldots\ldots\ldots\ldots\ldots\ldots\ldots (1')$$

would replace (1). Geometrically (1') would give approximately the volume of the solid defined in last article; the limit obtained by supposing the number of the areas δS to increase indefinitely, while the size of each area δS at the same time diminishes indefinitely, would give the volume of the solid and would be denoted by

$$\int f(x, y)\,dS, \dots\dots\dots\dots\dots\dots\dots\dots(9)$$

the integration being extended over the area $EFGH$.

It is easy to see, by Th. II., § 87, that (x, y) may be *any point* within or on the boundary of δS, so far as the *limit* (9) of the sum (1') is concerned; it is of great importance to bear this remark in mind, as the principle involved is constantly used (see for instance ex. 3, § 137).

If we take for δS the area bounded by two circular arcs of radii r and $r+\delta r$, and two radii making angles θ and $\theta+\delta\theta$ with the initial line, where r, θ are polar coordinates,

$$\delta S = \tfrac{1}{2}(r+\delta r)^2\delta\theta - \tfrac{1}{2}r^2\delta\theta = r\,\delta r\,\delta\theta + \tfrac{1}{2}(\delta r)^2\delta\theta,$$

so that $dS = r\,dr\,d\theta$.

If $f(x, y)$ becomes $F(r, \theta)$ when $r\cos\theta$, $r\sin\theta$ are put for x, y, we should get instead of (9), or the equivalents (5), (6),

$$\iint F(r, \theta)\,r\,dr\,d\theta, \dots\dots\dots\dots\dots(10)$$

the integration being extended over the area $EFGH$. In integrating with respect to θ, r is to be kept constant; the θ-integration would therefore give, in the geometrical representation, the area of a cylindrical section of the solid. Before evaluating an integral such as (10), the curve $EFGH$ should be drawn, and care has to be taken so that there may be no omission or inclusion of areas other than those belonging to the curve. The same remark applies to most integrations.

The reader will have little difficulty in extending these results to triple integrals,

$$\iiint f(x, y, z)\,dx\,dy\,dz \quad \text{or} \quad \int f(x, y, z)\,dv \dots\dots(11)$$

$dx\,dy\,dz$ or dv may be taken as an element of volume, and $f(x, y, z)$ might, for example, denote the density at (x, y, z). Integration with respect to z, keeping x, y constant, would give the mass of the column standing on the base $dx\,dy$; then the y-integration, keeping x constant, would give the mass of a slice of thickness dx perpendicular to the x-axis, and lastly the x-integration would give the total mass.

TRIPLE INTEGRALS. EXAMPLES.

Ex. 1. Find the volume of the tetrahedron bounded by the coordinate planes and the plane

$$x/a + y/b + z/c = 1$$

where a, b, c are positive.

The curve $EFGH$ is in this case the triangle OAB; the equation of AB is

$$y = b(1 - x/a)$$

and $MP' = b(1 - x/a)$,

while MP in § 135 is here zero.

$$SR = f(x, y) = z$$
$$= c(1 - x/a - y/b).$$

Hence using (5), the volume is

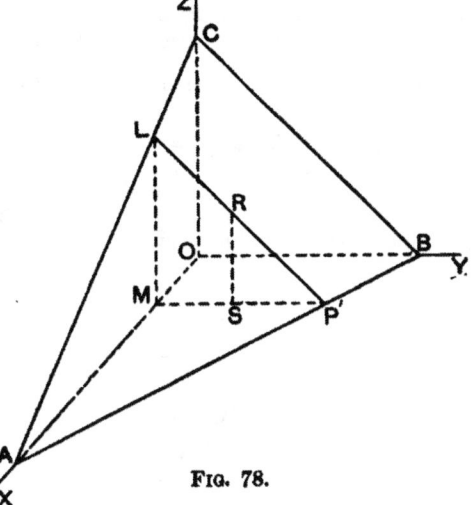

Fig. 78.

$$\int_0^a dx \int_0^{MP'} z\,dy = c \int_0^a dx \left[\left(1 - \frac{x}{a}\right) y - \frac{1}{2} \frac{y^2}{b} \right]_0^{MP'}$$
$$= \frac{1}{2} bc \int_0^a \left(1 - \frac{x}{a}\right)^2 dx = \frac{1}{6} abc.$$

Obviously $\frac{1}{2} bc (1 - x/a)^2$ is the area of the triangle LMP'.

Ex. 2. Find the value of $\int x^2 dv$ taken throughout the volume of the ellipsoid $x^2/a^2 + y^2/b^2 + z^2/c^2 = 1$.

$$\int x^2 dv = \int x^2 dx \int\int dy\,dz = \int_{-a}^{a} x^2 dx \left[\pi bc \left(1 - \frac{x^2}{a^2}\right) \right],$$

since, in integrating as to y and z, x is constant and $\int\int dy\,dz$ is the area of the section perpendicular to the x-axis. Integrate now as to x; the result is $4\pi a^3 bc/15$.

The *mean value* of the function x^2 throughout the volume of the ellipsoid is the above value divided by the volume, that is, $a^2/5$.

In general, *the mean value of a function $f(x, y)$ over an area $EFGH$* (Fig. 77) is the value of the integral (5) or (6) divided by the area; and a similar definition holds for the mean value throughout a volume. If, in the example, x^2 is the density at (x, y, z) of a mass occupying the volume of the ellipsoid, then $a^2/5$ is the *mean density* of the mass,

Ex. 3. If $f(x, y)$ is the product of a function $\phi(x)$ of x alone, and of a function $\psi(y)$ of y alone, it follows at once from § 135 that the integral of the product $\phi(x)\psi(y)$ taken over *the rectangle* $A_1B_1C_1D_1$ (Fig. 77) is equal to the product of the integrals

$$\int_a^b \phi(x)dx \text{ and } \int_{a'}^{b'} \psi(y)dy.$$

Now let $\phi(x) = e^{-x^2}$, $\psi(y) = e^{-y^2}$, and

$$U = \int_0^a e^{-x^2}dx = \int_0^a e^{-y^2}dy. \quad \ldots\ldots\ldots\ldots\ldots\text{(i)}$$

It follows that U^2, the product of these two integrals, is equal to the integral

$$\iint e^{-(x^2+y^2)}dxdy, \ \ldots\ldots\text{(ii)}$$

taken over the square $OABC$ of side $OA = a$ (Fig. 79).

Draw the arcs ADC, EBF from the centre O with the radii $OA = a$, $OB = a\sqrt{2}$. The integral (ii) is greater than the integral of the same function over the area $OADC$ and less than that over the area $OEBF$. These two integrals can be found by changing to polar coordinates; $dxdy$ is replaced by $rdrd\theta$ and $e^{-(x^2+y^2)}$ by e^{-r^2} and (ii) becomes, for the area $OADC$,

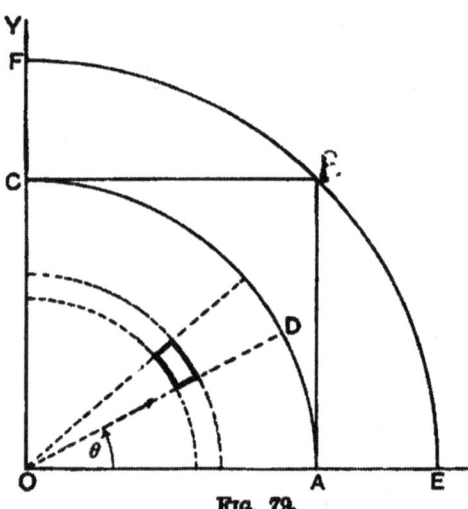

Fig. 79.

$$\iint e^{-r^2}rdrd\theta = \int_0^a e^{-r^2}rdr\int_0^{\frac{\pi}{2}} d\theta = \frac{\pi}{4}(1 - e^{-a^2})$$

since the integral of $e^{-r^2}r$ is $-\tfrac{1}{2}e^{-r^2}$. When the area is $OEBF$, the integral is $\frac{\pi}{4}(1 - e^{-2a^2})$.

U^2 lies between these two values; but when a tends to infinity both values tend to $\pi/4$; and therefore also U^2 tends to $\pi/4$, and U to $\tfrac{1}{2}\sqrt{\pi}$. Hence

$$\int_0^\infty e^{-x^2}dx = \underset{a=\infty}{\mathrm{L}} \int_0^a e^{-x^2}dx = \tfrac{1}{2}\sqrt{\pi}.$$

This example is a particular case of an integral of great importance (see Ex. XXX. 21), and the transformation is worthy of careful attention.

CENTRES OF INERTIA.

§ 137. Centres of Inertia. It is shown in works on mechanics that the coordinates $(\bar{x}, \bar{y}, \bar{z})$ of the centre of inertia of a set of n particles of masses $m_1, m_2, \ldots, m_n$ situated at the points $(x_1, y_1, z_1), (x_2, y_2, z_2), \ldots, (x_n, y_n, z_n)$ are given by the equations

$$\bar{x} = \frac{m_1 x_1 + m_2 x_2 + \ldots + m_n x_n}{m_1 + m_2 + \ldots + m_n} = \frac{\Sigma mx}{\Sigma m} \quad \ldots\ldots\ldots\ldots(1)$$

with similar expressions for $\bar{y}, \bar{z}$.

For a continuous distribution of matter the volume density ρ at the point (x, y, z) is the limit for $\delta v = 0$ of $\delta m/\delta v$ where δm is the mass of the volume δv surrounding the point; hence to the first order of infinitesimals

$$\delta m = \rho \delta v.$$

When the mass is supposed concentrated in a surface or in a line we have in a similar way $\delta m = \sigma \delta S$, $\delta m = \lambda \delta s$ where σ and λ are the surface density and the line density at a point and δS and δs elements of area and of length including the point.

A continuous mass may be supposed to be divided into n elements δm; if (x, y, z) are the coordinates of any point in the element δm then the coordinates of the centre of inertia of the mass are given by

$$\bar{x} = \operatorname*{L}_{n=\infty} \frac{\Sigma x \delta m}{\Sigma \delta m} = \frac{\int x dm}{\int dm} \quad \ldots\ldots\ldots\ldots\ldots(2)$$

with similar expressions for $\bar{y}, \bar{z}$. The integrations in (2) are to be extended through the total mass.

For volume, surface, and line distributions equations (2) take the forms

$$\frac{\int x \rho dv}{\int \rho dv}, \quad \frac{\int x \sigma dS}{\int \sigma dS}, \quad \frac{\int x \lambda ds}{\int \lambda ds} \quad \ldots\ldots\ldots\ldots(3)$$

respectively; the denominator is in each case the total mass.

The terms *mass-centre*, and *centroid* are sometimes used as equivalent to *centre of inertia*. The centroid of a volume, area, or line is the centre of inertia of a mass of uniform density occupying the volume, area, or line.

Ex. 1. A circular arc of uniform density, BAC (Fig. 80).

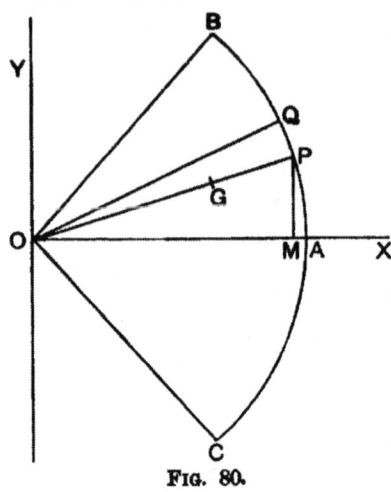

FIG. 80.

Let O be the centre of the circle; $OA = a$, $\angle COB = 2a$. Let OA bisect the angle COB, and take OA as the x-axis.

By symmetry, $\bar{y} = 0$.

Let $\angle XOP = \theta$;
$$OM = x = a\cos\theta$$
$$\text{arc } AP = s = a\theta;$$
$$ds = a\,d\theta.$$

The linear density λ is constant; hence the total mass is $2\lambda a a$.

Also
$$\int x\lambda\,ds = \lambda \int_{-a}^{a} a\cos\theta\, a\, d\theta = 2\lambda a^2 \sin a$$

and therefore
$$\bar{x} = 2\lambda a^2 \sin a / 2\lambda a a = a\sin a / a.$$

Ex. 2. A plane lamina of uniform density σ, in the form of a quadrant of an ellipse, $OAPB$ (Fig. 81).

In a case like this the use of a double integral may be avoided; for we may take a narrow strip $NPP'N'$, of breadth dy, parallel to OA as the element of mass. The centre of inertia of the strip is at its middle point, and therefore the moment of the strip about OB is

$\tfrac{1}{2}x \cdot \sigma x\,dy$ or $\tfrac{1}{2}\sigma x^2 dy$.

The total mass is
$$\pi\sigma ab/4,$$

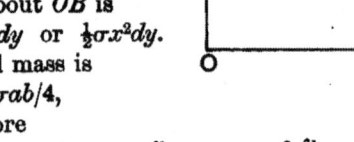

FIG. 81.

and therefore
$$\frac{\pi\sigma ab}{4}\bar{x} = \tfrac{1}{2}\sigma\int_0^b x^2 dy = \tfrac{1}{2}\sigma\frac{a^2}{b^2}\int_0^b (b^2 - y^2)dy = \tfrac{1}{3}\sigma a^2 b,$$

and therefore $\bar{x} = 4a/3\pi$.

In the same way, $\bar{y} = 4b/3\pi$, taking the strip $M'MPP'$ as element.

When the density is not uniform, the above method usually fails. Suppose $\sigma = kxy$ (k constant); the total mass M is
$$M = \iint kxy\,dx\,dy = k\int_0^a x\,dx\int_0^{ON} y\,dy = \tfrac{1}{2}k\int_0^a x \cdot ON^2 dx,$$

and since $ON^2 = y^2 = b^2(1 - x^2/a^2)$, we readily find $M = \tfrac{1}{8}ka^2b^2$.

Again,
$$M\bar{x} = \iint x \cdot kxy\,dx\,dy = k\int_0^a x^2\,dx \int_0^{ON} y\,dy = \tfrac{1}{15}ka^3b^2,$$
and therefore $\bar{x} = \tfrac{8}{15}a$. Similarly, $\bar{y} = \tfrac{8}{15}b$.

Ex. 3. A circular sector of uniform density.

Take the notation of ex. 1. We may take as element the small sector OPQ. The centre of inertia of OPQ may be taken as the point $(\tfrac{2}{3}a,\ \theta)$, and the moment about OY of the element is
$$\tfrac{2}{3}a\cos\theta \cdot \sigma\tfrac{1}{2}a^2 d\theta = \tfrac{1}{3}\sigma a^3 \cos\theta\,d\theta.$$
The total mass M is $\sigma a^2 \alpha$; $\bar{y}=0$ from symmetry; and $\bar{x}$ is given by
$$M\bar{x} = \tfrac{1}{3}\sigma a^3 \int_{-\alpha}^{\alpha} \cos\theta\,d\theta = \tfrac{2}{3}\sigma a^3 \sin\alpha,$$
so that
$$\bar{x} = \tfrac{2}{3}a\,\frac{\sin\alpha}{\alpha}.$$

When the density is not uniform, double integration will usually be required.

The centre of inertia of OPQ was taken *on OP*; as has been indicated several times, it does not matter for the *limit* whether we take the point as $(\tfrac{2}{3}a,\ \theta)$ or $(\tfrac{2}{3}a,\ \theta')$ where θ' is a value between θ and $\theta+\delta\theta$. Simplifications of this kind are of constant occurrence; a similar one was made in ex. 2 when the centre of inertia of $NPP'N'$ was taken at the middle point of NP.

Ex. 4. A uniform right circular cone.

From symmetry the centre of inertia is in the axis. Take a section perpendicular to the axis at a distance x from the vertex; if h is the height and A the area of the base of the cone, this section is $x^2 A/h^2$. We may take as element of mass the slice between this section and the parallel section at distance $x+\delta x$ from the vertex.

The total mass M is $\tfrac{1}{3}\rho h A$, and
$$M\bar{x} = \int_0^h x \cdot \rho\,\frac{x^2}{h^2}A\,dx = \tfrac{1}{4}\rho h^2 A,$$
and
$$\bar{x} = \tfrac{3}{4}h.$$

If the density is not uniform, double or triple integrals may be required since the element of mass could not be chosen as above. If, however, the density is a function of x alone, the method still applies; for example, if $\rho = kx$, the student may prove
$$M = \frac{A}{h^2}\int_0^h \rho x^2\,dx = \tfrac{1}{4}kAh^2;\quad \bar{x} = \tfrac{4}{5}h.$$

§ 138. Moments of Inertia. If $r_1, r_2, \ldots, r_n$ are the distances from an axis OR of n particles of masses $m_1, m_2, \ldots, m_n$ the sum
$$m_1 r_1^2 + m_2 r_2^2 + \ldots + m_n r_n^2,$$
or, in the notation of a sum, Σmr^2, is defined in works on

mechanics as the *moment of inertia* of the set of particles about the axis OR.

When the masses form a continuous body the summation is replaced by integration, as in the case of centres of inertia.

If the total mass of the system is M and if k is chosen so that
$$Mk^2 = \Sigma mr^2 \quad \text{or} \quad Mk^2 = \int r^2 dm,$$
the quantity k is called *the radius of gyration* of the system about the axis. The moment of inertia is often denoted by I.

The work of finding moments of inertia is simplified by the following theorems:

(i) If OX, OY, OZ are three rectangular axes, and if I_x, I_y, I_z are the moments of inertia about OX, OY, OZ respectively of a plane lamina lying in the plane XOY, then
$$I_z = I_x + I_y.$$

(ii) If I_R is the moment of inertia about any axis OR, I_G the moment about a parallel axis through the centre of inertia G and a the distance between these axes,
$$I_R = I_G + Ma^2,$$
where M is the total mass of the system.

The proofs of these theorems are very simple and may be left to the reader; they may be found in any work on mechanics.

Ex. 1. A thin straight rod of uniform density about an axis through one end perpendicular to the rod.

Let x be the distance from the axis of a point on the rod, λ the linear density, l the length of the rod. For the element of mass we may take $\lambda \delta x$; hence
$$I = \int_0^l x^2 \cdot \lambda dx = \tfrac{1}{3}\lambda l^3 = \tfrac{1}{3}Ml^2,$$
where $M = \lambda l$ is the mass of the rod. The radius of gyration k is therefore $l/\sqrt{3}$.

The moment about an axis through the *mid point* of the rod and perpendicular to it is $Ml^2/12$ as may be proved directly or by using theorem (ii.).

Ex. 2. A uniform rectangular lamina about an axis through its centre parallel to one side.

Let a, b be the lengths of the two sides and let the axis be parallel to the side a. Divide the lamina into thin strips parallel to the side b and let δm be the mass of a strip, M being the mass of the lamina.

EXAMPLES ON MOMENTS.

By ex. 1 the moment of inertia of δm about the axis is $\delta m\, b^2/12$ and therefore the moment of inertia of the whole rectangle is
$$\tfrac{1}{12}Mb^2.$$

In the same way the moment about an axis through the centre parallel to the side b is $Ma^2/12$ and therefore by Th. (i.) the moment about an axis through the centre perpendicular to its plane is
$$M(a^2+b^2)/12.$$

It is easy to deduce the moment of a uniform rectangular parallelepiped, whose edges are a, b, c, about an axis through its centre parallel to an edge. For, let the axis be parallel to the edge c and divide the solid into thin slices of mass δm by planes perpendicular to the edge c. The moment of one slice is, by the result just found,
$$\delta m(a^2+b^2)/12,$$
and therefore the moment of the solid is $M(a^2+b^2)/12$, M being the mass of the solid.

Ex. 3. *A uniform elliptic lamina about the major axis.*

Divide the lamina into strips of mass δm by lines parallel to the minor axis; then the moment of the strip is by ex. 1 $\delta m\,.\,(2y)^2/12$ or $\delta m\,.\,y^2/3$ where y is the ordinate of the strip.

If ρ is the density, δm is $2\rho y\,\delta x$; hence
$$I=\int \tfrac{1}{3}y^2 dm = \frac{2\rho}{3}\int_{-a}^{a} y^3 dx = \frac{4\rho}{3}\cdot\frac{b^3}{a^3}\int_0^a (a^2-x^2)^{\frac{3}{2}} dx.$$

But
$$\int_0^a (a^2-x^2)^{\frac{3}{2}} dx = a^4 \int_0^{\frac{\pi}{2}} \cos^4\theta\, d\theta = \frac{3\pi}{16}a^4,$$
by the substitution $x=a\sin\theta$. The total mass M is $\pi\rho ab$. Hence
$$I=\tfrac{1}{4}\pi\rho ab^3 = \tfrac{1}{4}Mb^2.$$

The moment about the minor axis is $Ma^2/4$ and about an axis through the centre perpendicular to its plane it is $M(a^2+b^2)/4$.

For a circle of radius a we get, by putting b equal to a, for the moment about a diameter $Ma^2/4$ and about an axis through the centre perpendicular to its plane $Ma^2/2$.

The last value may be found most simply by dividing the circle into thin concentric strips; then Theorem (i.) shows that the moment about a diameter, since all such moments are equal by symmetry, is half that about the axis perpendicular to the lamina.

Ex. 4. *A uniform ellipsoid about the axis OA.*

Divide the ellipsoid into thin slices by planes perpendicular to OA; the mass δm of a slice may be taken as
$$\pi \rho bc(1-x^2/a^2)\delta x,$$
and by the last example the moment of δm about OA is
$$\tfrac{1}{4}\delta m(a_1^2+b_1^2)$$
where $2a_1$, $2b_1$ are the axes of the section. But
$$a_1^2 = b^2(1-x^2/a^2),\ \ b_1^2 = c^2(1-x^2/a^2).$$

Substituting and integrating from $-a$ to a we get
$$I = \tfrac{1}{4}\pi\rho bc(b^2+c^2)\int_{-a}^{a}\left(1-\frac{x^2}{a^2}\right)^2 dx = M\frac{b^2+c^2}{5},$$
where $M = 4\pi\rho abc/3 =$ mass of ellipsoid.

The moments about the other axes may be found by symmetry.

§ 139. Polar Element of Volume. The expression for the element of volume dV in terms of the spherical polar coordinates r, θ, ϕ (§ 89a) of a point P is often required in physical applications.

Let a denote the plane through P and the axis OZ. *First*, keeping r and ϕ constant, let θ become $\theta + \delta\theta$; P thus describes an arc, PQ say, in the plane a and arc $PQ = r\delta\theta$. *Next*, let the plane a turn about OZ as an axis through the small angle $\delta\phi$, the coordinates r, θ being kept constant; P will describe an arc, PR say, equal to $r\sin\theta\delta\phi$ and, if $\delta\theta$ is kept constant, the arc PQ will describe an area, δS say, equal approximately to arc $PQ \times$ arc PR, that is, equal to $r^2\sin\theta\,\delta\theta\,\delta\phi$. *Finally*, keeping θ, ϕ, $\delta\theta$, $\delta\phi$ constant, let r become $r + \delta r$; the area δS will describe an element of volume δV equal approximately to $\delta S \times \delta r$, that is, equal to $r^2\sin\theta\,\delta r\,\delta\theta\,\delta\phi$. The limit of δV is the polar element of volume, so that

$$dV = r^2 \sin\theta\, dr\, d\theta\, d\phi.$$

The element of the surface of a sphere of radius r is

$$dS = r^2 \sin\theta\, d\theta\, d\phi.$$

If $r = f(\theta)$ is the polar equation of a curve lying in the plane ZOX, the initial line being OZ, we find by integrating dV from $\phi = 0$ to $\phi = 2\pi$ and then from $r = 0$ to $r = f(\theta)$ that the polar element of volume of a surface of revolution about the initial line is $\tfrac{2}{3}\pi r^3 \sin\theta\, d\theta$, where r now means $f(\theta)$.

Let P be the point (x, y, z) on a surface and let the rectangular parallelepiped standing on the rectangle $\delta x\, \delta y$ as base cut out of the surface the element of area $\delta\sigma$, and out of the tangent plane at P the element $\delta\sigma'$. If the normal to the tangent plane at P make with OZ the angle γ we have

$$\delta\sigma' \cos\gamma = \delta x\, \delta y, \quad \delta\sigma' = \delta x\, \delta y \sec\gamma.$$

POLAR ELEMENT OF VOLUME.

If we assume that the limit of $\delta\sigma'/\delta\sigma$ is unity we find
$$d\sigma = dx\, dy \sec \gamma.$$
The direction cosines of the normal can be found (§ 91) when the equation of the surface is known and thus $d\sigma$ can be expressed in terms of x, y, $\partial z/\partial x$, $\partial z/\partial y$.

Definitions. The terms *Line Integral, Surface Integral* occur so often that it may be worth while to define them, though we cannot find room for a consideration of their special properties and relations.

Let F denote a quantity such as a velocity or a force having direction as well as magnitude, and at the point P on a curve APQ let the angle between the direction of F and the tangent at P be ϵ. If s is the arc measured from a fixed point on the curve up to P, the integral
$$\int F \cos \epsilon\, ds \quad\quad\quad\quad\quad\quad\quad\quad\quad\quad\quad\quad (1)$$
taken from the value of s at a point A up to the value of s at another point B is called *the line integral of F along the curve AB.*

For example, in § 95, the work W is the line integral of the force F along the curve AP.

If X, Y, Z are the components of F parallel to the axes, the integral (1) may also (§ 95 (3)), be written
$$\int \left(X \frac{dx}{ds} + Y \frac{dy}{ds} + Z \frac{dz}{ds} \right) ds. \quad\quad\quad\quad\quad\quad (2)$$

Again, let δS be an element of surface, P a point on δS, and ϵ the angle between the normal to the surface at P and the direction of F. The integral
$$\int F \cos \epsilon\, dS \quad\quad\quad\quad\quad\quad\quad\quad\quad\quad\quad\quad (3)$$
taken over any portion of the surface is called *the surface integral of F over that portion.*

Thus if F is the electric intensity at P, then $F \cos \epsilon$ is the normal component N of the intensity, and the integral (3) is the surface integral of normal electric intensity over that portion of the surface.

EXERCISES XXX.

1. Find the mean value of y^2 over the range from 0 to π when
   (i) $y = a_1 \sin x + a_2 \sin 2x + \ldots + a_n \sin nx$.
   (ii) $y = b_1 \cos x + b_2 \cos 2x + \ldots + b_n \cos nx$.

2. If $\quad y = a_1 \sin x + b_1 \cos x + a_2 \sin 2x + b_2 \cos 2x$
   and $\quad z = A_1 \sin x + B_1 \cos x + A_2 \sin 2x + B_2 \cos 2x$,
find the mean value of the product yz over the range from 0 to 2π.

3. A particle falls freely from rest; show that the mean velocity with respect to the time is half the final but that the mean velocity with respect to the distance is two-thirds of the final velocity.

4. A particle of mass m describes a simple harmonic motion of amplitude a and period T; show that the mean kinetic energy is half the maximum kinetic energy.

5. Show that in a homogeneous liquid under gravity the mean pressure-intensity over a plane area immersed in the liquid is equal to the pressure-intensity at the centroid of the area.

6. If the density at a distance r from the centre of the earth is given by $\rho = (\rho_0 \sin kr)/kr$ where k is a constant, show that the mean density is

$$3\rho_0(\sin kR - kR \cos kR)/k^3 R^3$$

where R is the earth's radius. (Lamb's *Calculus*.)

Take as element of volume, δv, the shell between two spherical surfaces of radii r and $r + \delta r$; then $\delta v = 4\pi r^2 \delta r$ and $\delta m = \rho \delta v$. The total mass is found by integrating ρdv from $r = 0$ to $r = R$.

7. Find the centroid in the following cases:
(i) The area between the arc of a parabola, the axis, and the ordinate at the point (h, k).
(ii) The segment cut off from a parabola by the straight line joining the vertex and the point (h, k).
(iii) The *segment BAC* (Fig. 80).
(iv) The spherical sector formed by the revolution of the circular sector OAB (Fig. 80) about OA.
(v) The cardioid $r = a(1 + \cos \theta)$.

8. If the density of a hemisphere vary as the distance from the bounding plane, show that the distance from that plane of the centre of inertia is $8R/15$ where R is the radius.

9. Prove the *Theorems of Pappus*, namely,
(i) If an arc of a plane curve revolve about an axis in its plane which does not intersect it, the surface generated is equal to the length of the arc multiplied by the length of the path of the centroid of the arc.
(ii) If a plane area revolve about an axis in its plane which does not intersect it the volume generated is equal to the area multiplied by the length of the path of the centroid of the area.

Taking the x-axis as the axis of revolution, the theorems follow at once from the equations

$$\bar{y}\int ds = \int y\, ds\,;\quad \bar{y}\int dS = \int y\, dS = \tfrac{1}{2}\int(y_1^2 - y_2^2)dx$$

by multiplying by 2π. y_1, y_2 are the ordinates of the points in which a line perpendicular to the x-axis cuts the curve.

Deduce from (ii) the formula for the polar element of volume of a surface of revolution (§ 139.)

EXERCISES XXX. GAMMA-FUNCTION.

10. Find the moments of inertia in the following cases, the density being uniform :
 (i) A circular lamina of mass M and radius a about a tangent.
 (ii) A sphere of mass M and radius a about a tangent line.
 (iii) A triangular lamina of mass M and height h about its base.
 (iv) A right cone of mass M, height h, and radius of base a, (a) about its axis, (β) about an axis through its vertex parallel to the base.

11. A rectangle $ABCD$ revolves about an axis in its plane parallel to AB, and not intersecting the rectangle ; if a, b are the distances of AB, CD from the axis, show that the radius of gyration of the solid generated is given by
$$k^2 = \tfrac{1}{2}(a^2+b^2).$$

12. The moment of inertia of the anchor-ring (Exer. XXVIII., ex. 5) about its axis is $M(c^2+\tfrac{3}{4}a^2)$, the density of the solid being supposed uniform.

13. If $r^2 = x^2+y^2+z^2$, show that the mean value of r^2 throughout the volume of the ellipsoid $x^2/a^2+y^2/b^2+z^2/c^2=1$ is $(a^2+b^2+c^2)/5$.

14. The volume of the wedge intercepted between the cylinder
$$x^2+y^2=2ax$$
and the planes $\qquad z=x\tan a, \quad z=x\tan \beta$
is $\qquad \pi(\tan\beta - \tan a)a^3.$

15. If $n > 0$, the integral
$$\int_0^\infty e^{-x} x^{n-1} dx$$
has a definite value ; the integral is a function of n, usually called the *Gamma-function*, and denoted by $\Gamma(n)$. Show, by integrating by parts, that
$$\Gamma(n) = (n-1)\,\Gamma(n-1), \quad\ldots\ldots\ldots\ldots\ldots\ldots\ldots\text{(i)}$$
and that when n is an integer, $\Gamma(n) = (n-1)!$, $\Gamma(1)=1$.

If n is not an integer, let p be the integer next below n so that $(n-p)$ is a proper fraction, then (i) shows that
$$\Gamma(n) = (n-1)(n-2)\ldots(n-p)\,\Gamma(n-p) \ldots\ldots\ldots\ldots\ldots\text{(ii)}$$

16. Prove
 (i) $\Gamma(\tfrac{1}{2}) = \sqrt{\pi}$;
 (ii) $\Gamma(m+\tfrac{1}{2}) = \dfrac{2m-1}{2}\cdot\dfrac{2m-3}{2}\cdot\ldots\tfrac{1}{2}(\sqrt{\pi})\ldots$ (m integral).

Equation (i) follows from § 136, ex. 3, by putting $x=\sqrt{z}$ for
$$\frac{\sqrt{\pi}}{2} = \int_0^\infty e^{-x^2} dx = \tfrac{1}{2}\int_0^\infty e^{-z} z^{\tfrac{1}{2}-1} dz = \tfrac{1}{2}\Gamma(\tfrac{1}{2}).$$
Then (ii) follows from ex. 15 (ii).

17. Prove $\qquad \displaystyle\int_0^\infty e^{-ax} x^{n-1} dx = \dfrac{\Gamma(n)}{a^n}$ (a positive).

18. By the given substitutions, prove other formulae for $\Gamma(n)$:

$$x = r^2; \quad \Gamma(n) = 2\int_0^\infty e^{-r^2} r^{2n-1} dr \dots \dots \dots \dots \dots \dots \dots \text{(i)}$$

$$e^{-x} = z; \quad \Gamma(n) = \int_0^1 \left(\log \frac{1}{z}\right)^{n-1} dz \dots \dots \dots \dots \dots \dots \text{(ii)}$$

19. When m and n are both positive, the integral

$$\int_0^1 x^{m-1}(1-x)^{n-1} dx$$

has a definite value; it is a function of m and n usually called the *Beta-function*, and denoted by $B(m, n)$. Show that

$$B(m, n) = B(n, m).$$

20. By the given substitutions, prove other formulae for $B(m, n)$:

$$x = \cos^2\theta; \quad B(m, n) = 2\int_0^{\frac{\pi}{2}} \cos^{2m-1}\theta \sin^{2n-1}\theta \, d\theta \dots \dots \dots \dots \text{(i)}$$

$$x = \frac{1}{1+y}; \quad B(m, n) = \int_0^\infty \frac{y^{n-1} dy}{(1+y)^{m+n}} \dots \dots \dots \dots \dots \dots \text{(ii)}$$

21. Using form (i) of ex. 18, write

$$\Gamma(m) = 2\int_0^\infty e^{-x^2} x^{2m-1} dx, \quad \Gamma(n) = 2\int_0^\infty e^{-y^2} y^{2n-1} dy;$$

and then show, as in § 136, ex. 3, that

$$\Gamma(m) \times \Gamma(n) = 4\int_0^\infty e^{-r^2} r^{2(m+n)-1} dr \int_0^{\frac{\pi}{2}} \cos^{2m-1}\theta \sin^{2n-1}\theta \, d\theta;$$

and therefore, by exs. 18 (i) and 20 (i),

$$\Gamma(m)\Gamma(n) = \Gamma(m+n) B(m, n).$$

Thus the Beta-function can be expressed in terms of the Gamma-function.

22. Let $2m - 1 = p$, $2n - 1 = q$; then, from exs. 20 (i) and 21,

$$\int_0^{\frac{\pi}{2}} \cos^p\theta \sin^q\theta \, d\theta = \frac{B(m, n)}{2} = \frac{\Gamma\left(\frac{p+1}{2}\right)\Gamma\left(\frac{q+1}{2}\right)}{2\Gamma\left(\frac{p+q+2}{2}\right)},$$

where, since $m > 0$, $n > 0$, we have $\frac{1}{2}(p+1)$ and $\frac{1}{2}(q+1) > 0$, or p and q each greater than -1.

The student may test that this result includes the rule given in § 119. Tables of $\log \Gamma(n)$ for $1 \leq n \leq 2$ have been calculated (no wider range for n is necessary by ex. 15, (ii)), and many integrals can be expressed in terms of Gamma-functions.

23. Find the potential V at a point Q of a mass M distributed uniformly (density σ) over the surface of a sphere of radius a.

Take O, the centre of the sphere, as origin and OQ as z-axis; let dS be the surface element at P, and denote PQ by R and OQ by c. Then

$$V = \int \frac{\sigma dS}{PQ}; \quad dS = a^2 \sin\theta\, d\theta\, d\phi; \quad R^2 = a^2 + c^2 - 2ac\cos\theta.$$

The limits for ϕ are 0 and 2π, for θ they are 0 and π; in integrating as to ϕ, the other variable θ, and therefore in this case also PQ or R (which is a function of θ and not of ϕ) is to be kept constant. Hence

$$V = \sigma a^2 \int_0^\pi \frac{\sin\theta\, d\theta}{R} \int_0^{2\pi} d\phi = 2\pi\sigma a^2 \int_0^\pi \frac{\sin\theta\, d\theta}{R}.$$

Now change the variable from θ to R; we have $R\,dR = ac\sin\theta\, d\theta$. When $\theta = 0$, $R = \pm(a-c)$; R is a positive number, so that if Q is outside the sphere $R = c - a$, and if Q is inside $R = a - c$. When $\theta = \pi$, $R = a + c$ in both cases. Hence

$$V = 2\pi\sigma a^2 \left[\frac{R}{ac}\right] = \frac{4\pi\sigma a^2}{c} \quad (Q \text{ outside}) \dots\dots\dots\dots\dots(i)$$

$$= 4\pi\sigma a \quad (Q \text{ inside}). \dots\dots\dots\dots\dots(ii)$$

Thus $V = M/c$ when Q is outside, but $V = M/a = $ constant when Q is inside the sphere.

24. Same problem as in Ex. 23 for a solid sphere (density $\rho = $ constant).

Take as element of mass the shell bounded by radii r and $r + dr$, and use the results of Ex. 23, putting $\rho\, dr$ for σ, and r for a.

If Q is outside, the result (i) gives

$$V = \int_0^a \frac{4\pi\rho r^2 dr}{c} = \frac{4\pi\rho}{3} \frac{a^3}{c}. \dots\dots\dots\dots\dots(iii)$$

If Q is inside, V consists of two parts, V_1, V_2. V_1 is the potential due to the sphere of radius c, and by the result just found

$$V_1 = \frac{4\pi\rho}{3} \cdot \frac{c^3}{c} = \frac{4\pi\rho}{3} c^2.$$

V_2 is that due to the shell of radii c and a; by the result (ii) of ex. 23

$$V_2 = \int_c^a 4\pi\rho r\, dr = 2\pi\rho(a^2 - c^2).$$

Hence $\quad V = V_1 + V_2 = 2\pi\rho(a^2 - \tfrac{1}{3}c^2). \dots\dots\dots\dots\dots(iv)$

When $c = a$ the values given by (iii) and (iv) coincide.

CHAPTER XVI.

CURVATURE. ENVELOPES.

§ 140. Curvature. Let P and Q be two points on a plane curve, ϕ and $\phi + \delta\phi$ the angles which the tangents at P and Q make with the x-axis, s the arc measured from some fixed point on the curve up to P and δs the arc PQ. $\delta\phi$ will be the angle between the tangents at P and Q (Fig. 82, p. 354).

Definitions. (i) The angle $\delta\phi$ is called *the total curvature* of the arc PQ; (ii) the quotient $\delta\phi/\delta s$ is called *the average curvature* of the arc PQ; (iii) the limit of $\delta\phi/\delta s$ when Q approaches P as its limiting position, that is, $d\phi/ds$, is called *the curvature* of the curve at P.

For a circle of radius R, $\delta s = R\delta\phi$ and therefore

$$\frac{\delta\phi}{\delta s} = \frac{1}{R}; \quad \frac{d\phi}{ds} = \frac{1}{R} \quad \ldots\ldots\ldots\ldots\ldots\ldots\ldots(1)$$

that is, the average curvature of any arc of a circle is equal to the curvature at any point of that circle. In other words, a circle is a curve of constant curvature and its curvature is equal to the reciprocal of its radius.

Curvature is thus a magnitude of dimension -1 in length.

The curvature may be expressed in terms of the first and second derivatives of the ordinate at the point. For, since

$$\tan\phi = \frac{dy}{dx}, \quad \cos\phi = \frac{dx}{ds},$$

we get, by differentiating the first equation with respect to s,

$$\frac{d.\tan\phi}{d\phi} \frac{d\phi}{ds} = \frac{d}{dx}\left(\frac{dy}{dx}\right)\frac{dx}{ds},$$

CURVATURE OF PLANE CURVES.

that is, $\sec^2\phi \dfrac{d\phi}{ds} = \dfrac{d^2y}{dx^2} \div \sec\phi$;

and therefore, $\dfrac{d\phi}{ds} = \dfrac{d^2y}{dx^2} \div \sec^3\phi$ ....(2)

Hence, since $\sec^2\phi = 1 + (dy/dx)^2$ we find

$$\dfrac{d\phi}{ds} = \dfrac{d^2y}{dx^2} \div \left\{1 + \left(\dfrac{dy}{dx}\right)^2\right\}^{\frac{3}{2}} \quad \ldots\ldots\ldots\ldots(A)$$

Formula (A) may be considered fundamental.

COR. When the gradient dy/dx is so small that for all values of x within the range considered its square may be neglected, the curvature is approximately d^2y/dx^2. This approximate value is often used in Mechanics; for example, in the theory of the bending of beams.

Ex. 1. The parabola $y^2 = 4ax$.

$$\dfrac{dy}{dx} = \dfrac{2a}{y}; \quad \dfrac{d^2y}{dx^2} = -\dfrac{2a}{y^2}\dfrac{dy}{dx} = -\dfrac{4a^2}{y^3};$$

$$\dfrac{d\phi}{ds} = -\dfrac{4a^2}{y^3} \div \left\{1 + \dfrac{4a^2}{y^2}\right\}^{\frac{3}{2}} = \dfrac{-4a^2}{(y^2 + 4a^2)^{\frac{3}{2}}}.$$

If the normal at $P(x, y)$ meet the axis at G,

$$PG^2 = y^2 + 4a^2 \quad \text{and} \quad \dfrac{d\phi}{ds} = -\dfrac{4a^2}{PG^3}.$$

The meaning of the negative sign will be referred to in § 141.

Ex. 2. The ellipse $x^2/a^2 + y^2/b^2 = 1$.

$$\dfrac{dy}{dx} = -\dfrac{b^2x}{a^2y}; \quad \dfrac{d^2y}{dx^2} = -\dfrac{b^2}{a^2y} + \dfrac{b^2x}{a^2y^2}\left(-\dfrac{b^2x}{a^2y}\right) = \dfrac{-b^4}{a^2y^3},$$

since $b^2x^2 + a^2y^2 = a^2b^2$ by the equation of the ellipse. Hence

$$\dfrac{d\phi}{ds} = -\dfrac{a^4b^4}{(b^4x^2 + a^4y^2)^{\frac{3}{2}}}.$$

If p is the perpendicular from the centre on the tangent at (x, y),

$$p = \dfrac{a^2b^2}{(b^4x^2 + a^4y^2)^{\frac{1}{2}}} \quad \text{and} \quad \dfrac{d\phi}{ds} = -\dfrac{p^3}{a^2b^2}.$$

If PG is the normal at $P(x, y)$,

$$PG^2 = \dfrac{b^4x^2 + a^4y^2}{a^4} \quad \text{and} \quad \dfrac{d\phi}{ds} = -\dfrac{b^4}{a^2PG^3}.$$

A similar result holds for the hyperbola. Thus the curvature of a conic section varies inversely as the cube of the normal.

G.C.

§ 141. Circle, Radius, and Centre of Curvature. Let the normals at P and Q (Fig. 82) intersect at C'; when Q tends to P as its limiting position, C' will tend to a point C on the normal at P as its limiting position such that PC is equal to $ds/d\phi$.

For $\angle PC'Q = \delta\phi$, and

$$\frac{PC'}{\sin PQC'} = \frac{\text{chord } PQ}{\sin PC'Q} = \frac{\text{chord } PQ}{\text{arc } PQ} \cdot \frac{\delta s}{\delta\phi} \cdot \frac{\delta\phi}{\sin \delta\phi}.$$

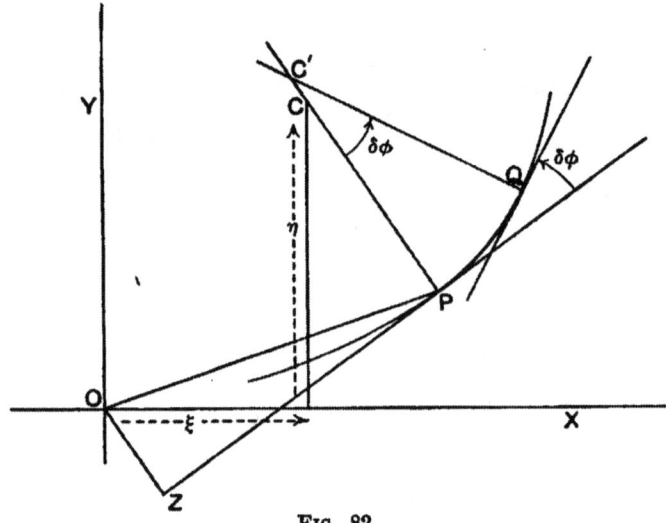

Fig. 82.

The limit of PQC' is 90° and the limits of the three fractions last written are 1, $ds/d\phi$, 1 respectively; hence the limit of PC' is $ds/d\phi$, as was to be proved.

The circle with centre C and radius PC has therefore the same tangent and the same curvature as the curve has at P. This circle is called *the circle of curvature*, its radius PC or $ds/d\phi$ *the radius of curvature*, and its centre C *the centre of curvature* at P. If any line through P meet the circle again at R, PR is called *a chord of curvature*.

If (x, y) are the coordinates of P, (ξ, η) those of C and ρ the radius of curvature PC or $ds/d\phi$ it is easy to prove

$$\xi = x - \rho \sin \phi, \quad \eta = y + \rho \cos \phi \dots\dots\dots\dots(1)$$

CIRCLE, RADIUS, CENTRE OF CURVATURE.

We will generally use ρ for the radius of curvature; the curvature will then be denoted by $1/\rho$.

If d^2y/dx^2 is zero at P then $1/\rho$ is zero by (A) and ρ or PC is infinite. Thus at a point of inflexion on a curve ρ is infinite.

We take Fig. 82 as the standard diagram. If we adhere to the convention that ϕ is always acute (§ 21) then dx/ds and $\sec\phi$ will be always positive and the root in (A) will have the positive sign. ρ and $1/\rho$ will therefore be positive or negative according as d^2y/dx^2 is positive or negative, that is according as the curve is concave upwards or convex upwards near P. Of course other conventions may be used but a little care, especially if a figure is drawn, will usually settle the question of sign. In many cases it is the numerical value alone that is important.

The limiting position C of the point C' is sometimes called the point of intersection of two *consecutive* normals. Of course there is no one normal that is the consecutive of another, but the phraseology is briefer than that used in the statement at the beginning of this article and is therefore sometimes useful.

It should be noticed that when the arc PQ is an infinitesimal of the first order the difference between PC' and QC' is of a higher order since the limit of $(QC' - PC')/\delta s$ is zero; for

$$QC' - PC' = QC'(1 - \cos\delta\phi) - PQ \cos QPC'.$$

§ 142. Other Formulae for the Curvature. Formula (A) is not very convenient unless the equation of a curve is in the form $y = f(x)$ or unless, as in the examples worked in § 140, the values of the derivatives can be easily calculated. We will therefore give one or two other formulae; the question of the *sign* of ρ usually needs special consideration.

(i) *Equation of form* $x = f(t)$, $y = F(t)$. The variable t need not, of course, represent *time* but we will for brevity use the fluxional notation.

Substitute in (A) the values of Dy, D^2y in terms of $\dot{x}, \ddot{x}, \dot{y}, \ddot{y}$, as given in § 98; we find

$$\frac{1}{\rho} = (\dot{x}\ddot{y} - \dot{y}\ddot{x})/(\dot{x}^2 + \dot{y}^2)^{\frac{3}{2}} \dots\dots\dots\dots\dots(B)$$

Since $D_x^2 y = (\dot{x}\ddot{y} - \dot{y}\ddot{x})/\dot{x}^3$, we can determine the sign of ρ when necessary in accordance with the convention of § 141.

(ii) *Polar Equations.* In (A) substitute the values of Dy, D^2y in terms of $D_\theta r$, $D_\theta^2 r$ and we get

$$\frac{1}{\rho} = \left\{ r^2 + 2\left(\frac{dr}{d\theta}\right)^2 - r\frac{d^2r}{d\theta^2} \right\} \div \left\{ r^2 + \left(\frac{dr}{d\theta}\right)^2 \right\}^{\frac{3}{2}} \quad \ldots\ldots(C)$$

Formula (C) is cumbrous. It is often simpler to find what is called *the p, r equation*, that is the relation between the perpendicular OZ from the origin O on the tangent at P (Fig. 82) and the radius OP (see ex. 2), and then to apply a formula we will now deduce.

In Fig. 82 we have, $OZ = p$, $OP = r$,

$$OC'^2 = OP^2 + PC'^2 - 2OP \cdot PC' \cos OPC'$$
$$= r^2 + PC'^2 - 2p \cdot PC'$$

since $p = OP \cos OPC' = r \sin \psi$ where ψ is, as usual, the angle between the tangent and the radius vector.

If $OQ = r + \delta r$ and if $p + \delta p =$ perpendicular from O on tangent at Q we find in the same way

$$OC'^2 = (r + \delta r)^2 + QC'^2 - 2(p + \delta p)QC'.$$

Equating the two values of OC'^2 we get

$$2r\delta r + (\delta r)^2 + (QC' - PC')(QC' + PC' - 2p) - 2QC'\delta p = 0.$$

But $(QC' - PC')$ and $(\delta r)^2$ are of order higher than the first, and therefore

$$\rho = \mathrm{L}\, QC' = \frac{r\,dr}{dp}. \quad \ldots\ldots\ldots\ldots\ldots\ldots\ldots\ldots\ldots(D)$$

Formula (D) may also be proved thus: since $p = r \sin \psi$ and $\phi = \theta + \psi$ we have (see § 88)

$$\frac{dp}{dr} = \sin \psi + r \cos \psi \frac{d\psi}{dr} = r\frac{d\theta}{ds} + r\frac{d\psi}{ds} = r\frac{d\phi}{ds},$$

and therefore $\qquad ds/d\phi = r\,dr/dp.$

An inspection of figures will show that when the curve is *concave* towards the origin (as in an ellipse with the centre as origin) p and r increase or decrease together, and therefore dr/dp and ρ are positive; when the curve is convex to the origin ρ is negative.

VARIOUS FORMULAE FOR CURVATURE.

We can now deduce (C) from (D); for (§ 88)
$$\tan \psi = r d\theta/dr$$
and
$$\frac{1}{p^2} = \frac{1}{r^2 \sin^2 \psi} = \frac{1}{r^2} + \left(\frac{1}{r^2}\frac{dr}{d\theta}\right)^2 \quad\ldots\ldots\ldots\ldots(i)$$

and by differentiating with respect to r we can find dp/dr. We will work out a slightly different formula that is of use in dynamics; namely, putting $r = 1/u$ we will find ρ in terms of u and θ.

Now
$$\frac{dr}{d\theta} = \frac{dr}{du}\frac{du}{d\theta} = -\frac{1}{u^2}\frac{du}{d\theta},$$

and therefore (i) becomes
$$\frac{1}{p^2} = u^2 + \left(\frac{du}{d\theta}\right)^2 \quad\ldots\ldots\ldots\ldots\ldots(ii)$$

Hence, differentiating with respect to u,
$$-\frac{2}{p^3}\frac{dp}{du} = 2u + 2\left(\frac{du}{d\theta}\right)\frac{d^2u}{d\theta^2}\frac{d\theta}{du}$$
$$= 2u + 2\frac{d^2u}{d\theta^2} \quad\ldots\ldots\ldots\ldots\ldots(iii)$$

But
$$\frac{dp}{du} = \frac{dp}{dr}\cdot\frac{-1}{u^2}; \quad \frac{1}{\rho} = \frac{1}{r}\frac{dp}{dr} = -u^3\frac{dp}{du},$$

and now by substitution in (iii), using (ii), we get
$$\frac{1}{\rho} = u^3\left(u + \frac{d^2u}{d\theta^2}\right) \div \left\{u^2 + \left(\frac{du}{d\theta}\right)^2\right\}^{\frac{3}{2}}$$
$$= \left(u + \frac{d^2u}{d\theta^2}\right) \div \left\{1 + \left(\frac{1}{u}\frac{du}{d\theta}\right)^2\right\}^{\frac{3}{2}}. \quad\ldots\ldots\ldots(E)$$

The root being taken positive, ρ will be positive or negative according as the arc is concave or convex to the origin.

(iii) *Intrinsic Equation.* Let s denote the arc of a curve measured from a fixed point on it up to the point P, and ϕ the angle which the tangent at P makes with a fixed tangent; the equation which expresses the relation between s and ϕ is called the *intrinsic equation* of the curve. This equation does not depend on any lines of reference outside the curve, such as the ordinary rectangular axes; hence the name.

When the intrinsic equation is given ρ is found at once by differentiation. In elementary work, however, the intrinsic equation is of comparatively small importance; it has usually to be deduced by integration from the ordinary equation, one of the coordinate axes being taken as the fixed tangent. The angle ϕ must not in this case be restricted to acute angles.

Ex. 1. $x^{\frac{2}{3}}+y^{\frac{2}{3}}=a^{\frac{2}{3}}$.

Let $x=a\cos^3 t$, $y=a\sin^3 t$, and use formula (B).

$$\dot{x}=-3a\cos^2 t\sin t;\qquad \ddot{x}=3a\cos t(2\sin^2 t-\cos^2 t);$$
$$\dot{y}=3a\sin^2 t\cos t;\qquad \ddot{y}=3a\sin t(2\cos^2 t-\sin^2 t);$$
$$\dot{x}^2+\dot{y}^2=9a^2\sin^2 t\cos^2 t;\quad \dot{x}\ddot{y}-\dot{y}\ddot{x}=-9a^2\sin^2 t\cos^2 t;$$
$$\rho=-3a\sin t\cos t=-3(axy)^{\frac{1}{3}}.$$

In this case $D_x^2 y = 1/3a\sin t\cos^4 t$ and ρ, if determined by the convention of § 141, will be $+3a\sin t\cos t$.

Ex. 2. $r^m = a^m \cos m\theta$.

Form the p, r equation and use formula (D).

$$\tan\psi = \frac{rd\theta}{dr} = -\cot m\theta = \tan\left(m\theta+\frac{\pi}{2}\right).$$

We will take $\psi = m\theta + \pi/2$; then

$$p = r\sin\psi = r\cos m\theta = r^{m+1}/a^m,$$

and therefore
$$\rho = \frac{rdr}{dp} = \frac{a^m}{(m+1)r^{m-1}}.$$

By giving different values to m we get several well-known equations. See Exercises XXXI. 10.

Ex. 3. Find the centre of curvature and the locus of the centre of curvature of an ellipse.

It is easy to show, with the notation of § 140, ex. 2, that

$$\sin\phi = -px/a^2,\quad \cos\phi = py/b^2,\quad \rho = -a^2 b^2/p^3,$$
$$\xi = x - \rho\sin\phi = x(1-b^2/p^2);\quad \eta = y(1-a^2/p^2).$$

Let θ be the eccentric angle of $P(x, y)$ and these values become

$$a\xi = (a^2-b^2)\cos^3\theta;\quad b\eta = -(a^2-b^2)\sin^3\theta.$$

To find the locus of the centre of curvature eliminate θ; thus
$$(a\xi)^{\frac{2}{3}}+(b\eta)^{\frac{2}{3}}=(a^2-b^2)^{\frac{2}{3}},$$
or, taking now x and y as current coordinates,
$$(ax)^{\frac{2}{3}}+(by)^{\frac{2}{3}}=(a^2-b^2)^{\frac{2}{3}}.$$
The curve is shown in Fig. 83, § 143.

Ex. 4. Show that the normal acceleration at a point P on a curve is v^2/ρ where v is the tangential velocity and $1/\rho$ the curvature at P.

At Q (Fig. 82) let the tangential velocity be $v+\delta v$; the components in the direction PC' of the velocity at P and at Q are 0 and $(v+\delta v)\sin\delta\phi$ respectively. Hence the normal acceleration at P is
$$\mathop{L}_{\delta t=0}\frac{(v+\delta v)\sin\delta\phi}{\delta t}=v\frac{d\phi}{dt}=v\frac{d\phi}{ds}\frac{ds}{dt}=v\cdot\frac{1}{\rho}\cdot v$$
as was to be proved.

EXERCISES XXXI.

1. The equation of any conic may be put in the form $y^2=2Ax+Bx^2$, where the x-axis is the focal axis and $2A$ is the latus rectum. If the normal at P meet the x-axis in G and if a is the angle between PG and the focal distance SP prove that
$$\rho=-PG^3/A^2=-PG/\cos^2 a.$$
Note that the projection of PG on SP is equal to the semi-latus rectum.

2. From the value of ρ in terms of a (ex. 1) prove the following construction for the centre of curvature K of any conic: Draw GH perpendicular to PG to meet SP at H, then draw HK perpendicular to HP to meet PG at K; K will be the centre of curvature.

3. For the rectangular hyperbola $xy=c^2$ show that
$$\rho=(x^2+y^2)^{\frac{3}{2}}/2c^2.$$

4. C is the centre of an ellipse, CD is a semi-diameter parallel to the tangent at P and θ is the eccentric angle of P; show that, numerically,
$$\rho=(a^2\sin^2\theta+b^2\cos^2\theta)^{\frac{3}{2}}/ab=CD^3/ab.$$
It may be shown that the eccentric angle of D is $\theta+\tfrac{1}{2}\pi$ or $\theta-\tfrac{1}{2}\pi$. CP, CD are called *conjugate* semi-diameters since, as may be readily proved, each diameter bisects all chords parallel to the other.

5. If r is the central radius of a point P on an ellipse, and p the perpendicular from the centre on the tangent at P, prove
$$a^2+b^2-r^2=a^2b^2/p^2;\quad \rho=a^2b^2/p^3.$$
For a hyperbola prove, with similar notation,
$$r^2-a^2+b^2=a^2b^2/p^2;\quad \rho=-a^2b^2/p^3.$$

6. For the curve $a^2y=x^3$ show that $\rho=(a^4+9x^4)^{3/2}/6a^4x$ and for the curve $ay^2=x^3$, $\rho=x^{1/2}(4a+9x)^{3/2}/6a$.

7. At the origin on the curve
$$ay = bx^2 + 2cxy + gy^2 + u_3 + u_4 + \ldots + u_n,$$
where u_n is of the nth degree and homogeneous in x and y, show that $Dy=0$, $D^2y=2b/a$, $\rho=a/2b$.

8. At the origin on the curve
$$y = 2x + 3x^2 - 2xy + y^2,$$
the radius of curvature is $5\sqrt{5}/6$.

9. Prove that the radius of curvature of the catenary
$$y = \frac{a}{2}\left(e^{\frac{x}{a}} + e^{-\frac{x}{a}}\right)$$
is y^2/a, and that of the catenary of uniform strength
$$y = c \log \sec(x/c)$$
is $c \sec(x/c)$.

10. Verify the general results given in ex. 2 § 142 for the particular cases :
  (i) Lemniscate $r^2=a^2\cos 2\theta$; $r^3=a^2p$; $\rho=a^2/3r$.
  (ii) Equilateral hyperbola $r^2\cos 2\theta=a^2$; $pr=a^2$; $\rho=r^3/a^2$.
  (iii) Parabola $r(1+\cos\theta)=2a$; $ar=p^2$; $\rho=2r^{3/2}/a^{1/2}$.
  (iv) Cardioid $r=a(1+\cos\theta)$; $r^3=2ap^2$; $\rho=4ap/3r$.

For the parabola $m=-1/2$; for the cardioid $m=1/2$, and $2a$ takes the place of a.

11. Show that the chord of curvature through the origin is $2p\,dr/dp$; for the curve $r^m=a^m\cos m\theta$, this chord is $2r/(m+1)$.

12. Show that for the equiangular spiral $r=ae^{\theta\cot\alpha}$ the radius of curvature is $r\cosec\alpha$; show also that the radius of curvature subtends a right angle at the origin.

13. If ψ is the angle between the focal radius of a conic and the tangent at P and α the angle between the focal radius and the normal, show by formula (E) that
$$\rho = l/\sin^3\psi = l/\cos^3\alpha,$$
the equation of the conic being $lu=1+e\cos\theta$.

Show also that if r and r' are the focal distances
$$rr'\cos^2\alpha = b^2 = al,$$
and that
$$\rho = \frac{(rr')^{3/2}}{ab} = \frac{rr'}{a\cos\alpha}.$$

14. If accents denote differentiation as to the arc s, show by differentiating the equations $\cos\phi = x'$, $\sin\phi = y'$ that, (ξ, η) being the centre of curvature,

$$1/\rho = -x''/y' = y''/x' \; ; \; 1/\rho^2 = (x'')^2 + (y'')^2,$$

and
$$\xi = x + \rho^2 x'', \quad \eta = y + \rho^2 y''.$$

15. Show from formula (E) that the condition for a point of inflexion is

$$u + \frac{d^2 u}{d\theta^2} = 0.$$

16. The circle $(x-a)^2 + (y-\beta)^2 = R^2$ and the curve $y = f(x)$ intersect at the point $P(a, b)$. If at P the values of Dy and $D^2 y$ are the same for the circle and the curve show that the circle is the circle of curvature at P.

The circle and the curve have the same tangent at P because P lies on both circle and curve, and the gradient of the circle at P is equal to that of the curve at P. Again, differentiate the equation of the circle twice and after differentiation put a, b (or $f(a)$), $f'(a)$, $f''(a)$, for x, y, Dy, $D^2 y$ respectively; we get

$$(a-a)^2 + (b-\beta)^2 = R^2 \ldots\ldots(\text{i}); \qquad (a-a) + (b-\beta)f'(a) = 0 \ldots\ldots(\text{ii});$$
$$1 + \{f'(a)\}^2 + (b-\beta)f''(a) = 0 \ldots\ldots\ldots\ldots\ldots\ldots(\text{iii})$$

From (ii) and (iii) we find

$$b - \beta = -[1 + \{f'(a)\}^2] \div f''(a); \quad a - a = f'(a)[1 + \{f'(a)\}^2] \div f''(a),$$

and therefore by substituting these values in (i)

$$R = [1 + \{f'(a)\}^2]^{\frac{3}{2}} \div f''(a)\ldots\ldots\ldots\ldots\ldots\ldots(\text{iv}).$$

But R as given by (iv) is the radius of curvature at P and (a, β) is the centre of curvature at P.

DEFINITION. Two curves $y = F(x)$, $y = f(x)$ which intersect at the point $P(a, b)$ are said to have *contact of the nth order* with each other at P if $F'(a) = f'(a)$, $F''(a) = f''(a)$, …… $F^{(n)}(a) = f^{(n)}(a)$ but $F^{(n+1)}(a)$ not equal to $f^{(n+1)}(a)$.

The circle of curvature has thus in general contact of the second order with the curve.

From Taylor's Theorem (§ 152) it will be seen that when the curves have contact of the nth order at (a, b) the difference $F(x) - f(x)$ between corresponding ordinates near (a, b) is an infinitesimal of order $n+1$ when $x-a$ is principal infinitesimal; for

$$F(x) - f(x) = \frac{(x-a)^{n+1}}{(n+1)!} \{F^{(n+1)}(a) - f^{(n+1)}(a) + R\},$$

where R is zero when $x = a$.

§ 143. Evolute. Involute. Parallel Curves.

DEFINITION. The locus of the centre of curvature of a given curve is called *the evolute* of that curve.

362   AN ELEMENTARY TREATISE ON THE CALCULUS.

The coordinates (ξ, η) of the centre of curvature C corresponding to the point $P(x, y)$ are given by
$$\xi = x - \rho \sin \phi, \quad \eta = y + \rho \cos \phi \quad \ldots \ldots \ldots \ldots (1)$$
The four quantities x, y, ϕ, ρ can all be expressed in terms of one quantity, for example x or s or t; the elimination of that quantity between the equations (1) will give a relation between ξ and η which will be the equation of the evolute.

The evolute of the ellipse is (§ 142, ex. 3) given by
$$(ax)^{\frac{2}{3}} + (by)^{\frac{2}{3}} = (a^2 - b^2)^{\frac{2}{3}},$$
and is shown in Fig. 83.

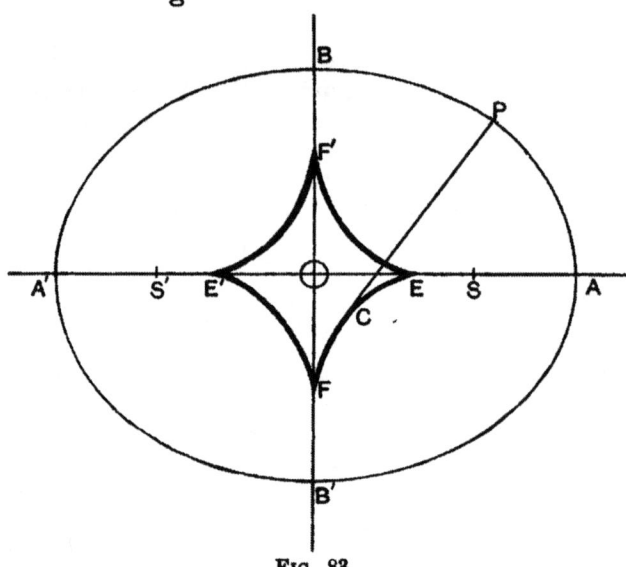

FIG. 83.

E, E', F, F' are the centres of curvature corresponding to the vertices A, A', B, B'; and
$$EA = A'E' = b^2/a, \quad FB = B'F' = a^2/b.$$

It is obvious how the radius of curvature may be utilised for graphing the curve.

The following are important properties of the evolute:

(i) The normal at P to the given curve is the tangent at C to the evolute. (ii) The length of an arc of the evolute is equal to the difference between the radii of curvature of

EVOLUTE AND INVOLUTE.

the given curve at the points corresponding to the ends of the arc.

(i) In equations (1) take s, an arc of the given curve, as independent variable; then

$$\frac{d\xi}{ds} = \frac{dx}{ds} - \rho \cos \phi \frac{d\phi}{ds} - \sin \phi \frac{d\rho}{ds} = -\sin \phi \frac{d\rho}{ds} \ldots \ldots (2)$$

since $dx/ds = \cos \phi$, $1/\rho = d\phi/ds$.

Similarly
$$\frac{d\eta}{ds} = \cos \phi \frac{d\rho}{ds} \ldots \ldots \ldots \ldots \ldots (3)$$

Therefore
$$\frac{d\eta}{d\xi} = \frac{d\eta}{ds} \bigg/ \frac{d\xi}{ds} = -\cot \phi.$$

Now the centre of curvature C (ξ, η) lies on the normal at P, and the gradient of the evolute at C is $d\eta/d\xi$, that is, $-\cot \phi$, which is the gradient of the normal to the given curve at P. Hence the normal PC coincides with the tangent to the evolute.

(ii) Let $d\sigma$ be the differential of an arc of the evolute; by (2) and (3)
$$d\xi = -\sin \phi \, d\rho, \quad d\eta = \cos \phi \, d\rho,$$
so that
$$d\sigma = \pm \sqrt{(d\xi^2 + d\eta^2)} = \pm d\rho \ldots \ldots \ldots \ldots (4)$$

The sign will be positive or negative according as σ increases or decreases as ρ increases. For the positive sign we have

$$\sigma = \rho + \text{const.} \ldots \ldots (5)$$

In Fig. 84 let σ be measured from C_1, and let P_1C_1, P_2C_2, P_3C_3 be ρ_1, ρ_2, ρ_3; then (5) gives

arc $C_1C_2 = \rho_2 + \text{const.}$
$= \rho_2 - \rho_1;$
arc $C_1C_3 = \rho_3 - \rho_1,$

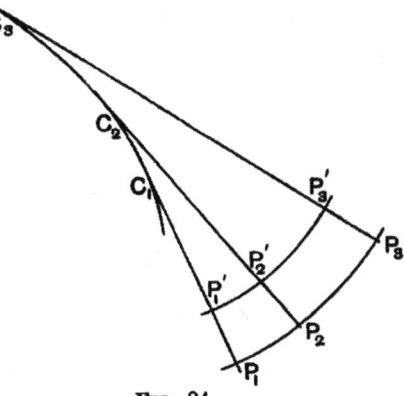

Fig. 84.

which proves the required result.

If a thread were wrapped round the curve C_1C_3 and one end fixed at C_3, the length of the thread being equal to ρ_3, it is clear that when the thread is unwound, and kept stretched in the process, the free end will describe the curve $P_1P_2P_3$. It is from this property that the evolute is named.

The curve P_1P_3 is said to be *an involute* of C_1C_3. Obviously any point on the thread will describe an involute, so that a given curve has an infinite number of involutes while it has but one evolute.

The two involutes P_1P_3, $P_1'P_3'$ are called *parallel curves*, since the distance between them measured along their common normals is constant.

§ 144. Envelopes.

The equation

$$y = ax + a/a \dots\dots\dots\dots\dots\dots(1)$$

where a, a are constants, represents a straight line. If we give a different constant value to a, say a_1, the equation will become

$$y = a_1 x + a/a_1 \dots\dots\dots\dots\dots\dots(2)$$

and will represent a different straight line. The coordinates of the point of intersection of (1) and (2) are

$$x = a/aa_1, \quad y = a/a + a/a_1 \dots\dots\dots\dots(3)$$

Suppose now that a_1 is taken closer and closer to a; the line (2) will therefore come closer and closer to the line (1), but the values (3) show that the point of intersection tends to a definite position when a_1 tends to a as its limit. The coordinates of the limiting position of the point of intersection are

$$x = a/a^2, \quad y = 2a/a \dots\dots\dots\dots\dots(4)$$

If we eliminate a between the two equations (4) we get

$$y^2 = 4ax \dots\dots\dots\dots\dots\dots(5)$$

so that, whatever value a may have, the limiting point lies on the parabola (5). It may be readily verified that whatever value a may have the line (1) is a tangent to the parabola.

In general the equation $f(x, y) = 0$ of a curve contains constants that determine the shape, size, and position of the curve. By assigning a series of different values to the constants we get a series of different curves. We will consider the case in which, as in the example just given, the series is determined by assigning different values to one constant, and we will speak of the series as a *family* of curves. The constant is then often called the *parameter* of the family; thus, in (1) a is the parameter of a family of straight lines.

Two curves of the family will, in general, intersect; if a and $a+\delta a$ are the values of the parameter for two curves C_1 and C_2 of the family, the point or points of intersection of C_1 and C_2 will tend to definite limiting positions as δa tends to zero, and the locus of these limiting positions is called the *envelope* of the family of curves.

Thus the parabola (5) is the envelope of the family (1); the evolute of a curve is the envelope of the family of straight lines composed of the normals of the curve (§§ 141, 143).

§ 145. Equation of Envelope.

Let the equation

$$f(x, y, a) = 0 \quad\quad\quad\quad\quad\quad (1)$$

represent a family of curves, the parameter a of the family being indicated in the functional symbol; a is constant for any one curve of the system. Let the equation

$$f(x, y, a+\delta a) = 0 \quad\quad\quad\quad\quad\quad (2)$$

represent another curve of the system. The coordinates of the points of intersection of (1) and (2) will satisfy

$$f(x, y, a+\delta a) - f(x, y, a) = 0,$$

and therefore also

$$\{f(x, y, a+\delta a) - f(x, y, a)\}/\delta a = 0 \quad\quad\quad\quad (3)$$

The limit of (3) for $\delta a = 0$ is

$$\partial f(x, y, a)/\partial a = 0, \quad\quad\quad\quad\quad\quad (4)$$

and therefore the coordinates of the points on the envelope satisfy equations (1), (4). The equation of the envelope is therefore got by eliminating a between these two equations. In forming (4) x and y are treated as constants, as is evident from the proof.

Thus, if $f(x, y, a) = -y + ax + a/a$,
$$\frac{\partial f(x, y, a)}{\partial a} = x - \frac{a}{a^2}.$$

Eliminate a between the equations
$$-y + ax + a/a = 0, \quad x - a/a^2 = 0$$
and we get $y^2 = 4ax$. The envelope is thus a parabola, as in § 144.

We saw in § 144 that the parabola (5) has each member of the family (1) as a tangent. We will now prove the

THEOREM. *In general, the envelope of a family of curves touches each member of the family.*

The gradient at a point (x, y) on (1) is given by
$$\frac{\partial f}{\partial x} + \frac{\partial f}{\partial y}\frac{dy}{dx} = 0 \ldots\ldots\ldots\ldots\ldots\ldots (5)$$
where, in differentiating, a must be kept constant.

On the other hand, to get the equation of the envelope, we have to eliminate a between (1) and (4); we may therefore take (1) for the equation of the envelope *provided we regard a as a function of x and y determined by* (4). The gradient at a point (x, y) on the envelope will therefore be found by taking *the total derivative* of (1); this total derivative is given by
$$\frac{\partial f}{\partial x} + \frac{\partial f}{\partial y}\frac{dy}{dx} + \frac{\partial f}{\partial a}\frac{da}{dx} = 0. \ldots\ldots\ldots\ldots (6)$$

Suppose now that the coordinates of the point (x, y) satisfy both (1) and (4); that point is therefore on the curve (1) and also on the envelope; and, by (4), equation (6) reduces to equation (5). Hence at the point (x, y) the gradient dy/dx is the same for the curve (1) and for the envelope, which proves the theorem.

It is assumed that $\partial f/\partial x$, $\partial f/\partial y$ are not both zero; if they are, the value of dy/dx given by (5) or (6) is not determinate and the theorem may not be true. The discussion of such cases, however, is beyond our limits.

Analytically, the problem of finding the envelope of the family (1) is equivalent to that of finding the turning values of the function $f(x, y, a)$ of the variable a, when x and y are regarded as constants.

THEOREM ON ENVELOPES.

The student should draw a few lines of the family $y = ax + a/a$ for both positive and negative values of a, and he will get a clear idea of a curve as the envelope of its tangents; the lines are easily drawn since the intercepts on the axes are $-a/a^2$ and a/a respectively.

Ex. 1. The evolute of the parabola $y^2 = 4ax$ considered as the envelope of its normals.

The normal at (h, k) is given by
$$2a(y-k) + k(x-h) = 0$$
or
$$8a^2y + 4a(x-2a)k - k^3 = 0, \quad\ldots\ldots\ldots\ldots\ldots\ldots\text{(i)}$$
since $h = k^2/4a$ by the equation of the parabola. Take k as the parameter of the family of straight lines (i), and find the envelope.

Differentiate (i) as to k; we get
$$4a(x-2a) - 3k^2 = 0. \quad\ldots\ldots\ldots\ldots\ldots\ldots\text{(ii)}$$
Eliminate k between (i) and (ii) and we get
$$27ay^2 = 4(x-2a)^3,$$
which is therefore the equation of the evolute.

Ex. 2. Find the envelope of the circles which pass through the origin and have their centres on the hyperbola
$$x^2 - y^2 = c^2.$$

Let (a, β) be the centre of any circle of the family; the equation of a circle is therefore
$$x^2 + y^2 - 2ax - 2\beta y = 0, \quad\ldots\ldots\ldots\ldots\ldots\ldots\text{(i)}$$
there being no constant term, since the circle goes through the origin.

Since the centre lies on the hyperbola, we have,
$$a^2 - \beta^2 = c^2. \quad\ldots\ldots\ldots\ldots\ldots\ldots\text{(ii)}$$

We might suppose (ii) solved for β in terms of a and the value inserted in (i); this shows that there is really but one parameter. It is simpler, however, to differentiate with respect to a, considering β as a function of a determined by (ii), and then to eliminate a, β and $d\beta/da$.

Differentiating (i) and (ii) with respect to a we get
$$x + y\frac{d\beta}{da} = 0, \quad a - \beta\frac{d\beta}{da} = 0. \quad\ldots\ldots\ldots\ldots\ldots\text{(iii)}$$

From (iii) $a/x = -\beta/y,$
and therefore by (ii) $a/x = -\beta/y = c/\sqrt{(x^2 - y^2)}.$

Substitute in (i) for a and β and reduce; we then get
$$(x^2 + y^2)^2 = 4c^2(x^2 - y^2),$$
which is the equation of a lemniscate.

It will be evident that the procedure is the same as that of finding maxima and minima.

§ 146. Cycloids.

As the cycloid is of some importance in dynamics, we will very briefly investigate its chief properties.

DEFINITION. The cycloid is the curve traced out by a point in the circumference of a circle (the *generating* circle) which rolls without slipping along a fixed straight line (the *base*).

Let OD (Fig. 85) be the base, P the tracing point on the generating circle LPI, and θ the angle between the radius SP and the radius SI, I being the point of contact with the base.

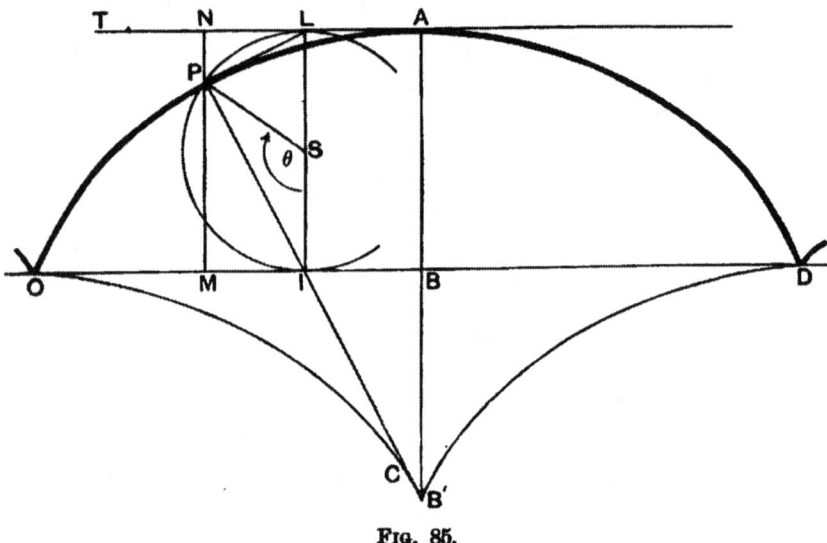

FIG. 85.

Suppose P to be at O when the circle begins to roll; draw PM perpendicular to OD, and let $OM = x$, $MP = y$. Then if a is the radius, we have

$$OI = \text{arc } PI = a\theta,$$
$$\left. \begin{array}{l} x = OI - SP \sin \theta = a(\theta - \sin \theta) \\ y = IS + SP \cos LSP = a(1 - \cos \theta) \end{array} \right\} \quad \ldots\ldots (1)$$

ISL being the diameter through I. Equations (1) are those of the cycloid.

THE CYCLOID.

When $\theta = \pi$, $x = \pi a = OB$ and P is at A, the greatest distance it can be from the base. A is called a *vertex*.

When $\theta = 2\pi$, $x = 2\pi a = OD$, and P is at D. The arch OAD is symmetrical about BA, and BA is called an *axis*.

If the circle were to continue rolling, P would trace out a series of arches congruent to OAD; when *the* cycloid is spoken of, it is usual to confine it to the one arch, and A, BA are then *the* vertex, *the* axis.

Properties. The following are easily established:

(i) $\tan\phi = D_x y = \cot\tfrac{1}{2}\theta = \tan(\tfrac{1}{2}\pi - \tfrac{1}{2}\theta) = \tan PIL$,

and therefore $\phi = \tfrac{1}{2}\pi - \tfrac{1}{2}\theta = \angle PIL$,

so that PL is the tangent and PI the normal at P.

(ii) $s = \text{arc } OP = 4a(1 - \cos\tfrac{1}{2}\theta)$; arc $OA = 4a$.

(iii) $\rho = PC = 4a\sin\tfrac{1}{2}\theta = 2PI$, *numerically*.

If the tangent AT and the normal AB are taken as axes, and PN drawn perpendicular to AT, we put $\theta' = \angle LSP = \pi - \theta$; then

$$x = AN = a(\theta' + \sin\theta'); \quad y = NP = a(1 - \cos\theta') \quad \ldots\ldots\ldots\ldots(1')$$

(ia) $\phi = \angle PLN = \tfrac{1}{2}\theta' = \angle PIL$.

(ii$a$) $s = \text{arc } AP = 4a\sin\tfrac{1}{2}\theta'$; $s^2 = 8a \cdot NP = 8ay$.

The coordinates of C, the centre of curvature, are

$$\xi = OM + 4a\sin\tfrac{1}{2}\theta\cos\tfrac{1}{2}\theta = a(\theta + \sin\theta),$$
$$\eta = -IC\sin\tfrac{1}{2}\theta = -a(1 - \cos\theta).$$

Hence, by equations (1'), the evolute of the cycloid OAD consists of the halves OCB', $B'D$ of an equal cycloid. In (1') the positive direction of y is *downwards*, but when O is origin the positive direction is *upwards*, so that η is negative.

B' is a cusp on the evolute; O, D are cusps on the original cycloid and vertices of the evolute.

Epicycloids and Hypocycloids. The curve traced out by a point on the circumference of a circle which rolls without slipping on the circumference of a fixed circle is called an *epicycloid* or a *hypocycloid* according as the rolling circle is outside or inside the fixed circle. When the rolling circle surrounds the fixed one the epicycloid is sometimes called a *pericycloid*.

Let Figure 86 represent the generation of an epicycloid, P being the tracing point and C the starting point. Let a and b be the radii of the fixed and of the rolling circles, θ and θ' the angles CAI and IBP; $AM=x$, $MP=y$. Then

arc PI = arc CI, so that, $b\theta' = a\theta$.
$$x = (a+b)\cos\theta - b\cos(\theta+\theta')$$
$$= (a+b)\cos\theta - b\cos[(a+b)\theta/b]$$
$$y = (a+b)\sin\theta - b\sin[(a+b)\theta/b] \quad\quad\quad\quad\quad (2)$$

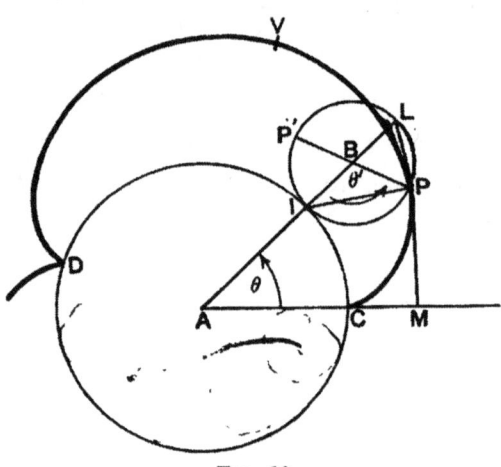

FIG. 86.

When the circles are on the same side of the tangent at I, that is, for the hypocycloids ($b<a$) and the pericycloids ($b>a$), it is only necessary to change the sign of b. Hence, the equations of the hypocycloid are of the form

$$x = (a-b)\cos\theta + b\cos[(a-b)\theta/b]$$
$$y = (a-b)\sin\theta - b\sin[(a-b)\theta/b] \quad\quad\quad\quad\quad (3)$$

When the ratio of b to a is a commensurable number the tracing point P will return to C after the circle B has rolled once or oftener round the fixed circle; when the ratio of b to a is incommensurable P will not return to C.

Trochoids. If the tracing point P is not on the circumference but on a radius or on a radius produced, the curve it describes is a *trochoid* or an *epitrochoid* or a *hypotrochoid*.

If the distance of P from the centre of the circle is to the radius in the ratio of λ to 1, the equations of the *trochoid* are got from equations (1) by multiplying $\sin\theta$ and $\cos\theta$ by λ, while the equations of the *epitrochoid* and the *hypotrochoid* are got from equations (2) and (3) respectively by multiplying the *coefficient* b of the second term by λ, as the student will easily prove.

EPICYCLOIDS. HYPOCYCLOIDS. 371

EXERCISES XXXII.

1. Show that for the parabola $y^2 = 4ax$,
$$\rho = -2a \csc^3 \phi, \quad \xi = 2a + 3a \cot^2 \phi, \quad \eta = -2a \cot^3 \phi,$$
and then find the equation of the evolute.

2. Show that for the hyperbola $x^2/a^2 - y^2/b^2 = 1$,
$$a^4 \xi = (a^2 + b^2) x^3, \quad b^4 \eta = -(a^2 + b^2) y^3,$$
and that the equation of the evolute is
$$(ax)^{\frac{2}{3}} - (by)^{\frac{2}{3}} = (a^2 + b^2)^{\frac{2}{3}}.$$

3. Show that for the rectangular hyperbola $xy = c^2$,
$$\xi = \frac{3}{2} x + \frac{y^3}{2c^2}, \quad \eta = \frac{3}{2} y + \frac{x^3}{2c^2},$$
and that the equation of the evolute is
$$(x+y)^{\frac{2}{3}} - (x-y)^{\frac{2}{3}} = (4c)^{\frac{2}{3}}.$$

4. Show that for the curve $x^{\frac{2}{3}} + y^{\frac{2}{3}} = a^{\frac{2}{3}}$ (see § 142, ex. 1),
$$\xi = a \cos^3 t + 3a \cos t \sin^2 t, \quad \eta = a \sin^3 t + 3a \sin t \cos^2 t,$$
and that the equation of the evolute is
$$(x+y)^{\frac{2}{3}} + (x-y)^{\frac{2}{3}} = 2a^{\frac{2}{3}}.$$

5. Prove that the envelope of the family of straight lines
$$x/a + y/\beta = 1,$$
  (i) when $a\beta = a^2$, is the hyperbola, $4xy = a^2$;
  (ii) when $a + \beta = a$, is the parabola $\sqrt{x} + \sqrt{y} = \sqrt{a}$;
  (iii) when $a^2 + \beta^2 = a^2$, is the curve $x^{\frac{2}{3}} + y^{\frac{2}{3}} = a^{\frac{2}{3}}$.

State the geometrical meaning of the conditions to which the parameters a, β are subject.

6. Prove that the envelope of the family of ellipses
$$x^2 a/^2 + y^2/\beta^2 = 1$$
  (i) when $a\beta = a^2$ is the two hyperbolas $2xy = \pm a^2$.
  (ii) when $a + \beta = a$ is the curve $x^{\frac{2}{3}} + y^{\frac{2}{3}} = a^{\frac{2}{3}}$.

State the geometrical meaning of the conditions to which the parameters a, β are subject.

7. The envelope of the circles described on the double ordinates of a parabola as diameters is an equal parabola.

8. If P, Q, R are functions of the coordinates of a point, and a a variable parameter, show that the envelope of
$$Pa^2 + 2Qa + R = 0$$
is
$$Q^2 - PR = 0,$$

and that the envelope of
$$P\cos a + Q\sin a = R$$
is
$$P^2 + Q^2 = R^2.$$

9. Show that whatever be the value of m, the straight line
$$y = mx + \sqrt{\{(a+bm^2)/ab\}}$$
touches the conic $ax^2 + by^2 = 1$.

10. A straight line moves so that (i) the product, (ii) the sum, of the squares of the perpendiculars drawn to it from two fixed points $(c, 0), (-c, 0)$ is constant; show that in each case the envelope is a central conic.

11. Show that the envelope of the circles described on the central radii of an ellipse as diameters is
$$(x^2 + y^2)^2 = a^2 x^2 + b^2 y^2.$$

12. The envelope of the ellipses $(x-a)^2/a^2 + (y-\beta)^2/b^2 = 1$ when the parameters a, β are connected by the equation
$$a^2/a^2 + \beta^2/b^2 = 1$$
is the ellipse $x^2/a^2 + y^2/b^2 = 4$. State the problem in geometrical language.

13. Show that the envelope of the family of straight lines
$$ax \sec a - by \csc a = a^2 - b^2$$
is the curve
$$(ax)^{\frac{2}{3}} + (by)^{\frac{2}{3}} = (a^2 - b^2)^{\frac{2}{3}}.$$

14. If in Fig. 82 $OZ = p$, show that the equations of the tangent and normal at P are
$$x \sin\phi - y \cos\phi = p \text{ (i)}; \quad x\cos\phi + y\sin\phi = \frac{dp}{d\phi} \text{ (ii)},$$
and show, from (ii), that $ZP = dp/d\phi$.

Consider the curve as the envelope of its tangents.

15. With the same notation as in ex. 14, show that the coordinates (ξ, η) of the centre of curvature are given by
$$\xi\cos\phi + \eta\sin\phi = \frac{dp}{d\phi}, \quad -\xi\sin\phi + \eta\cos\phi = \frac{d^2p}{d\phi^2},$$
or
$$\xi = \frac{dp}{d\phi}\cos\phi - \frac{d^2p}{d\phi^2}\sin\phi, \quad \eta = \frac{dp}{d\phi}\sin\phi + \frac{d^2p}{d\phi^2}\cos\phi.$$

16. With the same notation as in the last two examples, show that the projection of OC, where C is the centre of curvature, on PC, is
$$-\xi\sin\phi + \eta\cos\phi,$$
and that
$$\rho = p - \xi\sin\phi + \eta\cos\phi = p + \frac{d^2p}{d\phi^2}.$$

EXERCISES XXXII.

17. Show that the radius of curvature of the evolute of a curve is $\rho \, d\rho/ds$, where ρ is the radius of curvature at the corresponding point on the given curve.

Use § 143, (ii); $d\phi$ is the same for curve and evolute.

18. If A is the area between a curve, its evolute and two radii of curvature, show that

$$\frac{dA}{dx} = \frac{1}{2}\rho^2 \frac{d\phi}{dx} = \frac{1}{2}\left\{1 + \left(\frac{dy}{dx}\right)^2\right\}^2 \div \frac{d^2y}{dx^2}.$$

19. ABC is an arc of a circle whose centre is O and radius a; CP is the tangent at C and AP a part of an involute of the circle. Taking OA as the x-axis and putting ϕ for the angle AOC, show that the coordinates (x, y) of P are

$$x = a \cos \phi + a\phi \sin \phi, \quad y = a \sin \phi - a\phi \cos \phi,$$

and that the intrinsic equation of the involute is

$$s = \tfrac{1}{2} a \phi^2.$$

All the involutes of a circle are identically equal, so that we may speak of *the* involute of a circle.

20. Show that the p, r equation of the involute of a circle is

$$r^2 = p^2 + a^2.$$

21. The total length of the evolute of an ellipse is

$$4(a^3 - b^3)/ab.$$

22. The intrinsic equation of the cycloid, when the vertex A is the origin of s and the tangent AT the fixed tangent (Fig. 85) is $s = 4a \sin \phi$.

23. Show that for the epicycloid (Fig. 86) PL is the tangent and PI the normal at P.

For, $\quad \tan \phi = \dfrac{dy}{dx} = \dfrac{\cos \theta - \cos(\theta + \theta')}{\sin(\theta + \theta') - \sin \theta} = \tan(\theta + \tfrac{1}{2}\theta'),$

and PL makes with the x-axis the angle $\theta + \tfrac{1}{2}\theta'$. Similar results hold for the hypocycloid.

24. If s is the arc CP of an epicycloid, Fig. 86, show that

$$\frac{ds}{d\theta} = 2(a+b) \sin \frac{a\theta}{2b}; \quad s = \frac{4b(a+b)}{a}\left(1 - \cos \frac{a\theta}{2b}\right),$$

and that the length of CPD is $8b(a+b)/a$.

25. The intrinsic equation of an epicycloid is

$$s = \frac{4b(a+b)}{a}\left\{1 - \cos\left(\frac{a\phi}{a+2b}\right)\right\},$$

and the radius of curvature is

$$\rho = \frac{4b(a+b)}{a+2b} \sin\left(\frac{a\phi}{a+2b}\right).$$

Similar results hold for the hypocycloid, the sign of b being changed.

26. If $b = a/4$, show that the hypocycloid has four cusps, and that its equations are
$$x = a\cos^3\theta, \quad y = a\sin^3\theta.$$
Eliminating θ, we get $x^{\frac{2}{3}} + y^{\frac{2}{3}} = a^{\frac{2}{3}}$.

27. Show that if $b = a/2$ the hypocycloid becomes a diameter of the fixed circle.

28. Show that if $b = a$ the epicycloid becomes the cardioid
$$r = 2a(1 - \cos\theta),$$
the origin being at the point C; that is,
$$r\cos\theta = x - a, \quad r\sin\theta = y.$$

29. In ex. 25 put
$$\phi = \frac{(a+2b)\pi}{2a} + \phi'; \quad s = \frac{4b(a+b)}{a} + s';$$
that is, measure the arc from the middle point V of CPD (the vertex), and we get
$$s' = \frac{4b(a+b)}{a}\sin\frac{a\phi'}{a+2b}.$$
Show that the equation $s = l\sin n\phi$ will represent an epicycloid if n is less than unity, a hypocycloid if n is greater than unity.

30. If s, σ are corresponding arcs of a curve and its evolute
$$\sigma = \pm\frac{ds}{d\phi} + \text{const.}$$

Show from the result of ex. 29 that the evolute of an epicycloid is an epicycloid, and that of a hypocycloid is a hypocycloid.

31. Parallel rays fall on the circumference of a circle and are reflected, the angle of reflection being equal to that of incidence. If a is the radius of the circle, $(a\cos\theta, a\sin\theta)$ the point of incidence, the centre of the circle the origin of coordinates, and the x-axis parallel to the direction of the incident ray, show that the equation of the reflected ray is
$$y\cos 2\theta - x\sin 2\theta + a\sin\theta = 0,$$
and that the envelope of the reflected ray is an epicycloid
$$x = \frac{a}{4}(3\cos\theta - \cos 3\theta), \quad y = \frac{a}{4}(3\sin\theta - \sin 3\theta).$$

32. If v is the velocity of a particle describing a central orbit under an outward radial force F, then with the usual notation $v^2 = h^2/p^2$. Prove
$$F = \frac{d}{dr}(\tfrac{1}{2}v^2) = -h^2u^2\left(\frac{d^2u}{d\theta^2} + u\right).$$

This equation is the differential equation of the orbit. If $F = \pm\mu u^2$, show that the orbit is a conic, the centre of force being at a focus. (See §§ 169, 170).

CHAPTER XVII.

INFINITE SERIES.

§ 147. Infinite Series. For a thorough treatment of infinite series the student is referred to Chrystal's *Algebra*, vol. II.; an exceedingly good elementary account will be found in Osgood's *Introduction to Infinite Series* (Cambridge, U.S.A.; Harvard University). We will limit our discussion to those parts of the theory that are needed in the applications we make.

DEFINITION OF AN INFINITE SERIES. Let $u_1, u_2, u_3, \ldots$, be a set of quantities unlimited in number, u_n being a single-valued function of the integer n, and let s_n denote the sum of the first n terms,

$$s_n = u_1 + u_2 + \ldots + u_n \ldots\ldots\ldots\ldots\ldots(1)$$

When n increases indefinitely the sum (1) becomes an **Infinite Series**.

If as n increases indefinitely the sum s_n tends to a definite finite limit s, the infinite series is said to be **convergent**, and **to converge to the value** s, or **to have the value** s, or **to have the sum** s.

Ex. 1. Let $\quad s_n = 1 + \frac{1}{2} + \frac{1}{4} + \ldots + \frac{1}{2^{n-1}}$.

Here $\quad s_n = 2 - 1/2^{n-1}; \quad L s_n = 2 = s.$

If as n increases indefinitely s_n does not tend to a definite finite limit, the series is said to be **non-convergent**. In this case either s_n increases (numerically) beyond all bound, and then the series is said to be **divergent** or to **diverge**, or else s_n does not tend to a *definite* finite limit, and then the series is said to **oscillate**.*

* Some writers use *divergent* as equivalent to *non-convergent*.

Ex. 2. Let $s_n = 1+2+3+\ldots+n$.

Here s_n increases beyond all bound, and therefore the infinite series is divergent.

Ex. 3. Let $s_n = 1-1+1-1+\ldots+(-1)^{n-1}1$.

Here s_n is 0 or 1 according as n is even or odd, and though s_n does not become infinite, it does not tend to a *definite* finite limit. The series therefore oscillates.

It is obvious that if $u_1, u_2, \ldots$ are all of the same sign, the series cannot oscillate.

Notation. We will represent an infinite series by the notation
$$u_1 + u_2 + \ldots; \text{ or } \sum u; \text{ or } \sum_1^\infty u_n.$$

The following theorems are readily proved:

THEOREM I. *If* $u_1 + u_2 + \ldots$ *converges to the value* s, *the series* $cu_1 + cu_2 + \ldots$
where c is any finite quantity, converges to the value cs.

The proof is so simple that it may be left to the reader.

THEOREM II. *If* $u_1 + u_2 + \ldots$ *converges to the value* s, *and* $v_1 + v_2 + \ldots$ *to the value* t, *the series*
$$(u_1 + v_1) + (u_2 + v_2) + \ldots$$
converges to the value (s+t).

Let $s_n = u_1 + u_2 + \ldots + u_n$, $t_n = v_1 + v_2 + \ldots + v_n$, then $(u_1+v_1) + (u_2+v_2) + \ldots + (u_n+v_n) = s_n + t_n$ for every value of n, and the result follows at once.

The first theorem shows that the *product* of c and Σu is $\Sigma(cu)$, and the second shows that the *sum* of Σu and Σv is $\Sigma(u+v)$, and *sum* may obviously be considered as including *difference*.

In forming s_n it is to be understood that the terms are added on in the order in which they stand in the series, and it follows at once that when the series is convergent the law of association holds good; that is, we may group the terms as we please (so long as we do not change their order), and the value of the series will not be affected. It does not follow, however, that if we form a new series by writing the terms in a different order, the new series will converge to the same value as the old (see § 150).

DEFINITIONS. NOTATIONS.

The phrase "numerical value" or "absolute value" occurs so often that we will use the notation (now generally adopted)

$$|a|$$

to represent the numerical value of a. Thus,

$$|2|=2;\quad |-2|=2;\quad |-10+6|=4.$$

The following statements are easily proved:

(i) $$|a+b+c+\ldots| \leqq |a|+|b|+|c|+\ldots,$$

the equality holding only when $a, b, c \ldots$ have all the same sign.

(ii) If c is positive, the inequality

$$|a-b| < c$$

is equivalent to either of the inequalities

$$b-c < a < b+c;\quad a-c < b < a+c.$$

§ 148. Existence of a Limit. A function may be defined by an infinite series provided the series is convergent. Thus, the infinite geometrical progression

$$a + ax + ax^2 + \ldots$$

converges to the value $a/(1-x)$ so long as x is numerically less than 1, and we may say that if $-1 < x < 1$ the function $a/(1-x)$ is represented by the series, or that the series defines the function. But if x is greater than 1 the series is divergent and does not represent $a/(1-x)$ at all. *Convergent series alone* are of use in practice and, subject to certain restrictions, can be manipulated like expressions containing only a finite number of terms; non-convergent series can only be used under very special conditions.

It is not often, however, when a series is given, that we can, as in the case of the geometrical progression, actually assign the number which is the limit of s_n. It is necessary, therefore, to have a criterion for the *existence* of the limit, and we will now state three general theorems that will be of great service in leading to simple tests for the convergency of a series. The variable s_n is assumed to be a single-valued function of n, and n is to increase indefinitely; since all the limits are taken for $n = \infty$ we may omit the subscript "$n = \infty$."

THEOREM I. *If s_n is a function of n that* (i) *always increases as n increases, but* (ii) *always remains less than* *a definite quantity* a, *then as* n *increases indefinitely* s_n *will tend to a definite limit that is less than or equal to* a.

THEOREM II. *If* s_n *is a function of* n *that* (i) *always decreases as* n *increases, but* (ii) *always remains greater than a definite quantity* b, *then as* n *increases indefinitely* s_n *will tend to a definite limit that is greater than or equal to* b.

THEOREM III. *The necessary and sufficient condition that* s_n *should, as* n *increases indefinitely, tend to a definite limit is that the limit for* n *infinite of* $(s_{n+p} - s_n)$ *should be zero for every value of the integer* p; *or, in other words, given an arbitrarily small positive quantity* ϵ *it must be possible to choose* n, *say* n = m, *such that when* n ≧ m *the difference* $(s_{n+p} - s_n)$ *shall be numerically less than* ϵ, *whatever value the integer* p *may have.*

We do not propose to prove these theorems; the first and second have been given as exercises (Exer. VII., 14, 15), and the geometrical illustration there given affords some justification for assuming them. As to the third theorem it is easy to see that the condition stated is necessary. For, if s_n has a definite limit s, then since

$$s_{n+p} - s_n = (s_{n+p} - s) + (s - s_n),$$

we have

$$L(s_{n+p} - s_n) = L(s_{n+p} - s) + L(s - s_n) = 0.$$

To illustrate the sufficiency of the condition, take on the x-axis the points $A_1, A_2, A_3, \ldots$, which have $s_1, s_2, s_3, \ldots$ as abscissae. In this case A_{n+1} may be either to the right or to the left of A_n, since s_n does not necessarily either always increase or always decrease as n increases. But, by hypothesis, if $n \geqq m$,

$$|s_{n+p} - s_n| < \epsilon; \text{ that is, } s_m - \epsilon < s_{m+p} < s_m + \epsilon.$$

If P and Q are the points whose abscissae are $s_m - \epsilon$ and $s_m + \epsilon$, then the length of the segment PQ is 2ϵ, and every one of the points A_n for which n is greater than m lies within this segment. By assigning to ϵ smaller and smaller values we get shorter and shorter segments $P'Q', P''Q'', \ldots$, each lying within the one that precedes it. The ends $P, P', P'', \ldots$ move to the right, $Q, Q', Q'', \ldots$ to the left; by Theorem I. $P, P', P'', \ldots$ tend to a definite limit, and by Theorem II. $Q, Q', Q'', \ldots$ also tend to a definite limit, and since ϵ may be as small as we please, these two limits coincide, say at S. The points A_n therefore tend to S, and s, the abscissa of S, is the limit of s_n.

Examples 16, 17 of Set VII. illustrate Theorem I. To illustrate Theorem III. take

$$s_n = 1 - \tfrac{1}{2} + \tfrac{1}{3} - \tfrac{1}{4} + \ldots + (-1)^{n-1}\tfrac{1}{n}.$$

THEOREMS ON EXISTENCE OF A LIMIT.

Here $\quad s_{n+p} - s_n = \pm \left(\dfrac{1}{n+1} - \dfrac{1}{n+2} + \ldots \pm \dfrac{1}{n+p} \right),$

and $\quad \dfrac{1}{n+1} - \dfrac{1}{n+2} + \ldots \pm \dfrac{1}{n+p} = \dfrac{1}{n+1} - \left(\dfrac{1}{n+2} - \dfrac{1}{n+3} \right) - \ldots$
$$< 1/(n+1),$$
since each bracket is positive; if p is an even number the last bracket will contain but one term, $1/(n+p)$.

Again, $\dfrac{1}{n+1} - \dfrac{1}{n+2} + \ldots \pm \dfrac{1}{n+p} = \left(\dfrac{1}{n+1} - \dfrac{1}{n+2} \right) + \left(\dfrac{1}{n+3} - \dfrac{1}{n+4} \right) + \ldots$

and the expression on the right is *positive*; therefore $|s_{n+p} - s_n|$ lies between 0 and $1/(n+1)$.

Hence the limit of $s_{n+p} - s_n$ is zero, and s_n tends to a definite limit; the limit will be found later to be log 2 (p. 395), so that
$$\log 2 = 1 - \tfrac{1}{2} + \tfrac{1}{3} - \tfrac{1}{4} + \ldots.$$

It is clear that the Theorems I., II., III. hold even when the variable s_n is a continuous function, $f(x)$ say. If x tends to a finite limit x_1 we may put $x_1 \pm 1/n$ for x; when n tends to infinity x tends to x_1. If x tends to $-\infty$ we may put $-n$ for x.

§ 149. **Tests of Convergence.** The difference $s - s_n$ between the sum s_n of the first n terms and the value s of the series Σu is called *the remainder after* n *terms*; if we denote this remainder by r_n we have
$$s = s_n + r_n.$$

Clearly r_n is itself an infinite series $u_{n+1} + u_{n+2} + \ldots$ and the limit of r_n is zero. If the series is such that $|r_n|$ is small when n is small the series is said to be *rapidly* or *highly* convergent, because the calculation of a few terms will yield a good approximation to the value s. For purposes of calculation rapidity of convergence is valuable; but a series may yet be convergent though it require the calculation of a million terms to get a fair approximation.

Fundamental Test. Let $_p r_n$ denote $s_{n+p} - s_n$, that is,
$$_p r_n = u_{n+1} + u_{n+2} + \ldots + u_{n+p},$$
then $_p r_n$ is called a *partial remainder after* n *terms*. By § 148, Th. III., the necessary and sufficient condition that the series Σu should be convergent is that the limit of $_p r_n$ should be zero for every value of p.

If $p=1$ then $_pr_n = u_{n+1}$, and therefore a *necessary* condition of convergence is that u_{n+1} or, what amounts to the same thing, that u_n should converge to zero; but as we shall see (ex. 1) this condition is not *sufficient*.

This test is not easy of application; we therefore deduce one or two special tests that can be more readily used.

Comparison Test. Let $u_1 + u_2 + \ldots$ be a series of positive terms; if each term is less than or equal to the corresponding term of a series of positive terms $a_1 + a_2 + \ldots$ that is convergent, the series $u_1 + u_2 + \ldots$ is also convergent, but if each term is greater than or equal to the corresponding term of a series of positive terms $b_1 + b_2 + \ldots$ that is divergent, the series $u_1 + u_2 + \ldots$ is also divergent.

Let
$$s_n = \sum_1^n u_r, \quad t_n = \sum_1^n a_r, \quad t = L\, t_n,$$
then
$$s_n \leqq t_n; \quad t_n < t,$$
since all the terms of $a_1 + a_2 + \ldots$ are positive. Hence s_n, which increases as n increases, is always less than t; therefore (§ 148, Th. I.) s_n tends to a limit s that is less than or equal to t.

The proof for the case of divergence may be left to the reader.

Note. We may note here that in testing a series we are at liberty, when it is convenient, to disregard any *finite* number of terms; the rejection of such terms would affect the *value* but not the *existence* of the limit. Thus we need only suppose the terms of $u_1 + u_2 + \ldots$ to be *ultimately* positive.

Ex. 1. The series $1 + \frac{1}{2} + \frac{1}{3} + \ldots$ is called *the harmonic series*; show that it is divergent even though $L\, u_n = 0$.

Beginning with the third term take in succession 2 terms, then 4 or 2^2, then 8 or 2^3, and so on. Now
$$\frac{1}{3} + \frac{1}{4} > \frac{1}{4} + \frac{1}{4} \text{ or } \frac{1}{2}; \quad \frac{1}{5} + \frac{1}{6} + \frac{1}{7} + \frac{1}{8} > \frac{4}{8} \text{ or } \frac{1}{2},$$
and so on. Thus, the sum of 2^m terms is greater than
$$\left(1 + \frac{1}{2}\right) + \frac{1}{2} + \frac{1}{2} + \ldots \text{ to } m \text{ terms};$$
that is, greater than $1 + m/2$. We can therefore take n so large that s_n shall exceed any assigned number; that is, the series is divergent.

Ex. 2. The series
$$1 + \frac{1}{2^a} + \frac{1}{3^a} + \frac{1}{4^a} + \ldots$$
is convergent if $a > 1$, divergent if $a \leq 1$.

(i) $a > 1$. Group as in ex. 1, beginning with the *second* term.
$$\frac{1}{2^a} + \frac{1}{3^a} < \frac{1}{2^a} + \frac{1}{2^a} \text{ or } \frac{1}{2^{a-1}};$$
$$\frac{1}{4^a} + \frac{1}{5^a} + \frac{1}{6^a} + \frac{1}{7^a} < \frac{4}{4^a} \text{ or } \left(\frac{1}{2^{a-1}}\right)^2,$$
and so on. Hence the series is less than
$$1 + \frac{1}{2^{a-1}} + \left(\frac{1}{2^{a-1}}\right)^2 + \left(\frac{1}{2^{a-1}}\right)^3 + \ldots$$
which is a G.P. with common ratio less than 1 and therefore convergent. The given series is therefore also convergent.

(ii) $a \leq 1$. The case $a = 1$ is that of ex. 1. When $a < 1$ the terms are greater than the corresponding terms of the harmonic series; the series is therefore in this case divergent.

The Test Ratio. Let $u_1 + u_2 + \ldots$ be a series of positive terms, and let the limit for $n = \infty$ of u_{n+1}/u_n be ρ; the series will be convergent if $\rho < 1$, but divergent if $\rho > 1$. The test fails to discriminate if $\rho = 1$.

(i) $\rho < 1$. By the definition of a limit we can take n so large, say $n = m$, that when $n \geq m$ the ratio u_{n+1}/u_n shall differ from ρ by as little as we please and therefore shall be less than a proper fraction r. If m be so chosen we have
$$u_{m+1} < u_m r; \quad u_{m+2} < u_{m+1} r < u_m r^2; \quad u_{m+3} < u_m r^3$$
and so on. Hence, after the term u_m, each term of the series is less than the corresponding term of the G.P.
$$u_m r + u_m r^2 + u_m r^3 + \ldots.$$
Since r is less than 1 the G.P. and therefore also the given series is convergent.

(ii) $\rho > 1$. In the same way the series may be proved divergent when $\rho > 1$.

Cor. The remainder r_m of the given series is less than
$$u_m r + u_m r^2 + \ldots, \quad \text{that is,} \quad u_m r/(1-r).$$

Ex. 3.
$$1 + x + \frac{x^2}{2} + \frac{x^3}{3} + \ldots \quad (x \text{ positive}).$$
$$\frac{u_{n+1}}{u_n} = \frac{x^n}{n} \div \frac{x^{n-1}}{n-1} = \frac{n-1}{n} x; \quad \rho = x.$$

Hence the series is convergent if $x < 1$, divergent if $x > 1$; if $x = 1$ the series is the harmonic series and therefore divergent.

Ex. 4. $1+x+\dfrac{x^2}{2!}+\dfrac{x^3}{3!}+\ldots$ (x positive).

$$\frac{u_{n+1}}{u_n}=\frac{x}{n}\,;\quad \rho=0.$$

The series (the exponential series, § 49) is therefore convergent for every finite (positive) value of x. It will be seen immediately that we may suppose x to be either positive or negative.

§ 150. Absolute Convergence.   Power Series.

THEOREM I. *If a series which contains both positive and negative terms is convergent when all the negative terms have their signs changed, it is convergent as it stands.*

For the effect of restoring the negative signs is to diminish both $|s_n|$ and $|{}_pr_n|$.

DEFINITION 1. A series is said to be *absolutely* or *unconditionally* convergent when the series formed from it by making all its terms positive is convergent; that is, $u_1+u_2+\ldots$ is absolutely convergent when $|u_1|+|u_2|+\ldots$ is convergent. Any other convergent series is said to be *conditionally* convergent (sometimes *semi-convergent*).

The converse of Theorem I. is not true; the series $u_1+u_2+\ldots$ may be convergent, and the series $|u_1|+|u_2|+\ldots$ divergent (see ex. 1).

COR. A series is absolutely convergent if the limit of u_{n+1}/u_n is *numerically* equal to a proper fraction.

Absolutely convergent series are of special importance; no rearrangement of the terms affects the sum. It is possible, however, so to rearrange the terms of a conditionally convergent series that the series thus arising shall be convergent, but shall converge to a different value or even shall be divergent. Hence the words "conditional" and "unconditional." (See Chrystal's *Algebra*, vol. 2, chap. 26, § 13).

THEOREM II. *If* u_1, u_2, u_3, ... *are all positive, and each less than (or equal to) that which precedes it; if, further, the limit of* u_n *is zero, the series*

$$u_1-u_2+u_3-u_4+\ldots+(-1)^{n-1}u_n+\ldots$$

is convergent.

This series is called the *Alternating Series*.

We may write the sum of an even number of terms in the two forms

ABSOLUTE CONVERGENCE. POWER SERIES.

$$s_{2n} = (u_1 - u_2) + (u_3 - u_4) + \ldots + (u_{2n-1} - u_{2n})$$
$$= u_1 - (u_2 - u_3) - (u_4 - u_5) - \ldots - u_{2n}.$$

The first form shows that s_{2n} is positive and increases with n, while the second form shows that s_{2n} is less than u_1, because each difference is positive. Hence s_{2n} converges to a limit, s say.

Again, $s_{2n+1} = s_{2n} + u_{2n+1}$, and therefore, since $L\, u_{2n+1}$ is zero, s_{2n+1} and s_{2n} have the same limit; the series is therefore convergent.

COR. $|r_n|$ is less than u_{n+1}.

Ex. 1. $\quad 1 - \tfrac{1}{2} + \tfrac{1}{3} - \tfrac{1}{4} + \ldots$

The series satisfies all the conditions, and is therefore convergent, as was shown previously (§ 148); but the series $1 + \tfrac{1}{2} + \tfrac{1}{3} + \tfrac{1}{4} + \ldots$ is divergent.

THEOREM III. *If the series* $u_1 + u_2 + \ldots$ *is absolutely convergent, and if each of the quantities* $v_1, v_2, \ldots$ *is numerically less than a finite quantity* c, *the series* $u_1 v_1 + u_2 v_2 + \ldots$ *is absolutely convergent.*

For, the terms of $|u_1 v_1| + |u_2 v_2| + \ldots$ are less than the corresponding terms of

$$|u_1| c + |u_2| c + \ldots \text{ or } c\{|u_1| + |u_2| + \ldots\}.$$

Hence $|u_1 v_1| + |u_2 v_2| + \ldots$ is convergent, and therefore $u_1 v_1 + u_2 v_2 + \ldots$ is absolutely convergent.

Ex. 2. $\quad \dfrac{\sin x}{1} - \dfrac{\sin 2x}{2^2} + \dfrac{\sin 3x}{3^2} - \dfrac{\sin 4x}{4^2} + \ldots$

The series $\quad \dfrac{1}{1} - \dfrac{1}{2^2} + \dfrac{1}{3^2} - \dfrac{1}{4^2} + \ldots$

is absolutely convergent, and no sine is greater than 1; thus the series is absolutely convergent for every value of x.

DEFINITION 2. A series of ascending integral powers of a variable, x say, of the form

$$a_0 + a_1 x + a_2 x^2 + \ldots + a_n x^n + \ldots, \quad \ldots\ldots\ldots\ldots\ldots\text{(P)}$$

where the coefficients are constants, is called a **Power Series in** x.

It is with Power Series we chiefly have to deal; the following theorems are important.

THEOREM IV. *If the limit of a_{n+1}/a_n is numerically equal to $1/R$, the Power Series* (P) *converges absolutely when* x *is numerically less than* R, *but diverges when* x *is numerically greater than* R; *it may or may not converge when* x *is equal to* R.

For, disregarding the first term a_0, we have numerically
$$\frac{u_{n+1}}{u_n} = \frac{a_{n+1}}{a_n}x; \quad L\frac{u_{n+1}}{u_n} = xL\frac{a_{n+1}}{a_n} = \frac{x}{R};$$
and the result follows from Theorem I., Cor.

The following is a more general theorem:

THEOREM V. *If when* $x = R$ *none of the terms of the series* (P) *exceeds numerically a finite quantity* c, *the series* (P) *will be absolutely convergent so long as* x *is numerically less than* R.

For, if we write (P) in the form
$$a_0 + a_1 R(x/R) + a_2 R^2(x/R)^2 + \ldots$$
we see that the terms of (P) are numerically not greater than the corresponding terms of the geometrical progression
$$c + c(x/R) + c(x/R)^2 + \ldots,$$
and therefore the series is absolutely convergent so long as x/R is numerically less than unity.

The series (P) may or may not converge when $x = R$; if it does converge each term must, when $x = R$, be finite, and therefore it will converge absolutely when x is less than R numerically.

Interval of Convergence. When a series whose terms are functions of x is convergent when $a < x < b$, we may say the series converges within the interval (a, b). When the series converges for $a < x < b$, and diverges for $x < a$ and $x > b$, we may speak of (a, b) as *the* interval of convergence.

Ex. 3. The series $\quad x - \dfrac{x^2}{2} + \dfrac{x^3}{3} - \dfrac{x^4}{4} + \ldots$

converges (conditionally) when $x = 1$; therefore absolutely when $-1 < x < 1$. It diverges when $x = -1$ and when $|x| > 1$.

Ex. 4. The series $\quad x - \dfrac{x^2}{2^2} + \dfrac{x^3}{3^2} - \dfrac{x^4}{4^2} + \ldots$

converges absolutely when $-1 \leq x \leq 1$, diverges when $|x| > 1$.

For both series $(-1, 1)$ is the interval of convergence.

§ 151. **Uniform Convergence.** When the terms of a series are functions of a variable x and the series converges within a certain interval it will be possible, for a given value of x within the interval, to choose n so that the remainder r_n will be less than a given quantity. For different values of x, however, different values of n will usually be required to make the remainder less than the given quantity. Hence the

DEFINITION. A series, whose terms are functions of a variable x, is said to converge *uniformly* within an interval if it is possible to choose n, say $n=m$, so that for every value of n equal to or greater than m and *for every value of x within the interval* the remainder r_n shall be less than any given positive quantity ϵ.

We will indicate the variable by the notation $u_n(x)$, $s_n(x)$, $r_n(x)$, $s(x)$.

THEOREM I. *If the series* $u_1(x) + u_2(x) + \ldots$ *is uniformly convergent when* $a \leq x \leq b$, *and if each term is a continuous function of* x *for the same range, the sum* $s(x)$ *is also a continuous function for that range.*

Let x and x_1 be two values of the variable within the range; we have to show that, given ϵ, it is possible to take x_1 so near to x that the difference $|s(x_1) - s(x)|$ shall be less than ϵ. With the usual notation we have

$$s(x_1) - s(x) = s_n(x_1) - s_n(x) + r_n(x_1) - r_n(x),$$

and therefore

$$|s(x_1) - s(x)| \leq |s_n(x_1) - s_n(x)| + |r_n(x_1)| + |r_n(x)|.$$

First, since the series is uniformly convergent, we can choose m so that if $n \geq m$ both $|r_n(x_1)|$ and $|r_n(x)|$ shall be less than $\epsilon/3$. Suppose m so chosen.

Next, $s_m(x)$ is the sum of a finite number of continuous functions and therefore we can take x_1 so near x that $|s_m(x_1) - s_m(x)|$ shall be less than $\epsilon/3$.

Combining the two results, we can take x_1 so near x that $|s(x_1) - s(x)|$ shall be less than three times $\epsilon/3$, that is less than ϵ.

G.C. 2 B

The theorem is thus established when x is *within* the interval; the slight modifications required when $x=a$ or b may be left to the reader.

Theorem II. *A Power Series $a_0+a_1x+a_2x^2+\ldots$ represents a continuous function within its interval of convergence $(-R, R)$; the function may, however, become discontinuous at an end of the interval.*

We will show that if $-R<a\leq x\leq b<R$ the series is uniformly convergent; the result then follows from Theorem I.

Take ρ less than R but greater than $|b|$ or $|a|$; then by § 150, Theorem V., the series is absolutely convergent when $x=\rho$. Also if $a\leq x\leq b$

$$a_n x^n = a_n \rho^n (x/\rho)^n; \quad |a_n x^n| < |a_n \rho^n|$$

and therefore $|r_n(x)| < |a_n\rho^n| + |a_{n+1}\rho^{n+1}| + \ldots$

But, the series $a_0+a_1\rho+a_2\rho^2\ldots$ being absolutely convergent, we can choose m so that when $n\geq m$ the remainder $|a_n\rho^n|+|a_{n+1}\rho^{n+1}|+\ldots$ shall be less than ϵ, and therefore for this m we shall have $|r_n(x)|$ less than ϵ. But this is the condition for uniform convergency.

The proof requires x to be within the interval. We refer to Chrystal's *Algebra*, vol. 2, chap. 26, § 20, for the proof of the theorem (*Abel's Theorem*) that if the series is convergent when $x=R$ (or $-R$), the function represented by the series is continuous *up to and including* the value R (or $-R$); in other words, the value of the function when $x=R$ is the same as that of the series when $x=R$.

The method by which the uniform convergency of the power series was established is easily extended to prove

Theorem III. *If the terms of a series are continuous functions of x when $a\leq x\leq b$, and if they are numerically less than the corresponding terms of an absolutely convergent series, whose terms do not contain x, the series will be uniformly convergent for the same range.*

The student must not mix up uniform and absolute convergence; a series may be uniformly and yet not absolutely convergent, though such series are rather beyond our limits.

The theorems contained in Examples 9, 10, 11 of the following Exercises should be specially noted.

CONTINUITY OF SERIES. 387

EXERCISES XXXIII.

1. Show that the following series are convergent:

(i) $1 + 2^{-2} + 3^{-3} + 4^{-4} + \ldots$; (ii) $x + x^4 + x^9 + x^{16} + \ldots (0 < x < 1)$;

(iii) $1/(a+1)^a + 1/(a+2)^a + 1/(a+3)^a + \ldots (a > 0, a > 1)$.

2. Show that the following series are divergent:

(i) $\frac{1}{2} + \frac{1}{4} + \frac{1}{8} + \ldots$; (ii) $1 + \frac{1}{3} + \frac{1}{5} + \ldots$; (iii) $\Sigma 1/(a+n)$;

(iv) $\Sigma (n+1)/(n^2+1)$; (v) $\Sigma (an+b)/(cn^2+d) [a \neq 0]$.

3. If
$$\frac{1}{1^2} + \frac{1}{2^2} + \frac{1}{3^2} + \ldots = c,$$

prove (i) $\frac{1}{2^2} + \frac{1}{4^2} + \frac{1}{6^2} + \ldots = \tfrac{1}{4}c$; (ii) $\frac{1}{1^2} + \frac{1}{3^2} + \frac{1}{5^2} + \ldots = \tfrac{3}{4}c$.

The value of c is $\pi^2/6$ (Exercises XXXIV., 22).

4. Show that the series (*the Binomial Series*)
$$1 + mx + \frac{m(m-1)}{1 \cdot 2} x^2 + \frac{m(m-1)(m-2)}{1 \cdot 2 \cdot 3} x^3 + \ldots$$

is absolutely convergent for every value of m when $|x| < 1$, but divergent when $|x| > 1$.

For $\dfrac{u_{n+1}}{u_n} = \dfrac{m-n+1}{n} x = \left(\dfrac{m+1}{n} - 1\right) x$; $\mathrm{L} \left|\dfrac{u_{n+1}}{u_n}\right| = |x|$.

5. Show that if $f(n)$ is a rational integral function of n, the series $\Sigma f(n) x^n$ is absolutely convergent when $|x| < 1$, but divergent when $|x| > 1$.

Let $f(n) = an^r + bn^{r-1} + \ldots$, the degree of $f(n)$ being r; then
$$\frac{u_{n+1}}{u_n} = \frac{a(n+1)^r + \ldots}{an^r + \ldots} x; \quad \mathrm{L} \frac{u_{n+1}}{u_n} = x.$$

6. If the series Σa, Σb are absolutely convergent, show that the series

(i) $a_0 + a_1 \cos x + a_2 \cos 2x + a_3 \cos 3x + \ldots$,

(ii) $b_1 \sin x + b_2 \sin 2x + b_3 \sin 3x + \ldots$

are absolutely convergent for every value of x, and represent continuous functions. It follows that if (i) [or (ii)] represents a *discontinuous* function, Σa (or Σb) cannot be *absolutely* convergent.

7. Show that if $x \neq 0$, the series
$$e^{-x} \cos(x - a_1) + e^{-2x} \cos(2x - a_2) + e^{-3x} \cos(3x - a_3) + \ldots$$
represents a continuous function.

8. Show that if $x \geq 0$, and if Σa is absolutely convergent, the series
$$a_1 e^{-x} \cos(x - a_1) + a_2 e^{-2x} \cos(2x - a_2) + a_3 e^{-3x} \cos(3x - a_3) + \ldots$$
represents a continuous function.

9. If the power series $a_0+a_1x+a_2x^2+...$ is zero for every value of x in the interval $(-R, R)$, show that every coefficient is zero.

When $x=0$, the series reduces to the term a_0; therefore $a_0=0$. We now have
$$0=a_1x+a_2x^2+...=x(a_1+a_2x+...)=xf_1(x) \text{ say.}$$
Hence, either $x=0$ or $f_1(x)=0$. Suppose $x \neq 0$; therefore $f_1(x)=0$. But $f_1(x)$ is a continuous function, and therefore the *limit* of $f_1(x)$ for $x=0$ is equal to the *value* of $f_1(x)$ for $x=0$. Hence $a_1=0$. Similarly $a_2=0$, $a_3=0$, and so on.

10. Theorem of Identical Equality. If the two power series $a_0+a_1x+a_2x^2+...$, $b_0+b_1x+b_2x^2+...$ are equal for every value of x in the interval $(-R, R)$, show that $a_0=b_0$, $a_1=b_1$, ...

For we have $\quad 0=(a_0-b_0)+(a_1-b_1)x+(a_2-b_2)x^2+...,$
and the results follow from ex. 9.

11. Multiplication of Series. Suppose the two series
$$s=a_0+a_1x+a_2x^2+..., \quad t=b_0+b_1x+b_2x^2+...$$
to contain only positive terms, and to be convergent when $x \leq R$; let
$$w_n = a_0b_0+(a_0b_1+a_1b_0)x+(a_0b_2+a_1b_1+a_2b_0)x^2+...$$
$$+(a_0b_{n-1}+a_1b_{n-2}+...+a_{n-1}b_0)x^{n-1}$$
where the terms of w_n are formed by multiplying s_n and t_n, no term of degree higher than $n-1$ being placed in w_n. Show that the limit of w_n is st, the product of the two given series.

A little consideration shows that
$$s_nt_n < w_{2n} < s_{n+1}t_{n+1}; \quad s_{n+1}t_{n+1} = w_{2n}+\sigma_{2n}.$$
The inequalities show that w_{2n} or, what amounts to the same thing, that w_n converges to st.

Next, let s and t contain both positive and negative terms, and let them be absolutely convergent when $|x| \leq R$. Let σ'_{2n} be the value of σ_{2n} when all the terms are made positive; then by the first part, which holds when all the terms are made positive, the limit of σ'_{2n} is zero. But σ_{2n} is not greater than σ'_{2n}, and therefore the limit of σ_{2n} is zero. Hence the limit of w_{2n} is st. The rule may fail if the series are only conditionally convergent.

12. Determine $a_0, a_1, a_2, ...$ so that
$$\frac{\cos\theta+x}{1+2x\cos\theta+x^2} = a_0+a_1x+a_2x^2+....$$

Assuming convergency, multiply up by $1+2x\cos\theta+x^2$, and equate coefficients. We have
$$\cos\theta+x = a_0+(a_1+2a_0\cos\theta)x+(a_2+2a_1\cos\theta+a_0)x^2+....$$
Hence $\quad \cos\theta=a_0; \quad 1=a_1+2a_0\cos\theta; \quad 0=a_2+2a_1\cos\theta+a_0; \quad ....$

EXERCISES XXXIII.

Solving these equations we find
$$a_0 = \cos\theta, \quad a_1 = -\cos 2\theta, \quad a_2 = \cos 3\theta, \quad \ldots, \quad a_n = (-1)^n \cos(n+1)\theta,$$
and the series becomes
$$\cos\theta - x\cos 2\theta + x^2 \cos 3\theta - x^3 \cos 4\theta + \ldots.$$

The series is convergent when $|x|<1$, and therefore the assumption that there was a convergent power series is justified.

13. Deduce from ex. 12, or prove independently, that when $|x|<1$,
$$\frac{1-x^2}{1+2x\cos\theta+x^2} = 1 - 2x\cos\theta + 2x^2 \cos 2\theta - 2x^3 \cos 3\theta + \ldots.$$

14. Show that if θ is neither zero nor a multiple of 2π, the series
$$\cos\theta + \tfrac{1}{2}\cos 2\theta + \tfrac{1}{3}\cos 3\theta + \ldots$$
is convergent.

Multiply s_n by $2\sin\tfrac{1}{2}\theta$, express the product of cosine and sine as a difference of sines, and rearrange; we thus find
$$2s_n \sin\tfrac{1}{2}\theta = -\sin\tfrac{1}{2}\theta + \tfrac{1}{2}\sin\frac{3\theta}{2} + \frac{1}{2.3}\sin\frac{5\theta}{2} + \ldots + \frac{1}{(n-1)n}\sin\frac{2n-1}{2}\theta$$
$$+ \frac{1}{n}\sin\frac{2n+1}{2}\theta,$$
and therefore
$$2s_n \sin\tfrac{1}{2}\theta = -\sin\tfrac{1}{2}\theta + \frac{1}{n}\sin\frac{2n+1}{2}\theta$$
$$+ \left\{\frac{1}{2}\sin\frac{3\theta}{2} + \ldots + \frac{1}{(n-1)n}\sin\frac{2n-1}{2}\theta\right\}.$$

But the expression in the bracket has a definite limit for $n=\infty$, since the infinite series $\frac{1}{2} + \frac{1}{2.3} + \frac{1}{3.4} + \ldots$ is convergent. Hence, $2s_n \sin\tfrac{1}{2}\theta$ has a definite limit, and therefore also s_n unless $\sin\tfrac{1}{2}\theta$ is zero.

15. Show, with similar restrictions to those in ex. 14, that the series whose n^{th} terms are
$$\frac{1}{n}\sin n\theta, \quad (-1)^{n-1}\frac{1}{n}\cos n\theta, \quad (-1)^{n-1}\frac{1}{n}\sin n\theta$$
are convergent.

CHAPTER XVIII.

TAYLOR'S THEOREM.

§ 152. Taylor's Theorem. In § 72 we obtained the equation
$$f(x) = f(a) + (x-a)f'(a) + \tfrac{1}{2}(x-a)^2 f''(x_1),$$
and although all we know of x_1 is that it lies between a and x, yet when $x-a$ is small, the function $f(x)$ will be approximately represented by the quadratic function
$$f(a) + (x-a)f'(a) + \tfrac{1}{2}(x-a)^2 f''(a),$$
whose coefficients depend only on the values of $f(x), f'(x), f''(x)$ when $x = a$. We will now discuss the general theorem of which this is a particular case; we will first obtain a closed expression involving an undetermined number like x_1, and then, instead of a quadratic function, we shall get a Power Series. We will slightly modify the method used in § 72 so as to require only one application of Rolle's Theorem.

Let $f(x)$ and its first n derivatives be continuous from $x = a$ to $x = b$, and consider the quantity Q defined by the equation
$$f(b) - \left\{ f(a) + (b-a)f'(a) + \tfrac{1}{2}(b-a)^2 f''(a) + \ldots + \frac{(b-a)^{n-1}}{(n-1)!} f^{(n-1)}(a) \right\} = (b-a)^n Q. \quad \ldots (1)$$

By Rolle's Theorem we can find another expression for Q which, when substituted in (1), gives the general theorem sought.

Let $F(x)$ be a function of x defined by the equation
$$F(x) = f(b) - f(x) - (b-x)f'(x) - \tfrac{1}{2}(b-x)^2 f''(x) \ldots$$
$$- \frac{(b-x)^{n-1}}{(n-1)!} f^{(n-1)}(x) - (b-x)^n Q. \quad \ldots (2)$$

TAYLOR'S AND MACLAURIN'S THEOREM.

By equation (1) $F(a)=0$; also $F(b)=0$ identically. Further $F(x)$ and $F'(x)$ are continuous from $x=a$ to $x=b$, since $f(x)$ and its first n derivatives are so, by hypothesis. Hence $F'(x)$ is zero for a value of x, say x_1, between a and b. But if we differentiate (2) and reduce, we find

$$F'(x) = -\frac{(b-x)^{n-1}}{(n-1)!}f^{(n)}(x) + n(b-x)^{n-1}Q, \ldots\ldots(3)$$

and therefore, since $(b-x_1)$ is not zero,

$$Q = \frac{1}{n!}f^{(n)}(x_1) = \frac{1}{n!}f^{(n)}\{a+\theta(b-a)\} \ldots\ldots\ldots(4)$$

where $0 < \theta < 1$, because any number between a and b may be represented by $a+\theta(b-a)$.

Substitute in (1) the value of Q given by (4), and transpose the terms $f(a), (b-a)f'(a)\ldots$ to the other side of the equation; we then get

$$f(b) = f(a) + (b-a)f'(a) + \tfrac{1}{2}(b-a)^2 f''(a) + \ldots$$
$$+ \frac{(b-a)^{n-1}}{(n-1)!}f^{(n-1)}(a) + \frac{(b-a)^n}{n!}f^{(n)}\{a+\theta(b-a)\}.\ldots(5)$$

We may now write x instead of b, the only reason for using the symbol b instead of x in (1) being to prevent confusion when applying the Mean Value Theorem; thus, finally,

$$f(x) = f(a) + (x-a)f'(a) + \tfrac{1}{2}(x-a)^2 f''(a) + \ldots$$
$$+ \frac{(x-a)^{n-1}}{(n-1)!}f^{(n-1)}(a) + \frac{(x-a)^n}{n!}f^{(n)}\{a+\theta(x-a)\}.\ldots(6)$$

The theorem expressed by equation (6) is called **Taylor's Theorem**. The particular case of it for which $a=0$, namely,

$$f(x) = f(0) + xf'(0) + \frac{x^2}{2}f''(0) + \ldots$$
$$+ \frac{x^{n-1}}{(n-1)!}f^{(n-1)}(0) + \frac{x^n}{n!}f^{(n)}(\theta x).\ldots\ldots\ldots\ldots(7)$$

is called **Maclaurin's Theorem**.

The conditions under which Taylor's Theorem has been proved are that $f(x)$ and its first n derivatives are *continuous* (and therefore *finite*) over the range from $x=a$ to the particular value of x for which $f(x)$ is taken. In regard to the number θ, all that can be said is that it is a positive proper fraction; it will usually be different for different values of n and of x.

Remainder in Taylor's Theorem. In equation (6) denote the sum of the first n terms by $S_n(x)$ and the last term by $R_n(x)$, so that

$$f(x) = S_n(x) + R_n(x); \quad R_n(x) = \frac{(x-a)^n}{n!} f^{(n)}\{a + \theta(x-a)\}. \ldots (8)$$

If we suppose n to increase indefinitely the sum on the right of (6) becomes an infinite series, and if the limit of $R_n(x)$ is zero the series is convergent. Since $f(x)$ and its first n derivatives are by hypothesis continuous, every derivative must remain continuous in order that it may be possible to suppose n to become infinite. We therefore have the

THEOREM. *If $f(x)$ and all its derivatives are continuous for the range considered and if the limit of $R_n(x)$ is zero, the infinite series*

$$f(a) + (x-a)f'(a) + \frac{(x-a)^2}{2!} f''(a) + \ldots \ldots \ldots (9)$$

derived from (6) by making n infinite, is convergent and represents the function $f(x)$, that is, converges to the value $f(x)$.[*]

The series (9) is called *Taylor's Series for $f(x)$*; when it is necessary to draw a distinction between (6) and (9) the former may be called Taylor's *Formula*. Of course all that has been said about Taylor's series applies to the particular case of it, Maclaurin's series

$$f(0) + xf'(0) + \frac{x^2}{2!} f''(0) + \ldots \ldots \ldots \ldots (10)$$

The value of $R_n(x)$ given by (8) is called *Lagrange's form of the remainder in Taylor's series*. Another useful form of the remainder is obtained by writing $(b-a)Q$

---

[*] Cases may be constructed in which the series (9) is convergent and yet does not converge to the value $f(x)$; such cases, however, may be safely assumed not to occur in ordinary work.

instead of $(b-a)^n Q$ in equation (1). The last term of equation (3) becomes simply Q and $(b-a)Q$ becomes
$$\frac{(b-a)(b-x_1)^{n-1}}{(n-1)!}f^{(n)}(x_1) \text{ or } \frac{(b-a)^n(1-\theta)^{n-1}}{(n-1)!}f^{(n)}\{a+\theta(b-a)\}.$$
Hence
$$R_n(x) = \frac{(x-a)^n(1-\theta)^{n-1}}{(n-1)!}f^{(n)}\{a+\theta(x-a)\}\ldots\ldots(11)$$
This form is called *Cauchy's form of the remainder*.

If we put $(b-a)^p Q$ instead of $(b-a)^n Q$ in (1) we get
$$R_n(x) = \frac{(x-a)^n(1-\theta)^{n-p}}{(n-1)!\,p}f^{(n)}\{a+\theta(x-a)\},\ldots\ldots(12)$$
called the *Schlömilch-Roche form of the remainder*; $p=n$ gives Lagrange's form and $p=1$ gives Cauchy's.

In (5) put x for a and $x+h$ for b; we get
$$f(x+h) = f(x) + hf'(x) + \frac{h^2}{2!}f''(x) + \ldots + \frac{h^n}{n!}f^{(n)}(x+\theta h),\ldots(13)$$
a value of $f(x+h)$ that is often useful.

We will now apply these theorems to the expansion of functions, and will usually employ Maclaurin's Theorem; the two forms of remainder to be used are
$$R_n(x) = \frac{x^n}{n!}f^{(n)}(\theta x); \quad R_n(x) = \frac{x^n(1-\theta)^{n-1}}{(n-1)!}f^{(n)}(\theta x),$$
the first being Lagrange's and the second Cauchy's.

§ 153. Examples.
1. sin x.
$$f(x) = \sin x; \quad f'(x) = \cos x; \quad f''(x) = -\sin x; \quad f'''(x) = -\cos x;$$
$$f^{(iv)}(x) = \sin x; \quad f^{(n)}(x) = \sin\left(x + \frac{n\pi}{2}\right).$$
Hence
$$f(0) = 0; \quad f'(0) = 1; \quad f''(0) = 0; \quad f'''(0) = -1; \quad f^{(iv)}(0) = 0;$$
$$f^{(n)}(0) = \sin\frac{n\pi}{2}; \quad f^{(n)}(\theta x) = \sin\left(\theta x + \frac{n\pi}{2}\right).$$

Since $\sin(n\pi/2)$ is 0 or ± 1 according as n is even or odd, the coefficients of the even powers of x will be zero, and only odd powers of x will occur, the terms being alternately positive and negative. Thus
$$\sin x = x - \frac{x^3}{3!} + \frac{x^5}{5!} - \frac{x^7}{7!} + \ldots + \frac{x^n}{n!}\sin\left(\theta x + \frac{n\pi}{2}\right).$$

Again, $$R_n(x) = \frac{x^n}{n!}\sin\left(\theta x + \frac{n\pi}{2}\right),$$
and therefore is not greater numerically than $x^n/n!$, which has zero for limit. We thus get the series
$$\sin x = x - \frac{x^3}{3!} + \frac{x^5}{5!} - \frac{x^7}{7!} + \ldots,$$
which is absolutely convergent for every finite value of x.

2. cos x.

In the same way $\cos x = 1 - \frac{x^2}{2!} + \frac{x^4}{4!} - \frac{x^6}{6!} + \ldots,$

the series being absolutely convergent for every finite value of x.

3. e^x.

$f(x) = e^x$; $f^{(n)}(x) = e^x$; $f(0) = 1$; $f^{(n)}(0) = 1$ for every n.
$$e^x = 1 + x + \frac{x^2}{2!} + \frac{x^3}{3!} + \ldots,$$
the series being absolutely convergent for every finite value of x.

4. $(1+x)^m$.

$f(x) = (1+x)^m$; $f^{(n)}(x) = m(m-1)\ldots(m-n+1)(1+x)^{m-n}$.
$f(0) = 1$; $f^{(n)}(0) = m(m-1)\ldots(m-n+1)$.
$f^{(n)}(\theta x) = m(m-1)\ldots(m-n+1)(1+\theta x)^{m-n}$.

Hence
$$(1+x)^m = 1 + mx + \frac{m(m-1)}{1 \cdot 2}x^2 + \ldots + \frac{m(m-1)\ldots(m-n+2)}{(n-1)!}x^{n-1} + R_n(x).$$

If m is a positive integer the series stops with the $(m+1)^{th}$ term, since $f^{(n)}(x) = 0$ when $n > m$; if m is not a positive integer we have to consider $R_n(x)$. We take Cauchy's form,
$$R_n(x) = \frac{m(m-1)\ldots(m-n+1)}{(n-1)!}x^n(1-\theta)^{n-1}(1+\theta x)^{m-n}.$$

The infinite series
$$1 + mx + \frac{m(m-1)}{1 \cdot 2}x^2 + \ldots$$
converges absolutely if $|x| < 1$ and diverges if $|x| > 1$ (Exer. xxxiii, 4); we therefore need only consider values of x such that $|x| \leqq 1$.

(A) $|x| < 1$. $R_n(x)$ may be written as the product of the three factors,
$$mx(1+\theta x)^{m-1}; \quad \left(\frac{1-\theta}{1+\theta x}\right)^{n-1}; \quad \frac{(m-1)(m-2)\ldots(m-n+1)}{(n-1)!}x^{n-1}.$$

The first factor is finite for every n since $(1+\theta x)^{m-1}$ lies between 1 and $(1+x)^{m-1}$. The second factor cannot exceed unity. The third factor has zero for limit, since it is the nth term of the convergent series
$$1 + (m-1)x + \frac{(m-1)(m-2)}{1 \cdot 2}x^2 + \ldots.$$

EXAMPLES OF EXPANSIONS. 395

Hence the limit of $R_n(x)$ is zero, and the infinite series converges to $(1+x)^m$ for every value of m so long as $-1 < x < 1$.

(B) $x = \pm 1$. These cases are of less importance, and the investigation of $R_n(x)$ is tedious. We will therefore merely state the results, referring for proof to Chrystal's *Alg.*, vol. 2, chap. 26, § 6.

$x = +1$; series absolutely convergent if $m > 0$, but conditionally if $0 > m > -1$; oscillating if $m = -1$; divergent if $m < -1$.

$x = -1$; series absolutely convergent if $m > 0$; divergent if $m < 0$.

If $a \neq b$ the binomial $(a+b)^m$ may be written $a^m(1+b/a)^m$ or $b^m(1+a/b)^m$ and then x put for b/a or for a/b according as b is less or greater than a numerically.

5. log (1+x).

It is not possible to expand $\log x$ by Maclaurin's Theorem since $\log x$ is infinite when $x = 0$. We may expand $\log x$ in powers of $(x-a)$, if a is positive, using Taylor's Theorem, but it is simpler to take $\log(1+x)$.

$$f(x) = \log(1+x); \quad f'(x) = \frac{1}{(1+x)}; \quad f^{(n)}(x) = \frac{(-1)^{n-1}(n-1)!}{(1+x)^n}$$

$$f(0) = 0; \quad f'(0) = 1; \quad f^{(n)}(0) = (-1)^{n-1}(n-1)!$$

$$\log(1+x) = x - \frac{x^2}{2} + \frac{x^3}{3} - \frac{x^4}{4} + \ldots + R_n(x).$$

The infinite series diverges if $|x| > 1$ and if $x = -1$; we therefore consider the remainder for $-1 < x \leq 1$.

For x positive, Lagrange's form

$$R_n(x) = (-1)^{n-1} \frac{1}{n} \left(\frac{x}{1+\theta x} \right)^n$$

shows that the limit is zero, since $\{x/(1+\theta x)\}^n$ is never greater than unity and the limit of $1/n$ is zero.

For x negative, Cauchy's form

$$R_n(x) = (-1)^{n-1} x^n \cdot \frac{1}{1+\theta x} \cdot \left(\frac{1-\theta}{1+\theta x} \right)^{n-1}$$

shows that when $|x| < 1$ the limit is zero; for the limit of x^n is zero and the other factors are finite for every value of n.

Hence $\log(1+x) = x - \dfrac{x^2}{2} + \dfrac{x^3}{3} - \dfrac{x^4}{4} + \ldots$

where $-1 < x \leq 1$; the series is conditionally convergent when $x = 1$.

We may note that, putting $x = 1$, we get

$$\log 2 = 1 - \frac{1}{2} + \frac{1}{3} - \frac{1}{4} + \ldots.$$

6. Calculation of Logarithms.

The series just found is too slowly convergent for purposes of calculation; a more rapidly convergent series is got as follows. We have

$$\log(1+x) = x - \frac{x^2}{2} + \frac{x^3}{3} - \frac{x^4}{4} + \ldots, \quad \ldots\ldots\ldots\ldots\ldots(1)$$

and by writing $-x$ in place of x
$$\log(1-x) = -x - \frac{x^2}{2} - \frac{x^3}{3} - \frac{x^4}{4} - \ldots \ldots \ldots \ldots \ldots (2)$$
By subtraction we find, since $\log(1+x) - \log(1-x) = \log\{(1+x)/(1-x)\}$,
$$\log\left(\frac{1+x}{1-x}\right) = 2\left\{x + \frac{x^3}{3} + \frac{x^5}{5} + \ldots + \frac{x^{2n-1}}{2n-1} + \ldots\right\} \ldots \ldots \ldots (3)$$
Suppose x positive and let
$$(1+x)/(1-x) = (y+1)/y; \text{ so that } x = 1/(2y+1) < 1.$$
Equation (3) becomes
$$\log(y+1) = \log y + 2\left\{\frac{1}{2y+1} + \frac{1}{3}\left(\frac{1}{2y+1}\right)^3 + \frac{1}{5}\left(\frac{1}{2y+1}\right)^5 + \ldots\right\}, \ldots (4).$$
from which $\log(y+1)$ is found when $\log y$ is known. It may be noticed that (4) is not a *power series* in y.

With very little labour the logarithms of the prime numbers 2, 3, 5, 7, ..., may be found; thus
$$y = 1; \quad \log 2 = 2\left\{\frac{1}{3} + \frac{1}{3 \cdot 3^3} + \frac{1}{5 \cdot 3^5} + \ldots\right\};$$
$$y = 2; \quad \log 3 = \log 2 + 2\left\{\frac{1}{5} + \frac{1}{3 \cdot 5^3} + \frac{1}{5 \cdot 5^5} + \ldots\right\}.$$
Then $\log 4 = 2\log 2$; $\log 5$ is obtained by putting 4 for y; $\log 6 = \log 2 + \log 3$; and so on. Series (4) converges rapidly even when $y = 2$.

For particular numbers special artifices may be used. Thus, if $y = 49$ equation (4) would give $\log 7$ when $\log 2$ and $\log 5$ are known, the series being very rapidly convergent.

The student is referred to Chrystal's *Algebra*, vol. 2, chap. 28, § 11, for further information and references.

7. Huyghens' Rule for the Length of a Circular Arc.

If a is the chord of the whole arc and b the chord of half the arc, then the length (l) of the arc is, approximately, $(8b - a)/3$.

Let the arc subtend at the centre of the circle an angle of θ radians, and let the radius of the circle be r; then $l = r\theta$ and
$$a = 2r\sin\tfrac{1}{2}\theta = 2r\{\tfrac{1}{2}\theta - \tfrac{1}{6}(\tfrac{1}{2}\theta)^3 + \tfrac{1}{120}(\tfrac{1}{2}\theta)^5 - \ldots\} \ldots \ldots \ldots (i)$$
$$b = 2r\sin\tfrac{1}{4}\theta = 2r\{\tfrac{1}{4}\theta - \tfrac{1}{6}(\tfrac{1}{4}\theta)^3 + \tfrac{1}{120}(\tfrac{1}{4}\theta)^5 - \ldots\} \ldots \ldots \ldots (ii)$$

Multiply (ii) by 8 and then subtract (i); we thus eliminate θ^3. Therefore
$$8b - a = 2r\left\{\tfrac{3}{2}\theta - \frac{3\theta^5}{120 \times 2^7} + \ldots\right\}$$
$$= 3l\{1 - \theta^4/7680 + \ldots\}.$$
Hence, neglecting the fourth and higher powers of θ, we find $l = (8b - a)/3$. It may be shown that for an angle of 30° the relative error is less than 1 in 100,000, for an angle of 45° less than 1 in 20,000, and for an angle of 60° less than 1 in 6000.

§ 154. Calculation of the nth Derivative.

The practical difficulty in finding a power series by Maclaurin's Theorem lies in the calculation of $f^{(n)}(x)$; indeed, there are few cases besides those already treated in which the nth derivative can be expressed in a manageable form. The discussion of the remainder, $R_n(x)$, is impossible unless we know $f^{(n)}(x)$; in special cases, however, we can find $f^{(n)}(0)$, and the infinite series, if it converges, will (in general) represent $f(x)$ within its range of convergence.

In this connection Leibniz's Theorem (§ 68) will be found very serviceable.

As an example consider $f(x) = \sin(a \sin^{-1} x)$. It would be difficult to calculate $f^{(n)}(x)$ directly; we will, therefore, first calculate $f'(x)$ and $f''(x)$, and will then form a differential equation to which Leibniz's Theorem may be applied and which will lead to the value of $f^{(n)}(0)$.

$$f(x) = \sin(a \sin^{-1} x);$$
$$f'(x) = a \cos(a \sin^{-1} x)/\sqrt{(1-x^2)}. \quad \text{................(i)}$$
$$f''(x) = -a^2 \sin(a \sin^{-1} x)/(1-x^2) + a \cos(a \sin^{-1} x) \cdot x/(1-x^2)^{\frac{3}{2}},$$
$$= -a^2 f(x)/(1-x^2) + x f'(x)/(1-x^2) \quad \text{................(ii)}$$

and therefore
$$(1-x^2)f''(x) - x f'(x) + a^2 f(x) = 0 \quad \text{................(iii)}$$

By making x zero in $f(x)$, $f'(x)$, $f''(x)$, we find
$$f(0) = 0;\quad f'(0) = a;\quad f''(0) = 0.$$

The function on the left of (iii) is always zero and therefore its nth derivative is always zero. The function, being a sum of products, may be differentiated n times by applying Leibniz's Theorem to each of its terms and then adding the results. For the first term let $f''(x) = u$, $(1-x^2) = v$. Every derivative of v above the second is zero; the nth derivative of $f''(x)$ is $f^{(n+2)}(x)$, the $(n-1)^{th}$ is $f^{(n+1)}(x)$, and so on. Thus,
$$D^n\{(1-x^2)f''(x)\} = (1-x^2)f^{(n+2)}(x) + {}_nC_1(-2x)f^{(n+1)}(x) + {}_nC_2(-2)f^{(n)}(x),$$
$$= (1-x^2)f^{(n+2)}(x) - 2nx f^{(n+1)}(x) - (n^2-n)f^{(n)}(x).$$

In the same way
$$D^n\{x f'(x)\} = x f^{(n+1)}(x) + n f^{(n)}(x).$$

Also,
$$D^n\{a^2 f(x)\} = a^2 f^{(n)}(x).$$

Adding we find, after a slight reduction,
$$(1-x^2)f^{(n+2)}(x) - (2n+1)x f^{(n+1)}(x) - (n^2-a^2)f^{(n)}(x) = 0 \quad \text{..........(iv)}$$

and therefore when $x=0$
$$f^{(n+2)}(0) = (n^2-a^2)f^{(n)}(0) \quad \text{................(v)}$$

From (v) we find in succession all the derivatives above the second for $x=0$, since we know the first two.

$$f^{(4)}(0)=(2^2-a^2)f''(0)=0;$$
$$f^{(6)}(0)=(4^2-a^2)f^{(4)}(0)=0,$$

and so on; thus every *even* derivative is zero. Again,

$$f^{(3)}(0)=(1^2-a^2)f'(0)=(1^2-a^2)a;$$
$$f^{(5)}(0)=(3^2-a^2)f^{(3)}(0)=(3^2-a^2)(1^2-a^2)a,$$

and so on, the general value being

$$f^{(2n-1)}(0)=a(1^2-a^2)(3^2-a^2)\ldots\{(2n-3)^2-a^2\}.$$

Hence,

$$\sin(a\sin^{-1}x)=ax+\frac{a(1^2-a^2)}{3!}x^3+\frac{a(1^2-a^2)(3^2-a^2)}{5!}x^5+\ldots\text{(vi)}$$

The series (vi) will terminate if a is an odd integer; in all other cases it will not terminate. The ratio of the term in x^{2n+1} to the term before it is

$$\frac{(2n-1)^2-a^2}{2n(2n+1)}x^2,$$

and since the limit of this ratio is x^2 the series (vi) is absolutely convergent so long as $-1<x<1$.

For many purposes only a few terms of the development of a function are required, and the calculation of a small number of derivatives may always be effected with more or less labour. Thus, the first three or four derivatives of $\log(1+\sin x)$ are easily calculated and the first three terms of the expansion obtained, $x-x^2/2+x^3/6$.

It is usually simpler, however, in cases like this to proceed as follows: suppose

$$y=a_1x+a_2x^2+\ldots;\quad f(y)=b_0+b_1y+b_2y^2+\ldots.$$

Substitute for y in the series $b_0+b_1y+b_2y^2+\ldots$ its value in terms of x and rearrange in powers of x; the series obtained will be convergent for sufficiently small values of x.

For example,

$$y=\sin x=x-\frac{x^3}{6}+\ldots;\quad \log(1+y)=y-\frac{y^2}{2}+\frac{y^3}{3}+\ldots,$$

and therefore

$$\log(1+\sin x)=\left(x-\frac{x^3}{6}+\ldots\right)-\frac{1}{2}\left(x-\frac{x^3}{6}+\ldots\right)^2+\frac{1}{3}\left(x-\frac{x^3}{6}+\ldots\right)^3-\ldots$$
$$=x-\frac{1}{2}x^2+\frac{1}{6}x^3-\ldots.$$

The proof of the method cannot be gone into here.

DIFFERENTIATION AND INTEGRATION OF SERIES.

§ 155. Differentiation and Integration of Series. The properties of a function are often most simply investigated by using an infinite series which represents the function; we must, therefore, see under what conditions a series may be differentiated or integrated term by term. The rules for differentiating and integrating a sum have been proved with the express limitation that the number of terms is *finite*; their extension to infinite series requires justification.

We begin with the theorem in integration; ϵ denotes as usual a given arbitrarily small positive quantity.

THEOREM I. *If the series* $u_1(x) + u_2(x) + \ldots$ *is* **uniformly convergent** *from* $x = a$ *to* $x = b$ *and converges to* $f(x)$, *then the series*
$$\int_c^x u_1(x)\,dx + \int_c^x u_2(x)\,dx + \ldots,$$
where $a \leq c < x \leq b$ *is also convergent and converges to the value*
$$\int_c^x f(x)\,dx.$$

Let $f(x) = s_n(x) + r_n(x)$ and let
$$\sigma_n(x) = \int_c^x s_n(x)\,dx; \quad \rho_n(x) = \int_c^x r_n(x)\,dx;$$
then
$$\sigma_n(x) = \int_c^x u_1(x)\,dx + \int_c^x u_2(x)\,dx + \ldots + \int_c^x u_n(x)\,dx,$$
and $\int_c^x f(x)\,dx = \sigma_n(x) + \rho_n(x).$

Now, since the series is uniformly convergent, we can choose m so that, if $n \geq m$, the remainder $r_n(x)$ will, for *every* value of x from $x = a$ to $x = b$, be less than ϵ; therefore, m being so chosen, if $n \geq m$ the quantity $\rho_n(x)$ is numerically less than
$$\int_c^x \epsilon\,dx, \text{ that is, } \epsilon(x-c).$$

Hence, if $n \geq m$, the difference
$$\int_c^x f(x)\,dx - \sigma_n(x),$$
is numerically less than $\epsilon(x-c)$; that is, the limit for $n = \infty$ of the difference is zero, and therefore
$$\int_c^x f(x)\,dx = \lim_{n=\infty} \sigma_n(x) = \int_c^x u_1(x)\,dx + \int_c^x u_2(x)\,dx + \ldots.$$

THEOREM II. *If the series* $u_1(x) + u_2(x) + \ldots$ *is convergent and converges to* $f(x)$ *when* $a \leq x \leq b$, *then the derivative of* $f(x)$ *is obtained by differentiating the series term by term, that is*
$$f'(x) = u_1'(x) + u_2'(x) + \ldots,$$
provided the series $u_1'(x) + u_2'(x) + \ldots$ *is* **uniformly** *convergent from* $x = a$ *to* $x = b$.

Let $\quad F(x) = u_1'(x) + u_2'(x) + \ldots;$
then by Theorem I., since $u_1'(x) + u_2'(x) + \ldots$ is uniformly convergent,
$$\int_c^x F(x)\, dx = \int_c^x u_1'(x)\, dx + \int_c^x u_2'(x)\, dx + \ldots,$$
$$= \{u_1(x) - u_1(c)\} + \{u_2(x) - u_2(c)\} + \ldots,$$
$$= \{u_1(x) + u_2(x) + \ldots\} - \{u_1(c) + u_2(c) + \ldots\},$$
$$= f(x) - \text{constant}.$$
Therefore
$$\frac{d}{dx}\int_c^x F(x)\, dx = f'(x); \text{ that is, } F(x) = f'(x).$$

By § 151 Theorem II. we see that a power series may be integrated term by term if x is *within* the interval of convergence.

We will now show that the series obtained by differentiating the power series is uniformly convergent when x is *within* the interval of convergence, and that the derivative of the series is therefore got by differentiating it term by term. For, in the notation of § 151, the series $\Sigma a_n \rho^n$ is absolutely convergent, and therefore $|a_n \rho^n|$ is finite, less than c say, for every n.

The series obtained by differentiation is
$$a_1 + 2a_2 x + 3a_3 x^2 + \ldots + n a_n x^{n-1} + \ldots.$$
Numerical values alone being considered, we have
$$n a_n x^{n-1} = n a_n \rho^{n-1} \left(\frac{x}{\rho}\right)^{n-1} < n \frac{c}{\rho}\left(\frac{x}{\rho}\right)^{n-1},$$
and therefore the terms of the series of derivatives are numerically less than the corresponding terms of the series
$$\frac{c}{\rho}\left\{1 + 2\left(\frac{x}{\rho}\right) + 3\left(\frac{x}{\rho}\right)^2 + \ldots\right\}$$

EXPANSIONS BY INTEGRATION OF SERIES.

But this series is absolutely convergent since the test ratio is x/ρ and x/ρ is numerically less than unity. Hence, the series of derivatives is uniformly convergent when x is *within* the interval of convergence of the power series $\Sigma a_n x^n$. (§ 151, Th. II.)

Ex. $\quad \log(1+x) = x - \tfrac{1}{2}x^2 + \tfrac{1}{3}x^3 - \ldots \quad (-1 < x \leqq 1)$.

By differentiation we find
$$1/(1+x) = 1 - x + x^2 - \ldots$$
This equation is true if $-1 < x < 1$, but not if $x = 1$.

§ 156. Examples. We will give two examples of the development of a function as a power series by the integration of a known series.

1. $\tan^{-1}x$.

If $-1 < x < 1$, we have
$$\frac{1}{1+x^2} = 1 - x^2 + x^4 - x^6 + \ldots + (-1)^n x^{2n} + \ldots, \quad \ldots\ldots\ldots\ldots(1)$$
and therefore, integrating from 0 to x,
$$\tan^{-1}x = x - \frac{x^3}{3} + \frac{x^5}{5} - \frac{x^7}{7} + \ldots + (-1)^n \frac{x^{2n+1}}{2n+1} + \ldots \quad \ldots\ldots\ldots(A)$$

The expansion (A) is proved for $|x| < 1$. The series (1) oscillates when $x = \pm 1$, but (A) is convergent for $x = \pm 1$; we may therefore apply Abel's theorem (p. 386), and deduce that (A) remains true even when $x = \pm 1$.

If $x = 1$ we find, since $\tan^{-1}1 = \pi/4$,
$$\frac{\pi}{4} = 1 - \frac{1}{3} + \frac{1}{5} - \frac{1}{7} + \ldots \quad \ldots\ldots\ldots\ldots\ldots(A_1)$$

The series (A_1) is called *Gregory's* (sometimes *Leibniz's*) series for π; it is too slowly convergent, however, to be suitable for calculation. A better series is got by using *Machin's* formula, namely
$$\frac{\pi}{4} = 4\tan^{-1}\left(\frac{1}{5}\right) - \tan^{-1}\left(\frac{1}{239}\right).$$
It will be a good exercise to calculate π from this formula by using the expansion (A); the series for $\tan^{-1}(1/5)$ and $\tan^{-1}(1/239)$ converge rapidly and give π to 5 or 6 decimals with little labour.

2. $\sin^{-1}x$.

If $-1 < x < 1$, we get by the binomial expansion
$$\frac{1}{\sqrt{(1-x^2)}} = (1-x^2)^{-\frac{1}{2}} = 1 + \frac{1}{2}x^2 + \frac{1.3}{2.4}x^4 + \frac{1.3.5}{2.4.6}x^6 + \ldots,$$
and therefore, integrating from 0 to x,
$$\sin^{-1}x = x + \frac{1}{2}\frac{x^3}{3} + \frac{1.3}{2.4}\frac{x^5}{5} + \frac{1.3.5}{2.4.6}\frac{x^7}{7} + \ldots.$$

The following example shows how we may obtain an approximate value of an integral by means of a series:

3. The time of a complete oscillation of a simple pendulum of length l, oscillating through an angle a on each side of the vertical is $4K\sqrt{(l/g)}$ where

$$K = \int_0^{\frac{\pi}{2}} \frac{d\phi}{\sqrt{(1-k^2\sin^2\phi)}}, \quad k = \sin\tfrac{1}{2}a \, ;$$

to find a series for K.

Expand $(1-k^2\sin^2\phi)^{-\frac{1}{2}}$ by the binomial theorem, and then integrate term by term. We have

$$\frac{1}{\sqrt{(1-k^2\sin^2\phi)}} = 1 + \frac{1}{2}k^2\sin^2\phi + \frac{1\cdot 3}{2\cdot 4}k^4\sin^4\phi + \dots .$$

The integrals of $\sin^2\phi$, $\sin^4\phi$, ... are given on page 285; therefore

$$K = \frac{\pi}{2}\left\{ 1 + \left(\frac{1}{2}\right)^2 k^2 + \left(\frac{1\cdot 3}{2\cdot 4}\right)^2 k^4 + \left(\frac{1\cdot 3\cdot 5}{2\cdot 4\cdot 6}\right)^2 k^6 + \dots \right\}.$$

When a is so small that k^2, k^4, ... may be neglected, $K = \pi/2$ and the period is $2\pi\sqrt{(l/g)}$.

4. To evaluate $\displaystyle\int_0^\pi \frac{\cos rx\, dx}{1 - 2a\cos x + a^2}$ (r a positive integer)

If $|a| < 1$ we have by ex. 13 Exer. XXXIII.

$$\frac{1}{1 - 2a\cos x + a^2} = \frac{1}{1-a^2}\{1 + 2a\cos x + 2a^2\cos 2x + 2a^3\cos 3x + \dots\}.$$

Also
$$\int_0^\pi \cos nx \cos rx\, dx = 0 \text{ if } n \neq r$$
$$= \frac{\pi}{2} \text{ if } n = r.$$

Therefore, when the series is multiplied by $\cos rx$ and integrated, every term will vanish except that arising from $2a^r\cos rx\cos rx$; we thus get

$$\int_0^\pi \frac{\cos rx\, dx}{1 - 2a\cos x + a^2} = \frac{\pi a^r}{1-a^2}. \quad\quad\dots\dots\dots\dots(i)$$

If $|a| > 1$ we have

$$\frac{1}{1 - 2a\cos x + a^2} = \frac{1}{a^2}\cdot\frac{1}{1 - 2\frac{1}{a}\cos x + \frac{1}{a^2}},$$

and we may expand in powers of $1/a$; or, we may write $1/a$ for a in (i) and then multiply by $1/a^2$. We find for the integral $\pi a^{-r}/(a^2-1)$.

EXERCISES XXXIV.

1. Prove that the following expansions hold for every finite value of x:

(i) $\sin(x+a) = \sin a + x\cos a - \dfrac{x^2}{2}\sin a - \dfrac{x^3}{3!}\cos a + \ldots,$

(ii) $e^x \cos x = 1 + x - \dfrac{2x^3}{3!} - \dfrac{2^2 x^4}{4!} - \ldots + 2^{\frac{n}{2}} \cos \dfrac{n\pi}{4} \dfrac{x^n}{n!} + \ldots,$

(iii) $e^x \sin x = x + x^2 + \dfrac{2x^3}{3!} - \dfrac{2^2 x^5}{5!} - \ldots + 2^{\frac{n}{2}} \sin \dfrac{n\pi}{4} \dfrac{x^n}{n!} + \ldots,$

(iv) $e^{x\cos a}\cos(x\sin a) = 1 + x\cos a + \dfrac{x^2}{2!}\cos 2a + \dfrac{x^3}{3!}\cos 3a + \ldots.$

Show that $D^n e^{x\cos a}\cos(x\sin a) = e^{x\cos a}\cos(x\sin a + na).$

2. From 1 (ii), (iii) derive the expansions of $\cosh x \cos x$, $\sinh x \sin x$, $\cosh x \sin x$, $\sinh x \cos x$.

3. Prove that if $|x| < 1$,
$$\log(1 + x + x^2) = x + \dfrac{x^2}{2} - \dfrac{2x^3}{3} + \dfrac{x^4}{4} + \ldots.$$

4. Show that, as far as the terms stated,

(i) $\sec x = 1 + \dfrac{x^2}{2} + \dfrac{5x^4}{24} + \dfrac{61x^6}{720};$

(ii) $\tan x = x + \dfrac{x^3}{3} + \dfrac{2x^5}{15} + \dfrac{17x^7}{315};$

(iii) $x \cot x = 1 - \dfrac{x^2}{3} - \dfrac{x^4}{45} - \dfrac{2x^6}{945}.$

These expansions may be obtained by division, replacing $\cos x$ and $\sin x$ by the equivalent series. Can $\cot x$ be developed by Maclaurin's Theorem?

5. If x is so small that squares and higher powers may be neglected, show that
$$\{\sqrt{(4+x)} + \sqrt[3]{(1+x)}\} \div \{\sqrt[4]{(1+4x)} + \sqrt[5]{(1-x)}\} = \dfrac{3}{2} - \dfrac{37x}{120}.$$

6. If $f(x) = x/(e^x - 1)$, show that the limits of $f(x)$ and of $f'(x)$ for $x = 0$ are 1 and $-\tfrac{1}{2}$ respectively. Show also by differentiating n times the equation
$$e^x f(x) - f(x) - x = 0$$
that $\quad e^x\{f^{(n)}(x) + {}_nC_1 f^{(n-1)}(x) + \ldots + {}_nC_1 f'(x) + f(x)\} = f^{(n)}(x),$

and therefore that if $n > 1$,
$${}_nC_1 f^{(n-1)}(0) + {}_nC_2 f^{(n-2)}(0) + \ldots + {}_nC_1 f'(0) + f(0) = 0,$$
the limits of the functions for $x = 0$ being taken as the values for $x = 0$.

7. If
$$\frac{x}{e^x - 1} = 1 - \tfrac{1}{2}x + \frac{B_1}{2!}x^2 - \frac{B_2}{4!}x^4 + \frac{B_3}{6!}x^6 - \ldots$$
show that $B_1 = 1/6$, $B_2 = 1/30$, $B_3 = 1/42 \ldots$.

The numbers $B_1, B_2, \ldots$ are called *Bernoulli's numbers* (see Chrystal's *Alg.*, vol. 2, chap. 28, § 6).

8. Show that
$$\frac{x}{e^x + 1} = \frac{x}{2} - (2^2 - 1)\frac{B_1}{2!}x^2 + (2^4 - 1)\frac{B_2}{4!}x^4 - (2^6 - 1)\frac{B_3}{6!}x^6 + \ldots$$

9. If $f(x) = (\sin^{-1} x)/\sqrt{(1-x^2)}$, show that
$$(1 - x^2) f'(x) - x f(x) = 1,$$
and that if $|x| < 1$,
$$\frac{\sin^{-1} x}{\sqrt{(1-x^2)}} = x + \frac{2}{3}x^3 + \frac{2 \cdot 4}{3 \cdot 5}x^5 + \frac{2 \cdot 4 \cdot 6}{3 \cdot 5 \cdot 7}x^7 + \ldots$$

10. Show from ex. 9 that
  (i) $\theta = \sin \theta \cos \theta \left(1 + \frac{2}{3}\sin^2\theta + \frac{2 \cdot 4}{3 \cdot 5}\sin^4\theta + \ldots\right)$.
  (ii) $\tan^{-1} z = \frac{z}{1 + z^2}\left\{1 + \frac{2}{3}\frac{z^2}{1+z^2} + \frac{2 \cdot 4}{3 \cdot 5}\left(\frac{z^2}{1+z^2}\right)^2 + \ldots\right\}$.

Put $x = \sin \theta$, $\tan \theta = z$.

11. Deduce from ex. 9 by integration that, if $|x| < 1$,
$$\tfrac{1}{2}\{\sin^{-1} x\}^2 = \frac{x^2}{2} + \frac{2}{3}\frac{x^4}{4} + \frac{2 \cdot 4}{3 \cdot 5}\frac{x^6}{6} + \ldots$$

12. Show that $\cos(a \sin^{-1} x)$ satisfies equation (iii), § 154, and prove the expansion ($|x| < 1$),
$$\cos(a \sin^{-1} x) = 1 - \frac{a^2}{2!}x^2 - \frac{a^2(2^2 - a^2)}{4!}x^4 - \frac{a^2(2^2 - a^2)(4^2 - a^2)}{6!}x^6 - \ldots$$

13. Prove from the series for $\sin(a \sin^{-1} x)$ and $\cos(a \sin^{-1} x)$ that
  (i) $\sin m\theta = m \sin \theta - \frac{m(m^2 - 1^2)}{3!}\sin^3\theta + \frac{m(m^2 - 1^2)(m^2 - 3^2)}{5!}\sin^5\theta - \ldots$.
  (ii) $\cos m\theta = 1 - \frac{m^2}{2!}\sin^2\theta + \frac{m^2(m^2 - 2^2)}{4!}\sin^4\theta - \ldots$.

Series for $\cos m\theta / \cos \theta$ and $\sin m\theta / \cos \theta$ may be obtained by differentiating $\sin(a \sin^{-1} x)$ and $\cos(a \sin^{-1} x)$.

14. Show that if $|x| < 1$,
  (i) $\log\{x + \sqrt{(1 + x^2)}\} = x - \frac{1}{2}\frac{x^3}{3} + \frac{1 \cdot 3}{2 \cdot 4}\frac{x^5}{5} - \frac{1 \cdot 3 \cdot 5}{2 \cdot 4 \cdot 6}\frac{x^7}{7} + \ldots$;
  (ii) $\tfrac{1}{2}\{\log(1+x)\}^2$
$$= \frac{x^2}{2} - \left(1 + \frac{1}{2}\right)\frac{x^3}{3} + \left(1 + \frac{1}{2} + \frac{1}{3}\right)\frac{x^4}{4} - \left(1 + \frac{1}{2} + \frac{1}{3} + \frac{1}{4}\right)\frac{x^5}{5} + \ldots$$

15. Prove that if $|x|<1$,

(i) $\quad e^{k\sin^{-1}x}=1+kx+\dfrac{k^2}{2!}x^2+\dfrac{k(k^2+1^2)}{3!}x^3+\dfrac{k^2(k^2+2^2)}{4!}x^4+\ldots$

(ii) $\quad \{x+\sqrt{(1+x^2)}\}^k=1+kx+\dfrac{k^2}{2!}x^2+\dfrac{k(k^2-1^2)}{3!}x^3+\dfrac{k^2(k^2-2^2)}{4!}x^4+\ldots$

To prove convergency, note that both in (i) and in (ii) the series formed of the odd terms and the series formed of the even terms are *separately* convergent or divergent according as $|x|$ is less than or greater than unity.

16. Show that, with the usual notation, the perimeter of an ellipse is

$$4a\int_0^{\frac{\pi}{2}}\sqrt{(1-e^2\sin^2\phi)}d\phi$$

$$=2\pi a\left\{1-\left(\frac{1}{2}\right)^2\frac{e^2}{1}-\left(\frac{1.3}{2.4}\right)^2\frac{e^4}{3}-\left(\frac{1.3.5}{2.4.6}\right)^2\frac{e^6}{5}-\ldots\right\}$$

17. Prove (i) the perimeter of an ellipse of small eccentricity e exceeds that of a circle of the same area in the ratio $1+3e^4/64$ approximately; (ii) the surface of an ellipsoid of revolution (either prolate or oblate) of small eccentricity e exceeds that of a sphere of equal volume by the fraction $2e^4/25$ of itself.

18. Show by integrating $(\cos\theta+x)/(1+2x\cos\theta+x^2)$ first with respect to x, and next with respect to θ (see Exer. XXXIII., 12), that if $|x|<1$,

(i) $\quad \frac{1}{2}\log(1+2x\cos\theta+x^2)=x\cos\theta-\dfrac{x^2}{2}\cos 2\theta+\dfrac{x^3}{3}\cos 3\theta-\ldots;$

(ii) $\quad \tan^{-1}\left(\dfrac{x\sin\theta}{1+x\cos\theta}\right)=x\sin\theta-\dfrac{x^2}{2}\sin 2\theta+\dfrac{x^3}{3}\sin 3\theta-\ldots$

19. Deduce from ex. 18, by taking the limit for $x=1$, that if $-\pi<\theta<\pi$,

(i) $\cos\theta-\frac{1}{2}\cos 2\theta+\frac{1}{3}\cos 3\theta-\ldots=\log(2\cos\frac{1}{2}\theta)$.

(ii) $\sin\theta-\frac{1}{2}\sin 2\theta+\frac{1}{3}\sin 3\theta-\ldots=\frac{1}{2}\theta$.

Show that the series (ii) does not represent the function $\theta/2$, except when $-\pi<\theta<\pi$, and that the *value* of the series when $\theta=\pi$ is zero, but that the *limit* of the series for $\theta=\pi$ is $\pi/2$. Show also that neither series can be differentiated term by term, although both are convergent (Exer. XXXIII., 15).

20. Deduce from 19 by putting $\theta=\pi-x$ that if $0<x<2\pi$,

(i) $\cos x+\frac{1}{2}\cos 2x+\frac{1}{3}\cos 3x+\ldots=-\log(2\sin\frac{1}{2}x)$,

(ii) $\sin x+\frac{1}{2}\sin 2x+\frac{1}{3}\sin 3x+\ldots=\dfrac{\pi}{2}-\dfrac{x}{2}$.

21. By integrating 20 (ii) show that if $0 \leq x \leq 2\pi$,
$$\frac{x^2}{4} - \frac{\pi x}{2} = \left(\frac{\cos x}{1^2} + \frac{\cos 2x}{2^2} + \frac{\cos 3x}{3^2} + \ldots\right) - C$$
where
$$C = \frac{1}{1^2} + \frac{1}{2^2} + \frac{1}{3^2} + \ldots$$

The series is *uniformly* convergent for every value of x, and we may therefore give to x the values 0 and 2π after integration. The series, however, is periodic, and does not represent the function $x^2/4 - \pi x/2$ outside the interval $(0, 2\pi)$.

22. Deduce from 21 that

(i) $\dfrac{1}{1^2} + \dfrac{1}{3^2} + \dfrac{1}{5^2} + \ldots = \dfrac{\pi^2}{8}$; (ii) $\dfrac{1}{1^2} + \dfrac{1}{2^2} + \dfrac{1}{3^2} + \ldots = \dfrac{\pi^2}{6}$;

(iii) $\dfrac{1}{1^2} - \dfrac{1}{2^2} + \dfrac{1}{3^2} - \dfrac{1}{4^2} + \ldots = \dfrac{\pi^2}{12}$.

To get (i) put $x = \pi$ in 21; (ii), (iii) readily follow. (Exercises XXXIII., 3.)

23. Show that

(i) $\displaystyle\int_0^1 \frac{1}{x} \log(1+x)\,dx = \frac{1}{1^2} - \frac{1}{2^2} + \frac{1}{3^2} - \ldots = \frac{\pi^2}{12}$,

(ii) $\displaystyle\int_0^1 \frac{1}{x} \log(1-x)\,dx = -\left(\frac{1}{1^2} + \frac{1}{2^2} + \frac{1}{3^2} + \ldots\right) = -\frac{\pi^2}{6}$,

(iii) $\displaystyle\int_0^{\frac{\pi}{4}} \tan\theta \log(\cot\theta)\,d\theta = \frac{1}{4}\left(\frac{1}{1^2} - \frac{1}{2^2} + \frac{1}{3^2} - \frac{1}{4^2} + \ldots\right) = \frac{\pi^2}{48}$.

To get (iii) put $\tan\theta = x$; note that $\underset{x=0}{L}\, x^n \log x = 0$ (Exercises VII., 10).

24. Prove

(i) $\dfrac{\cos x}{1^2} - \dfrac{\cos 2x}{2^2} + \dfrac{\cos 3x}{3^2} - \dfrac{\cos 4x}{4^2} + \ldots = \dfrac{\pi^2}{12} - \dfrac{x^2}{4}$,

(ii) $\dfrac{\cos x}{1^2} + \dfrac{\cos 3x}{3^2} + \dfrac{\cos 5x}{5^2} + \ldots = \dfrac{\pi^2}{8} - \dfrac{\pi x}{4}$.

In (i) $-\pi \leq x \leq \pi$; in (ii) $0 \leq x \leq \pi$.

25. Show that for every finite value of x

(i) $\dfrac{1}{\pi}\displaystyle\int_0^\pi \cos(x\cos\theta)\,d\theta = 1 - \dfrac{x^2}{2^2} + \dfrac{x^4}{2^2 \cdot 4^2} - \dfrac{x^6}{2^2 \cdot 4^2 \cdot 6^2} + \ldots$;

(ii) $\dfrac{1}{\pi}\displaystyle\int_0^\pi \cos(x\cos\theta)\sin^{2r}\theta\,d\theta$

$= \dfrac{(2r)!}{(r!\,2^r)^2}\left\{1 - \dfrac{x^2}{2(2r+2)} + \dfrac{x^4}{2\cdot 4(2r+2)(2r+4)} - \ldots\right\}.$

EXERCISES XXXIV.

26. If y denote the series (or integral) in 25 (i) show that
$$\frac{d^2y}{dx^2}+\frac{1}{x}\frac{dy}{dx}+y=0.$$

27. If u denote the series (or integral) in 25 (ii) and if $y=x^r u$ show that
$$x^2\frac{d^2y}{dx^2}+x\frac{dy}{dx}+(x^2-r^2)y=0.$$

y is called a *Bessel Function* of order r, and is (but for a numerical factor) usually denoted by $J_r(x)$; the function in 26 is $J_0(x)$. (See Gray and Mathews, *Bessel Functions*).

28. Show that if n is a positive integer
$$\sin x(1+2\cos 2x+2\cos 4x+\ldots+2\cos 2nx)=\sin(2n+1)x,$$
and then prove
$$\int_0^{\frac{\pi}{2}}\frac{\sin(2n+1)x}{\sin x}dx=\frac{\pi}{2}.$$

29. Prove the following results, a being positive and r a positive integer:

(i) $\int_0^{\pi}\log(1-2a\cos x+a^2)dx=0$ if $a<1$
$$=2\pi\log a\text{ if }a>1;$$

(ii) $\int_0^{\pi}\frac{x\sin x\,dx}{1-2a\cos x+a^2}=\frac{\pi}{a}\log(1+a)\text{ if }a<1$
$$=\frac{\pi}{a}\log\left(1+\frac{1}{a}\right)\text{ if }a>1;$$

(iii) $\int_0^{\pi}\cos rx\log(1-2a\cos x+a^2)dx=-\pi a^r/r\text{ if }a<1$
$$=-\pi a^{-r}/r\text{ if }a>1;$$

(iv) $\int_0^{\pi}\frac{\sin x\sin rx\,dx}{1-2a\cos x+a^2}=\frac{\pi}{2}a^{r-1}\text{ if }a<1$
$$=\frac{\pi}{2}a^{-(r+1)}\text{ if }a>1.$$

30. Prove

(i) $\int_0^x\frac{\sin x}{x}dx=x-\frac{x^3}{3!\,3}+\frac{x^5}{5!\,5}-\ldots$

(ii) $\int_0^1 x^x dx=1-\frac{1}{2^2}+\frac{1}{3^3}-\frac{1}{4^4}+\ldots$

To obtain (ii), put x^x in the form $e^{x\log x}$ and expand.

CHAPTER XIX.

TAYLOR'S THEOREM FOR FUNCTIONS OF TWO OR MORE VARIABLES. APPLICATIONS.

§ 157. Taylor's Theorem for Functions of two or more Variables. We will consider very briefly the expansions corresponding to Taylor's Theorem when there are more variables than one. The expressions for the remainder are very complicated and will not be written down although the form they would take can easily be gathered from the proof; any adequate discussion of the remainder, however, would take us too far into the theory of algebraic forms. It is, of course, assumed that the functions and their derivatives up to and including those that would appear in the remainder are all continuous.

We will take first the expansion of $f(x+h, y+k)$ in powers of h and k; this expansion corresponds to (13) of § 152.

$f(x+h, y+k)$ is the value of $f(x+ht, y+kt)$ when $t=1$; the latter function, considered as a function of t, can be expanded by Maclaurin's Theorem. For brevity denote $f(x+ht, y+kt)$ by $F(t)$ and let accents indicate derivatives with respect to t; then

$$F(t) = F(0) + tF'(0) + \frac{t^2}{2!}F''(0) + \ldots + R_n(t). \ldots\ldots(1)$$

We will now show how to express the t-derivatives of $F(t)$ in terms of the partial derivatives with respect to x and y of $F(t)$.

Let
$$x+ht = \alpha;\ y+kt = \beta, \ldots\ldots\ldots\ldots\ldots\ldots(2)$$

then
$$F'(t) = \frac{\partial F}{\partial \alpha}\frac{d\alpha}{dt} + \frac{\partial F}{\partial \beta}\frac{d\beta}{dt} = h\frac{\partial F}{\partial \alpha} + k\frac{\partial F}{\partial \beta}. \ldots\ldots(3)$$

But $\dfrac{\partial F}{\partial x} = \dfrac{\partial F}{\partial a}\dfrac{\partial a}{\partial x} = \dfrac{\partial F}{\partial a}$, since, by (2), $\dfrac{\partial a}{\partial x} = 1$,

and, similarly, $\dfrac{\partial F}{\partial y} = \dfrac{\partial F}{\partial \beta}$.

Thus (3) becomes

$$F'(t) = h\frac{\partial F}{\partial x} + k\frac{\partial F}{\partial y}. \quad\ldots\ldots\ldots\ldots\ldots(4)$$

The student will perhaps see the meaning of (4) more clearly by taking a particular case, say $F(t) = (x+ht)^m(y+kt)^n$. The example will also illustrate the fact that $F'(t)$ is a function of $x+ht$ and $y+kt$, and that $F''(t)$ may therefore be found in the way we now state.

Next $F''(t)$ is the t-derivative of $F'(t)$ and will be obtained by replacing $F(t)$ in (4) by $F'(t)$; thus

$$F''(t) = h\frac{\partial F'}{\partial x} + k\frac{\partial F'}{\partial y}$$

$$= h\left\{h\frac{\partial^2 F}{\partial x^2} + k\frac{\partial^2 F}{\partial x \partial y}\right\} + k\left\{h\frac{\partial^2 F}{\partial x \partial y} + k\frac{\partial^2 F}{\partial y^2}\right\}$$

$$= h^2\frac{\partial^2 F}{\partial x^2} + 2hk\frac{\partial^2 F}{\partial x \partial y} + k^2\frac{\partial^2 F}{\partial y^2}. \quad\ldots\ldots\ldots\ldots(5)$$

Similarly,

$$F'''(t) = h^3\frac{\partial^3 F}{\partial x^3} + 3h^2k\frac{\partial^3 F}{\partial x^2 \partial y} + 3hk^2\frac{\partial^3 F}{\partial x \partial y^2} + k^3\frac{\partial^3 F}{\partial y^3}\ldots(6)$$

The law of formation of the derivatives is now clear; we will show immediately how the value of $F^{(m)}(t)$ can be written down in a more compact form. We first consider the values of $F'(0)$, $F''(0)$, $F'''(0)$.

$F(0) = f(x, y)$ and the values of $F'(0)$, $F''(0)$, $F'''(0)$ are got by simply replacing the function $F(t)$ in (4), (5), (6) by $f(x, y)$. To get the Lagrangian form of the remainder, we must in $F^{(n)}(t)$ replace t by θt; if $n=3$, then in (6) $F(t)$ would be replaced by $f(x+h\theta t, y+k\theta t)$. Thus (1) becomes

$$f(x+ht, y+kt) = f(x, y) + t\left(h\frac{\partial f}{\partial x} + k\frac{\partial f}{\partial y}\right)$$

$$+ \frac{t^2}{2!}\left(h^2\frac{\partial^2 f}{\partial x^2} + 2hk\frac{\partial^2 f}{\partial x \partial y} + k^2\frac{\partial^2 f}{\partial y^2}\right) + \ldots \quad (7)$$

To get $f(x+h, y+k)$ put 1 for t in (7); therefore,

$$f(x+h, y+k) = f(x, y) + h\frac{\partial f}{\partial x} + k\frac{\partial f}{\partial y}$$
$$+ \frac{1}{2!}\left(h^2\frac{\partial^2 f}{\partial x^2} + 2hk\frac{\partial^2 f}{\partial x \partial y} + k^2\frac{\partial^2 f}{\partial y^2}\right) + \ldots \quad (8)$$

Equation (8) gives the required expansion; the expansion (7) is however a form that is useful.

The values of $F''(t)$, $F'''(t)$, in (5), (6) may be written more compactly in the *symbolical forms*

$$\left(h\frac{\partial}{\partial x} + k\frac{\partial}{\partial y}\right)^2 F, \quad \left(h\frac{\partial}{\partial x} + k\frac{\partial}{\partial y}\right)^3 F \ldots \ldots \ldots (9)$$

if these are interpreted as follows:—Let the binomial be expanded as if $h\frac{\partial}{\partial x}$ and $k\frac{\partial}{\partial y}$ were single quantities; after expansion place F as the last factor of each term and then replace a term like

$$3\left(h\frac{\partial}{\partial x}\right)^2\left(k\frac{\partial}{\partial y}\right)F$$

first by $3h^2 k \dfrac{\partial^3}{\partial x^2 \partial y}F$, then by $3h^2 k \dfrac{\partial^3 F}{\partial x^2 \partial y}$.

In this notation the $(m+1)^{\text{th}}$ term of (7) would be

$$\frac{t^m}{m!}\left(h\frac{\partial}{\partial x} + k\frac{\partial}{\partial y}\right)^m f$$
$$= \frac{t^m}{m!}\left(h^m \frac{\partial^m f}{\partial x^m} + mh^{m-1}k\frac{\partial^m f}{\partial x^{m-1}\partial y} + \ldots + k^m\frac{\partial^m f}{\partial y^m}\right).$$

The form (8) may be easily adapted to the expansion of $f(x+h, y+k)$ in powers of x and y; we have merely to interchange x with h and y with k. Using the suffix notation, we get

$$f(x+h, y+k) = f(h, k) + xf_h + yf_k$$
$$+ \frac{1}{2!}(x^2 f_{hh} + 2xy f_{hk} + y^2 f_{kk}) + \ldots \quad \ldots \ldots (10)$$

To form $f_h, f_{hh} \ldots$ we may differentiate $f(x, y)$ with respect to x and y, and then replace x by h and y by k.

SYMBOLICAL FORM OF DERIVATIVES.

In (10) we may of course suppose, if we please, $h=0$, $k=0$; we should thus get the expansion of $f(x, y)$ corresponding to Maclaurin's Theorem.

When there are three or more variables the expansions are similar. Thus for three variables

$$f(x+h,\ y+k,\ z+l) = f(x,\ y,\ z) + h\frac{\partial f}{\partial x} + k\frac{\partial f}{\partial y} + l\frac{\partial f}{\partial z}$$
$$+ \frac{1}{2!}\left(h\frac{\partial}{\partial x} + k\frac{\partial}{\partial y} + l\frac{\partial}{\partial z}\right)^2 f + \dots \quad (11)$$

where the symbolical expression is to be interpreted in the same way as before.

§ 158. Examples.

1. To find the equation of the tangent plane at the point $P(h, k, l)$ on the surface $f(x, y, z) = 0$.

The equations of the straight line through P, with the direction cosines λ, μ, ν, are

$$(x-h)/\lambda = (y-k)/\mu = (z-l)/\nu = r, \dots\dots\dots\dots\dots\dots(i)$$

where r is the distance from (h, k, l) to (x, y, z). Let (x, y, z) be the point Q on the surface; then

$$x = h + \lambda r,\ y = k + \mu r,\ z = l + \nu r\ ;\quad f(x, y, z) = 0.$$

In $f(x, y, z)$ put for x, y, z the values just written, and expand by Taylor's Theorem; therefore

$$0 = f(h, k, l) + r(\lambda f_h + \mu f_k + \nu f_l) + Ar^2 + \dots \quad\quad\quad \text{(ii)}$$

But $f(h, k, l) = 0$, since the point P is on the surface; therefore one value of r given by (ii) is zero. The other roots of (ii) are the distances from P to the several points in which the line (i) meets the surface. Let $r_1 = PQ$; then (ii) becomes, since r_1 is not zero,

$$0 = \lambda f_h + \mu f_k + \nu f_l + Ar_1 + \dots \quad\quad\quad\quad \text{(iii)}$$

As r_1 tends to zero the line (i) tends towards the position of a tangent line; but (iii) shows that as r_1 tends to zero, so does

$$\lambda f_h + \mu f_k + \nu f_l.$$

Hence the line (i) will be a tangent line if λ, μ, ν satisfy the equation
$$\lambda f_h + \mu f_k + \nu f_l = 0 \dots\dots\dots\dots\dots\dots\dots\dots\dots\dots\text{(iv)}$$

If we eliminate λ, μ, ν from equations (i) and (iv), we shall obtain an equation which is true for the coordinates of any point on any tangent line through P. The result of the elimination is

$$(x-h)f_h + (y-k)f_k + (z-l)f_l = 0,$$

the same equation, except for the notation, as was found in § 91.

2. Euler's Theorems of Homogeneous Functions.

DEFINITION. A function u of two or more variables is said to be homogeneous and of degree n if, when the variables x, y, ... are replaced by λx, λy, ... respectively, the function u becomes $\lambda^n u$ whatever the quantity λ may be.

Let $u = f(x, y)$ be a homogeneous function of degree n in two variables x, y. Then

$$xu_x + yu_y = nu, \quad \ldots\ldots\ldots\ldots\ldots\ldots\ldots\ldots\ldots\text{(i)}$$

$$x^2 u_{xx} + 2xy u_{xy} + y^2 u_{yy} = n(n-1)u. \quad \ldots\ldots\ldots\ldots\ldots\text{(ii)}$$

Replace x and y by $(1+t)x$ and $(1+t)y$, that is, by $x+xt$ and $y+yt$; then u becomes $(1+t)^n u$, that is,

$$f(x+xt, y+yt) = (1+t)^n u.$$

Expand the function on the left by Taylor's Theorem and that on the right by the Binomial Theorem; therefore

$$f(x, y) + t(xf_x + yf_y) + \frac{t^2}{2}(x^2 f_{xx} + 2xy f_{xy} + y^2 f_{yy}) + \ldots$$

$$= u + ntu + \frac{n(n-1)}{2} t^2 u + \ldots .$$

Equating coefficients of the same powers of t, we get equations (i), (ii).

It is easy to see that

$$\left(x \frac{\partial}{\partial x} + y \frac{\partial}{\partial y} \right)^m u = n(n-1) \ldots (n-m+1) u,$$

and that the theorems may be extended to homogeneous functions of three or more variables. For example,

$$xu_x + yu_y + zu_z = nu \ldots\ldots\ldots\ldots\ldots\ldots\ldots\ldots\ldots\text{(iii)}$$

Ex. Let $u = \tan^{-1}(y/x)$; then u is of zero-degree.

$$u_x = \frac{1}{1 + y^2/x^2} \times \frac{-y}{x^2} = \frac{-y}{x^2 + y^2}; \quad u_y = \frac{x}{x^2 + y^2};$$

$$xu_x + yu_y = \frac{-xy}{x^2 + y^2} + \frac{xy}{x^2 + y^2} = 0.$$

§ 159. Maxima and Minima of a function of two or more Variables.

DEFINITION. $f(a, b)$ is said to be a *maximum* value of $f(x, y)$ if $f(a+h, b+k)$ is *less* than $f(a, b)$ for all values of h and k, positive or negative, that lie between zero and certain finite values however small; $f(a, b)$ is said to be a *minimum* value of $f(x, y)$ if $f(a+h, b+k)$ is *greater* than $f(a, b)$ for all such values of h and k.

Similar definitions hold for functions of more than two variables.

We will assume the continuity of the functions and their derivatives for all values of the independent variables considered.

A *necessary* condition that $f(a, b)$ should be a maximum or a minimum (*a turning value*) is that *both* f_x *and* f_y *should be zero when* x=a, y=b. For $f(a, b)$ cannot be a turning value of $f(x, y)$ unless it is a turning value of the function $f(x, b)$ of x alone when $x=a$ and also a turning value of the function $f(a, y)$ of y alone when $y=b$; therefore $f_x(x, b)$ vanishes when $x=a$ and $f_y(a, y)$ when $y=b$.

To investigate *sufficient* conditions expand $f(a+h, b+k)$; we get

$$f(a+h, b+k) - f(a, b) = \tfrac{1}{2}(h^2 f_{aa} + 2hk f_{ab} + k^2 f_{bb}) + R \ldots (1)$$

where the terms hf_a, kf_b are omitted since $f_a=0, f_b=0$ when $f(a, b)$ is a turning value.

If $f(a, b)$ is a turning value the expression on the right of (1) must retain the same sign for all small values of h and k, the negative sign for a maximum value and the positive sign for a minimum value. Now R contains h and k in the *third* degree, if we suppose R to be the remainder in Taylor's Theorem; it seems natural therefore to assume that, for sufficiently small values of h and k, the sign will be that of the quadratic expression in h and k. Yet this assumption is not sound as the following example, given by Peano, will show.

Let $f(x, y) = 3x^2 - 6xy^2 + y^4$; then $a=0$, $b=0$, $f(a, b)=0$, and equation (1) becomes

$$f(h, k) = 3h^2 + (-6hk^2 + k^4). \ldots\ldots\ldots\ldots\ldots (2)$$

Here we have R exactly, $R = -6hk^2 + k^4$. The terms of second degree reduce to $3h^2$, and are therefore positive so long as h is not zero. Yet $f(h, k)$ is not of the same sign for all small values of h and k. For let $k = \sqrt{(\lambda h)}$, and we find

$$f(h, k) = (\lambda - 2)(\lambda - 4)h^2.$$

Hence $f(h, k)$ is positive or negative according as λ does not or does lie between 2 and 4. In other words, $f(0, 0)$ is not a minimum value of $f(x, y)$, even though the terms of second degree are positive except when $h=0$.

The difficulty just noticed would require a fuller consideration of the remainder in Taylor's Theorem than we have room to give. We therefore simply state that $f(a, b)$ will be a turning value if

$$f_{aa}f_{bb} > (f_{ab})^2,$$

and the value will be a maximum if f_{aa} (or f_{bb}) is negative, a minimum if f_{aa} (or f_{bb}) is positive.

It may be seen that a *necessary* condition that $f(a, b, c)$ should be a turning value of $f(x, y, z)$ is that f_x, f_y, f_z should all vanish when $x=a, y=b, z=c$.

In many cases it is known that a turning value of a function must exist; it is usual to assume without further proof that the values of the variables that make the first derivatives vanish are those that give the turning value.

§ 160. Examples. The most important cases are those in which the function whose turning values are required is given as a function of two or three or more variables, the variables being connected by one or more equations of condition. The best method of proceeding in such cases is usually the following. Let the function be u and let there be, say, four variables with two equations of condition,

$$u = f(x, y, z, w)\ (1);$$
$$\phi(x, y, z, w) = 0\ (2);\quad \psi(x, y, z, w) = 0\ (3).$$

Suppose for the moment that z and w are found from (2) and (3) in terms of x, y, and that these values are substituted in (1) which thus becomes a function of two independent variables x, y; let $D_x u, D_y u$ denote the first derivatives on the supposition that the substitutions have been made. For a turning value $D_x u$ and $D_y u$ must both be zero. Now

$$D_x u = f_x + f_z \frac{\partial z}{\partial x} + f_w \frac{\partial w}{\partial x}, \dots\dots\dots\dots\dots(4)$$

and $\partial z/\partial x, \partial w/\partial x$ are found by differentiating (2), (3); thus

$$\phi_x + \phi_z \frac{\partial z}{\partial x} + \phi_w \frac{\partial w}{\partial x} = 0 \dots (5);\quad \psi_x + \psi_z \frac{\partial z}{\partial x} + \psi_w \frac{\partial w}{\partial x} = 0.\dots(6)$$

Instead of solving (5), (6) for $\partial z/\partial x$ and $\partial w/\partial x$, multiply (5) by λ, (6) by μ and add to (4); therefore

$$D_x u = f_x + \lambda\phi_x + \mu\psi_x$$
$$+ (f_z + \lambda\phi_z + \mu\psi_z)\frac{\partial z}{\partial x} + (f_w + \lambda\phi_w + \mu\psi_w)\frac{\partial w}{\partial x}. \quad\ldots\ldots(7)$$

In exactly the same way we find
$$D_y u = f_y + \lambda\phi_y + \mu\psi_y$$
$$+ (f_z + \lambda\phi_z + \mu\psi_z)\frac{\partial z}{\partial y} + (f_w + \lambda\phi_w + \mu\psi_w)\frac{\partial w}{\partial y}. \quad\ldots\ldots(8)$$

It will be noticed that the coefficients of $\partial z/\partial x$ and $\partial w/\partial x$ in (7) are respectively equal to those of $\partial z/\partial y$ and $\partial w/\partial y$ in (8); therefore choose the multipliers λ, μ (and this is in general possible) so that these coefficients are zero, and the values of $D_x u$, $D_y u$ will reduce to the first three terms of (7), (8) respectively.

For the turning values of u the derivatives $D_x u$, $D_y u$ are zero; therefore for the turning values we have the four equations,
$$\left.\begin{array}{l} f_x + \lambda\phi_x + \mu\psi_x = 0, \quad f_y + \lambda\phi_y + \mu\psi_y = 0, \\ f_z + \lambda\phi_z + \mu\psi_z = 0, \quad f_w + \lambda\phi_w + \mu\psi_w = 0, \end{array}\right\} \ldots\ldots(9)$$
and these four equations together with equations (2), (3) are just sufficient to determine λ, μ and the values of x, y, z, w that give the turning values of u.

The equations (9) are symmetrical in x, y, z, w, and this method, called the method of *undetermined multipliers*, is specially simple when the functions f, ϕ, ψ are homogeneous. We have taken four variables and two equations of condition, but it is clear that the reasoning is quite general. We may state the rule for writing down the equations (9) thus:

Form $\qquad\qquad df + \lambda d\phi + \mu d\psi$

and equate to zero the coefficients of dx, dy, dz, dw.

Of course df means $f_x dx + f_y dy + f_z dz + f_w dw$ and $d\phi$, $d\psi$ have like meanings.

Ex. 1. $\quad u = x^2 + y^2 + z^2$ (1); $\phi = ax + by + cz - k = 0$ (2)

Clearly u has a minimum value; for, by (2), x, y, z cannot be simultaneously zero and u is always positive. Now
$$du + \lambda d\phi = (2x + \lambda a)dx + (2y + \lambda b)dy + (2z + \lambda c)dz,$$

and therefore, equating to zero the coefficients of dx, dy, dz, we find for the values of x, y, z that make u a minimum
$$x/a = -\lambda/2 = y/b = z/c.$$
By (2) each of these fractions is equal to $k/(a^2+b^2+c^2)$, and then by substitution for x, y, z in (1) we see that the minimum value of u is
$$k^2/(a^2+b^2+c^2).$$
The student may also solve the example by replacing z in (1) by its value $(k-ax-by)/c$ deduced from (2); he must be on his guard against confounding the value of u_x in this method with the value of u_x in the first method.

Ex. 2. Find the turning values of u when
$$u = a^2x^2 + b^2y^2 + c^2z^2, \quad \text{...(1)}$$
and
$$x^2 + y^2 + z^2 = 1 \; ; \quad \text{...(2)}$$
$$lx + my + nz = 0 \; . \quad \text{...(3)}$$
In this case u is really a function of only one variable, but the method of undetermined multipliers is equally applicable.

To get rid of the factor 2 we take λ, 2μ as the multipliers; then we readily find
$$a^2x + \lambda x + \mu l = 0, \quad b^2y + \lambda y + \mu m = 0, \quad c^2z + \lambda z + \mu n = 0. \quad \text{........(4)}$$
Multiply the first of equations (4) by x, the second by y, the third by z, and add; then taking note of (2), (3), we find
$$a^2x^2 + b^2y^2 + c^2z^2 + \lambda = 0, \text{ that is, } \lambda = -u,$$
where u *is now a turning value*, since the values of x, y, z that satisfy (4) are those that determine the turning values.

Put $-u$ for λ in (4), and we get
$$x = \mu l/(u - a^2), \quad y = \mu m/(u - b^2), \quad z = \mu n/(u - c^2).$$
If we now put these values of x, y, z in (3) the factor μ divides out and we get a quadratic equation for u,
$$l^2/(u - a^2) + m^2/(u - b^2) + n^2/(u - c^2) = 0. \quad \text{..................(5)}$$
One root of (5) will be the *maximum* value of u, and the other the *minimum*.

EXERCISES XXXV.

1. Verify Euler's theorem on Homogeneous Functions (taking first derivatives only) in the following cases.

 (i) $ax^2 + 2bxy + cy^2$; (ii) $ax^3 + by^3 + cz^3$; (iii) $\sqrt{x} + \sqrt{y} + \sqrt{z}$
 (iv) $(x+y)/(x^2+y^2)$; (v) $(x+y+z)/(x^2+y^2+z^2)$;
 (vi) $\tan^{-1}(r/z)$ where $r = \sqrt{(x^2+y^2+z^2)}$; (vii) $1/r$.

2. If u is homogeneous of degree n, prove

 (i) $xu_{xx} + yu_{xy} = (n-1)u_x$; (ii) $xu_{xy} + yu_{yy} = (n-1)u_y$.

3. Show that if a is positive $3axy - x^3 - y^3$ is a maximum when $x = a$, $y = a$, but neither a maximum nor a minimum when $x = 0$, $y = 0$.

EXERCISES XXXV.

4. The function $x^3y^2(6-x-y)$ is a maximum when $x=3$, $y=2$, but is neither a maximum nor a minimum when $x=0$, $y=0$.

5. Show that if a, b, c are positive, and if
$$a/x + b/y + c/z = 1,$$
the sum $x+y+z$ is a minimum, when
$$x/\sqrt{a} = y/\sqrt{b} = z/\sqrt{c} = \sqrt{a} + \sqrt{b} + \sqrt{c}.$$
Show also that if p, q, r are positive, the product $x^p y^q z^r$ is a minimum, when
$$px/a = qy/b = rz/c = p+q+r.$$

6. If $u = x^2 + y^2$, and if $ax^2 + 2hxy + by^2 = 1$, find the maximum and the minimum value of u, and interpret the result geometrically.

7. If $u = x^2 + y^2 + z^2$, and if
$$x^2/a^2 + y^2/b^2 + z^2/c^2 = 1, \text{ and } lx + my + nz = 0,$$
find the maximum and the minimum value of u, and interpret the result geometrically.

8. If
$$u = x_1^2 + x_2^2 + \ldots + x_n^2,$$
and if
$$a_1 x_1 + a_2 x_2 + \ldots + a_n x_n = k,$$
show that the minimum value of u is $k^2/(a_1^2 + a_2^2 + \ldots + a_n^2)$.

9. If x, y, z are the perpendiculars from any point P on the sides a, b, c of a triangle of area Δ, show that the minimum value of $x^2+y^2+z^2$ is
$$4\Delta^2/(a^2+b^2+c^2).$$

10. Show that the minimum value of
$$(a_1 x + b_1 y + c_1)^2 + (a_2 x + b_2 y + c_2)^2 + \ldots + (a_n x + b_n y + c_n)^2$$
is given by the values of x and y which satisfy the equations
$$(\Sigma a_1^2)x + (\Sigma a_1 b_1)y + (\Sigma a_1 c_1) = 0;$$
$$(\Sigma a_1 b_1)x + (\Sigma b_1^2)y + (\Sigma b_1 c_1) = 0.$$

11. Show that the centroid of n given points is the point, the sum of the squares of whose distances from the n points is a minimum.

12. Apply the method of undetermined multipliers to find the evolute of an ellipse considered as the envelope of the normals.

The normal is $\quad a^2 x/\alpha - b^2 y/\beta = a^2 - b^2$
where $\quad\quad\quad\quad \alpha^2/a^2 + \beta^2/b^2 = 1.$

Hence $\quad -\dfrac{a^2 x}{\alpha^2} + \lambda \dfrac{2\alpha}{a^2} = 0, \quad \dfrac{b^2 y}{\beta^2} + \lambda \dfrac{2\beta}{b^2} = 0,$

and therefore $\quad \lambda = \tfrac{1}{2}(a^2 - b^2), \quad \alpha^3 = a^4 x/(a^2-b^2), \text{ etc.}$

13. Show that the envelope of
$$x\alpha^m + y\beta^m = a^{m+1} \text{ where } \alpha^n + \beta^n = b^n$$
is
$$x^{\tfrac{n}{n-m}} + y^{\tfrac{n}{n-m}} = \left(\dfrac{a^{m+1}}{b^m}\right)^{\tfrac{n}{n-m}}.$$

G.C.   2 D

§ 161. Indeterminate Forms.

A function $f(x)$, that is in general well defined for a certain range of values of its argument, may for a particular value, a say, of its argument take a form (such as $0/0$) that has no meaning. It is possible, however, that $f(x)$ may have a definite *limit A* when x converges to a. Although $f(x)$ is really undefined, has no value that can be calculated by the ordinary rules of algebra, when $x=a$, yet it has become the established practice to call $f(a)$ in such a case an *indeterminate form*, and *to define A as the value of $f(x)$ when $x=a$*. The value thus assigned by the definition is usually called *the true value* of $f(x)$ when $x=a$.

If it be clearly understood that this "true value" is assigned by definition and is therefore arbitrary, there is a certain advantage from the procedure, namely, $f(x)$ becomes continuous up to and including the value a, it being supposed that $f(x)$ is in general continuous.

The typical indeterminate forms are

$$0/0 \ ; \ \infty/\infty \ ; \ \infty - \infty \ ; \ 0 \times \infty \ ; \ 0^0 \ ; \ \infty^0 \ ; \ 1^\infty.$$

We have already had some important cases of such forms; the derivative of $f(x)$ is a case of $0/0$.

$0 \times \infty$ is seen in $x \log x$ when $x=0$; the true value is zero.

$x^n e^{-x}$ or x^n/e^x when $x = +\infty$ gives $0 \times \infty$ or ∞/∞ and the limit is zero. (See Exercises VII., 8, 9.) It is easy to see that the result holds whether n is integral or fractional.

1^∞ is the case of $(1+x)^{\frac{1}{x}}$ when $x=0$; in this case the limit or true value is e (§ 48 Cor.).

In many cases the limits are found most simply by algebraical transformations and the use of series. We will take one or two examples before indicating the general theorems.

Ex. 1. $\quad \dfrac{x^{\frac{3}{2}}-1+(x-1)^{\frac{3}{2}}}{(x^2-1)^{\frac{3}{2}}-x+1}$ when $x=1$; form $\dfrac{0}{0}$.

Divide numerator and denominator by $(x^{\frac{1}{2}}-1)$; we see at once that the limit is $-3/2$. The "true value" of the fraction when $x=1$ is therefore $-3/2$.

Ex. 2. $\quad (\sin^{-1}x - x)/x^3$ when $x=0$; form $0/0$.

Expand $\sin^{-1}x \, (=x+x^3/6+ \dots)$; x cancels in numerator, and after dividing numerator and denominator by x^3 we get $1/6$ as the limit.

Ex. 3. $\sec x/\sec 3x$ when $x = \pi/2$; form ∞/∞.

Let $x = \tfrac{1}{2}\pi - u$; then
$$\underset{x=\frac{1}{2}\pi}{L} \frac{\sec x}{\sec 3x} = \underset{u=0}{L} \frac{-\sin 3u}{\sin u} = -3.$$

Ex. 4. $\dfrac{1}{x^2} - \cot^2 x$ when $x=0$; form $\infty - \infty$.

$$\frac{1}{x^2} - \cot^2 x = \left(1 + \frac{x}{\sin x}\cos x\right)\left(\frac{x}{\sin x}\right)\left(\frac{\sin x - x \cos x}{x^3}\right).$$

The limit of the first factor is 2 and of the second factor 1; also
$$\sin x - x\cos x = x - \frac{x^3}{6} + \ldots - x\left(1 - \frac{x^2}{2} + \ldots\right) = \frac{x^3}{3} + \ldots,$$
so that the limit of the third factor is 1/3.

Hence the limit or true value is 2/3.

Ex. 5. x^x when $x=0$; form 0^0.

Let $u = x^x$; then $\log u = x \log x$. The limit of $x \log x$ or $\log u$ is 0, as we have just seen, so that the limit of u or x^x is 1.

Ex. 6. $(1/x)^{\tan x}$ when $x=0$; form ∞^0.

The logarithm of the function is
$$-\tan x \log x = -\frac{\tan x}{x}(x \log x)$$
and has therefore 0 for limit; the limit of the function is therefore 1.

§ 162. Method of the Calculus.
We will now prove the general theorem for the evaluation of indeterminate forms, the continuity of the functions near the critical values being assumed.

THEOREM. *If $\phi(a)$ and $\psi(a)$ are either both zero or both infinite, and if $\phi'(x)/\psi'(x)$ converges to a limit when x converges to a, then $\phi(x)/\psi(x)$ converges to the same limit.*

It will save repetition to observe at once that if $\phi'(x)/\psi'(x)$ is indeterminate when $x=a$, the theorem shows that if $\phi''(x)/\psi''(x)$ converges to a limit then $\phi'(x)/\psi'(x)$ and therefore also $\phi(x)/\psi(x)$ converges to the same limit; and so on.

We need the following extension of the Mean Value Theorem of § 72; if $\phi(x)$, $\phi'(x)$, $\psi(x)$, $\psi'(x)$ are continuous for the range $a \leq x \leq b$ and if $\psi'(x)$ is not zero so long as $a < x < b$, then
$$\frac{\phi(b) - \phi(a)}{\psi(b) - \psi(a)} = \frac{\phi'(x_1)}{\psi'(x_1)}, \quad\ldots\ldots\ldots\ldots\ldots(1)$$
where $a < x_1 < b$ (*Generalised Theorem of Mean Value*).

The proof is obtained at once by considering $F(x)$ where

$$F(x) = \frac{\phi(b)-\phi(a)}{\psi(b)-\psi(a)}\{\psi(x)-\psi(a)\} - \{\phi(x)-\phi(a)\},$$

because $F(a)=0$, $F(b)=0$, and therefore $F'(x_1)=0$ and we can divide by $\psi'(x_1)$, for $\psi'(x_1)$ is not zero since x_1 lies between a and b.

I. **Form 0/0.** Let $\phi(a)=0$, $\psi(a)=0$, and in (1) put x for b;

therefore $\qquad \dfrac{\phi(x)}{\psi(x)} = \dfrac{\phi'(x_1)}{\psi'(x_1)} \quad (a < x_1 < x),$

and $\qquad \underset{x=a}{L}\dfrac{\phi(x)}{\psi(x)} = \underset{x_1=a}{L}\dfrac{\phi'(x_1)}{\psi'(x_1)} = \underset{x=a}{L}\dfrac{\phi'(x)}{\psi'(x)}.$

If $a = \infty$ the substitution $x = 1/z$ reduces the problem to the evaluation of the limit for $z=0$, and therefore the theorem holds in this case also.

II. **Form ∞/∞.** *First*, let $\phi(x)$, $\psi(x)$ be infinite when x is infinite. Let c be a large but finite value of x; then, by (1), putting x for b and c for a

$$\frac{\phi(x)-\phi(c)}{\psi(x)-\psi(c)} = \frac{\phi'(x_1)}{\psi'(x_1)} \quad (c < x_1 < x).$$

We may also write

$$\frac{\phi(x)-\phi(c)}{\psi(x)-\psi(c)} = \frac{\phi(x)}{\psi(x)} \times \frac{1-\phi(c)/\phi(x)}{1-\psi(c)/\psi(x)},$$

and therefore by equating values

$$\frac{\phi(x)}{\psi(x)} = \frac{\phi'(x_1)}{\psi'(x_1)} \times \frac{1-\psi(c)/\psi(x)}{1-\phi(c)/\phi(x)}.$$

Now, let c be taken so large that $\phi'(x_1)/\psi'(x_1)$ differs from its limit A by less than ϵ_1 and let c be then kept fixed; $\phi(c)$, $\psi(c)$ will, though large, be finite. Then let x be taken so large (and this choice is possible since $\phi(x)$, $\psi(x)$ tend to infinity) that the second fraction on the right shall differ from 1 by less than ϵ_2. The fraction $\phi(x)/\psi(x)$ is now the product of two factors, the first of which differs from A by less than ϵ_1 and the second of which differs from 1 by

EVALUATION BY METHOD OF CALCULUS.

less than ϵ_2 where ϵ_1, ϵ_2 may be as small as we please. Hence the limit of $\phi(x)/\psi(x)$ is A; that is

$$\mathop{L}_{x=\infty} \frac{\phi(x)}{\psi(x)} = \mathop{L}_{x=\infty} \frac{\phi'(x)}{\psi'(x)}.$$

Second, let $\phi(a)$, $\psi(a)$ be infinite, a being finite. The substitution $x = a + 1/z$ reduces the problem to the evaluation of the limit for $z = \infty$, and therefore the theorem holds in this case also.

The above proof is that given in the *Calculus* of Gennochi-Peano (German Translation, Leipzig: Teubner).

III. Other Forms. If $\phi(a) = 0$, $\psi(a) = \infty$, we may write

$$\phi(x) \times \psi(x) = \phi(x) \div \{1/\psi(x)\},$$

and the case reduces to case I.

The forms 0^0, ∞^0, 1^∞, are reduced by taking logarithms as illustrated in § 161, ex. 5, 6.

The form $\infty - \infty$ may be treated as in § 161, ex. 4; or expansion in series may be used.

Of course the method of differentiation may be combined with that of expansion in series.

Ex. 1. If n is positive, $(\log x)/x^n$ converges to zero when x becomes infinite; for

$$\mathop{L}_{x=\infty} \frac{\log x}{x^n} = \mathop{L}_{x=\infty} \frac{\frac{1}{x}}{nx^{n-1}} = \mathop{L}_{x=\infty} \frac{1}{nx^n} = 0.$$

Ex. 2. Find the limit for $x=0$, $y=0$ of the function of *two independent variables* $(x-y)/(x+y)$.

We take this example to illustrate the arbitrariness of the definition of a "true value," and also to show the great difference between limits for a function of one variable and limits for a function of two variables.

The above function may be made to tend to *any value whatever*; for let $y = \lambda x$ and we get

$$\mathop{L}\frac{x-y}{x+y} = \mathop{L}\frac{x-\lambda x}{x+\lambda x} = \frac{1-\lambda}{1+\lambda}.$$

By proper choice of λ we can make $(1-\lambda)/(1+\lambda)$ equal to any number whatever.

Geometrically, the z-axis lies on the surface

$$z(x+y) = x-y,$$

and as x and y tend to zero the point (x, y, z) may be made to approach any point on the z-axis.

EXERCISES XXXVI.

Find the limits (the "true values") of the functions in examples 1–15 for the given values of the argument.

1. $\{x-(n+1)x^{n+1}+nx^{n+2}\}/(1-x)^2$ when $x=1$.
2. $\{a-\sqrt{(a^2-x^2)}\}/x^2$ when $x=0$.
3. $x-\sqrt{(x^2-2ax)}$ when $x=\infty$.
4. $\sqrt[n]{\{(x+a_1)(x+a_2)...(x+a_n)\}}-x$ when $x=\infty$.

Put $x=1/z$ and expand by the binomial theorem.

5. $(1+1/x)^{x^2}$ and $(1+1/x^2)^x$ when $x=\infty$.
6. $\dfrac{\log x}{1-x}$ and $\dfrac{1-x+\log x}{1-\sqrt{(2x-x^2)}}$ when $x=1$.
7. $\dfrac{\tan x - x}{x - \sin x}$ and $\dfrac{\tan nx - n\tan x}{n\sin x - \sin nx}$ when $x=0$.
8. $\left(\dfrac{\pi}{2}-x\right)\tan x$ and $x\tan x - \dfrac{\pi}{2}\sec x$ when $x=\dfrac{\pi}{2}$.
9. $\log(1+ax)/\log(1+bx)$ and $(e^{ax}-e^{-ax})/\log(1+bx)$ when $x=0$.
10. $\dfrac{1}{x}-\dfrac{1}{x^2}\log(1+x)$ when $x=0$.
11. $(a^x-b^x)/(c^x-g^x)$ when $x=0$.
12. $\dfrac{\log\tan ax}{\log\tan bx}$ and $\dfrac{\log\tan ax - \log\tan bx}{\log\sin ax - \log\sin bx}$ when $x=0$.
13. $\left(\dfrac{a_1^x+a_2^x+...+a_n^x}{n}\right)^{\frac{n}{x}}$ when $x=0$.
14. $\dfrac{\sinh x - \sin x}{x^3}$ and $\dfrac{\cosh x - \cos x}{x^2}$ when $x=0$.
15. $(\cos ax)^{\operatorname{cosec}^2 bx}$ and $(\cos ax)^{\operatorname{cosec}^3 bx}$ when $x=0$.
16. If the equation of a curve is

$$u_2+u_3+u_4+\ldots=0$$

where $u_2, u_3, u_4, \ldots,$ are homogeneous of degrees 2, 3, 4, ..., in the coordinates, show that when the factors of u_2 are real, the equation $u_2=0$ gives the tangents at the origin.

Put $x=r\cos\theta$, $y=r\sin\theta$, and let $u_2, u_3, \ldots,$ become $r^2v_2, r^3v_3, \ldots$; then two values of r are zero since r^2 is a factor of $u_2+u_3+\ldots$. If, then, θ be chosen so that v_2 tends to zero, another value of r will tend to zero. The equation $v_2=0$ is a quadratic for $\tan\theta$, and therefore when its roots are real and different we get two gradients; when they are real and equal we get one gradient; when they are imaginary the values of $\tan\theta$ are imaginary, and the origin is then an isolated point.

EXERCISES XXXVI.

Definition. A point on a curve at which there are two distinct tangents is called a *node*.

At a node two branches of the curve cross each other, intersecting at a finite angle. In Fig. 61, p. 312, and in Fig. 63, p. 313, the origin is a node.

17. If $x^7 + 2x^6y + 5x^3 + a^2x^2 - b^2y^2 = 0$, find the value of dy/dx when $x = 0$, $y = 0$.

18. If, when x becomes infinite, $\phi(x)$ converges to zero, show that when x becomes infinite $\phi'(x)$, if it converges to a finite limit at all, will converge to zero.

Suppose $\phi'(x)$ has the limit A different from zero; the equation
$$\phi(x) = \phi(c) + (x-c)\phi'(x_1), \quad (c < x_1 < x)$$
shows that $\phi(x)$ must tend to infinity, because the term $(x-c)\phi'(x_1)$ tends to $(x-c)A$, that is, to infinity. But this is contrary to the hypothesis that $\phi(x)$ tends to zero, so that if A is finite, it must be zero.

19. Show that the series
$$\frac{1}{(\log 2)^a} + \frac{1}{(\log 3)^a} + \frac{1}{(\log 4)^a} + \dots$$
is divergent for every positive value of a.

Compare with $1/2 + 1/3 + 1/4 + \dots$; the limit for $n = \infty$ of
$$\frac{1}{(\log n)^a} \div \frac{1}{n}, \text{ that is, of } \left(n^{\frac{1}{a}}/\log n\right)^a$$
is infinite (§ 162, ex. 1). Hence the given series is divergent since the harmonic series is divergent. The series is obviously divergent when a is negative.

CHAPTER XX.

DIFFERENTIAL EQUATIONS.

§ 163. Differential Equations. We propose in this chapter to discuss a few differential equations that occur in elementary work. Nothing beyond the merest outline can be given; the student will find ample treatment in Forsyth's *Differential Equations* (Macmillan) or Murray's *Differential Equations* (Longmans).

An ordinary differential equation is an equation between one independent variable, one dependent variable and one or more derivatives of the dependent variable.

A partial differential equation is an equation between two or more independent variables, one dependent variable and partial derivatives of the dependent variable.

We deal only with ordinary differential equations.

The *order* of a differential equation is that of the highest derivative contained in it; the *degree* is that of the highest derivative when the equation is cleared of fractions and the powers of the derivatives are positive integers.

Thus the equation
$$x^2 y'' + xy' + (x^2 - r^2)y = 0$$
is of the second order and of the first degree. The equation
$$xy'^2 - yy' + a = 0$$
is of the first order and of the second degree.

By the theory of elimination explained in algebra we can eliminate *one* quantity from *two* equations, *two* quantities from *three* equations, n quantities from $(n+1)$ equations. Hence if an equation containing x, y and constants is differentiated once the new equation will contain x, y, y' and constants, and from the two equations *one* constant may be eliminated; the resulting equation will be a differential equation of the *first* order and will contain *one* constant fewer than the given equation.

Similarly, if the given equation is differentiated twice, we shall have three equations from which two constants may be eliminated; the resulting equation will be of the *second* order and will contain *two* constants fewer than the given equation; and so on.

The given equation is in each case called *the complete primitive* of the resulting differential equation and we see that the complete primitive contains *one, two* ... constants that do not occur in the differential equation when that equation is of the *first, second* ... order. In the process of elimination no account is taken of the particular value of the constants; these constants may therefore be called *arbitrary*.

Ex. 1. Let the given equation be
$$y = Ax^2 + B, \quad \quad \quad \quad (1)$$
and differentiate twice; we find
$$Dy = 2Ax; \quad \quad \quad \quad (2)$$
$$D^2y = 2A \quad \quad \quad \quad (3)$$

The first differentiation eliminates B; we can eliminate A from (2) and (3), getting
$$xD^2y - Dy = 0 \quad \quad \quad \quad (4)$$

Whatever be the value of B, equation (1) represents a parabola with a given latus rectum $1/A$, and with its axis lying along the y-axis; hence (2) is *the differential equation* of all such parabolas. Equation (4) again is the differential equation of all parabolas whose axes lie along the y-axis.

Ex. 2. Let the given equation be
$$(x-a)^2 + (y-b)^2 = c^2, \quad \quad \quad \quad (1)$$
and differentiate twice; we find
$$(x-a) + (y-b)Dy = 0; \quad \quad \quad \quad (2)$$
$$1 + (Dy)^2 + (y-b)D^2y = 0 \quad \quad \quad \quad (3)$$

If we eliminate a and b from equations (1), (2), (3), we find
$$c^2(D^2y)^2 = \{1 + (Dy)^2\}^3 \quad \quad \quad \quad (4)$$

Equation (4) is the differential equation of all circles with radius c; equation (2) is the differential equation of all circles whose centre is the point (a, b); equation (3) is that of all circles whose centres are on the line $y = b$.

§ **164. Complete Integral.** If in example 1 of last article we suppose equation (4) to be given, and if we pass from the differential equation to equation (1) we are said to *integrate* or *solve* the differential equation. From this point of view (1) is called *the complete integral* of (4), and A and B are called the arbitrary constants of integration.

Equation (4) is of the *second* order and (1) contains *two* arbitrary constants. It is proved in works on Differential Equations that a complete integral exists for every differential equation and that when the equation is of the nth order the integral contains n arbitrary constants.

A *particular integral* is one obtained by assigning a definite value to one or more of the arbitrary constants in the complete integral. Thus $y = x^2 - 1$, $y = 2x^2$, $y = x^2$ are particular integrals of (4) in Ex. 1 of last article.

Another way of considering the integration of a differential equation may be illustrated thus:—Find a function y (i) that shall satisfy the equation

$$x D^2 y - Dy = 0.$$

(ii) that shall be equal to b when $x = a$ and (iii) that shall have its first derivative equal to c when $x = a$.

Since the complete integral $y = Ax^2 + B$ contains *two* arbitrary constants A, B we can determine them to satisfy conditions (ii), (iii). These conditions give

$$b = Aa^2 + B;\ c = 2Aa.$$

so that $\qquad A = c/2a,\ B = b - \tfrac{1}{2}ac,$

and the function $\qquad y = \dfrac{c}{2a}x^2 + \dfrac{2b - ac}{2}$

satisfies conditions (i), (ii), (iii).

For another illustration of a similar kind see § 69, exs. 1, 2.

The student should work through the following set of Exercises; several of the differential equations occur frequently in physical applications. The primitive, considered as the integral of the differential equation, is in each case the *complete integral*. It will be noticed (see examples 7, 8) that the one differential equation may arise from different primitives into which the constants enter in different forms.

EXERCISES XXXVII.

1. If $y = Ax + B$, then $D^2y = 0$. What is the geometrical meaning of the equations $Dy = A$, $D^2y = 0$?

2. If $y = Ax^{n-1} + Bx^{n-2} + \ldots + Kx + L$, a rational integral function of degree $(n-1)$, prove $D^ny = 0$.

3. If $y = Ax^2 + Bx + C$, then $D^3y = 0$. Interpret geometrically

4. If $y = Ax^3 + Bx^2 + Cx$, then
$$D^3\left(\frac{y}{x}\right) = 0 \text{ or } x^3D^3y - 3x^2D^2y + 6xDy - 6y = 0.$$

5. If $y = A/x + B$, then
$$x^2Dy = -A, \quad D^2y + \frac{2}{x}Dy = 0.$$

6. If $y = A \log x + B$, then
$$xDy = A, \quad D^2y + \frac{1}{x}Dy = 0.$$

7. If $y = A \cos nx + B \sin nx$ or if $y = C \cos(nx - E)$ then
$$D^2y + n^2y = 0.$$

8. If $y = Ae^{nx} + Be^{-nx}$ or if $y = C \cosh nx + E \sinh nx$, then
$$D^2y - n^2y = 0.$$

9. If $y = A/x + B + x^2$, then
$$D^2y + \frac{2}{x}Dy = 6.$$

10. If $y = A \cos nx + B \sin nx + E \cos px + F \sin px$ where A, B are arbitrary and n, p unequal, prove
$$D^2y + n^2y = (n^2 - p^2)E \cos px + (n^2 - p^2)F \sin px.$$

11. If $y = e^{-\frac{1}{2}kx}(A \cos nx + B \sin nx)$, then
$$D^2y + k\,Dy + (n^2 + \tfrac{1}{4}k^2)y = 0.$$

12. If $y = e^{-\frac{1}{2}kx}(Ae^{nx} + Be^{-nx})$, then
$$D^2y + k\,Dy - (n^2 - \tfrac{1}{4}k^2)y = 0.$$

13. If $y = Ae^{mx} + Be^{nx}$, then
$$D^2y - (m+n)Dy + mny = 0.$$

14. If $y = (A + Bx)e^{nx}$, then
$$D^2y - 2n\,Dy + n^2y = 0.$$
(Compare 13 and 14.)

15. If $y = (A + Bx) \cos nx + (E + Fx) \sin nx$, then
$$D^4y + 2n^2 D^2y + n^4y = 0.$$

16. If $y = (A \cos nx + B \sin nx)/x$, then
$$D^2(xy) + n^2xy = 0, \text{ or } D^2y + \frac{2}{x}Dy + n^2y = 0.$$

428 AN ELEMENTARY TREATISE ON THE CALCULUS.

17. If $y = (Ae^{nx} + Be^{-nx})/x$, then
$$D^2y + \frac{2}{x}Dy - n^2y = 0.$$

18. If $y = mx + a/m$, m being arbitrary, then
$$x(Dy)^2 - y\,Dy + a = 0.$$

19. If $x^2/(a^2+k) + y^2/(b^2+k) = 1$, k being arbitrary, then
$$xy(Dy)^2 + (x^2 - y^2 - a^2 + b^2)Dy - xy = 0.$$

The primitive represents a family of central conics having the same foci (confocal conics).

20. Show that the complete integral of equation (iii) § 154 is
$$f(x) = A\sin(a\sin^{-1}x) + B\cos(a\sin^{-1}x).$$

§ 165. Equations of the First Order and of the First Degree. We will now state one or two types of equations which can be readily integrated; at any rate their integration can be reduced to the evaluation of an ordinary integral. So far as the theory of differential equations is concerned, the solution may be considered to be obtained when the equation is reduced to either of the forms
$$\frac{dy}{dx} = f(x), \quad \frac{dy}{dx} = F(y),$$

for these equations give at once
$$y = \int f(x)\,dx + C; \quad x = \int \frac{dy}{F(y)} + C,$$

and the rest of the work is ordinary integration.

Type I. Variables Separable. The variables are said to be separable when the equation may be written
$$f(x)\,dx + F(y)\,dy = 0,$$
where $f(x)$ is a function of x alone and $F(y)$ a function of y alone. The solution is
$$\int f(x)\,dx + \int F(y)\,dy = C.$$

Ex. 1. $\qquad n(x+a)Dy + m(y+b) = 0.$

We have $\qquad \dfrac{n\,dy}{y+b} + \dfrac{m\,dx}{x+a} = 0;$

therefore $\qquad n\log(y+b) + m\log(x+a) = \text{const.},$
or $\qquad \log\{(y+b)^n(x+a)^m\} = \text{const.},$
or $\qquad (y+b)^n(x+a)^m = C.$

TYPES OF EQUATIONS OF FIRST ORDER. 429

Any one of the last three equations may be taken as the solution, but the last form is usually the most convenient since the integral is algebraic.

Type II. Homogeneous Equations. An equation is called homogeneous when it is of the form
$$Dy = f(x, y)/F(x, y),$$
where $f(x, y)$, $F(x, y)$ are homogeneous and of the same degree in x and y.

To solve, change the dependent variable by the substitution $y = vx$; the equation becomes
$$xDv + v = f(1, v)/F(1, v) = \phi(v),$$
and the variables are now separable.

Ex. 2. $\quad 2xy Dy = x^2 + y^2.$
We find $\quad 2x^2 v(xDv + v) = x^2(1 + v^2),$
whence $\quad \dfrac{dx}{x} - \dfrac{2v\,dv}{1 - v^2} = 0;$
therefore, $\quad \log\{x(1 - v^2)\} = \text{const.} = \log C,$
or $\quad x^2 - y^2 = Cx.$

The equation $(ax + by + c)Dy = a'x + b'y + c'$ may be made homogeneous by the substitutions
$$\xi = ax + by + c, \quad \eta = a'x + b'y + c',$$
provided $ab' - a'b$ is not zero. (See Exer. XXXVIII., 6, 7.)

Type III. Linear Equations. An equation is said to be *linear* if the dependent variable and its derivatives occur in it *only in the first degree*. The linear equation of the first order is therefore of the form
$$Dy + Py = Q,$$
where P, Q are functions of x (or constants).

Let $P_1 = \int P\,dx$ and multiply by e^{P_1};

then since $\quad De^{P_1} = e^{P_1} DP_1 = e^{P_1} P,$
we find $\quad e^{P_1} Dy + e^{P_1} Py = D(e^{P_1} y).$
Hence, $\quad D(e^{P_1} y) = e^{P_1} Q;$
and therefore $\quad e^{P_1} y = \int e^{P_1} Q\, dx + C.$

Cor. The equation $Dy + Py = Qy^n$ may be reduced to the linear form by putting $v = y^{-n+1}$ and taking v as the dependent variable.

Ex. 3. $(1-x^2)Dy + xy = ax.$

Here $$Dy + \frac{x}{1-x^2}y = \frac{ax}{1-x^2},$$

and $$P_1 = \int \frac{x\,dx}{1-x^2} = -\tfrac{1}{2}\log(1-x^2) = \log\frac{1}{\sqrt{(1-x^2)}},$$

$$e^{P_1} = 1/\sqrt{(1-x^2)}.$$

Therefore $$\frac{1}{\sqrt{(1-x^2)}}y = \int \frac{ax\,dx}{(1-x^2)^{\frac{3}{2}}} + C = \frac{a}{\sqrt{1-x^2}} + C,$$

and $$y = a + C\sqrt{(1-x^2)}.$$

Ex. 4. When an electric current of strength x is flowing in a circuit of inductance L and resistance R subject to an extraneous electromotive force E, the equation of the current at time t is

$$L\dot{x} + Rx = E.$$

First suppose E constant, equal to E_0, L and R being constant. We have
$$\dot{x} + \frac{R}{L}x = \frac{E_0}{L},$$

and therefore $$e^{\frac{Rt}{L}}x = \frac{E_0}{L}\int e^{\frac{Rt}{L}}dt + C = \frac{E_0}{R}e^{\frac{Rt}{L}} + C,$$

and $$x = \frac{E_0}{R} + Ce^{-\frac{Rt}{L}}.$$

When $t=0$ the current $x=0$, and therefore $C = -E_0/R$; hence

$$x = \frac{E_0}{R}\left(1 - e^{-\frac{Rt}{L}}\right).$$

The part $E_0 e^{-\frac{Rt}{L}}/R$ is the extra or induced current and dies away to zero as the total current attains its steady value E_0/R.

Next suppose $E = E_0 \cos(pt - a)$; then since

$$\int e^{\frac{Rt}{L}}\cos(pt-a)\,dt = \frac{Le^{\frac{Rt}{L}}}{R^2+p^2L^2}\{R\cos(pt-a) + pL\sin(pt-a)\},$$

we find $$x = Ce^{-\frac{Rt}{L}} + \frac{E_0}{R^2+p^2L^2}\{R\cos(pt-a) + pL\sin(pt-a)\}.$$

As t increases, the term $Ce^{-Rt/L}$ becomes of less and less importance; the other term gives the steady oscillation. The steady oscillation may be put in the form

$$x = \frac{E_0}{\sqrt{(R^2+p^2L^2)}}\cos(pt - a - a_1)$$

where $\tan a_1 = pL/R$. The quantity $\sqrt{(R^2+p^2L^2)}$ is called the *impedance* of the circuit.

LINEAR EQUATIONS. EXACT EQUATIONS.

Type IV. Exact Equations. The equation
$$M + NDy = 0, \text{ or, } Mdx + Ndy = 0,$$
where M, N are functions of x and y, is called an *exact* equation if $Mdx + Ndy$ is a complete differential, that is, if $\partial M/\partial y$ is equal to $\partial N/\partial x$ (§ 94). In this case there exists a function u such that
$$du = Mdx + Ndy,$$
and, obviously, the integral is $u = $ constant.

Ex. 5. $2xy - y^2 + 2x + (x^2 - 2xy + 2y)Dy = 0.$
Here $M = 2xy - y^2 + 2x, \; N = x^2 - 2xy + 2y,$
and $\partial M/\partial y = 2x - 2y = \partial N/\partial x,$

so that the equation is exact. Knowing that the equation is exact, we can readily arrange $Mdx + Ndy$ as a sum of complete differentials; we find
$$(2xy\,dx + x^2 dy) - (y^2 dx + 2xy\,dy) + 2x\,dx + 2y\,dy,$$
that is, $d(x^2 y) - d(xy^2) + d(x^2) + d(y^2),$
so that $u = x^2 y - xy^2 + x^2 + y^2,$
and the integral is $x^2 y - xy^2 + x^2 + y^2 = C.$

Ex. 6. $x^3 - 2y^2 + 2xy\,Dy = 0.$

This equation is not exact, but it becomes exact when multiplied by $1/x^3$. We find
$$\frac{x^3 - 2y^2}{x^3} + \frac{2y}{x^2} Dy = D\left(\frac{x^3 + y^2}{x^2}\right),$$
and the integral is $(x^3 + y^2)/x^2 = C$, or, $x^3 + y^2 = Cx^2$.

The factor $1/x^3$ which makes the equation exact is called an *integrating factor*; when an equation is not exact it may be possible to guess an integrating factor and thus integrate it.

§ 166. Equations of First Order but not of First Degree.

Let Dy be denoted by p; the equation, when of the nth degree, will have the form
$$Ap^n + Bp^{n-1} + \ldots + Kp + L = 0 \ldots\ldots\ldots\ldots(1)$$
where $A, B, \ldots$ are functions of x and y (or constants).

If possible, solve for p; there will be in general n values
$$p = p_1, \; p = p_2, \ldots$$
and each of these equations when integrated will give a relation between x and y that will satisfy (1).

Ex. 1. $xyp^2 - (x^2 + y^2)p + xy = 0.$
Therefore $p = y/x$ or $p = x/y,$
and these equations have as integrals
$$y = Cx, \quad y^2 - x^2 = C.$$

Ex. 2. *Clairaut's Equation,*
$$y = xp + f(p). \quad \text{...........................(i)}$$
This equation is of a special form and is integrated thus:
Differentiate (i) as to x, and we find
$$p = p + x\frac{dp}{dx} + f'(p)\frac{dp}{dx}; \quad \text{or} \quad \{x + f'(p)\}\frac{dp}{dx} = 0. \quad \text{............(ii)}$$
Hence *either* $dp/dx = 0$, that is, $p = \text{constant} = C$; or
$$x + f'(p) = 0 \quad \text{...........................(iii)}$$
The substitution of C for p in (i) gives *the complete integral*
$$y = Cx + f(C). \quad \text{...........................(iv)}$$

On the other hand, if p is eliminated between (i) and (iii), we shall get a relation between x and y that will satisfy (i). This relation is not obtained by assigning a particular constant value to C in (iv), and is called a *Singular Solution*.

The singular solution is in fact the envelope of the family of lines (iv); for if we eliminate C between (iv) and $x + f'(C) = 0$, we clearly must get the same equation as that called the Singular Solution (we have simply interchanged C and p). As we have seen (§ 145), the gradient of the envelope is the same as that of the family (iv) at their points of meeting.

For example, the complete integral of $y = xp + a/p$ is
$$y = Cx + a/C,$$
and the Singular Solution is given by
$$y^2 = 4ax.$$

§ 167. Equations of the Second Order.

Type I. $D^2y = f(x)$, a function of x alone.

Integrate twice with respect to x; two constants will be introduced.

Type II. $D^2y = f(y)$, a function of y alone.

Multiply by Dy; then since $Dy\, D^2y = D\{\tfrac{1}{2}(Dy)^2\}$
$$\tfrac{1}{2}(Dy)^2 = \int f(y) Dy\, dx + C = \int f(y)\, dy + C.$$

It may now be possible to integrate this equation of the first order.

Ex. 1. The equation of motion of a simple pendulum of length l is $l\ddot{\theta} = -g\sin\theta$. To integrate, multiply by $\dot{\theta}$, then
$$\tfrac{1}{2}l(\dot{\theta})^2 = g\cos\theta + C.$$
When $t = 0$, let $\theta = a$, $\dot{\theta} = 0$; then
$$C = -g\cos a$$

and $\theta = -\sqrt{\left(\dfrac{2g}{l}\right)}\sqrt{(\cos\theta - \cos a)} = -2\sqrt{\left(\dfrac{g}{l}\right)}\sqrt{\left(\sin^2\dfrac{a}{2} - \sin^2\dfrac{\theta}{2}\right)},$

the negative sign being taken because θ decreases as t increases.

If we put $\sin\tfrac{1}{2}\theta = \sin\tfrac{1}{2}a \sin\phi$, we get after reduction

$$\frac{dt}{d\phi} = -\sqrt{\left(\frac{l}{g}\right)}\frac{1}{\sqrt{(1 - \sin^2\tfrac{1}{2}a \sin^2\phi)}}.$$

The integration cannot be carried further by means of the elementary functions, but t may be expressed by an infinite series. The value of t for the quarter period is $K\sqrt{(l/g)}$ [§ 156, ex. 3]. In general,

$$t = \sqrt{\left(\frac{l}{g}\right)}\int_\phi^{\tfrac{\pi}{2}} \frac{d\phi}{\sqrt{(1 - \sin^2\tfrac{1}{2}a \sin^2\phi)}}.$$

Type III. $D^2y = f(Dy)$, a function of Dy alone.

Let $Dy = v$ and we get $Dv = f(v)$ and it may be possible to find v, and then y.

Ex. 2. The equation $cD^2y = \{1 + (Dy)^2\}^{\tfrac{3}{2}}$ gives (p. 276)

$$x = cv/(1 + v^2)^{\tfrac{1}{2}} + a \text{ (constant)}.$$

Then
$$Dy = v = \pm (x - a)/\sqrt{\{c^2 - (x - a)^2\}},$$
$$y = \mp \sqrt{\{c^2 - (x - a)^2\}} + b \text{ (constant)},$$
or
$$(x - a)^2 + (y - b)^2 = c^2.$$

§ 168. Linear Equations. The typical equation of the second order is

$$D^2y + PDy + Qy = R \quad\quad\quad\quad (1)$$

where P, Q, R are functions of x alone (or constants).

The complete integral of *all linear equations* is the sum of two functions:—

I. *The Complementary Function* (C.F.) which is the complete integral of the equation when R (or in general the term independent of y and its derivatives) is zero. This function will contain *two* (when the equation is of the nth order, n) arbitrary constants.

II. *The Particular Integral* (P.I.) which is any solution whatever of the equation as it stands. This function contains no arbitrary constant.

We prove the proposition for equations of the second order, but it is easy to see that the reasoning is general; for the equation of the nth order there will be n functions like u, v, and n constants.

If $y = u$ and $y = v$ satisfy

$$D^2y + PDy + Qy = 0, \quad\quad\quad\quad (2)$$

so does $y = Au + Bv$ where A, B are constants. For if
$$D^2u + PDu + Qu = 0, \quad D^2v + PDv + Qv = 0,$$
then also $\quad D^2(Au + Bv) + PD(Au + Bv) + Q(Au + Bv) = 0,$
and therefore $Au + Bv$ satisfies (2), and since it contains two constants it is the complete integral of (2).

Next, if $y = w$ is the particular integral, that is, if w verifies equation (1) and if $Au + Bv$ is the complementary function, then $Au + Bv + w$ will satisfy (1). For when $y = Au + Bv + w$,
$$D^2y + PDy + Qy = D^2(Au + Bv) + PD(Au + Bv) + Q(Au + Bv)$$
$$+ D^2w + PDw + Qw.$$

The first line on the right is zero, and, since w satisfies (1), the second line is equal to R. This value of y therefore satisfies (1), and since it contains two constants it is the complete integral of (1).

The only equations we consider are those in which P, Q are *constants*.

§ **169. Complementary Function.** The equation to be integrated is
$$D^2y + aDy + by = 0. \dots\dots\dots\dots\dots\dots(3)$$

I. Let $y = e^{\lambda x}$ (λ constant); then
$$(\lambda^2 + a\lambda + b)e^{\lambda x} = 0.$$

If therefore λ is a root of the equation (the *auxiliary equation*)
$$\lambda^2 + a\lambda + b = 0 \dots\dots\dots\dots\dots\dots(4)$$
$e^{\lambda x}$ will satisfy (3). The two roots λ_1, λ_2 of (4) are
$$\lambda_1 = -\tfrac{1}{2}a + \sqrt{(\tfrac{1}{4}a^2 - b)}, \quad \lambda_2 = -\tfrac{1}{2}a - \sqrt{(\tfrac{1}{4}a^2 - b)}$$
and $e^{\lambda_1 x}$, $e^{\lambda_2 x}$ are two solutions of (3). Hence the complete integral of (3) is
$$y = Ae^{\lambda_1 x} + Be^{\lambda_2 x} = e^{-\frac{1}{2}ax}(Ae^{nx} + Be^{-nx})\dots\dots\dots(5)$$
where $n = \sqrt{(\tfrac{1}{4}a^2 - b)}$.

We must however consider special cases.

II. If $a^2 = 4b$ equation (4) has *two equal roots*, namely $\lambda_1 = \lambda_2 = -\tfrac{1}{2}a$. In this case (5) becomes
$$y = (A + B)e^{-\frac{1}{2}ax},$$
and there is only *one distinct constant*, for we might obviously replace $A + B$ by C.

When $a^2 = 4b$ let $y = e^{-\frac{1}{2}ax}u$ and (3) becomes, after rejecting the factor $e^{-\frac{1}{2}ax}$,
$$D^2u = 0$$
of which the complete integral is $u = A + Bx$.

THE COMPLEMENTARY FUNCTION.

Hence the complete integral of (3) when the auxiliary equation has two equal roots, each $= -\frac{1}{2}a$, is
$$y = (A + Bx)e^{-\frac{1}{2}ax}. \quad\quad\quad\quad\quad\quad (6)$$

III. If $a^2 < 4b$ the roots of (4) are imaginary. Again, let $y = e^{-\frac{1}{2}ax}u$ and equation (3) becomes
$$D^2u + m^2u = 0. \quad\quad\quad\quad\quad\quad (7)$$
where $\frac{1}{4}a^2 - b = -m^2$ and m is real. Now (7) is satisfied by $u = \cos mx$, $u = \sin mx$; its complete integral is thus
$$u = A \cos mx + B \sin mx,$$
and therefore the complete integral of (3) when $a^2 < 4b$ is
$$y = e^{-\frac{1}{2}ax}u = e^{-\frac{1}{2}ax}(A \cos mx + B \sin mx). \quad\quad (8)$$

We shall now show how to write down (5) and (8) when the roots of (4) are known.

Let i denote as usual $\sqrt{(-1)}$. When the roots of (4) are real let $\frac{1}{4}a^2 - b = n^2$; the roots then are
$$-\tfrac{1}{2}a + n, \quad -\tfrac{1}{2}a - n,$$
and the solution is $y = e^{-\frac{1}{2}ax}(Ae^{nx} + Be^{-nx})$.

When the roots of (4) are imaginary let $\frac{1}{4}a^2 - b = -n^2$; the roots then are
$$-\tfrac{1}{2}a + ni, \quad -\tfrac{1}{2}a - ni,$$
and the solution is
$$y = e^{-\frac{1}{2}ax}(A \cos nx + B \sin nx),$$
so that instead of e^{nix}, e^{-nix} we have $\cos nx$, $\sin nx$.

It should be noticed that the auxiliary equation is obtained by replacing D by λ and rejecting y.

Ex. 1. $D^2y + 7Dy - 8y = 0.$
Aux. Eq. $\lambda^2 + 7\lambda - 8 = 0$; $\lambda_1 = 1$, $\lambda_2 = -8$.
Solution $y = Ae^x + Be^{-8x}.$

Ex. 2. $D^2y + 2Dy + 10y = 0.$
Aux. Eq. $\lambda^2 + 2\lambda + 10 = 0$; $\lambda_1 = -1 + 3i$, $\lambda_2 = -1 - 3i$.
Solution $y = e^{-x}(A \cos 3x + B \sin 3x).$

Ex. 3. $D^4y - 2D^3y + 5D^2y - 8Dy + 4y = 0.$
Aux. Eq. $\lambda^4 - 2\lambda^3 + 5\lambda^2 - 8\lambda + 4 = 0$;
             $\lambda_1 = 1 = \lambda_2$, $\lambda_3 = 2i$, $\lambda_4 = -2i.$

The equal roots λ_1, λ_2 give $(A + Bx)e^x$; the imaginary roots $2i$, $-2i$ give $E \cos 2x + F \sin 2x$. Hence the
Solution $y = (A + Bx)e^x + E \cos 2x + F \sin 2x.$

§ 170. Particular Integral.

The most important practical cases are those in which R is a sum of terms of the form Le^{ax}, $L \sin ax$, $L \cos ax$; the simplest method of finding a particular solution is by substitution. Equation (1) is now

$$D^2y + aDy + by = R \quad \quad \quad (9)$$

I. $R = Le^{ax}$. Let $y = Ce^{ax}$ and try to find C so that equation (9) shall be verified. We find

$$C(a^2 + aa + b)e^{ax} = Le^{ax},$$

and Ce^{ax} will satisfy (9) if $C = L/(a^2 + aa + b)$.

There are exceptional cases, however.

I. (a). If a is a root of the auxiliary equation (4) then

$$a^2 + aa + b = 0,$$

and the value of C is infinite. In this case try Cxe^{ax} or Cx^2e^{ax} according as a is a single or a double root of the auxiliary equation.

Ex. 1. $\quad\quad D^2y - 2Dy + y = e^x + e^{2x}.$

Aux. Eq. $\quad \lambda^2 - 2\lambda + 1 = 0$; $\lambda = 1$ twice.

To find P.I. take e^x and e^{2x} separately; that is, since the coefficient of x in e^x is 1, and 1 is a double root of Aux. Eq., try Cx^2e^x, for P.I. corresponding to e^x, and Ee^{2x} for P.I. corresponding to e^{2x}. Hence we put

$$y = Cx^2e^x + Ee^{2x},$$

and the equation becomes

$$2Ce^x + Ee^{2x} = e^x + e^{2x},$$

so that $C = \tfrac{1}{2}$, $E = 1$, and therefore

$$\text{P.I.} = \tfrac{1}{2}x^2e^x + e^{2x}.$$

The part corresponding to e^{2x} may be obtained at once by direct application of I. The complete integral is now

$$y = \text{C.F.} + \text{P.I.}$$
$$= (A + Bx)e^x + \tfrac{1}{2}x^2e^x + e^{2x}.$$

II. $R = L \sin ax + M \cos ax$.

Take as trial solution

$$y = E \sin ax + F \cos ax.$$

We find

$$(-a^2E - aaF + bE)\sin ax + (-a^2F + aaE + bF)\cos ax$$
$$= L \sin ax + M \cos ax,$$

THE PARTICULAR INTEGRAL.

and the equation will be satisfied if
$$(b-a^2)E - aaF = L\ ;\quad aaE + (b-a^2)E = M\ ;$$
or $E = \dfrac{(b-a^2)L + aaM}{(b-a^2)^2 + a^2a^2}\ ;\quad F = \dfrac{-aaL + (b-a^2)M}{(b-a^2)^2 + a^2a^2}.$

If $a = 0$ we get
$$E = L/(b-a^2)\ ;\quad F = M/(b-a^2)\ ;$$
but this solution fails if $a^2 = b$, that is, when the complementary function is $A\cos ax + B\sin ax$. We have then

II. (a). If $a = 0$ and $a^2 = b$ it will be found on trial that
$$\text{P.I.} = -\frac{L}{2a}x\cos ax + \frac{M}{2a}x\sin ax,$$
when $R = L\sin ax + M\cos ax$.

Ex. 2. The equation
$$\ddot{x} + k\dot{x} + \mu x = a\cos(nt - \alpha) \dots\dots\dots\dots\dots\text{(i)}$$
is typical in dynamical and electrical theory.

C.F. is easily found. To find P.I. try
$$x = E\cos(nt - \alpha) + F\sin(nt - \alpha),\ \dots\dots\dots\dots\text{(ii)}$$
and we find by substitution in (i)
$$(-n^2E + knF + \mu E)\cos(nt - \alpha)$$
$$+(-n^2F - knE + \mu F)\sin(nt - \alpha) = a\cos(nt - \alpha).$$
Hence (ii) will satisfy (i) if
$$(\mu - n^2)E + knF = a\ ;\quad -knE + (\mu - n^2)F = 0.$$
Therefore $\quad E = \dfrac{(\mu - n^2)a}{(\mu - n^2)^2 + k^2n^2}\ ;\quad F = \dfrac{kna}{(\mu - n^2)^2 + k^2n^2}.$

Hence $\quad\text{P.I.} = \dfrac{a\{(\mu - n^2)\cos(nt - \alpha) + kn\sin(nt - \alpha)\}}{\{(\mu - n^2)^2 + k^2n^2\}}$
$$= a\cos(nt - \alpha - \alpha_1)/\sqrt{\{(\mu - n^2)^2 + k^2n^2\}}$$
where $\tan\alpha_1 = kn/(\mu - n^2)$.

If $k = 0$ and $n^2 = \mu$, we have II. (a). In this case
$$\text{P.I.} = \frac{a}{2n}t\sin(nt - \alpha).$$

III. If R is a rational integral function of x we may put for y a rational integral function and try to determine the coefficients so as to satisfy the equation.

§ 171. **Simultaneous Equations.** We will illustrate, by solving one or two examples, some methods of integrating *simultaneous* ordinary differential equations, the number of equations being equal to the number of dependent variables. We take t as the independent variable and restrict ourselves to two dependent variables, x and y.

Ex. 1. $\quad\quad\quad\quad\quad\quad\quad \dot{x} = -\omega y;$ ....(i)
$$\dot{y} = \omega x. \quad\quad\quad\quad\quad\quad\quad\quad\quad\quad\quad\quad\quad\quad\text{(ii)}$$

Differentiate (i) and substitute for $\dot{y}$ its value ωx given by (ii); we thus get an ordinary equation with *one* dependent variable, namely, $\ddot{x} + \omega^2 x = 0$, of which the integral is

$$x = A\cos\omega t + B\sin\omega t \quad \text{or} \quad x = C\cos(\omega t - E). \quad\text{......(iii)}$$

The value of y is now found from (i); we get

$$y = A\sin\omega t - B\cos\omega t \quad \text{or} \quad y = C\sin(\omega t - E). \quad\text{......(iv)}$$

It should be noticed that although A and B are arbitrary, yet the constants in y are determinate as soon as those in x are chosen. If, however, (i) contains x alone, and (ii) y alone, the constants in x do not condition those of y. Thus the equations

$$\ddot{x} + \omega^2 x = 0, \quad \ddot{y} + \omega^2 y = 0$$

give $\quad\quad\quad x = A\cos\omega t + B\sin\omega t, \quad y = E\cos\omega t + F\sin\omega t,$
and there is no relation between A, B and E, F.

Ex. 2. $\dot{x} + 5x - 3y = 0$ ....(i); $\dot{y} + 15x - 7y = 0.$ ....(ii)
Differentiate (i); $\quad\quad\quad \ddot{x} + 5\dot{x} - 3\dot{y} = 0.$ ....(iii)
From (i), (ii), (iii) we can eliminate y and $\dot{y}$; we find
$$\ddot{x} - 2\dot{x} + 10x = 0, \quad\quad\quad\quad\quad\quad\quad\quad\text{(iv)}$$
of which the integral is
$$x = e^t(A\cos 3t + B\sin 3t). \quad\quad\quad\quad\quad\quad\text{(v)}$$

Equation (i) now determines y, namely,

$$y = e^t\{(2A + B)\cos 3t + (2B - A)\sin 3t\}. \quad\quad\text{(vi)}$$

If (i), (ii) had each contained both $\dot{x}$ and $\dot{y}$, we should have differentiated both (i) and (ii), and from the *four* equations we should have eliminated the *three* quantities y, $\dot{y}$, $\ddot{y}$.

Ex. 3. As the last example we take the equations

$$L\dot{x} + M\dot{y} + Rx = P, \quad\quad\quad\quad\quad\quad\quad\quad\text{(i)}$$
$$M\dot{x} + N\dot{y} + Sy = Q, \quad\quad\quad\quad\quad\quad\quad\quad\text{(ii)}$$

which connect two mutually influencing electric circuits. x and y denote the currents, L and N the self-inductances, M the mutual inductance, R and S the resistances, and P and Q the extraneous electromotive forces. The product LN is greater than M^2.

We may proceed as in example 2 by differentiating (i) and (ii) and eliminating y, $\dot{y}$, $\ddot{y}$; but we will illustrate another method. The principle of complementary function and particular integral evidently holds for simultaneous linear equations; P, Q are either constants or

functions of t, and we may apply the principle to (i) and (ii). The complementary function is thus obtained from
$$L\dot{x} + M\dot{y} + Rx = 0; \quad\quad\quad\quad\quad\quad\quad\quad\text{(iii)}$$
$$M\dot{x} + N\dot{y} + Sy = 0. \quad\quad\quad\quad\quad\quad\quad\quad\text{(iv)}$$

Let $x = Ae^{\lambda t}$, $y = Be^{\lambda t}$ where A, B are constants, and substitute in (iii), (iv). We find
$$(L\lambda + R)A + M\lambda B = 0; \quad\quad\quad\quad\quad\quad\text{(v)}$$
$$M\lambda A + (N\lambda + S)B = 0. \quad\quad\quad\quad\quad\quad\text{(vi)}$$

If we eliminate the ratio $A : B$ from (v) and (vi) we get the condition that (v) and (vi) should be simultaneously satisfied, namely
$$(L\lambda + R)(N\lambda + S) - M^2\lambda^2 = 0,$$
or
$$(LN - M^2)\lambda^2 + (LS + NR)\lambda + RS = 0. \quad\quad\quad\text{(vii)}$$

The roots of (vii) are real; for
$$(LS + NR)^2 - 4(LN - M^2)RS = (LS - NR)^2 + 4M^2 RS,$$
so that the discriminant of (vii) is positive. Also, since $LN > M^2$, the roots of (vii) have the same sign; both are negative. If we call them $-\lambda_1$, $-\lambda_2$ and take the constants as A_1, A_2 and B_1, B_2 we get for the solutions of (iii), (iv),
$$x = A_1 e^{-\lambda_1 t} + A_2 e^{-\lambda_2 t}, \quad y = B_1 e^{-\lambda_1 t} + B_2 e^{-\lambda_2 t}. \quad\quad\text{(viii)}$$
B_1 is connected with A_1, and B_2 with A_2 by equation (v) or (vi), that is,
$$B_1 = A_1(R - L\lambda_1)/M\lambda_1, \quad B_2 = A_2(R - L\lambda_2)/M\lambda_2.$$

If P, Q are constants, the particular integrals are clearly
$$x = P/R, \quad y = Q/S,$$
and these added to (viii) give the complete integrals of (i), (ii).

The only other important case is that in which $P = E_0 \cos(nt - a)$ $Q = 0$, and the particular integral is found by assuming as a trial solution,
$$x = E\cos(nt - a) + F\sin(nt - a),$$
$$y = G\cos(nt - a) + H\sin(nt - a),$$
and determining the constants E, F, G, H.

The equations $\ddot{x} + k\dot{y} + c^2 x = 0$, $\ddot{y} - k\dot{x} + c^2 y = 0$, are the equations for the small motions of the bob of a gyrostatic pendulum (gyroscope axis along suspension), and also the elementary equations of motion of an electron in a magnetic field in the theory of the Zeeman effect. (See Gray, *Magnetism and Electricity*, Vol. I., § 565. In Chapter X. of this work will be found several instructive examples.)

EXERCISES XXXVIII.

Integrate equations 1-16.

1. $(1 + x^2)Dy = 1 + y^2;$
2. $\sqrt{(1 - x^2)}Dy = \sqrt{(1 - y^2)};$
3. $y - xDy = m(y^2 + Dy);$
4. $(xy + x^2)Dy + y^2 = 0;$
5. $xDy - y = \sqrt{(x^2 + y^2)};$
6. $(2x + 18y - 14)Dy = 6x + 5y - 7;$
7. $(ax + by + c)Dy = m(ax + by) + g;$ Replace y by the

substitution $\eta = ax + by$.

8. $(ax + by + c)Dy = fx - ay + g$;
9. $Dy + y = e^{-x}$;
10. $xDy + y = x$;
11. $(1 - x^2)Dy - xy = 1$;
12. $(1 + x^2)Dy + 2xy = x^3$;
13. $Dy + ay = \cos(bx + c)$;
14. $xDy + y = x^3y^6$;
15. $x^2Dy + y^2 = xy$;
16. $y(x^2 - y^2 - b^2)Dy + x(x^2 + y^2 - a^2) = 0$.

Find the complete integral and the singular solutions (where they exist) of equations 17-19.

17. $(y - px)^2 = a^2p^2 + b^2$;
18. $y = px + p^3$;
19. $x^2(y - px) = yp^2$.

Solve equations 20-27.

20. $D^2y - (a+b)Dy + aby = 0$;
21. $D^3y - 5D^2y + 6Dy = 0$;
22. $D^2y - 6Dy + 10y = \sin 2x$;
23. $D^2y - 3Dy + 2y = e^x$;
24. $D^2y + n^2y = a \cos nx + b \sin nx$;
25. $D^2y - n^2y = ae^{nx} + be^{-nx}$;
26. $D^2y - 6Dy + 13y = x^2$;
27. $D^4y + 2D^2y + y = 0$.

Integrate the simultaneous equations 28-31.

28. $\dot{x} - 7x + y = 0$, $\dot{y} - 2x - 5y = 0$;
29. $\dot{x} + \dot{y} + 2x + y = 0$, $\dot{y} + 5x + 3y = 0$;
30. $\dot{x} + 2x - 3y = t$, $\dot{y} - 3x + 2y = e^{2t}$;
31. $\ddot{x} - 3x - 4y = 0$, $\ddot{y} + x + y = 0$.

32. Integrate the equations $\ddot{x} = 0$, $\ddot{y} = -g$, determining the constants so that $x = 0$, $y = 0$, $\dot{x} = V\cos\alpha$, $\dot{y} = V\sin\alpha$ when $t = 0$.

33. Integrate the equations $\ddot{x} = -\mu x$, $\ddot{y} = -\mu y$, choosing the constants so that $x = a$, $y = 0$, $\dot{x} = 0$, $\dot{y} = b\sqrt{\mu}$ when $t = 0$.

34. Integrate the equation $\ddot{x} = -\mu/x^2$, choosing the constants so that $x = a$, $\dot{x} = 0$ when $t = 0$.

35. The equation $BD^4y = w$ occurs in the theory of the bending of beams, B being the flexural rigidity and w the weight per unit length; integrate the equation under the conditions:

(i) $y = 0$, $D^2y = 0$ when $x = 0$ and when $x = l$;

(ii) $y = 0$, $Dy = 0$ when $x = 0$ and when $x = l$;

(iii) $y = 0$, $Dy = 0$ when $x = 0$ and $D^2y = 0$, $D^3y = 0$ when $x = l$.

36. The plates of a charged condenser of capacity C are connected by a wire of self-inductance L and resistance R; if at time t the difference of potential between the plates is V, then V satisfies the equation
$$CL\ddot{V} + RC\dot{V} + V = 0,$$
and the current γ is $-C\dot{V}$. Show that the discharge will be oscillatory if $CR^2 < 4L$, that the period T is given by
$$T = 4\pi L/\sqrt{\{4L/C - R^2\}},$$
and that the logarithmic decrement of the potential is $RT/4L$.

37. Integrate the equation $D^2y + \dfrac{2}{x}Dy + n^2y = 0$ by changing the dependent variable from y to u where $u = xy$.

Give the complete integral, and also the integral which remains finite as x converges to zero.

38. Show that the complete integral of
$$x^2D^2y + axDy + by = 0$$
is
$$y = Ax^{\lambda_1} + Bx^{\lambda_2}$$
where λ_1, λ_2 are the roots of the equation
$$\lambda(\lambda - 1) + a\lambda + b = 0.$$
Take as trial solution $y = x^\lambda$ and proceed as in § 169.

39. Integrate
  (i) $xD^2y + 2Dy = 6x$; (ii) $x^3D^3y - 3x^2D^2y + 6xDy - 6y = x^2$;
  (iii) $x^2D^2y - 2y = x$.

40. Integrate the equation
$$x^2D^2y + xDy + n^2y = 0$$
by changing the independent variable from x to θ where $x = e^\theta$. The equation for λ, corresponding to that of example 38, has in this case imaginary roots.

41. Integrate $\dfrac{d}{dr}\left(\dfrac{dw}{dr} + 2\dfrac{w}{r}\right) = 0.$

42. Integrate $\dfrac{d^2V}{dr^2} + \dfrac{1}{r}\dfrac{dV}{dr} = 0.$

43. Find Dy from the equation
$$a^2D^2y = y\{1 + (Dy)^2\}^{\frac{3}{2}}.$$

44. If $y = uv$, where u, v, are functions of x, show that the linear equation
$$D^2y + PDy + Qy = R \quad\dotfill\text{(i)}$$
becomes, the accents denoting x-derivatives,
$$vu'' + (2v' + Pv)u' + (v'' + Pv' + Qv)u = R. \quad\dotfill\text{(ii)}$$
It follows that if v is any solution of (i) when $R = 0$, the value of u (and therefore of y) can be found; for the coefficient of u is zero, and (ii) is linear, and of the first order when u' is the dependent variable.

45. Integrate $\quad x^2D^2y + xDy - y = x^3.$
Put $y = xu$.

ANSWERS.

CHAPTER I.

§ 5, p. 6. 1. $3\frac{1}{2}$; 2; -3; 4·56. 3. (i) +; (ii) −

§ 6, p. 9. 2. The locus in each case is a straight line: in cases (i), (ii), (iv) the line is perpendicular to the axis of abscissae, and in (iii) the line is the axis of ordinates. When the ordinate is given the lines are parallel to or coincident with the axis of abscissae.

4. (i) +; (ii) −.

§ 7, p. 10. 2. (i) $\sqrt{5}$; (ii) $\sqrt{17}$; (iii) $\sqrt{5}$; (iv) $2\sqrt{13}$; (v) $3\pi\sqrt{2}/2$ or (i) 2·24; (ii) 4·12; (iii) 2·24; (iv) 7·21; (v) 6·66

Set I., p. 19.

1. 1, 1, 1. 2. $(ax+b)^2 - (ax+b) - 2$.
3. $x^4 - 5x^2 + 1$; $x^6 - 5x^3 + 1$; $\sin^2 x - 5\sin x + 1$; $-2·04$.
9. $ay^2 + byx + c$; $ax^2 + bx^2 + c$; $ay^2 + by^2 + c$.

CHAPTER II.

Set II., p. 29.

1. A, C, D on curve; B, E not on curve.
2. $Y'OY$ is an axis of symmetry for (i), (iii), (vi), (vii). Point $(1, -1)$ lies on (i), (ii). $a = 0$.
3. Turning points. (i) $(0, -1)$; (ii) $(0, -1)$; (iii) $(0, 1)$; (iv) $(\frac{3}{4}, \frac{9}{8})$; (v) $(-\frac{1}{2}, \frac{5}{4})$; (vi) $(\frac{3}{4}, \frac{1}{8})$. Abscissae (i) $-1, 1$; (ii) $-\frac{1}{\sqrt{2}}, \frac{1}{\sqrt{2}}$; (iii) $-\frac{1}{\sqrt{2}}, \frac{1}{\sqrt{2}}$; (iv) $0, \frac{3}{2}$; (v) $\frac{1}{2}(-1 \pm \sqrt{5})$; (vi) $\frac{1}{2}, 1$.
6. (i), (ii), (iv). In (iii) y is imaginary when x is negative.

Set III., p. 32.

1. (i) $-1, 2$; (ii) $\frac{3}{2}, -1$; (iii) $-\frac{3}{2}, -1$. 3. C lies on line.
6. (i) $x+y=3$; (ii) $x+y=1$; (iii) $x+y=0$; (iv) $3x-2y+6=0$.
7. $y = 2x - 5$. 8. $y - b = c(x - a)$. 10. $(1, 1), (-3, 9)$.

Set IV., p. 41.

8. (i) $-2·84, ·44, 2·40$; (ii) $-3·14$. 9. $-1·98, -·06, 2·06, 3·98$.

ANSWERS. 443

CHAPTER III.

Set V., p. 54.

11. (i) $\frac{\sqrt{3}}{2}$, (ii) $\frac{1}{3}\sqrt{15}$. **14.** (i) $\frac{1}{\sqrt{2}}$; (ii) $\frac{1}{\sqrt{2}}$.

15. The equation is equivalent to $y+1 = \pm \frac{1}{2}(x-3)$.

Set VI., p. 60.

12. (i) an ellipse, $(3, -4)$, $2a=6$, $2b=4$; (ii) a hyperbola, $(-11, 5)$, $2a=4$, $2b=2\sqrt{3}$. **19.** (i) $\frac{7}{5}$; (ii) $\frac{27}{\sqrt{313}}$.

20. (i) $ab/\sqrt{(a^2\sin^2\theta + b^2\cos^2\theta)}$ which may be written $b/\sqrt{(1-e^2\cos^2\theta)}$ when $a^2 - b^2 = e^2 a^2$;
(ii) $(1 - e\cos\theta)(ab/\sqrt{(a^2\sin^2\theta + b^2\cos^2\theta)}) = b(1-e\cos\theta)/\sqrt{(1-e^2\cos^2\theta)}$;
(iii) $b(1+e\cos\theta)/\sqrt{(1-e^2\cos^2\theta)}$.

CHAPTER IV.

§ 32, p. 67. 4. The values of $\delta y_1/\delta x_1$ are in order 331, 315·25, 303·01, 300·3001, 300·030001.

5. The values of $\delta y_1/\delta x_1$ are
 (i) ·015038, ·015077, ·015100, ·015107;
 (ii) ·008594, ·008661, ·008701, ·008713.

6. The values of $\delta y_1/\delta x_1$ are
 (i) ·001332, ·001334, ·001335, ·001336;
 (ii) ·005950, ·005990, ·006028, ·00603.

§ 37, p. 74. 1. $0, \frac{1}{2}g, g, 2g$. **2.** $-a/v_1^2$.

CHAPTER VI.

§ 53, p. 106. $-10, -4, 0, 2, 8$.

§ 57, p. 111. 2. $5t^4, -4/t^3, 3/2\sqrt{t}, -2t^{-\frac{3}{2}}$.

4. $x^5, \frac{2}{3}x^{\frac{3}{2}}, 4\sqrt{x}, -1/x, -3/4x^4$.

Set VIII., p. 115.

1. $21x^2 + 10x + 4$. **2.** $112x - 10$. **3.** $3x^2 - 4x - 5$.

4. $1/(5-2x)^2$. **5.** $\frac{1}{2\sqrt{x}} - \frac{1}{2\sqrt{x^3}}$. **6.** $\frac{3}{2\sqrt{x}}\left(\sqrt{x} - \frac{1}{\sqrt{x}}\right)^2\left(\frac{x+1}{x}\right)$.

7. $n(x^{n-1} - x^{-n-1})$. **8.** $m(ax^{m-1} - bx^{-m-1})$.

9. $5x^{\frac{1}{4}} + \frac{1}{2}x^{-\frac{3}{4}} + \frac{1}{2}x^{-\frac{4}{4}} - \frac{5}{4}x^{-\frac{5}{4}}$. **10.** $2(x+1)(x^3-1)/x^3$.

11. $(ad-bc)/(ct+d)^2$. **12.** $-ac/(b+ct)^2$.

13. $2\{(aB-bA)t^2 + (aC-cA)t + (bC-cB)\}/(At^2 + 2Bt + C)^2$.

14. $\dfrac{6(t^2-2)}{(t+1)^2(t+2)^2}$. **16.** $vw\dot{u} + wu\dot{v} + uv\dot{w}$.

17. Abscissae of turning points (a) $\frac{1}{2}$, (b) ± 1, (c) 0, ± 1.

18. (i) $x^2 - x + C$; (ii) $\frac{3}{2}x^2 + \frac{1}{x} + C$; (iii) $\frac{1}{3}ax^3 + \frac{1}{2}bx^2 + cx + C$.

19. $y = \frac{1}{3}x^3 - \frac{1}{2}x^2 + x$. **21.** $Vt - \frac{1}{2}gt^2$ feet.

Set IX., p. 119.

1. $\dfrac{-1}{2\sqrt{(1-x)}}$. **2.** $\dfrac{2-x}{2(1-x)^{\frac{3}{2}}}$. **3.** $\tfrac{1}{2}(4x-3)/\sqrt{(2x+1)(x-2)}$.

4. $a^2(a^2-x^2)^{-\frac{3}{2}}$. **5.** $a^2(a^2+x^2)^{-\frac{3}{2}}$. **6.** $(ax+\tfrac{1}{2}b)/\sqrt{(ax^2+bx+c)}$.

7. $\dfrac{-2x}{(x^2-1)\sqrt{(x^4-1)}}$. **8.** $\dfrac{(aB-bA)x^2+(aC-cA)x+(bC-cB)}{(Ax^2+2Bx+C)\sqrt{(ax^2+2bx+c)}\sqrt{(Ax^2+2Bx+C)}}$.

9. $-\dfrac{(x+7)(x+1)^2}{(x-1)^5}$. **10.** $\dfrac{(mx-nx+mb-na)(x+a)^{m-1}}{(x+b)^{n+1}}$.

11. $\dfrac{2-5x}{x^3(x-1)^4}$. **14.** $-2x/3y$, $-\tfrac{2}{3}, \tfrac{2}{3}, -\tfrac{2}{3}, \tfrac{2}{3}$.

15. $-(2x+2y-5)/(2x+2y+1)$, grad. $=1$. **16.** a, $b-ct$; $\tan\phi=(b-ct)/a$.

17. (i) $-\dfrac{x-a}{y-b}$; (ii) $\dfrac{A+2Bx}{2y}$; (iii) $-\dfrac{c^2}{x^2}$; (iv) $-\dfrac{my}{nx}$.

18. $\dfrac{1}{3a}\sqrt{u}$. **19.** $\tfrac{1}{2}u^n$. **20.** $\dfrac{1}{a}f(u)$.

Set X., p. 125.

3. $yy_1=2a(x+x_1)$; $(y-y_1)2a+(x-x_1)y_1=0$.

CHAPTER VII.

Set XI., p. 131.

1. $3(\cos 3x - \sin 3x)$. **2.** $\dfrac{2\pi}{a}\cos\dfrac{2\pi}{a}(x+b)$.

3. $m\cos mx \cos nx - n\sin mx \sin nx$. **4.** $x\cos x$. **5.** $x\sin x$. **6.** $\sin^2 x$.

7. $\cos^2 x$. **8.** $\cos^3 x$. **9.** $\sin^3 x$. **10.** $\tfrac{1}{3}(\sin 3x + \cos 3x)$.

11. $\dfrac{1}{a}\sin(ax+b)$. **12.** $\dfrac{1}{a}\tan(ax+b)$. **13.** $\tfrac{1}{2}x+\tfrac{1}{4}\sin 2x$.

14. $\tfrac{1}{2}x - \tfrac{1}{4}\sin 2x$. **15.** $-\tfrac{1}{12}\cos 6x - \tfrac{1}{4}\cos 2x$.

16. $-2a\cos(ax+b)\sin(ax+b)$. **17.** $\tan(\tfrac{1}{2}x+1)\sec^2(\tfrac{1}{2}x+1)$.

18. $\cos 2x/\sqrt{\sin 2x}$. **19.** $\sin x (3-\cos^2 x)/\cos^4 x$.

20. $\sin x/(1+\cos x)^2$. **21.** $2\sin x/(1+\cos x)^2$.

22. $(\cos x - \sin x \tan^2 x)/(1+\tan x)^2$.

25. (i) $t=\dfrac{1}{n}\{(N+\tfrac{1}{2})\pi+e\}$, $s=0$; (ii) $t=\dfrac{1}{n}(N\pi+e)$, $s=\pm a$ where N is any integer.

26. $-a\sin t$, $b\cos t$; $\tan\phi = -(b/a)\cot t$.

28. $b\tan(x/b)$; $(a^2/2b)\sin(2x/b)$.

36. For $\cos x$ the inequalities are not changed; for $\sin x$ put $>$ in place of $<$.

ANSWERS. 445

Set XII., p. 134.

1. $\dfrac{3}{\sqrt{(1-9x^2)}}$. 2. $\dfrac{1}{\sqrt{(2+x-x^2)}}$. 3. $\dfrac{3}{5-2x+2x^2}$.

4. $\dfrac{-1}{\sqrt{(2x-x^2)}}$. 5. $\sin^{-1}x + \dfrac{x}{\sqrt{(1-x^2)}}$. 6. $\tan^{-1}x + \dfrac{x}{1+x^2}$.

7. $\sin^{-1}\left(\dfrac{x}{\sqrt{3}}\right)$. 8. $\dfrac{1}{\sqrt{3}}\tan^{-1}\left(\dfrac{x}{\sqrt{3}}\right)$. 9. $\sin^{-1}\left(\dfrac{x-1}{2}\right)$.

Set XIII., p. 139.

1. $1 + \log x$. 2. $x^{n-1}(1 + n\log x)$. 3. $\cot x$. 4. $-\tan x$.
5. $1/\sin x$. 6. $2/\cos x$. 7. $2/\sin x$.
8. $a/(a^2 - x)\sqrt{x}$. 9. $\dfrac{1}{2\sqrt{(x^2 - a^2)}}$. 10. $(x+1)e^x$.
11. $x^{n-1}(x+n)e^x$. 12. $-2e^{-x}\sin x$. 13. $\dfrac{xe^x}{(1+x)^2}$.
14. $\tfrac{1}{3}\log(3x+4)$. 15. $\dfrac{1}{2a}\log\dfrac{x-a}{x+a}$. 16. $\dfrac{1}{12}\log\dfrac{2x-3}{2x+3}$.
17. $\log(x+\sqrt{x^2+1})$. 18. $\dfrac{1}{a}e^{ax}$. 21. a; $\dfrac{c^2}{a}e^{\frac{2x}{a}}$.
22. $a\left(e^{\frac{x}{a}}+e^{-\frac{x}{a}}\right)\Big/\left(e^{\frac{x}{a}}-e^{-\frac{x}{a}}\right)$, $\tfrac{1}{4}a\left(e^{\frac{2x}{a}}-e^{-\frac{2x}{a}}\right)$, y^2/a.

Set XIV., p. 146.

1. $28x^3 - 6x^2$; $84x^2 - 12x$; $168x - 12$; 168.
2. $(x^2+1)^{-\frac{3}{2}}$. 3. $12x^2 - 12ax + 2a^2$; $12(2x-a)$.
4. $y' = -(x-1)^{-2} - (x+1)^{-2} + (x+2)^{-2}$;
   $y'' = 2(x-1)^{-3} + 2(x+1)^{-3} - 2(x+2)^{-3}$;
   $y^{(n)} = (-1)^n(n!)(x-1)^{-n-1} +$ two similar terms.
5. $y^{(n)} = -2^{n-1}\cos(2x + n\pi/2)$.
6. $y^{(n)} = x^2\cos\theta + 2nx\sin\theta - n(n-1)\cos\theta$ where $\theta = x + n\pi/2$.
7. $y = \tfrac{1}{4}\sin 2x + \tfrac{1}{8}\sin 4x$; $y^{(n)} = 2^{n-2}\sin(2x + n\pi/2) + 2^{2n-3}\sin(4x + n\pi/2)$.
8. $\dfrac{1}{x}$; $\dfrac{(-1)^n(n-2)!}{x^{n-1}}$. 9. $e^x(x+n)$.
10. $e^x\{x^2 + 2nx + n(n-1)\}$. 12. Ex. 1. $x = \tfrac{3}{14}$; Ex. 2. $x = 0$.
13. $y' = 0$ when $x = -1/\sqrt{3}$ or $+1/\sqrt{3}$. 23. $-\dfrac{1}{2at^3}$. 24. $-\dfrac{b}{a^2\sin^3 t}$.

CHAPTER VIII.

§ 70, p. 157. -67×10^{-6}, 50×10^{-6}.

Set XV., p. 159.

1. aV/OP^2.   3. $-dp/dx = gp$.
4. $-dN/dt$ is the time-rate of *decrease* of the number of lines that pass through the circuit; or, the time-rate at which lines are *withdrawn from* the circuit.
5. $E = RC + LdC/dt$.   6. $X = -dE/dx$.
7. (i) $k\log(v_2/v_1)$; (ii) $k(v_2^{1-\gamma} - v_1^{1-\gamma})/(1-\gamma)$ or $(p_1v_1 - p_2v_2)/(\gamma - 1)$.

CHAPTER IX.

Set XVI. a., p. 176.

1. $x = -1$, max.; $x = 2$, min.   2. $x = 1$, max.; $x = 3$, min.
3. $x = 0$, min.; $x = -4/7$ max.
4. $x = -a$, max.; $x = -\frac{1}{2}a$, min.; $x = \frac{1}{3}a$, max.
5. $x = -1$, min.; $x = 1$, max.   6. $x = -1$, max.; $x = \frac{1}{3}$, min.
7. $x = -1$, min.; $x = 1$, max. if $a > 0$.   8. $x = -1$ min.; $x = 1$ max.
9. $x = \frac{1}{2}a$, max.   10. $x = -\frac{1}{2}$, min.; $x = \frac{1}{2}$, max.
11. $x = c$, min. if $b > 0$.   12. No max. or min.
13. $m^m n^n \{k/(m+n)\}^{m+n}$.   15. $(a+b)^2/c$; $4ab/c^2$.
16. $(m_1 x_1 + m_2 x_2 + \ldots)/(m_1 + m_2 + \ldots)$.   19. $3abc$.
20. $abc/3\sqrt{3}$; $3d^2/(abc)^{\frac{2}{3}}$.   20. $2ab$.

Set XVI. b., p. 177.

1. $\tan QAB = b/a\sqrt{2}$.   2. $\frac{1}{2}(a + \sqrt{a^2 + 8b^2})$.
3. $\frac{1}{6}(a + b - \sqrt{a^2 - ab + b^2})$.   10. $x = \frac{1}{2}a$; $x = \frac{1}{3}a$.
11. $\frac{\sqrt{3}}{3}d$; $\frac{\sqrt{6}}{3}d$.   12. $\frac{1}{2}d$; $\frac{\sqrt{3}}{2}d$.   13. $\dfrac{ua}{\sqrt{(v^2 - u^2)}}$.
14. $a/\sqrt{2}$.   15. $AP : PB = a : b$.

Set XVI. c., p. 179.

10. $\dfrac{3\sqrt{3}}{4}a$.   12. $AP^2 : PB^2 = a^3 : b^3$.
13. (i) $\tan\theta = \sqrt[3]{\left(\dfrac{b}{a}\right)}$; (ii) $\sqrt{\left(\dfrac{b}{a}\right)}$; (iii) $\dfrac{b}{a}$.   17. $\dfrac{1}{e}$.   18. $-\dfrac{1}{e}$.
19. e.   20. $\dfrac{1}{2e}$.   21. $\left(\dfrac{b}{a}\right)^{\frac{a}{a-b}} - \left(\dfrac{b}{a}\right)^{\frac{b}{a-b}}$.

ANSWERS. 447

Set XVII., p. 182.

1. Origin a point of inflexion on (i), (iii), (iv).
2. $x = \pm \dfrac{1}{\sqrt{3}}$.
3. x for points of inflexion (i) $\pm a/\sqrt{3}$; (ii) 0, $\pm a\sqrt{3}$; (iii) $\pm a/\sqrt{3}$; (iv) 0, $\pm a\sqrt{3}$.
5. The origin.
7. (i) 0, π; (ii) $\dfrac{\pi}{2}, \dfrac{3\pi}{2}$; (iii) 0, π.
9. (i) $x=2$, (ii) $x = \pm \dfrac{1}{\sqrt{2}}$.
10. $x = \dfrac{2}{b-a} \log\left(\dfrac{b}{a}\right)$.
11. $bx + c - 2\theta = n\pi$ (§ 75, Ex. 4).

CHAPTER X.

Set XVIII., p. 201.

9. (1) $\tfrac{1}{6}a^2(\theta_2^3 - \theta_1^3)$; (2) $\tfrac{1}{2}a^2\left(\dfrac{1}{\theta_1} - \dfrac{1}{\theta_2}\right)$; (3) $\tfrac{1}{2}a^2 \log \dfrac{\theta_2}{\theta_1}$;
(4) $\tfrac{1}{4}a^2 \tan\alpha \left(e^{2\theta_2 \cot\alpha} - e^{2\theta_1 \cot\alpha}\right)$; (5) $\tfrac{1}{8}a^2[6\theta - 8\sin\theta + \sin 2\theta]_{\theta_1}^{\theta_2}$.
10. $\tfrac{1}{2}a^2$.

CHAPTER XI.

§ 91. p. 218.
2. $ax_1 x + by_1 y + cz_1 z = 1$; $(x - x_1)/ax_1 = (y - y_1)/by_1 = (z - z_1)/cz_1$.
3. $by_1 y + cz_1 z = x + x_1$; $-(x - x_1) = (y - y_1)/by_1 = (z - z_1)/cz_1$.
4. $ax_1 x + by_1 y + cz_1 z = 0$; $(x - x_1)/ax_1 = (y - y_1)/by_1 = (z - z_1)/cz_1$.

Set XIX., p. 239.

2. $\dot{r}^2 + 2(D_\theta r)^2 - r D_\theta^2 r = 0$.
3. $-\dfrac{1 + n \cos t}{a \sin^3 t}$.

CHAPTER XII.

Set XX., p. 253.

1. $2\cdot 137\,812$.
2. $\cdot 226\,074$.
3. $2\cdot 188\,920$.
4. $2\cdot 588\,968$.
5. $\cdot 057\,014$; $1\cdot 467\,65$.
6. $1\cdot 895\,494$.
7. $\cdot 739\,085$.
8. $1\cdot 165\,6$; $4\cdot 604\,2$.
9. $91\cdot 964$.
10. (i) $4\cdot 730\,04$; (ii) $1\cdot 875\,1$.
11. $5\cdot 600\,257$, or in deg., $320°\,52'\,16''$.
12. $x = 1\cdot 996$, $y = \cdot 909$.

Set XXI., p. 260. 2. $1\cdot 57$ in.

CHAPTER XIII.

§ 111, p. 266. 1. $\tfrac{2}{3}x^{3/2}$; $2\sqrt{x}$; $\tfrac{2}{9}(3x-4)^{3/2}$; $\tfrac{2}{3}\sqrt{(3x-4)}$; $\sin^{-1}\dfrac{x}{\sqrt{3}}$.
2. 2; 0; 1; $\log(b^2/a^2)$; $-\log 3$.

Set XXII., p. 269.

1. $\frac{3}{4}x^4 + \frac{5}{3}x^3 + \frac{17}{2}x^2 + 51x + 150\log(x-3)$. 2. $x + \log(2x-1)$.
3. $\log(x^2 - 3x + 2)$. 4. $\log\dfrac{(x-2)(x-3)}{(x-1)^2}$. 5. $\dfrac{1}{2\sqrt{21}}\log\dfrac{\sqrt{7}+x\sqrt{3}}{\sqrt{7}-x\sqrt{3}}$.
6. $\dfrac{1}{\sqrt{21}}\tan^{-1}\left(x\sqrt{\dfrac{3}{7}}\right)$. 7. $\frac{1}{2}\sin^{-1}\left(\dfrac{2x}{\sqrt{3}}\right)$.
8. $\frac{1}{2}\log(2x + \sqrt{4x^2+3})$. 9. $\frac{1}{2}x + \frac{1}{4}\sin 2x$.
10. $\frac{1}{12}\sin 3x + \frac{3}{4}\sin x$. 11. $\frac{1}{2}x + \dfrac{1}{4a}\sin 2(ax+b)$.
12. $\frac{3}{8}x - \frac{1}{4}\sin 2x + \frac{1}{32}\sin 4x$. 13. $\frac{1}{2}\sin x - \frac{1}{14}\sin 7x$.
14. $\frac{1}{2}\cos(x+1) - \frac{1}{14}\cos(7x+5)$. 15. $\frac{1}{4}x + \frac{1}{8}\sin 2x + \frac{1}{16}\sin 4x + \frac{1}{24}\sin 6x$.
16. $\pi/4$. 17. $\pi/4$. 18. $\pi/8$. 19. $\frac{1}{4}\log 3$.
20. $\frac{1}{4}\log(\frac{3}{5})$. 21. $\pi/6$. 22. $\pi/2$. 23. (iii) $4\pi abc/3$.

Set XXIII., p. 280.

1. $\dfrac{2}{\sqrt{23}}\tan^{-1}\left(\dfrac{4x+3}{\sqrt{23}}\right)$. 2. $\sin^{-1}\left(\dfrac{2x-a}{a}\right)$. 3. $\log(x - \frac{1}{2}a + \sqrt{x^2 - ax})$.
4. $\sin^{-1}\left(\dfrac{2x-a-b}{b-a}\right)$. 5. $\frac{1}{2}\log(a^2 + x^2)$. 6. $\sqrt{(a^2 + x^2)}$.
7. $\frac{1}{6}\log\left(\dfrac{x^3-1}{x^3+1}\right)$. 8. $\dfrac{1}{\sqrt{3}}\tan^{-1}\left(\dfrac{2x^2+1}{\sqrt{3}}\right)$. 9. $\sqrt{(x^2 + 2x - 3)}$.
10. $\log\sin x$. 11. $\log(1 + \sin x)$. 12. $\log(x + \sin x)$.
13. $\frac{1}{3}\tan^3 x - \tan x + x$. 14. $-\frac{1}{4}\cot^4 x + \frac{1}{2}\cot^2 x + \log\sin x$.
15. $\dfrac{1}{ab}\tan^{-1}\left(\dfrac{b}{a}\tan x\right)$. 16. $-\cos x + \cos^3 x - \frac{3}{5}\cos^5 x + \frac{1}{7}\cos^7 x$.
17. $-\frac{1}{5}\cos^5 x + \frac{2}{7}\cos^7 x - \frac{1}{9}\cos^9 x$. 18. $\tan x - \cot x$.
19. $\frac{1}{4}\sec^4 x$. 20. $2\sqrt{(a-x)}\{\frac{1}{5}(a-x)^2 - \frac{1}{3}a(a-x)\}$.
21. $-\frac{2}{3}(x+2a)\sqrt{(a-x)}$. 22. $\log(x + \sqrt{x-1}) - \dfrac{2}{\sqrt{3}}\tan^{-1}\left(\dfrac{2\sqrt{(x-1)}+1}{\sqrt{3}}\right)$.
23. (i) $8/15$; (ii) $8/315$; (iii) π/ab; (iv) $\frac{1}{4}\log 3$;
    (v) $\frac{1}{2}\log 2$; (vi) $\pi/3\sqrt{3}$; (vii) $2\pi/3\sqrt{3}$; (viii) $\pi/2$.
24. $\frac{1}{2}\log(x^2 + x + 1) + \dfrac{1}{\sqrt{3}}\tan^{-1}\left(\dfrac{2x+1}{\sqrt{3}}\right)$. 25. $x - 2\tan^{-1} x$.
26. $\frac{1}{3}(x-1)^3 + 2\log(x^2 + 2x + 3)$. 27. $\frac{3}{8}\log(x^2 - 2) + \frac{1}{8}\log(x^2 + 2)$.
28. $x + 4\log(x-1) - 4/(x-1)$. 29. $\sin^{-1} x - \sqrt{(1-x^2)}$.
30. $\sqrt{(x^2-1)} + \log(x + \sqrt{x^2-1})$. 31. $\sqrt{(x^2 + ax)} + \dfrac{a}{2}\log(x + \dfrac{a}{2} + \sqrt{x^2 + ax})$.
32. $\sqrt{(ax - x^2)} + \dfrac{a}{2}\sin^{-1}\left(\dfrac{2x-a}{a}\right)$. 33. $\sin^{-1}\left(\dfrac{x-1}{3x}\right)$.

ANSWERS. 449

$\frac{\sqrt{(x^2-1)}}{x+1}$.   **35.** $-\frac{\sqrt{(x^2-1)}}{x-1}$.   **36.** $-\sqrt{\left(\frac{1-x}{1+x}\right)}$.

$-\sqrt{\left(\frac{1+x}{1-x}\right)}$.   **38.** $\frac{1}{n}\log\left(\frac{x^n}{x^n+1}\right)$.   **39.** $\frac{x+1}{2\sqrt{(x^2+2x+3)}}$.

$\frac{1}{2}x + \frac{1}{2}\log(\sin x + \cos x)$.   **41.** $\frac{3}{5}x - \frac{1}{5}\log(\sin x + 2\cos x)$.

(i) $\pi/4$; (ii) $\pi/2$; (iii) $\pi/4$; (iv) $\pi/(1-r^2)$ or $\pi/(r^2-1)$ according as $r^2 < 1$ or $r^2 > 1$; (v) $a/\sin a$; (vi) $\pi/2\sqrt{(1-k^2)}$; (vii) $\frac{1}{4}\log 3$.

(i) $\pi/2$; (ii) $-\pi/2$.   **45.** $8a^2/15$.   **46.** Each $= 2a^2/3$.   **47.** $\frac{1}{2}(2a^2+b^2)\pi$.

Set XXIV., p. 288.

$-(x+1)e^{-x}$.   **2.** $-(x^3+3x^2+6x+6)e^{-x}$.   **3.** $\sin x - x\cos x$.

$x \sin x + \cos x$.   **5.** $-\frac{1}{4}x\cos 2x + \frac{1}{8}\sin 2x$.

$-x^2 \cos x + 2x \sin x + 2\cos x$.   **7.** $\frac{x^{n+1}}{n+1}\log x - \frac{x^{n+1}}{(n+1)^2}$.

$\frac{1}{2}(\log x)^2$.   **9.** $-\frac{1}{2}e^{-x} + \frac{1}{10}e^{-x}(\cos 2x - 2\sin 2x)$.

$e^x/(1+x)$.   **11.** $-\frac{1}{2}e^{-x^2}$.   **12.** $x\sin^{-1}x + \sqrt{(1-x^2)}$.

$x\tan^{-1}x - \frac{1}{2}\log(1+x^2)$.   **14.** $\frac{1}{2}x^2\sin^{-1}x - \frac{1}{4}\sin^{-1}x + \frac{1}{4}x\sqrt{(1-x^2)}$.

$\frac{1}{2}(1+x^2)\tan^{-1}x - \frac{1}{2}x$.   **16.** $\frac{1}{2}(x-1)\sqrt{(3+2x-x^2)} + 2\sin^{-1}\frac{x-1}{2}$.

$\frac{1}{2}(x+1)\sqrt{(x^2+2x+3)} + \log(x+1+\sqrt{x^2+2x+3})$.

$\frac{1}{2}(x-a)\sqrt{(2ax-x^2)} + \frac{1}{2}a^2\sin^{-1}\frac{x-a}{a}$.

$\frac{1}{2}(x+a)\sqrt{(2ax+x^2)} - \frac{1}{2}a^2\log(x+a+\sqrt{2ax+x^2})$.

$\frac{1}{2}\sin^{-1}x - \frac{1}{2}x\sqrt{(1-x^2)}$.   **21.** $\frac{x\sin x}{1+\cos x}$.   **22.** $\frac{e^{-3x}}{25}(4\sin 4x - 3\cos 4x)$.

$\frac{1}{2}(\cosh x \sin x + \sinh x \cos x)$.   **24.** $\frac{1}{2}(\cosh x \sin x - \sinh x \cos x)$.

$35\pi/256$, $5\pi/16$, $3\pi/256$, $4/35$, $7\pi/256$, $13/15 - \pi/4$.

$\pi a^4/16$, $\pi a^3/2$, $5\pi a^4/8$.   **27.** $(\pi-2)a^2/4$.

$m!\, n!/(m+n+1)!$.   **33.** $m!\, n!/(m+n)!$.

$\pi a^6/32$, $(21\pi/32 - 28/15)a^6$.   **36.** $32\sqrt{2}/15$.   **37.** πa^2.

$4a(1-\cos\theta/2)$.   **40.** $\frac{1}{2}a\theta\sqrt{(1+\theta^2)} + \frac{1}{2}a\log(\theta+\sqrt{1+\theta^2})$.   **41.** $(r-a)\sec a$.

Set XXV., p. 296.

$\log(x+2) - \frac{1}{2}\log(2x+1) - \frac{1}{3}\log(3x+2)$.

$15x^2 - 5\log(x^2-1) + 80\log(x^2-4)$.   **3.** $\Sigma\frac{a^2}{(a-b)(a-c)}\log(x-a)$.

$-\frac{1}{2}\frac{1}{x+1} + \frac{1}{4}\log\frac{x-1}{x+1}$.   **5.** $\frac{1}{2x^2} + \frac{1}{x} + \log\frac{x-1}{x}$.

$\frac{1}{16}\frac{1}{(x+1)^2} - \frac{1}{16}\frac{1}{(x-1)^2} + \frac{3}{16}\left(\frac{1}{x+1} + \frac{1}{x-1} + \log\frac{x-1}{x+1}\right)$.

$-\frac{1}{4}\frac{1}{(x^2-1)^2}$.   **8.** $\frac{2}{3}\log(x+1) + \frac{1}{6}\log(x^2-x+1) + \frac{1}{\sqrt{3}}\tan^{-1}\frac{2x-1}{\sqrt{3}}$.

G.C.   2 F

9. $\frac{1}{3}x^3 - \frac{1}{3}\tan^{-1}x^3$. 10. $x + \log(x^2 - x + 1) + \frac{2}{\sqrt{3}}\tan^{-1}\frac{2x-1}{\sqrt{3}}$.

11. $\frac{\sqrt{2}}{3}\tan^{-1}\frac{x}{\sqrt{2}} + \frac{1}{6}\log\frac{x-1}{x+1}$. 12. $\frac{1}{a^2-b^2}\left(\frac{1}{b}\tan^{-1}\frac{x}{b} - \frac{1}{a}\tan^{-1}\frac{x}{a}\right)$.

13. $\frac{1}{2(a^2-b^2)}\log\frac{x^2+b^2}{x^2+a^2}$. 14. $\frac{1}{a^2-b^2}\left(a\tan^{-1}\frac{x}{a} - b\tan^{-1}\frac{x}{b}\right)$.

15. $\frac{1}{4}\log(x-1) - \frac{1}{8}\log(x^2+1) - \frac{1}{2}\tan^{-1}x + \frac{1}{4}\frac{1-x}{1+x^2}$.

16. $\frac{1}{4\sqrt{2}}\log\frac{x^2+x\sqrt{2}+1}{x^2-x\sqrt{2}+1} + \frac{1}{2\sqrt{2}}\tan^{-1}(x\sqrt{2}+1) + \frac{1}{2\sqrt{2}}\tan^{-1}(x\sqrt{2}-1)$.

17. $\frac{5}{16}\tan^{-1}\frac{x+1}{2} + \frac{1}{x^2+2x+5} - \frac{3}{8}\frac{x+1}{x^2+2x+5}$.

18. $-\frac{1}{3bx^3} + \frac{a}{2b^2x^2} - \frac{a^2}{b^3x} + \frac{a^3}{b^4}\log\frac{ax+b}{x}$.

19. $\frac{1}{12}\tan^{-1}\frac{3\sin x}{4}$. 20. $\frac{1}{12}\log\frac{2+\cos x}{2-\cos x} + \frac{1}{6}\log\frac{1-\cos x}{1+\cos x}$.

21. $-\frac{1}{8}\frac{x+4}{(x^2+4x+6)^2} - \frac{3}{32}\frac{x+2}{x^2+4x+6} - \frac{3\sqrt{2}}{64}\tan^{-1}\frac{x+2}{\sqrt{2}}$.

22. $\frac{1}{2a}\log\frac{x^2-ax+a^2}{x^2+ax+a^2}$.

23. $\frac{1}{25}\log(x+1) + \frac{1}{150}\log(x^2+4) - \frac{2}{75}\log(4x^2+1)$
$\quad + \frac{8}{75}\tan^{-1}2x - \frac{1}{150}\tan^{-1}\frac{1}{2}x$.

24. $-\frac{1}{x} - \frac{1}{2}\frac{x}{1+x^2} - \frac{3}{2}\tan^{-1}x$. 25. $\frac{1}{(a-b)^{m+n-1}}\int\left(\frac{1}{u}-1\right)^{m+n-2}du$.

26. $2\sqrt{x} - 2\tan^{-1}\sqrt{x}$. 27. $2x^{\frac{1}{2}} - 3x^{\frac{1}{3}} + 6x^{\frac{1}{6}} - 6\log(1+x^{\frac{1}{6}})$.

28. $2\sqrt{(x-1)}\left\{x + \frac{3}{5}(x-1)^2 + \frac{1}{7}(x-1)^3\right\}$.

29. $\frac{2}{b^2}\frac{2a+bx}{\sqrt{(a+bx)}}$. 30. $\frac{1}{\sqrt{2}}\tan^{-1}\frac{x\sqrt{2}}{\sqrt{(1-x^2)}}$.

31. $\frac{1}{\sqrt{2}}\tanh^{-1}\frac{x\sqrt{2}}{\sqrt{(1+x^2)}} = \frac{1}{2\sqrt{2}}\log\frac{\sqrt{(x^2+1)}+x\sqrt{2}}{\sqrt{(x^2+1)}-x\sqrt{2}}$.

32. $\frac{x^2}{2a} - \frac{x\sqrt{(x^2-a)}}{2a} + \frac{1}{2}\log(x+\sqrt{x^2-a})$.

33. $\frac{1}{b^2}\sqrt{(a+bx^2)}\left\{\frac{1}{5}(a+bx^2)^2 - \frac{1}{3}a(a+bx^2)\right\}$. 34. $\frac{4}{25}(1+x^{\frac{5}{2}})^{\frac{5}{2}} - \frac{4}{15}(1+x^{\frac{5}{2}})^{\frac{3}{2}}$.

35. $(2x^2-1)\sqrt{(1+x^2)}/3x^3$. 36. $-(a-x)^{\frac{3}{2}}/(a+x)^{\frac{1}{2}}$.

37. $\frac{1}{2\sqrt{2}}\log\frac{1+\tan x - \sqrt{2\tan x}}{1+\tan x + \sqrt{2\tan x}} + \frac{1}{\sqrt{2}}\tan^{-1}(\sqrt{2\tan x}-1)$
$\quad + \frac{1}{\sqrt{2}}\tan^{-1}(\sqrt{2\tan x}+1)$.

ANSWERS. 451

CHAPTER XIV.

Set XXVI., p. 306.

1. $a/(a^2+b^2)$.   2. $b/(a^2+b^2)$.   3. π.   4. $3\pi/16a^5$.
5. $5\pi a^3/16$.   6. πa.   7. π.   8. $\pi(b-a)^2/8$.
9. $\pi/\sqrt{(a^2-b^2)}$.   10. $\pi/2ab$.   11. $\pi(a^2+b^2)/4a^3b^3$.
12. $(e-\tan^{-1}e)/e^3$.   13. $\frac{1}{2e^2}\sqrt{(1+e^2)}-\frac{1}{2e^3}\log(e+\sqrt{1+e^2})$.   14. 0.
15. -1.   16. $-1/9$.   18. $n!$ 19. $\pi^2/4$.   20. $\dfrac{2\pi}{\sqrt{(1-e^2)}}\tan^{-1}\sqrt{\dfrac{1-e}{1+e}}$

Set XXVII., p. 312.

1. $2\pi ah^2$; $\dfrac{8\pi}{3}\sqrt{a}\left[(a+h)^{\frac{3}{2}}-a^{\frac{3}{2}}\right]$.   2. $\pi h^2\sqrt{(bc)}$.   3. $\frac{4}{3}ab$; $4\pi ab^2/15$.
4. $\pi(b-a)^2(b+a)/4c$.   5. $\pi(a^2+b^2)/2$.   7. $ab/30$.
9. $(4-\pi)a^2/2$; $(4+\pi)a^2/2$.   10. $3\pi a^2$; $2\pi^2 a^3$.   11. πa^2; $\pi^2 a^3/2$.
12. $(\pi-2)a^2/2$.   13. $\pi a^3(10-3\pi)/6$.   16. 2.
18. $3\pi a^2$, $4a(1-\cos a/2)$, $5\pi^2 a^3$, $\pi^2 a^3$.   19. $abc/6$.   20. $\pi ac^2/2$.
22. $\dfrac{l^2}{4}\left(\tan\dfrac{a}{2}+\dfrac{1}{3}\tan^3\dfrac{a}{2}\right)$;

$$\dfrac{l^2 e}{2(1-e^2)}\left\{-\dfrac{\sin a}{1+e\cos a}+\dfrac{2}{e\sqrt{(1-e^2)}}\tan^{-1}\left(\sqrt{\dfrac{1-e}{1+e}}\tan\dfrac{a}{2}\right)\right\}.$$

23. $\pi a^2/12$.   24. $\pi a^2/4$, $\pi a^2/2$.   25. $(3-2\log 2)a^2/4$.   26. $(4-\pi)a^2/2$.

Set XXVIII., p. 322.

3. $16a^2/3$.   6. 11π.   9. $\dfrac{\pi}{2mn}\left(\dfrac{a^2}{m^2}+\dfrac{b^2}{n^2}\right)$.

CHAPTER XV.

Set XXX., p. 347.

1. (i) $(a_1^2+a_2^2+\ldots+a_n^2)/2$.   (ii) $(b_1^2+b_2^2+\ldots+b_n^2)/2$.
2. $(a_1A_1+b_1B_1+a_2A_2+b_2B_2)/2$.
7. (i) $3h/5$, $3k/8$.   (ii) $2h/5$, $k/2$.   (iii) $\frac{2}{3}a\sin^3 a/(a-\sin a\cos a)$, 0.
   (iv) $\bar{x}=\frac{3}{8}a(1+\cos a)$.   (v) $5a/6$, 0.
10. (i) $5Ma^2/4$.   (ii) $7Ma^2/5$.   (iii) $Mh^2/6$.
   (iv) $3Ma^2/10$, $3M(a^2+4h^2)/20$.

CHAPTER XIX.

Set XXXV., p. 416.

6. $(1-au)(1-bu)-h^2u^2=0$; u_1, u_2 are the squares of the semi-axes of the conic $ax^2+2hxy+by^2=1$.

7. $l^2a^2/(a^2-u)+m^2b^2/(b^2-u)+n^2c^2/(c^2-u)=0$; u_1, u_2 are the squares of the semi-axes of the conic in which the plane cuts the ellipsoid.

Set XXXVI., p. 422.

1. $\frac{1}{2}n(n+1)$. **2.** $1/2a$. **3.** a. **4.** $(a_1+a_2+\ldots a_n)/n$.

5. ∞ and 1. **6.** -1 and -1. **7.** 2 and 2.

8. 1 and -1. **9.** a/b and $2a/b$. **10.** $1/2$.

11. $(\log a - \log b)/(\log c - \log g)$. **12.** 1 and 1. **13.** $a_1 a_2 \ldots a_n$.

14. $\frac{1}{3}$ and 1. **15.** $e^{-a^2/2b^2}$ and 0. **17.** $\pm(a/b)$.

CHAPTER XX.

Set XXXVIII., p. 439.

1. $y-x=C(1+xy)$. **2.** $\sin^{-1}y-\sin^{-1}x=C$.

3. $Cy=(1-my)(x+m)$. **4.** $xy^2=C(2y+x)$.

5. $y=Cx^2-1/4C$. **6.** $(2x-3y+1)^2(x+2y-2)=C$.

7. $x=C+\int \frac{\eta+c}{R}d\eta$, where $R=(a+mb)\eta+ac+bg$.

8. $by^2+2axy-fx^2-2gx+2cy=C$. **9.** $y=(x+C)e^{-x}$.

10. $y=\frac{1}{2}x+C/x$. **11.** $y=(\sin^{-1}x+C)/\sqrt{(1-x^2)}$. **12.** $(1+x^2)y=\frac{1}{3}x^3+C$.

13. $y=Ce^{-ax}+\{a\cos(bx+c)+b\sin(bx+c)\}/(a^2+b^2)$.

14. $1/y^5=\frac{5}{2}x^2+Cx^5$. **15.** $x/y=C+\log x$.

16. $x^4+2x^2y^2-y^4-2a^2x^2-2b^2y^2=C$.

17. $(y-Cx)^2=a^2C^2+b^2$; $x^2/a^2+y^2/b^2=1$.

18. $y=Cx+C^3$; $27y^2+4x^3=0$. **19.** $y^2=Cx^2+C^2$.

20. $y=Ae^{ax}+Be^{bx}$. **21.** $y=A+Be^{2x}+Ce^{3x}$.

22. $y=e^{3x}(A\cos x+B\sin x)+(2\cos 2x+\sin 2x)/30$.

23. $y=Ae^x+Be^{2x}-xe^x$.

24. $y=A\cos nx+B\sin nx+x(a\sin nx-b\cos nx)/2n$.

25. $y=Ae^{nx}+Be^{-nx}+x(ae^{nx}-be^{-nx})/2n$.

26. $y=e^{2x}(A\cos 2x+B\sin 2x)+(169x^2+156x+46)/2197$.

27. $y=(A+Bx)\cos x+(E+Fx)\sin x$.

ANSWERS. 453

28. $x = e^{6t}(A\cos t + B\sin t)$,
$y = e^{6t}\{(A - B)\cos t + (A + B)\sin t\}$.

29. $x = A\cos t + B\sin t$,
$y = \tfrac{1}{2}(B - 3A)\cos t - \tfrac{1}{2}(A + 3B)\sin t$.

30. $x = Ae^t + Be^{-5t} - \tfrac{2}{5}t - \tfrac{13}{25} + \tfrac{3}{7}e^{2t}$,
$y = Ae^t - Be^{-5t} - \tfrac{3}{5}t - \tfrac{12}{25} + \tfrac{4}{7}e^{2t}$.

31. $x = (A + Bt)e^t + (E + Ft)e^{-t}$,
$y = \tfrac{1}{2}(B - A - Bt)e^t - \tfrac{1}{2}(E + F + Ft)e^{-t}$.

32. $x = Vt\cos a$, $y = Vt\sin a - \tfrac{1}{2}gt^2$.

33. $x = a\cos nt$, $y = b\sin nt$ where $n = \sqrt{\mu}$.

34. $\dfrac{dx}{dt} = -\sqrt{(2\mu)}\sqrt{\left(\dfrac{1}{x} - \dfrac{1}{a}\right)}$; $t\sqrt{\left(\dfrac{2\mu}{a^3}\right)} = \theta + \sin\theta\cos\theta$, where $x = a\cos^2\theta$.

35. (i) $By = \tfrac{1}{24}w(x^4 - 2lx^3 + l^3x)$;
(ii) $By = \tfrac{1}{24}wx^2(l - x)^2$;
(iii) $By = \tfrac{1}{24}wx^2(x^2 - 4lx + 6l^2)$.

37. $y = (A\cos nx + B\sin nx)/x$; $y = B(\sin nx)/x$.

39. (i) $y = A + B/x + x^2$; (ii) $y = Ax^3 + Bx^2 + Cx - x^2\log x$;
(iii) $y = Ax^2 + B/x - \tfrac{1}{2}x$.

40. $y = A\cos(n\log x) + B\sin(n\log x)$. **41.** $w = Ar + B/r^2$.

42. $V = A\log r + B$. **43.** $(Dy)^2 + 1 = 4a^4/(C - y^2)^2$.

45. $y = \tfrac{1}{6}x^3 + Ax + B/x$.

INDEX.

Abdank-Abakanowicz, 192.
Abel's Theorem, 386.
Abscissa, 4, 7.
Acceleration, 150.
   angular, 153.
   normal, 359.
   radial, 239.
Adiabatic curves, 127.
   expansion, 230.
Algebraic functions, 43.
Amsler's planimeter, 321.
Anchor-ring, 322, 349.
Angle, 31, 219.
   between two lines, 42, 207.
Appell, 322.
Approximations, 196, 244-269.
   rule for, in expansions, 249.
   to areas and volumes, 328.
   to integrals, 299, 308, 328.
   to roots of equations, 244.
Arc, derivative of, 124, 201.
   of circle, Huyghens' approximation, 396.
Area, approximations to, 328.
   derivative of, 185, 201.
   interpretation of, 187.
   of closed curves, 316.
   of surfaces, 193, 338.
   of some common curves and surfaces, 309.
   sign of, 186.
   swept out by moving line, 319.
Argument of function, 14.
Asymptote, 38, 250.
Auxiliary circle, 54.
   equation, 434.
Attraction, 151, 154, 241.
Axes, change of, 52.
   rectangular, 6, 205.

Bernoulli's numbers, 404.
Bessel Function, 407.
Beta Function, 350.
Binomial Theorem, 394.

Cardioid, 202, 360.
Catenary, 139, 360.
Cauchy, 121.
Cauchy's form of remainder, 393.
Centre of curvature, 354.
   of gravity, or inertia, or mass, 341.
Centroid, 341.
Chrystal's Algebra, 173, 250, 290, 375, 382, 386, 395, 396, 404.
   Elementary Algebra, 20.
Circle, Area of, 85.
   of curvature, 354.
   involute of, 373.
   perimeter of, 85.
Cissoid, 314.
Clairaut's equation, 432.
Commutative property of derivatives, 221.
Complementary function, 433.
Complete differential, 213, 224.
   integral, 426.
Compound interest law, 97.
Concavity, 180.
Cone, surface and volume of, 86, 309.
   moments of, 349.
Confocal conics, 428.
Conic section, definition, equation and properties of, 47, 54, 61.
   polar equation of, 63.
   tangent properties of, 124-128.
   confocal, 428.
Conical point, 218.

The numbers refer to pages.

INDEX.

Conicoid, 218.
Consecutive normals, 355.
Constant, 13.
   arbitrary, 262, 425.
   elimination of, 424.
Contact of curves, 361.
Continuity, 12, 87.
   of elementary functions, 90.
   of series, 385.
Convergence of series, 375.
   absolute or unconditional, 382.
   conditional, 382.
   uniform, 385.
Convexity, 180.
Coordinate geometry—
   of two dimensions, 27.
   of three dimensions, 205.
Coordinates—
   cylindrical, 210.
   polar, 10.
   rectangular, 7.
   spherical polar, 210.
Corrections, small, 258.
$\cos x$, expansion of, 394.
Curvature, 352.
   centre of, 354.
   chord of, 354, 360.
   circle of, 354.
   formulae for, 353, 355.
   radius of, 354.
Curves—
   contact of, 361.
   derived, 183.
   equation of, 23, 209.
   family of, 365.
   integral, 190.
   tracing of, 311.
Cusp, 46.
   of second kind, 261.
Cycloid, 368.
   properties of, 369, 373.
Cylinder, surface and volume of, 86, 309.

Decreasing function, 104.
Definite integral, *see* 'Integral.'
Definite value, 15.
Density, 341.
Derivatives, 101.
   geometrical interpretation of, 105.
   not definite, 107.

Derivatives of sum, product etc., 112-114.
   of a function of a function and of inverse functions, 116.
   of implicit functions, 119, 214.
   of arc, 124, 201.
   of area, 185, 201.
   of surface and volume, 193, 346.
   successive or higher, 142.
Derivatives, partial, 204.
   commutative property of, 221.
   geometrical illustrations of, 214.
   of higher orders, 220.
Derivatives, total, 212.
Derived curve, 183.
   function, 102.
Differential, 120.
   complete or total, 213, 224.
   higher, 234.
Differential coefficient, 102 *see* 'Derivatives.'
Differential Equations, 424.
   degree of, 424.
   exact, 431.
   homogeneous, 429.
   linear, 429, 433.
   order of, 424.
   ordinary, 424.
   partial, 424.
   simultaneous, 437.
Differentiation, 101.
   logarithmic, 113.
   of series, 400.
   see 'Derivatives.'
Dimensions of magnitudes, 68.
Direction cosines, 207.
Directrix of conic, 47.
Discontinuity, 88, 154, 387.
Divergent series, 375.
Durand, 193.
Dynamics, 149-155, 225, 341-347.

Eccentric angle, 55.
Eccentricity of a conic, 47.
Elasticity, coefficient of, 156, 230.
Electric current equations, 159, 430, 438.
Elimination of constants, 424.
Ellipse, definition and simpler properties of, 49, 54, 61.
   area of 281, 310.
   curvature of, 353, 359.

The numbers refer to pages.

Ellipse, evolute of, 362.
  perimeter of, 405.
  tangent properties of, 124-128.
Ellipsoid, moments of inertia of, 345.
  volume of, 270, 310.
  of revolution, *see* 'spheroid.'
Elliptic lamina.
  centroid of quadrant of, 342.
  moments of inertia of, 345.
Energy, kinetic, 150.
Envelopes, 364.
  contact-property of, 366.
Epicycloid, 369.
  properties of, 373.
Epitrochoid, 370.
Equations, of a curve, 23, 209.
  of a surface, 209.
  theory of, 242-254.
  differential, 424.
Errors, superposition of small, 258.
Euler, 253.
  theorems of, on homogeneous functions, 412.
Everett, 70.
Evolute, 361.
Expansion, coefficient of, 156, 230.
Expansions of functions, 390, 408.
Explicit function, 16.
Exponential function, 96, 394.
  graph of, 58.
Extension, 152.

Fluent, fluxion, 109.
Focus of a conic, 47.
Forms, indeterminate, 418.
Forsyth's Differential Equations, 424.
Function, algebraic, 43.
  definition of, 14.
  explicit, 16.
  graphical representation of, 20.
  homogeneous, 412.
  implicit, 17.
  inverse, 18.
  multiple-valued, 17.
  notation for, 16.
  of a function, 90.
  periodic, 56, 303.
  single-valued, 17.
  transcendental, 56.

Gamma Function, 349.
Gennochi-Peano's Calculus, 421.
Geometry, coordinate—
  of two dimensions, 27.
  of three dimensions, 205.
Gradient, 32, 102.
Graphical integration, 192.
Graphs, 20, 311.
  general observations on, 59.
  of inverse functions, 44.
Gray's, Absolute Measurements, 70, 175.
  Magnetism, and Electricity, 439.
  Physics, 154, 160.
Gray and Mathews, Bessel Functions, 407.
Gregory's series for π, 401.
Gyration, radius, of 344.

Harmonic motion, 152, 160.
Heat, conduction of, 157.
Henrici's Report on Planimeters, 322.
Hobson's Trigonometry, 257.
Holditch's Theorem, 323.
Homogeneous functions,
  Euler's theorems on, 412.
Huyghens' rule for circular arc, 396.
Hyperbola, definition and simpler properties of, 50, 54, 61.
  area of sector of, 289.
  curvature of, 359.
  evolute of, 371.
  rectangular, referred to asymptotes, 54.
  tangent properties of, 124-128.
Hyperbolic functions, 139-142.
Hypocycloid, 369, 373.
Hypotrochoid, 370.

Identical Equality, theorem of, 388.
Impedance, 430.
Implicit function, 17.
  differentiation of, 119, 214.
Increasing function, 104.
Increment, 65.
Indeterminate forms, 418.
Inductance, 159, 430, 438.
Inertia, centre of, 341.
  moment of, 343.

The numbers refer to pages.

INDEX. 457

Infinite, 60, 80, 195.
  series, see 'Series.'
Infinitesimals, 195-200.
Inflexion, point of, 35, 180, 239.
Inflexional tangent, 35.
Integral curve, 190.
  function, 188.
Integral, complete, 426.
  definite, 263, 298-309.
  double, 334.
  general, 189, 262.
  geometrical representation of, 188, 263.
  indefinite, 262.
  limit of a sum, 324.
  line, 347.
  particular, 426, 433.
  related, 301.
  standard forms, 265, 278.
  surface, 347.
  triple, 338.
  see 'Approximations.'
Integrand, 262.
  infinite, 304.
Integraph, 192.
Integrating factor, 431.
Integration, 262, 295.
  by algebraic and trigonometric transformations, 267.
  by change of variable, 271, 340.
  by partial fractions, 268, 290.
  by parts, 281.
  by successive reduction, 284.
  of quadratic functions, 274.
  of trigonometric functions, 278.
  of irrational functions, 294.
  of rational functions, 292.
  of series, 399.
  along a curve, 318, 347.
  over an area, 337.
  through a volume, 338.
Intercept, 31, 33.
Intrinsic equation, 357.
Inverse function, 17.
  differentiation of, 116.
  graph of, 44.
Involute, 361.
Isolated point, 313.

Lagrange's remainder, 392, 409.
Lamb's Calculus, 348.
Laplace's Equation, 223, 235.

Leibniz, 121.
  series for π, 401.
  theorem on derivative of product, 144.
Limits, 74-86.
  distinction between limit and value, 81, 405.
  theorems on existence of, 100, 377.
  of a definite integral, 263.
Line integral, 347.
Linear differential equations, 429, 433.
  function, 31.
Lituus, 202.
Lodge's Mensuration, 331.
Logarithmic differentiation, 113.
  function, 57.
  series, 395.
Logarithms, calculation of, 395.
  derivative of, 136.
  graph of 58.
Lüroth, 257.

Maclaurin's Theorem, 391, 411.
Maclean's Physical Units, 70.
Magnitudes—
  dimensions of, 68.
  directed, 13.
  geometrical representation of, 13.
Mass-centre, 341.
Maxima and Minima, 166.
  elementary methods, 171.
  of functions of several variables, 412.
Maxwell's Heat, 232.
Mean-Value Theorems—
  Derivative, 162, 419.
  Integral, 300, 309.
Mean value of a function, 332, 339.
Mechanics, see "Dynamics."
Minima, see "Maxima."
Moment of differential, 121.
Moment of inertia, 343.
Momentum, 150.
Multipliers, undetermined, 415.
Multiple-valued function, 17.
Murray's Differential Equations, 424.

The numbers refer to pages.

Napier's base, 59, 92.
Newton, 109.
  his method of approximating to the roots of equations, 244.
Node, 423.
Normal, 123, 201, 216.
Number e, 92.
  π, 85, 401.

Order of differential equation, 424.
  of infinitesimals, 195.
Ordinate, 7.
Origin of coordinates, 8.
  change of, 52.
Oscillating series, 375.
Osgood on Infinite Series, 375.

Pappus' Theorems, 348.
Parabola, definition and simpler properties of, 48, 54, 61.
  arc of, 127, 314.
  curvature of, 353.
  evolute of, 367, 371.
  semi-cubical, 127.
  tangent properties of, 124-128.
Parallel curves, 361.
Parameter, 365.
Partial Derivatives, see "Derivatives, partial."
Peano, 413, 421.
Pendulum, period of oscillation of, 402, 432.
Pericycloid, 369.
Period of a function, 56, 303.
Perpendicular, length of, 63.
Plane, equation of, 209.
  tangent, 215, 411.
Planimeter, 321.
Plotting of points, 9.
Points, conical, 218.
  distance between two, 9, 206.
  isolated, 313.
  turning, 24, 167.
Polar formulae, 200.
  tangent, normal, etc., 201.
Potential, 153, 223, 351.
Power, fundamental limit, 91.
  derivative of, 111.
Power series, 383.
  continuity of, 386.
  differentiation and integration of, 400.

Primitive of differential equation, 425.
Prismoid, 332.
Proportional parts, 255.

Radius of curvature, 354.
  of gyration, 344.
Rates, 65-73, 101.
Rational fractions, integration of, 290.
Rational function, 34.
  integration of, 292.
Reduction, successive, 284.
Remainder in Taylor's and Maclaurin's Theorems, 392, 409.
Ring, see "Anchor-ring."
Robin's Tracts, 121.
Rolle's Theorem, 161.
Roots, see 'Equations.'

Schlömilch-Roche's form of remainder, 393.
Segments, directed, 1.
  addition and subtraction of, 2, 3.
  measure of, 5, 12.
  symmetric, 3.
Series, infinite, 375.
  alternating, 382.
  differentiation of, 400.
  integration of, 399.
  multiplication of, 388.
  semi-convergent, 382.
  See 'Convergence of series,' 'Power-series.'
Sign of area, 186.
Simpson's Rules, 330, 332.
Simultaneous differential equations, 437.
$\sin x$, $\sin^{-1}x$, expansion of, 393, 401.
Slope, 102.
Solution of a differential equation, 426.
  singular, 432.
Space-rate of change, 103, 150.
Sphere, surface and volume of, 194, 309.
Spheroid, oblate and prolate, 310.
  surface and volume of, 310.
Spiral, of Archimedes, 201.

The numbers refer to pages.

Spiral, equiangular, 202, 360.
   reciprocal, 202.
Stationary value, 105.
Step, *see* 'Segments.'
Subnormal, 123, 201.
Subtangent, 123, 201.
Surface, equation of, 209.
   of revolution, 193.
     areas and volumes of, 309, 312-315.
   integral, 347.
Symmetry, 9, 23.
   centre of, 29.

$\operatorname{Tan}^{-1}x$, expansion of, 401.
Tangent, definition of, 78.
   length of, 123, 201.
   inflexional, 35.
   plane, 216, 411.
Taylor's Theorem and Series—
   for function of one variable, 390-398.
   for function of several variables, 408-412.
Thermodynamics, 228-233.
Time-rate of change, 103.
Tore, 322, 349.
Total derivative, 211.
   differential, 213, 224.

Trapezoidal rule, 329.
Trigonometric functions, direct and inverse, 56.
   differentiation of, 129, 133.
   integration of, 265, 278, 284.
Trochoid, 370.
True value, 418.
Turning value, 24, 166.

Ultimately equal, 199.
Uniform convergence, 385.
Units, 26, 28.

Value, stationary, 105.
   true, 418.
   turning, 24, 166.
Variable, dependent and independent, 12.
   change of, 233, 271.
Variation, near a turning value, 174.
   in a given direction, 218.
Velocity, 149.
   angular, 153.
   components of, 110.
Volumes, 193, 309, 331, 335.
   polar element of, 346.

Wallis's value of π, 307.
Work, 150, 225.

The numbers refer to pages.

The borrower must return this item on or before the last date stamped below. If another user places a recall for this item, the borrower will be notified of the need for an earlier return.

*Non-receipt of overdue notices does **not** exempt the borrower from overdue fines.*

Harvard College Widener Library
Cambridge, MA 02138   617-495-2413

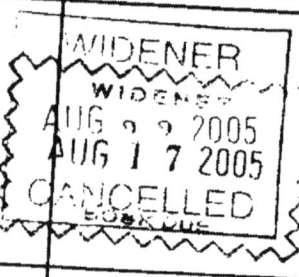

Please handle with care.
Thank you for helping to preserve library collections at Harvard.

www.ingramcontent.com/pod-product-compliance
Lightning Source LLC
LaVergne TN
LVHW012154200226
832130LV00014B/792